T0137406

Emerging Technologies for Biorefineries, Biofuels, and Value-Added Commodities

Zhi-Hua Liu • Art Ragauskas
Editors

Emerging Technologies for Biorefineries, Biofuels, and Value-Added Commodities

Editors
Zhi-Hua Liu
School of Chemical Engineering and Technology
Tianjin University
Jinnan District, Tianjin, China

Art Ragauskas
Chemical & Bio-molecular Engineering
University of Tennessee
Knoxville, TN, USA

ISBN 978-3-030-65586-0 ISBN 978-3-030-65584-6 (eBook)
https://doi.org/10.1007/978-3-030-65584-6

This Springer imprint is published by the registered company Springer Nature Switzerland AG
The registered company address is: Gewerbestrasse 11, 6330 Cham, Switzerland

Preface

Lignocellulosic biorefinery has gained much attention worldwide as a potential solution to reduce our dependence on fossil energy, decrease greenhouse gas emission, and eventually mitigate climate change. However, there are still some technical barriers in developing a competitive biorefinery industry. As intrinsic recalcitrance of lignocellulose associates with the presence of lignin-carbohydrate complex (LCC), it has been recognized that the sustainability and profitability of biorefineries highly depended on deconstructing LCC and valorizing both carbohydrate and lignin polymer of the plant cell wall to produce biofuels and value-added commodities. Biorefinery involves several technologies and processes to fractionate lignocellulose into primary building blocks—cellulose, hemicellulose, and lignin, which could be further converted to biofuels, chemicals, and materials. To improve bioeconomy in a biorefinery concept, it is necessary to develop the integrated conversion processes including pretreatment, saccharification, fermentation, and lignin valorization.

This book has thereby summarized the emerging technologies of biorefining, biofuel, and value-added commodities, which have potential to make biorefineries economically and environmentally sustainable. Chapter 1 presents the challenges and perspectives of biorefineries along with basic concepts. Chapters 2, 3, 4, 5 and 6 focus on the leading and emerging pretreatment technology for deconstructing LCC. Chapters 7, 8, 9 and 10 give a detailed summary of the advanced saccharification and fermentation technology. Chapters 11, 12, 13, 14 and 15 introduce the emerging lignin valorization technologies to valuable co-products of biorefineries. These chapters have been organized from the viewpoint of process engineering, product engineering, material science, bioconversion, and systems and synthetic biology. The above information could provide comprehensive understanding of lignocellulosic biorefinery and a directional guidance of advance biorefinery scenario design.

Overall, this book could have a broad impact in the fields of biorefining, biofuel, bio-based products, biomaterials, environmental waste utilization, biotechnology, industrial engineering, and others. Authors offer a few commonplace remarks by way of introduction in the hope that others may come up with valuable opinions for the better development of biorefinery industry.

Tianjin, China
April 2021

Zhi-Hua Liu

Contents

Chapter 1
Challenges and Perspectives of Biorefineries

Zhi-Hua Liu

1.1 Introduction

Increased demand for energy, shortage of petroleum sources, and change in global climate are the three main challenging issues hindering the sustainable development of society across the world [1–4]. To solve these problems and address the societal needs, it is urgent to develop alternative, sustainable, and cheap energy sources to displace fossil fuel and products. Lignocellulosic biomass (LCB) mainly containing cellulose, hemicellulose, and lignin is a carbon-neutral renewable resource on earth [5–7]. LCB has been recognized as a potential energy resource of mixed sugars and aromatics for conversion to biofuels, and chemicals and materials (Fig. 1.1) [1, 8–11]. The enhanced utilization of LCB holds many potential advantages for reducing the greenhouse gas emissions and meeting the energy demand and the societal needs.

First of all, a wide variety of LCB, mainly including agricultural residues (e.g., corn cobs, wheat straw), energy crops (e.g., switchgrass, sweet sorghum, sugarcane bagasse), and forestry residues (e.g., spruce trees, lopping wastes), can be used as feasible feedstock for the production of bio-based products. LCB is produced through photosynthesis, by which light energy is converted to chemical energy and CO_2 is stored in the form of carbohydrates. A yearly production of LCB is about 2.0×10^{11} tons, which is about ten times the world usage [12], while more than 1.3 billion tons of agricultural and forestry residues could be produced annually alone in the United States [7, 13]. LCB has multicomponent and macromolecular characteristics, which makes it as a potential feedstock for the production of various bio-based products (Fig. 1.1). Besides those, lignin in LCB is the largest renewable

Z.-H. Liu (✉)
School of Chemical Engineering and Technology, Tianjin University, Tianjin, China

Department of Plant Pathology and Microbiology, Texas A&M University, College Station, Texas, USA
e-mail: zhliu@tju.edu.cn

Z.-H. Liu, A. Ragauskas (eds.), *Emerging Technologies for Biorefineries, Biofuels, and Value-Added Commodities*,
https://doi.org/10.1007/978-3-030-65584-6_1

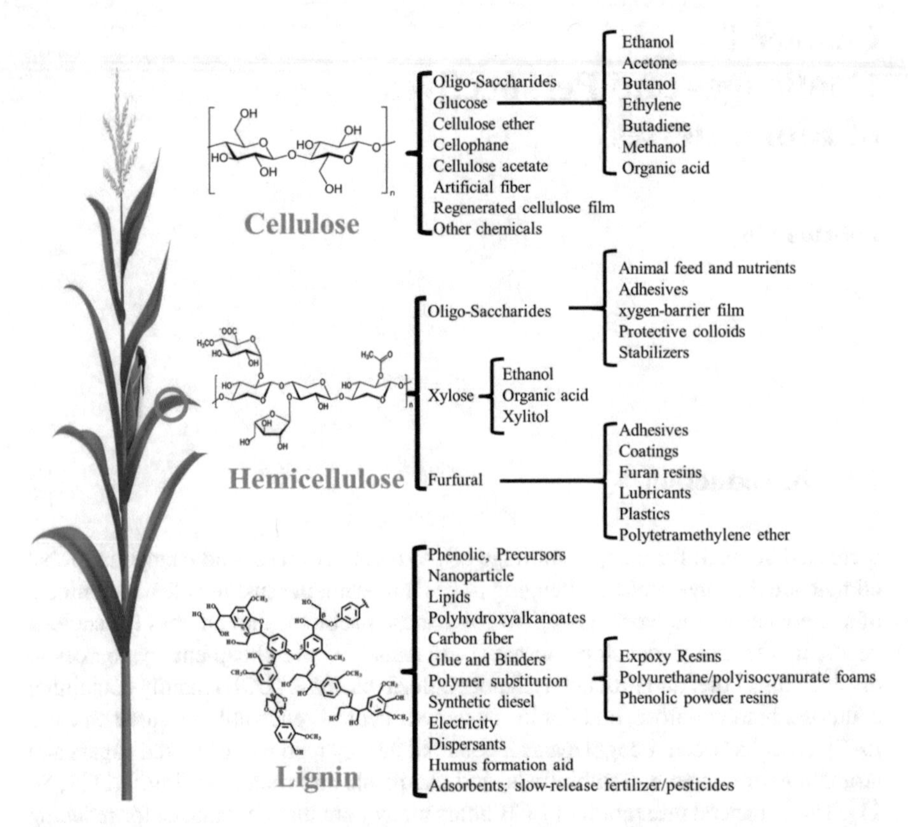

Fig. 1.1 Lignocellulosic biomass used as renewable sources for biofuels, chemicals, and materials

aromatic carbon source and the world's second abundant terrestrial polymer after cellulose [1, 5, 14]. Lignin represents a potential source for the production of biofuels, chemicals, and materials [8–10, 15–17]. Biofuels derived from renewable LCB can be used for the generation of thermal energy and offer a key solution to the current challenge in energy. Across the full life cycle, biofuels can be carbon neutral, which is beneficial for reducing greenhouse gas emissions [14, 18]. Overall, LCB is widely recognized as a potential renewable and alternative feedstock for the production of biofuels and high-value bioproducts.

1.2 Intrinsic Properties of Lignocellulosic Biomass (LCB)

In the past few decades, although breakthroughs have provided a foundation and potential for efficient utilization of LCB, biorefinery to biofuels and bio-based products is yet to be fully commercialized [16, 17, 19]. Several issues still hinder the industrial implementation of biorefinery, which need to be addressed to enhance the yields and properties of targeted products and reduce the process capital cost toward commercial relevance. One of the most significant challenges is how to overcome the intrinsic properties of LCB for maximizing the output and improving the purity of multicomponents for their high-value utilization.

LCB has multicomponent and macromolecular characteristics and is generally composed of cellulose, hemicellulose, lignin, protein, ash, extractives, and others (Fig. 1.1) [20–25]. The three main polymers in cell wall form a large and very intricate three-dimensional (3D) lignin-carbohydrate complex (LCC) structure by intertwining with each other in a different degree.

Cellulose (30–50% of LCB) is the main constituent of plant cell wall of LCB. It is a polymer of β-D-glucopyranose linked together by β-(1, 4) glycosidic bonds, which is unbranched and homopolymer and has a crystalline structure [23–25]. The disaccharide cellobiose is used as the repeating unit to form the cellulose chain, which is grouped together to form microfibrils and hence cellulose fibers [26]. Hemicellulose (20–40% of LCB) is another most abundant sugar polymer, which locates in secondary cell walls. It has a lower molecular weight compared with cellulose. In many LCB, it is heteropolysaccharide with backbone chains of β-1,4-linked β-D-xylopyranose units and is composed of mainly pentoses (xylose and arabinose) and hexoses (glucose, mannose, and galactose) [27]. Unlike cellulose, hemicellulose has side chain branching and is an inhomogeneous polymer. Among the key compositions of LCB, hemicellulose is the most thermo-chemically sensitive and easily degraded into by-products in pretreatment with higher severity parameters [28]. Lignin (15–25% of LCB) is aromatic polymer with an amorphous heterogeneous network of phenyl propane units consisting primarily of syringyl, guaiacyl, and *p*-hydroxy phenol and presents mainly in secondarily thickened plant cell walls [5, 29]. Lignin is generally considered as the "glue," by which the different components of LCB are bound together [26]. Multicompositions and macromolecules are two of the intrinsic characteristics of LCB. Isolation of cellulose, hemicellulose, and lignin singly and/or completely, for single bio-based product production, would consume much more solvent and high energy. This situation is also contrary to the principle of energy conservation and should be economically unfeasible. Taking petroleum refinery as an example, in order to achieve high economic efficiency, it never produces a single product, but produces multiple products due to the fact that a single product production will have high risks and poor performance in the face of market fluctuations [30]. Based on the multicompositions and macromolecules, biorefining of LCB to produce multiple products such as ethanol, materials, and chemicals has become a dynamic research area, which will lead to improved economic feasibility of biorefinery.

During its evolution, the plant biomass has evolved complex rigid and compact structure to protect its structural components from attack and degradation by microbes and chemicals. These structural and chemical mechanisms for resisting the degradation of LCB are known as "biomass recalcitrance" [7, 31–34], which is mainly constructed by following key factors: (1) the complex tissue or organ structure of the LCB; (2) the arrangement, density, and size of the vascular bundles; (3) 3D spatial network as a protective matrix polymer at the cell-wall level; (4) the complexity of the multiple cell-wall constituents; (5) the type of lignin and the degree of lignification; (6) the properties of the cellulose macromolecules (e.g., crystallinity and degree of polymerization) at the molecular level; and (7) the compositional and structural barriers generated during the conversion process. Another intrinsic characteristic of LCB is the heterogeneity, which is mostly constructed by various morphological fractions and tissues, the complexes of cell-wall constituents, and the multiple monomers of polymers [35, 36]. In biorefinery, these intrinsic physicochemical features of LCB create mass-transport limitations and affect liquid penetration and/or enzyme accessibility and activity in the fractionation, purification, and conversion of LCB. Therefore, these intrinsic properties of LCB constitute the barriers in biorefinery and thus contribute to the high costs of the process. Due to these complex structures, there is still a substantial knowledge gap in understanding of intrinsic properties and their relationship with the conversion of LCB. Thus, the comprehensive understanding of intrinsic properties and effective technologies to modify LCB and overcome these limitations are needed and necessary for achieving sustainable production of bio-based products from LCB in biorefinery.

1.3 Emerging Pretreatment Technologies

To overcome the complex heterogeneous structure and recalcitrance of LCB to release the monosaccharides, the pretreatment of LCB needs to be carried out for biomass utilization [13, 23, 37]. The aim of the pretreatment is to remove the physical and chemical impediments that hinder the subsequent enzymatic hydrolysis, and thus to convert the complex LCB into simple components such as cellulose, hemicellulose, and lignin and increase the productivity of monosaccharides. Pretreatment is one of the most important units in the conversion of LCB, which currently presents the most critical challenge and determines the capital cost of the process [37, 38]. Over the past few decades, various pretreatments such as physical, chemical, physicochemical, or biological methods have been developed and evaluated to deconstruct different kinds of LCB for subsequent conversion [26, 31, 36–43]. Although some of these approaches have successfully made the transition from lab-scale platform to the industrial demonstration, significant challenges to pretreatment still remain. Each conventional pretreatment approach has its own specific advantages and disadvantages, which are summarized in Table 1.1. For example, physical pretreatment generally requires high amount of energy and leads to low enzymatic hydrolysis yield, which will affect its feasibility and cost-effectiveness. Most of chemical pretreatment approaches employ excessive amount of expensive and hazardous chemicals, which are environmentally unfriendly and may result in the corrosion of equipment. Biological pretreatment can be carried out at mild pretreatment conditions and is environmentally friendly, but it needs long processing time and generally obtains low yields of sugars.

Table 1.1 The modification of physicochemical properties of lignocellulosic biomass (LCB) by conventional leading pretreatments

Leading pretreatments	Cellulose structure	Removal of hemicellulose	Delignification	Lignin structure	Increase in ASA	Inhibitor generation	Disadvantages	Advantages
Mechanical comminution	++	–	–	+	+	–	1) High energy consumption 2) Low enzymatic hydrolysis yield	1) Alters the inherent structure of LCB 2) Reduces size of biomass particles 3) Decreases the crystallinity of cellulose
Dilute sulfuric acid	–	++	–	+	++	++	1) Equipment corrosion and high cost of process 2) Sugar degradation and loss 3) Inhibitor generation 4) Conditioning of pretreated slurry	1) High glucose yield 2) High removal of hemicellulose 3) Increased ASA
Liquid hot water	–	++	–	+	+	+	1) High holding temperature 2) Low hydrolysis rate 3) C5 sugar degradation and loss 4) Generation of inhibitory compounds	1) No chemical use 2) Low initial and maintenance costs
Steam explosion	–	+	–	+	++	+	1) Incomplete disruption of LCB matrix 2) Generation of toxic components 3) Degradation of hemicellulose sugar	1) Cost-effective 2) Hemicellulose removal 3) High yield of glucose 4) No chemical use and low environmental impact

(continued)

Table 1.1 (continued)

Leading pretreatments	Cellulose structure	Removal of hemicellulose	Delignification	Lignin structure	Increase in ASA	Inhibitor generation	Disadvantages	Advantages
Alkali pretreatment	–	+	+ +	+ +	+ +	–	1) Long residence time 2) Alteration of lignin structure 3) Need for neutralization of slurry	1) Efficient removal of lignin 2) Low inhibitor formation 3) Low energy demands
AFEX	++	–	–	++	+ +	–	1) Incomplete disruption of LCB matrix 2) High enzyme dose 3) Need for solvent recovery 4) Less effective for softwood	1) Mild pretreatment conditions 2) Increased ASA 3) Change in cellulose structure 4) High efficiency and selectivity for lignin reaction
Organosolv	–	–	+ +	+ +	+ +	–	1) Need for solvent recovery 2) High cost	1) Purified solid and liquid stream 2) High-quality lignin
Biological pretreatment (fungus)	+	–	+	+ +	+	–	1) Low hydrolysis rate 2) Large space requirement 3) Long processing time	1) Mild pretreatment conditions 2) Environmentally friendly 3) Low process cost

+ + represents high effect; + represents moderate effect; – represents low effect; *AFEX* ammonia fiber expansion; *ASA* accessible surface area

Pretreatment should be simple, eco-friendly, cost-effective, and economically feasible. An effective pretreatment must comprehensively considerate various situations to facilitate the LCB utilization and reduce the process costs. Generally, main objectives of an effective pretreatment include the following aspects [13, 23, 26, 44]:

1. Increase the accessible surface area (ASA) of the carbohydrates to enzymes.
2. Improve the sugar yields in the subsequent enzymatic hydrolysis.
3. Preserve the hemicellulose and reduce the degradation of the carbohydrates.
4. Limit the generation of inhibitors for hydrolysis and fermentation.
5. Consider the lignin reactivity and yield.
6. Ensure reasonable operational conditions such as operating pressure, temperature, and residence time.
7. Minimize the need of chemicals and avoid the corrosion of equipment.
8. Employ green solvent and consider its cost, safety, and recovery.
9. Reduce energy demands and equipment costs.
10. Minimize the process costs.

Besides these conventional pretreatment approaches, various emerging technologies of the pretreatment, including combinatorial pretreatment, deep eutectic solvents, supercritical fluids, ionic liquids, and microwave irradiation, have been developed and investigated to deconstruct the conversion of LCB in recent years [44–50]. These innovative emerging pretreatment approaches hold great promise of being feasible, cost-effective, and commercialized in the future. However, these technologies are still limited to the lab-scale production of biofuels and bio-based products. Comparative performance of these emerging pretreatment technologies on different LCBs has also not been fully evaluated. The feasibility of emerging pretreatment technologies is generally limited by the high capital cost of process operation and equipment employed. Further, for industrial application of the emerging pretreatment technologies in biorefinery, green, simplified, and eco-friendly technology solutions are needed to meet the challenges in the pretreatment of LCB and thus to the large-scale industrial production requirements in a sustainable way (Table 1.2).

1.4 Enzymatic Hydrolysis Technologies

Enzymatic hydrolysis is a process that is employed to convert pretreated LCB to produce glucose from cellulose, and xylose, arabinose, mannose, and galactose from hemicellulose. Cellulases involved in the enzymatic hydrolysis process are multiple enzyme systems, usually including endoglucanase, cellobiohydrolase, and β-glucosidase [51, 52]. Endoglucanase can attack the low-crystallinity region of the cellulose fiber and create free chain-ends. Cellobiohydrolase can degrade the free cellulose chain to release cellobiose units. β-glucosidase can hydrolyze cellobiose to generate glucose. Additionally, various ancillary enzymes also involve in the

Table 1.2 Emerging pretreatment technologies for the deconstruction and conversion of lignocellulosic biomass (LCB)

Pretreatments	Cellulose structure	Hemicellulose structure	Delignification	Lignin structure	Increase in ASA	Inhibitor generation	Disadvantages	Advantages
Combinatorial pretreatment	++	++	++	++	++	+	1) Complexity of process operation	1) Effective deconstruction 2) Maximizes the component output 3) Increases the processibility of lignin 4) Mild operation conditions
Deep eutectic solvents	–	++	++	++	++	–	1) Poor stability under higher holding temperature	1) Green solvent 2) Biodegradable and biocompatible
Supercritical fluids	++	+	++	++	++	–	1) High energy needed 2) High costs of utilities and process	1) Green solvent 2) Low degradation of sugars 3) Suitable for mobile biomass processor
γ-Valerolactone (GVL)								
Ionic liquids (ILs)	+ +	+	+	+	+ +	–	1) High cost of ionic liquid 2) Need for solvent recovery and recycling 3) Complexity of synthesis and purification	1) Lignin and hemicellulose hydrolysis 2) Mild processing conditions
Microwave irradiation	–	+	–	–	+	–	1) High energy consumption 2) Low enzymatic hydrolysis yield 3) Need to combine with other approach	1) Fast heat transfer, short residence time 2) Selectivity and uniform volumetric heating performance 3) Low formation of by-products

+ + represents high effect; + represents moderate effect; – represents low effect; *ASA* accessible surface area enzymatic hydrolysis process to attack hemicelluloses, including xylanase, β-xylosidase, galactomannanase, glucuronidase, and glucomannanase. All of these types of enzymes synergistically hydrolyze complex cellulose and hemicellulose to free sugars in enzymatic hydrolysis.

Despite the progresses, there are still several factors affecting the enzymatic hydrolysis performance, mainly relating to the properties of pretreated solids, the cost of enzymes, inhibitory effect, and hydrolysis conditions [36, 53–55]. For example, each pretreatment with its unique feature deconstructs the LCB, alters the accessible surface area (ASA) of carbohydrate to enzymes, and thus defines the hydrolysis performance [32, 36, 56]. The changes in ASA by pretreatment are generally related to several factors, including the removal of chemical composition (hemicelluloses, lignin, acetyl group, and pectin), the change in cellulose structure (crystallinity and degree of polymerization), and the modification of porous structure (pore size and volume, particle size, and specific surface area) [32, 33, 57–59]. To enhance the hydrolysis performance, the synergistic effects of these factors are needed to be understood and innovative technologies should be designed and developed to eliminate the barriers of biomass recalcitrance and improve the ASA of carbohydrate to enzymes.

Another factor determining the hydrolysis performance is the solid substrate content, which will obviously affect the economic feasibility of the conversion of LCB [53, 60–62]. To make a feasible conversion process, more than 4% ethanol concentration is needed by increasing the product titer, improving the utilization efficiency of equipment, and reducing the capital cost of the process. Correspondingly, the initial solid loading and sugar levels should be more than 20% and 8%, respectively. Enzymatic hydrolysis carried out at more than 15% solid loading is generally called high solid enzymatic hydrolysis. It offers many advantages, including low energy input, high overall productivity, low water consumption, and low operating costs, and thus shows a potential for the industrial application [61–63]. However, several technical challenges, including insufficient mass and heat transfer, inhibition effects, water constraint, change in rheological characteristics, and difficulty with handling, hinder the industrial implementation of high solid enzymatic hydrolysis [60–62, 64, 65]. Therefore, various emerging strategies are needed to overcome these limitations and improve the performance and sustainability of high solid enzymatic hydrolysis [66–68]. For example, fed-batch operation mode shows the advantages of reducing the initial viscosity of the system and maintaining a lot of free water, which are helpful for the reduction of inhibition effect and the improvement of the diffusion and mixing limitations. Novel reactor systems and process intensification approaches have shown great promise for improving the performance of high solid enzymatic hydrolysis and the feasibility of LCB conversion [60–62, 64, 65].

Besides these, inhibitory effect is one of the most important factors determining the enzymatic hydrolysis performance [69–71]. One of the inhibitory effects is end-product feedback inhibition of cellulase activity [72, 73]. For example, cellulase activity is inhibited by cellobiose produced in enzymatic hydrolysis. Several strategies have been evaluated to remove the feedback inhibition, such as the supplementation of β-glucosidases, the use of high enzyme loading, and the removal of sugars during enzymatic hydrolysis. Previous studies also confirmed that lignin in pretreated LCB restricts the enzymatic hydrolysis due to the physical prevention of the enzymes to access the carbohydrates and the nonproductive adsorption of enzymes onto the lignin surface [59, 71, 74, 75]. The main inhibitory mechanism of lignin on enzymes needs to be comprehensively understood and effective strategies are required to remove the inhibitory effect of lignin for enhancing the enzymatic hydrolysis performance.

Overall, to improve the productivity of monosaccharides from carbohydrates in the enzymatic hydrolysis, studies are required to focus on improving the accessibility of LCB to enzymes, reducing the enzyme cost, removing the inhibitory effect, optimizing the hydrolysis process, and developing novel strategies.

1.5 Fermentation Technologies

Fermentation in the LCB conversion process is employed to convert the monosaccharides in hydrolysate released from cellulose and hemicelluloses into biofuels (e.g., ethanol, butanol, acetone, isobutanol, and lipids) and other bio-based products (e.g., organic acids) via yeasts or bacteria. Figure 1.2 shows the several biological conversion process strategies of LCB, including separate hydrolysis and fermentation (SHF), simultaneous saccharification and fermentation (SSF); simultaneous saccharification and co-fermentation (SScF), and consolidated bioprocess (CBP). The fermentation capacity of hemicellulose sugars, the inhibitory effect of degradation products from pretreatment, the properties of pretreated LCB, and the fermentation conditions define the fermentation performance of LCB to high-value products [62, 67, 76–78].

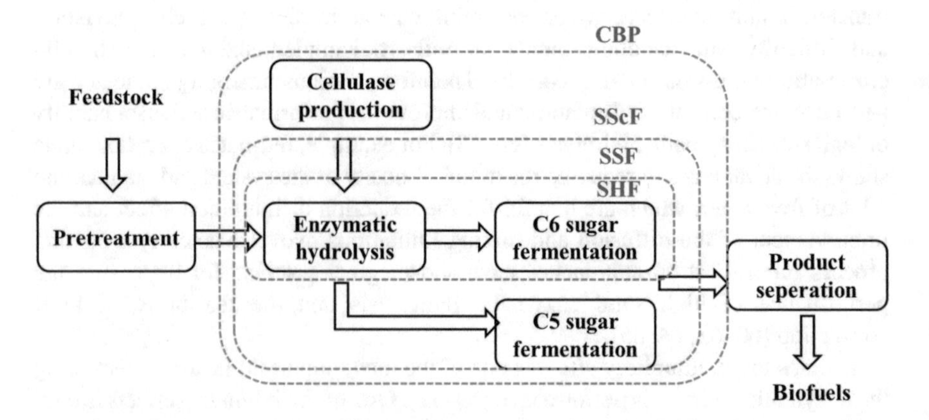

Fig. 1.2 Biological conversion process of lignocellulosic biomass (LCB). *SHF* separate hydrolysis and fermentation; *SSF* simultaneous saccharification and fermentation; *SScF* simultaneous saccharification and co-fermentation; *CBP* consolidated bioprocess

The complete co-fermentation of hexose and pentose sugars to produce biofuels is required to make the conversion process economically feasible [67, 78–80]. Most strains cannot naturally consume xylose or other C5 sugars from LCB, impeding the carbon utilization efficiency and hence resulting in the high cost of conversion process. However, in the past decades, many challenging efforts have been made to improve the xylose fermentation. One of the strategies is to optimize the metabolic pathway through genetic modifications of microbes to improve their metabolic capabilities. Some microbes like yeasts after genetic modification are capable of producing ethanol or accumulating lipids by consuming both glucose and xylose [81–83]. Even the conversion of LCB for biofuels through current fermentation technology is still not economical; emerging synthetic biotechnology combined with the production of high-value co-products such as fatty acids could potentially make the biofuel conversion process profitable [84–86]. Advancements in the fields of systems and synthetic biology will increase the ability to implement different strategies of genetically modified microbes to produce a wide range of products.

The product yield and productivity in fermentation of the hydrolysates is dependent on the amount of inhibitors formed or released from pretreatment and hydrolysis, such as weak acids, furans, and phenolic compounds [76, 79, 87]. Biological, physical, and chemical detoxification methods have been employed to remove inhibitors from the hydrolysates prior to fermentation. As different microbes show different capacities to degrade and tolerate various inhibitors, strain screening is important to enhance the productivity of fermentation by increasing the inhibitor tolerance. Additionally, the genetic and evolutionary engineering strategies are powerful tools to obtain tolerant strains for improving their inhibitor tolerance. Recently, systems and synthetic biology have also provided advanced and powerful approaches for improving tolerance of strains for the biofuel production from LCB.

For fermentation condition and mode, various fermentation strategies, such as batch, fed-batch, and continuous fermentation, have been employed and evaluated to improve the product productivity in fermentation [76, 88–90]. Other promising approaches, such as high cell density, immobilized cell, and cell recirculation, have been exploited to enhance the capacity of the cells to effectively ferment the hydrolysate from LCB.

As the bioconversion of LCB is still in a developmental phase, above limitations need to be overcome and extensive research will be required to make LCB fermentation profitable. For example, new microbes need to be screened and identified and the available microbes need to be engineered for improving the utilization of a broad range of sugars and LCB sources and tolerating the inhibitors generated from pretreatment. Innovative fermentation strategies are required to improve the productivity, facilitate the conversion process, and improve the overall economics of the production process.

1.6 Lignin Valorization

Lignin (15–30%, dry weight of LCB), an abundant renewable alkyl-aromatic polymer found in the cell walls of terrestrial plants, provides structure and rigidity to plant cell walls and offers support and protection. It also enables water and nutrient transport through plant tissues. Lignin is a natural and highly effective barrier against the attack from microbes and chemicals and contributes to biomass recalcitrance [1, 91, 92]. Depending on the plant species, lignin constituents vary considerably, which results in the substantial diversity in the chemistry and structure of lignin polymer.

Lignin is a heterogeneous and recalcitrant polymer (Fig. 1.3). It consists of phenylpropane units synthesized from *p*-hydroxyphenylpropane (H), guaiacyl propane (G), and syringyl propane (S), which are conjugated together via radical coupling reactions to form a variety of ether and C–C chemical bonds, including β-*O*-4-aryl ether, β-5, β–β, and α-*O*-4-aryl ether linkages [1, 92, 93]. Besides these, lignin units also contain various functional groups, including phenolic hydroxyl group, aliphatic hydroxyl group, carbonyl group, and benzyl alcohol group. These varied physicochemical properties contribute to the heterogeneity and recalcitrance of lignin, which impede the deconstruction of LCB, and the depolymerization, separation, purification, and processing of lignin polymer. Therefore, the degree of lignification of LCB will define its deconstruction performance and the costs of the production of biofuels and bio-based products via the biological conversion approaches.

Despite these challenges in its processibility, lignin represents a unique potential feedstock used in the production of fine chemicals and the fabrication of functional

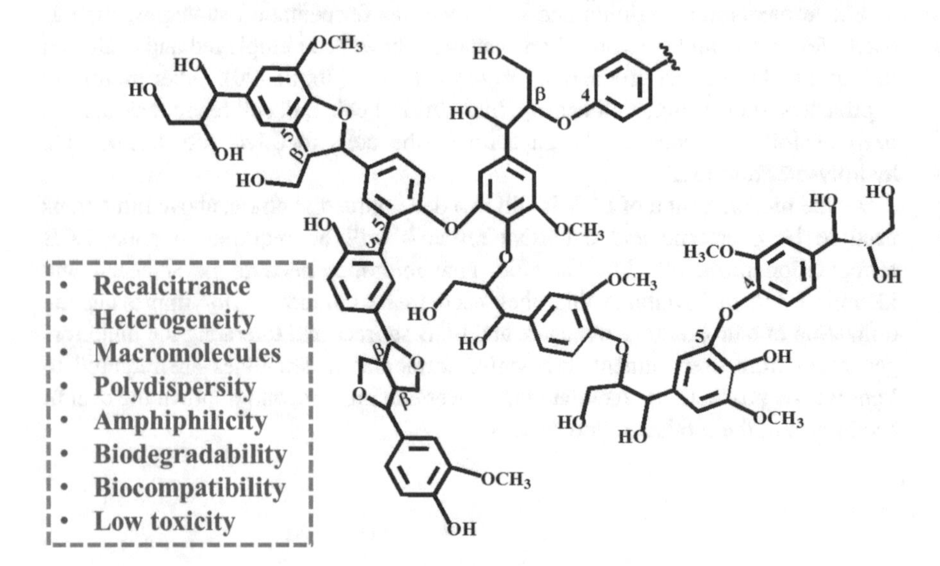

Fig. 1.3 Schematic structures and intrinsic properties of lignin polymer in lignocellulosic biomass

materials [1, 14, 90–92, 94, 95]. As lignin is a renewable reservoir of high-molecular-weight aromatic polymer as well as the aromatic building blocks, it can create pathway and avenues via the chemical and biological conversion of lignin to chemicals and fuels. Lignin also possesses numerous other eco-friendly properties, including biodegradability, biocompatibility, and low toxicity, and thus it can be valorized to produce environmentally friendly materials applied in the fields of drug delivery, tissue engineering, molecular imaging, and others [8, 96–98]. Besides these, the chemistry, reactivity, and processibility of lignin can be modified and altered during the processing of LCB to facilitate the lignin valorization.

Despite these potentials, the high-value conversion of lignin to biofuels, chemicals, and materials has been a challenging task due to the intrinsic heterogeneity and recalcitrance, the condensation and degradation of the lignin macromolecules, and the generation of relatively intractable solid residues during the processing. Therefore, effective strategies are needed to be designed for advanced applications of chemicals and materials using lignin as renewable sources.

1.7 Modern Biorefinery Concept with Composition Valorization for Multiproducts

A biorefinery is generally an overall concept of the processing of LCB as contrasting with the petrochemical refinery. In a biorefinery, LCB feedstocks are converted into a multiplicity of valuable products (biofuels, chemical, materials) and energy with the integration of various processes, technologies, and equipment [99–102]. Currently, most biorefineries with integrated biological conversion processes comprise five major sections: biomass harvest and storage, pretreatment, enzymatic hydrolysis, fermentation of sugar to fuels, and product separation. The development and implementation of biorefineries is vital to meet the perspective toward a sustainable economy from renewable LCB. Besides that, biorefineries produce renewable fuels and products from LCB, which have become increasingly urgent in the face of current global problems and shown the potential to meet the increasing energy demand and reduce the greenhouse gas emissions. To achieve the goals of sustainable development of human society, biorefineries have to play a dominant role in the twenty-first century. Despite nearly a century of research and development attempting to convert LCB into valuable products, conversional biorefinery scenarios still remain uncommercialized.

An emerging concept of modern biorefineries is the conversion of the multicomponents of renewable LCB to generate various biofuels and valuable commodities with green processes and sustainable technologies (Figs. 1.4 and 1.5). As the success of biorefineries is dependent on the full conversion of the three components (cellulose, hemicellulose, and lignin) of LCB, the drive toward lignin valorization offers unique opportunities to make biorefineries economically feasible and has witnessed a significant resurgence in recent years [1]. Lignin is an abundant

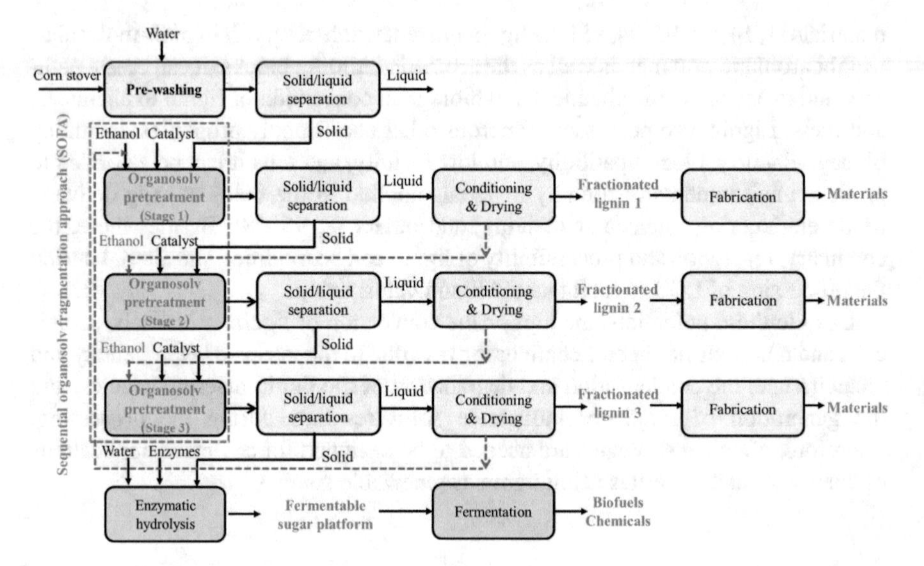

Fig. 1.4 The frame example of a modern biorefinery concept for the co-production of biofuels, chemicals, and materials from lignocellulosic biomass

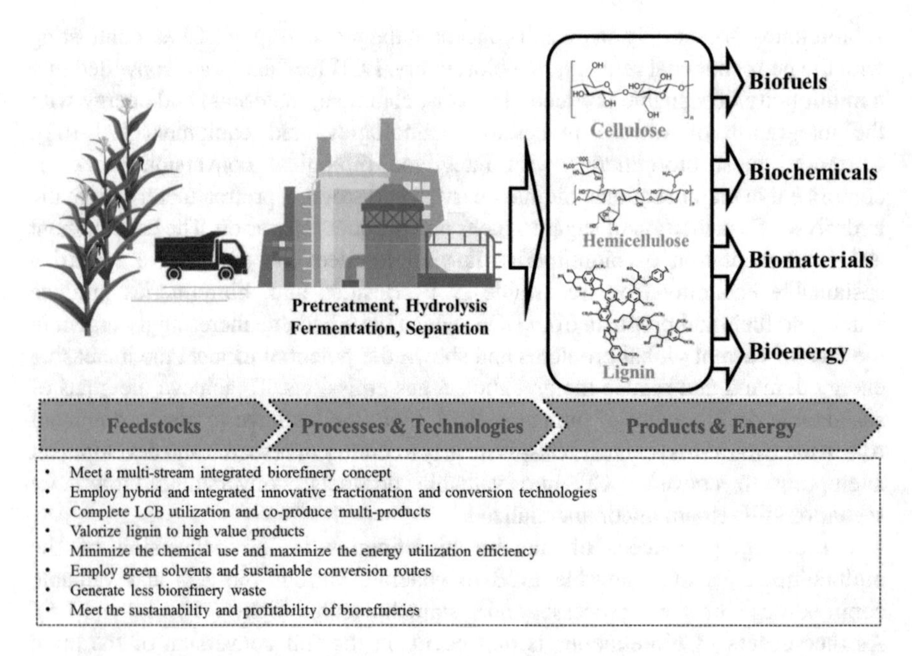

Fig. 1.5 Biorefinery concept for the production of biofuels, chemicals, and materials from lignocellulosic biomass

terrestrial polymer on Earth and the only large-volume renewable feedstock that is composed of aromatics. The first consideration of conventional biorefineries generally focuses on the conversion of carbohydrates into fuels and chemicals, whereas lignin-enriched biorefinery solid is considered as wastes to be burned and only utilized with a low-value way. The advent of biorefineries that convert plant biomass into ethanol will generate substantially more lignin. Most importantly, as biomass recalcitrance is associated with the presence of LCC, the inherent value of LCB highly depends on deconstructing and valorizing lignin polymer in a biorefinery, and not just exploiting the more uniform carbohydrates. Therefore, efforts are still needed to develop novel process and biorefinery scenarios to transform lignin into higher-value co-products in a biorefinery. To achieve sustainability and feasibility, ideal innovative modern biorefineries could meet the following objectives:

1. Meet a multistream integrated biorefinery concept.
2. Employ hybrid and integrated innovative fractionation and conversion technologies.
3. Complete LCB utilization and co-produce multiproducts.
4. Valorize lignin to high-value products.
5. Minimize the chemical use and maximize the energy utilization efficiency.
6. Employ green solvents and sustainable conversion routes.
7. Generate less biorefinery waste.
8. Meet the sustainability and profitability of biorefineries.

The effective utilization of LCB feedstocks in large-scale applications requires to develop and implement the conversion process with multitechnologies for multiproducts and facilitate the profitability of innovative modern biorefinery. In other words, the approach adapting a multistream integrated concept is of utmost importance to the realization of a profitable biorefinery industry. Figure 1.4 shows an example of the biorefinery scenario for biofuels and co-production of valuable lignin-derived product, which would be the main frame of a modern biorefinery concept. To address the challenge in biorefinery and facilitate the full conversion of LCB, an innovative sequential organosolv fragmentation approach (SOFA) by using green solvent plus different-stage catalysts was developed to deconstruct the corn stover biomass, overcome the recalcitrance of biomass and lignin, improve the hydrolysis performance, and selectively dissolve lignin for producing multiple uniform lignin streams [47, 98]. In a biorefinery, this approach applied the integrated multistream concept by using integrated innovative conversion technologies and co-production of multiproducts. SOFA significantly improved the carbohydrate output and obtained an increased monomer-sugar release. The sugar platform can be fermented to convert into biofuels and chemicals. In this biorefinery scenario, the chemistry and reactivity of lignin have been tailored by SOFA to produce lignin with suitable properties for fabricating high-value product, lignin nanoparticles. Lignin nanoparticles have exhibited high quality with a spherical shape, small effective diameter, and good stability. The process holds promise for improving biorefinery cost-effectiveness. It synergistically improved the carbohydrate and lignin output from LCB and enhanced the lignin processability for the fabrication of high-value materials. By employing the low holding temperature and green solvent, it minimized the energy input and made the process clean and sustainable. Besides these, SOFA led to less generation of biorefinery waste and reduced the post-treatment cost of wastes. Overall, this scenario provided a frame example of a modern biorefinery concept for the co-production of biofuels, chemicals, and materials from LCB. It represents a sustainable means for upgrading the low-value lignin in a biorefinery concept and could contribute to the profitability of biorefineries.

Although it is likely that many biofuels and chemicals will be produced from LCB as alternatives to those currently produced from petroleum, several challenges still remain to accelerate the industrial implementation of biorefineries (Fig. 1.5). The biorefinery concept is still not fully developed, so it is important to formulate standards for the biorefineries and develop green sustainable processes and technologies to make a biorefinery profitable. The effective integrated process configurations are required for the advances in the cost-effective conversion of LCB. To improve long-term sustainability, modern biorefineries should consider how to develop microbial strains and accompanying processes for the production of building block and/or platform chemical derivatives from LCB and their further commercialization. Recent breakthroughs have been made to provide a foundation for lignin valorization; however, the studies on the effective fractionation of lignin polymer and the synthesis process of targeted products still need to be addressed to facilitate the lignin valorization toward commercial relevance. In addition, lignin valorization needs to be geared toward producing an entirely new line of chemicals and products with high-value utilization, and these processes and products need to be integrated into existing refineries or chemical plants for improving the sustainability and profitability of modern biorefinery.

References

1. Ragauskas, A. J., et al. (2014). Lignin valorization: Improving lignin processing in the biorefinery. *Science, 344*, 709.
2. Lynd, L. R. (2017). The grand challenge of cellulosic biofuels. *Nature Biotechnology, 35*, 912–915.
3. Cantero, D., et al. (2019). Pretreatment processes of biomass for biorefineries: Current status and prospects. *Annual Review of Chemical and Biomolecular Engineering, 10*, 289–310.
4. Alonso, D. M., et al. (2017). Increasing the revenue from lignocellulosic biomass: Maximizing feedstock utilization. *Science Advances, 3*, e1603301.
5. Liu, Z. H., et al. (2019). Identifying and creating pathways to improve biological lignin valorization. *Renewable and Sustainable Energy Reviews, 105*, 349–362.
6. Jin, M. J., Gunawan, C., Uppugundla, N., Balan, V., & Dale, B. E. (2012). A novel integrated biological process for cellulosic ethanol production featuring high ethanol productivity, enzyme recycling and yeast cells reuse. *Energy & Environmental Science, 5*, 7168–7175.
7. Himmel, M. E., et al. (2007). Biomass recalcitrance: Engineering plants and enzymes for biofuels production. *Science, 315*, 804–807.
8. Mohanty, A. K., Vivekanandhan, S., Pin, J. M., & Misra, M. (2018). Composites from renewable and sustainable resources: Challenges and innovations. *Science, 362*, 536–542.

9. Seidl, P. R., & Goulart, A. K. (2016). Pretreatment processes for lignocellulosic biomass conversion to biofuels and bioproducts. *Current Opinion in Green and Sustainable Chemistry, 2*, 48–53.
10. Zhang, X., Tu, M. B., & Paice, M. G. (2011). Routes to potential bioproducts from Lignocellulosic biomass lignin and hemicelluloses. *Bioenergy Research, 4*, 246–257.
11. Liu, Z. H., et al. (2013). Effects of biomass particle size on steam explosion pretreatment performance for improving the enzyme digestibility of corn stover. *Industrial Crops and Products, 44*, 176–184.
12. Kumar, R., Singh, S., & Singh, O. V. (2008). Bioconversion of lignocellulosic biomass: Biochemical and molecular perspectives. *Journal of Industrial Microbiology & Biotechnology, 35*, 377–391.
13. Kumari, D., & Singh, R. (2018). Pretreatment of lignocellulosic wastes for biofuel production: A critical review. *Renewable and Sustainable Energy Reviews, 90*, 877–891.
14. Ragauskas, A. J., et al. (2006). The path forward for biofuels and biomaterials. *Science, 311*, 484–489.
15. Mohanty, A. K., Misra, M., & Drzal, L. T. (2002). Sustainable bio-composites from renewable resources: Opportunities and challenges in the green materials world. *Abstracts of Papers of the American Chemical Society, 223*, D70–D70.
16. Ozdenkci, K., et al. (2017). A novel biorefinery integration concept for lignocellulosic biomass. *Energy Conversion and Management, 149*, 974–987.
17. Kawaguchi, H., Hasunuma, T., Ogino, C., & Kondo, A. (2016). Bioprocessing of bio-based chemicals produced from lignocellulosic feedstocks. *Current Opinion in Biotechnology, 42*, 30–39.
18. Tilman, D., Hill, J., & Lehman, C. (2006). Carbon-negative biofuels from low-input high-diversity grassland biomass. *Science, 314*, 1598–1600.
19. Hassan, S. S., Williams, G. A., & Jaiswal, A. K. (2019). Lignocellulosic biorefineries in Europe: Current state and prospects. *Trends in Biotechnology, 37*, 231–234.
20. Rinaldi, R., et al. (2016). Paving the way for lignin valorisation: Recent advances in bioengineering, Biorefining and Catalysis. *Angewandte Chemie, International Edition, 55*, 8164–8215.
21. Renders, T., Van den Bosch, S., Koelewijn, S. F., Schutyser, W., & Sels, B. F. (2017). Lignin-first biomass fractionation: The advent of active stabilisation strategies. *Energy & Environmental Science, 10*, 1551–1557.
22. Amiri, M. T., Dick, G. R., Questell-Santiago, Y. M., & Luterbacher, J. S. (2019). Fractionation of lignocellulosic biomass to produce uncondensed aldehyde-stabilized lignin. *Nature Protocols, 14*, 921–954.
23. Mood, S. H., et al. (2013). Lignocellulosic biomass to bioethanol, a comprehensive review with a focus on pretreatment. *Renewable and Sustainable Energy Reviews, 27*, 77–93.
24. Balat, M. (2011). Production of bioethanol from lignocellulosic materials via the biochemical pathway: A review. *Energy Conversion and Management, 52*, 858–875.
25. Ding, S. Y., et al. (2012). How does plant Cell Wall nanoscale architecture correlate with enzymatic digestibility? *Science, 338*, 1055–1060.
26. Agbor, V. B., Cicek, N., Sparling, R., Berlin, A., & Levin, D. B. (2011). Biomass pretreatment: Fundamentals toward application. *Biotechnology Advances, 29*, 675–685.
27. Saha, B. C. (2003). Hemicellulose bioconversion. *Journal of Industrial Microbiology & Biotechnology, 30*, 279–291.
28. Girio, F. M., et al. (2010). Hemicelluloses for fuel ethanol: A review. *Bioresource Technology, 101*, 4775–4800.
29. Liu, Z. H., et al. (2019). Cooperative valorization of lignin and residual sugar to polyhydroxyalkanoate (PHA) for enhanced yield and carbon utilization in biorefineries. *Sustainable Energy & Fuels, 3*, 2024–2037.
30. Zhang, Y. H. P. (2008). Reviving the carbohydrate economy via multi-product lignocellulose biorefineries. *Journal of Industrial Microbiology & Biotechnology, 35*, 367–375.

31. Chen, H. Z., & Liu, Z. H. (2015). Steam explosion and its combinatorial pretreatment refining technology of plant biomass to bio-based products. *Biotechnology Journal, 10*, 866–885.
32. Zhao, X. B., Zhang, L. H., & Liu, D. H. (2012). Biomass recalcitrance. Part I: the chemical compositions and physical structures affecting the enzymatic hydrolysis of lignocellulose. *Biofuels, Bioproducts and Biorefining, 6*, 465–482.
33. Holwerda, E. K., et al. (2019). Multiple levers for overcoming the recalcitrance of lignocellulosic biomass. *Biotechnol Biofuels, 12*, 1–12.
34. Tarasov, D., Leitch, M., & Fatehi, P. (2018). Lignin-carbohydrate complexes: Properties, applications, analyses, and methods of extraction: A review. *Biotechnology for Biofuels, 11*, 269.
35. Liu, Z. H., & Chen, H. Z. (2016). Mechanical property of different corn Stover morphological fractions and its correlations with high solids enzymatic hydrolysis by periodic peristalsis. *Bioresource Technology, 214*, 292–302.
36. Liu, Z. H., Qin, L., Li, B. Z., & Yuan, Y. J. (2015). Physical and chemical characterizations of corn Stover from leading pretreatment methods and effects on enzymatic hydrolysis. *ACS Sustainable Chemistry & Engineering, 3*, 140–146.
37. Mosier, N., et al. (2005). Features of promising technologies for pretreatment of lignocellulosic biomass. *Bioresource Technology, 96*, 673–686.
38. Yang, B., & Wyman, C. E. (2008). Pretreatment: The key to unlocking low-cost cellulosic ethanol. *Biofuels, Bioproducts and Biorefining, 2*, 26–40.
39. Wyman, C. E., et al. (2005). Coordinated development of leading biomass pretreatment technologies. *Bioresource Technology, 96*, 1959–1966.
40. Sindhu, R., Binod, P., & Pandey, A. (2016). Biological pretreatment of lignocellulosic biomass - an overview. *Bioresource Technology, 199*, 76–82.
41. Liu, Z. H., & Chen, H. Z. (2017). Two-step size reduction and post-washing of steam exploded corn stover improving simultaneous saccharification and fermentation for ethanol production. *Bioresource Technology, 223*, 47–58.
42. Liu, Z. H., et al. (2013). Evaluation of storage methods for the conversion of corn stover biomass to sugars based on steam explosion pretreatment. *Bioresource Technology, 132*, 5–15.
43. Feng, L., et al. (2014). Combined severity during pretreatment chemical and temperature on the Saccharification of wheat straw using acids and alkalis of differing strength. *BioResources, 9*, 24–38.
44. Shen, X. J., et al. (2019). Facile fractionation of lignocelluloses by biomass-derived deep eutectic solvent (DES) pretreatment for cellulose enzymatic hydrolysis and lignin valorization. *Green Chemistry, 21*, 275–283.
45. Hassan, S. S., Williams, G. A., & Jaiswal, A. K. (2018). Emerging technologies for the pretreatment of lignocellulosic biomass. *Bioresource Technology, 262*, 310–318.
46. Liu, Z. H., et al. (2017). Synergistic maximization of the carbohydrate output and lignin processability by combinatorial pretreatment. *Green Chemistry, 19*, 4939–4955.
47. Liu, Z. H., et al. (2019). Codesign of combinatorial Organosolv pretreatment (COP) and lignin nanoparticles (LNPs) in biorefineries. *ACS Sustainable Chemistry & Engineering, 7*, 2634–2647.
48. Zhang, K., Pei, Z. J., & Wang, D. H. (2016). Organic solvent pretreatment of lignocellulosic biomass for biofuels and biochemicals: A review. *Bioresource Technology, 199*, 21–33.
49. Kim, K. H., Dutta, T., Sun, J., Simmons, B., & Singh, S. (2018). Biomass pretreatment using deep eutectic solvents from lignin derived phenols. *Green Chemistry, 20*, 809–815.
50. Serna, L. V. D., Alzate, C. E. O., & Alzate, C. A. C. (2016). Supercritical fluids as a green technology for the pretreatment of lignocellulosic biomass. *Bioresource Technology, 199*, 113–120.
51. Zhang, J. W., Zhong, Y. H., Zhao, X. N., & Wang, T. H. (2010). Development of the cellulolytic fungus Trichoderma reesei strain with enhanced beta-glucosidase and filter paper activity using strong artificial cellobiohydrolase 1 promoter. *Bioresource Technology, 101*, 9815–9818.

52. Mota, T. R., de Oliveira, D. M., Marchiosi, R., Ferrarese, O., & dos Santos, W. D. (2018). Plant cell wall composition and enzymatic deconstruction. *AIMS Bioengineering, 5*, 63–77.
53. Chen, H. Z., & Liu, Z. H. (2017). Enzymatic hydrolysis of lignocellulosic biomass from low to high solids loading. *Engineering in Life Sciences, 17*, 489–499.
54. Alvira, P., Tomas-Pejo, E., Ballesteros, M., & Negro, M. J. (2010). Pretreatment technologies for an efficient bioethanol production process based on enzymatic hydrolysis: A review. *Bioresource Technology, 101*, 4851–4861.
55. Chandra, R. P., et al. (2007). Substrate pretreatment: The key to effective enzymatic hydrolysis of lignocellulosics? *Advances in Biochemical Engineering/Biotechnology, 108*, 67–93.
56. Hall, M., Bansal, P., Lee, J. H., Realff, M. J., & Bommarius, A. S. (2010). Cellulose crystallinity – A key predictor of the enzymatic hydrolysis rate. *The FEBS Journal, 277*, 1571–1582.
57. Kumar, R., et al. (2018). Cellulose-hemicellulose interactions at elevated temperatures increase cellulose recalcitrance to biological conversion. *Green Chemistry, 20*, 921–934.
58. Van Dyk, J. S., & Pletschke, B. I. (2012). A review of lignocellulose bioconversion using enzymatic hydrolysis and synergistic cooperation between enzymes-factors affecting enzymes, conversion and synergy. *Biotechnology Advances, 30*, 1458–1480.
59. Yang, Q., & Pan, X. J. (2016). Correlation between lignin physicochemical properties and inhibition to enzymatic hydrolysis of cellulose. *Biotechnology and Bioengineering, 113*, 1213–1224.
60. Jorgensen, H., Vibe-Pedersen, J., Larsen, J., & Felby, C. (2007). Liquefaction of lignocellulose at high-solids concentrations. *Biotechnology and Bioengineering, 96*, 862–870.
61. Modenbach, A. A., & Nokes, S. E. (2012). The use of high-solids loadings in biomass pretreatment-a review. *Biotechnology and Bioengineering, 109*, 1430–1442.
62. Nguyen, T. Y., Cai, C. M., Kumar, R., & Wyman, C. E. (2017). Overcoming factors limiting high-solids fermentation of lignocellulosic biomass to ethanol. *Proceedings of the National Academy of Sciences of the United States of America, 114*, 11673–11678.
63. Jin, M. J., et al. (2017). Toward high solids loading process for lignocellulosic biofuel production at a low cost. *Biotechnology and Bioengineering, 114*, 980–989.
64. Liu, Z. H., & Chen, H. Z. (2016). Periodic peristalsis releasing constrained water in high solids enzymatic hydrolysis of steam exploded corn stover. *Bioresource Technology, 205*, 142–152.
65. Liu, Z. H., & Chen, H. Z. (2016). Periodic peristalsis enhancing the high solids enzymatic hydrolysis performance of steam exploded corn stover biomass. *Biomass and Bioenergy, 93*, 13–24.
66. da Silva, A. S., et al. (2016). High-solids content enzymatic hydrolysis of hydrothermally pretreated sugarcane bagasse using a laboratory-made enzyme blend and commercial preparations. *Process Biochemistry, 51*, 1561–1567.
67. Liu, Z. H., & Chen, H. Z. (2016). Simultaneous saccharification and co-fermentation for improving the xylose utilization of steam exploded corn stover at high solid loading. *Bioresource Technology, 201*, 15–26.
68. Hu, J. G., et al. (2015). The addition of accessory enzymes enhances the hydrolytic performance of cellulase enzymes at high solid loadings. *Bioresource Technology, 186*, 149–153.
69. Xue, S. S., et al. (2015). Sugar loss and enzyme inhibition due to oligosaccharide accumulation during high solids-loading enzymatic hydrolysis. *Biotechnology for Biofuels, 8*, 195.
70. Qin, L., et al. (2016). Inhibition of lignin-derived phenolic compounds to cellulase. *Biotechnology for Biofuels, 9*, 1–10.
71. Li, X., et al. (2018). Inhibitory effects of lignin on enzymatic hydrolysis: The role of lignin chemistry and molecular weight. *Renewable Energy, 123*, 664–674.
72. Binod, P., Janu, K. U., Sindhu, R., & Pandey, A. (2011). Hydrolysis of Lignocellulosic biomass for bioethanol production. *Bioscience, Biotechnology, and Biochemistry*, 229–250, Academic press.

73. Teugjas, H., & Valjamae, P. (2013). Product inhibition of cellulases studied with C-14-labeled cellulose substrates. *Biotechnology for Biofuels, 6*, 1–14.
74. Sun, S. L., Huang, Y., Sun, R. C., & Tu, M. B. (2016). The strong association of condensed phenolic moieties in isolated lignins with their inhibition of enzymatic hydrolysis. *Green Chemistry, 18*, 4276–4286.
75. Yoo, C. G., Li, M., Meng, X. Z., Pu, Y. Q., & Ragauskas, A. J. (2017). Effects of organosolv and ammonia pretreatments on lignin properties and its inhibition for enzymatic hydrolysis. *Green Chemistry, 19*, 2006–2016.
76. Liu, Z. H., Qin, L., Zhu, J. Q., Li, B. Z., & Yuan, Y. J. (2014). Simultaneous saccharification and fermentation of steam-exploded corn stover at high glucan loading and high temperature. *Biotechnology for Biofuels, 7*, 1–16.
77. Zeng, Y. N., Zhao, S., Yang, S. H., & Ding, S. Y. (2014). Lignin plays a negative role in the biochemical process for producing lignocellulosic biofuels. *Current Opinion in Biotechnology, 27*, 38–45.
78. Zhang, G. C., Liu, J. J., Kong, I. I., Kwak, S., & Jin, Y. S. (2015). Combining C6 and C5 sugar metabolism for enhancing microbial bioconversion. *Current Opinion in Chemical Biology, 29*, 49–57.
79. Raud, M., Kikas, T., Sippula, O., & Shurpali, N. J. (2019). Potentials and challenges in lignocellulosic biofuel production technology. *Renewable and Sustainable Energy Reviews, 111*, 44–56.
80. Nogue, V. S., & Karhumaa, K. (2015). Xylose fermentation as a challenge for commercialization of lignocellulosic fuels and chemicals. *Biotechnology Letters, 37*, 761–772.
81. Lau, M. W., & Dale, B. E. (2009). Cellulosic ethanol production from AFEX-treated corn stover using Saccharomyces cerevisiae 424A(LNH-ST). *Proceedings of the National Academy of Sciences of the United States of America, 106*, 1368–1373.
82. Ko, J. K., Um, Y., Woo, H. M., Kim, K. H., & Lee, S. M. (2016). Ethanol production from lignocellulosic hydrolysates using engineered Saccharomyces cerevisiae harboring xylose isomerase-based pathway. *Bioresource Technology, 209*, 290–296.
83. Jin, M. J., Lau, M. W., Balan, V., & Dale, B. E. (2010). Two-step SSCF to convert AFEX-treated switchgrass to ethanol using commercial enzymes and Saccharomyces cerevisiae 424A(LNH-ST). *Bioresource Technology, 101*, 8171–8178.
84. Lee, S. K., Chou, H., Ham, T. S., Lee, T. S., & Keasling, J. D. (2008). Metabolic engineering of microorganisms for biofuels production: From bugs to synthetic biology to fuels. *Current Opinion in Biotechnology, 19*, 556–563.
85. Jullesson, D., David, F., Pfleger, B., & Nielsen, J. (2015). Impact of synthetic biology and metabolic engineering on industrial production of fine chemicals. *Biotechnology Advances, 33*, 1395–1402.
86. Qi, H., Li, B. Z., Zhang, W. Q., Liu, D., & Yuan, Y. J. (2015). Modularization of genetic elements promotes synthetic metabolic engineering. *Biotechnology Advances, 33*, 1412–1419.
87. Palmqvist, E., & Hahn-Hagerdal, B. (2000). Fermentation of lignocellulosic hydrolysates. II: inhibitors and mechanisms of inhibition. *Bioresource Technology, 74*, 25–33.
88. Brethauer, S., & Wyman, C. E. (2010). Review: Continuous hydrolysis and fermentation for cellulosic ethanol production. *Bioresource Technology, 101*, 4862–4874.
89. Puligundla, P., Smogrovicova, D., Mok, C., & Obulam, V. S. R. (2019). A review of recent advances in high gravity ethanol fermentation. *Renewable Energy, 133*, 1366–1379.
90. Liu, Z. H., Xie, S. X., Lin, F. R., Jin, M. J., & Yuan, J. S. (2018). Combinatorial pretreatment and fermentation optimization enabled a record yield on lignin bioconversion. *Biotechnology for Biofuels, 11*, 21.
91. Li, M., Pu, Y. Q., & Ragauskas, A. J. (2016). Current understanding of the correlation of Lignin structure with biomass recalcitrance. *Frontiers in Chemistry, 4(45)*, 1–8.
92. Yang, H. B., et al. (2019). Overcoming cellulose recalcitrance in woody biomass for the lignin-first biorefinery. *Biotechnol Biofuels, 12*, 171.

93. Giummarella, N., Pu, Y. Q., Ragauskas, A. J., & Lawoko, M. (2019). A critical review on the analysis of lignin carbohydrate bonds. *Green Chemistry, 21*, 1573–1595.
94. Wu, X. Y., et al. (2020). Lignin-derived electrochemical energy materials and systems. *Biofuels, Bioproducts and Biorefining, 14*, 650–672.
95. Huang, C., et al. (2019). Bio-inspired nanocomposite by layer-by-layer coating of chitosan/ hyaluronic acid multilayers on a hard nanocellulose-hydroxyapatite matrix. *Carbohydrate Polymers, 222(115036)*, 1–7.
96. Yiamsawas, D., Beckers, S. J., Lu, H., Landfester, K., & Wurm, F. R. (2017). Morphology-controlled synthesis of lignin Nanocarriers for drug delivery and carbon materials. *ACS Biomaterials Science & Engineering, 3*, 2375–2383.
97. Figueiredo, P., et al. (2017). In vitro evaluation of biodegradable lignin-based nanoparticles for drug delivery and enhanced antiproliferation effect in cancer cells. *Biomaterials, 121*, 97–108.
98. Liu, Z. H., et al. (2019). Defining lignin nanoparticle properties through tailored lignin reactivity by sequential organosolv fragmentation approach (SOFA). *Green Chemistry, 21*, 245–260.
99. Jenkins, R., & Alles, C. (2011). Field to fuel: Developing sustainable biorefineries. *Ecological Applications, 21*, 1096–1104.
100. Valdivia, M., Galan, J. L., Laffarga, J., & Ramos, J. L. (2016). Biofuels 2020: Biorefineries based on lignocellulosic materials. *Journal of Microbial Biotechnology, 9*, 585–594.
101. Cherubini, F. (2010). The biorefinery concept: Using biomass instead of oil for producing energy and chemicals. *Energy Conversion and Management, 51*, 1412–1421.
102. Fernando, S., Adhikari, S., Chandrapal, C., & Murali, N. (2006). Biorefineries: Current status, challenges, and future direction. *Energy & Fuels, 20*, 1727–1737.

Chapter 2
Deconstruction of Lignocellulose Recalcitrance by Organosolv Fractionating Pretreatment for Enzymatic Hydrolysis

Ziyuan Zhou, Dehua Liu, and Xuebing Zhao

2.1 Introduction of Lignocellulosic Biomass

Generally, lignocellulosic biomass (LCB) mainly exists in the form of forestry; agricultural residues such as corn stover, sugarcane bagasse, wheat straw, etc.; and industrial wastes (e.g., furfural residues) [81, 149, 161]. In nature, LCB is an abundant feedstock with a yearly production of higher than 10^{10} million tons [82]. Nowadays, LCB has become an attractive raw material in biorefinery for production of biofuels, chemicals, and materials due to its low cost and widespread availability [117].

LCB consists of three main polymeric compositions, namely, cellulose, hemicellulose, and lignin, with a relative low portion of extractives, ash, proteins, and so on [58]. The structure of lignocellulosic biomass is schematically shown in Fig. 2.1. Cellulose is the major component of LCB with repeating disaccharide cellobiose as monomer units [53]. Its structure is composed of massive intramolecular and intermolecular hydrogen bonding [71, 137], which makes it difficult to hydrolyze [82]. Hemicelluloses are the second abundant carbohydrates in LCB [169]. Being different from cellulose, hemicelluloses are a group of heteropolysaccharides consisting of several units like D-xylose, D-glucose, D-mannose, etc. [17]. They bind to cellulose via hydrogen bonds and to lignin via covalent bond [17], increasing the strength of cell wall. Unlike cellulose, hemicelluloses with variable and amorphous structures are easy to degrade by enzymes or chemicals [46]. Lignin is another major component formed by three types of structural units (guaiacyl, sringyl, and

Z. Zhou · D. Liu · X. Zhao (✉)
Key Laboratory of Industrial Biocatalysis, Ministry of Education, Tsinghua University, Beijing, China

Institute of Applied Chemistry, Department of Chemical Engineering, Tsinghua University, Beijing, China
e-mail: zhaoxb@mail.tsinghua.edu.cn

Z.-H. Liu, A. Ragauskas (eds.), *Emerging Technologies for Biorefineries, Biofuels, and Value-Added Commodities*,
https://doi.org/10.1007/978-3-030-65584-6_2

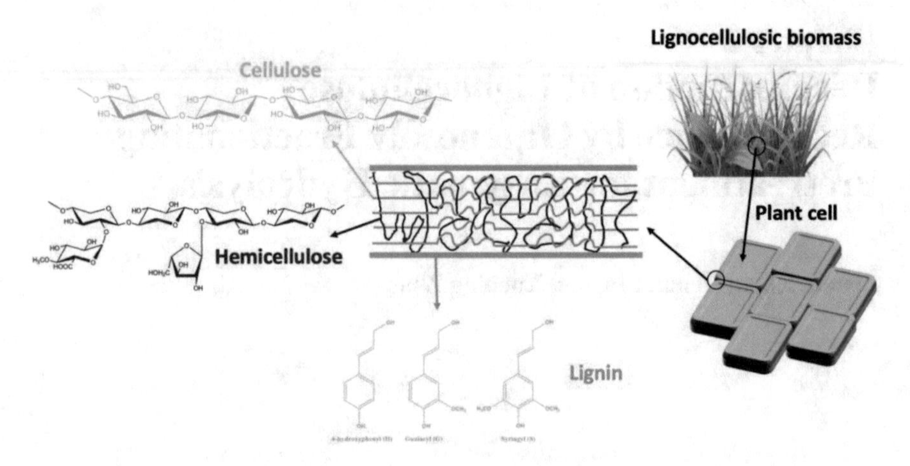

Fig. 2.1 Structure of lignocellulosic biomass [53, 67]

p-hydroxyphenyl propane) [14]. The existence of lignin could improve the compressive strength and stiffness of cell wall [140]. Cellulose, hemicellulose, and lignin nonuniformly distribute in the cell wall layers. Depending on the species, tissues, and maturity of cell walls, these polymers are organized into varying structure and quantity [23]. Different types of LCB have different contents of the major compositions. Table 2.1 summarizes the chemical compositions of different types of LCB, including woody and nonwoody biomass and other wastes.

2.2 The Role of Pretreatment on Lignocellulose Bioconversion

The poly-hierarchical structure and multicomponent laminate compositions of cell wall greatly contribute to the recalcitrance of lignocellulose for efficient enzymatic deconstruction and saccharification [126]. The main factors affecting enzymatic digestibility of biomass can be divided into two types, namely, the enzyme-related and substrate-related factors. Enzyme-related factors generally refer to the enzyme activity and synergism of the enzyme cocktail components [158], while the substrate-related factors mainly refer to the cellulose accessibility to enzymes, which is affected by contents of hemicelluloses and lignin, lignin structure and distribution in cell wall, cellulose crystallinity, specific surface area, porosity, and pore structure of the substrates.

Cellulose crystallinity has been considered as an important factor affecting biomass digestibility. In general, cellulose in crystalline state is more difficult to hydrolyze by chemical and biological methods than cellulose in amorphous state. Cellulase with a size of 3–8 nm is only able to digest crystalline cellulose from its surface; thus, the enzymatic hydrolysis efficiency is relatively low [2, 159]. It has been found that the regenerated cellulose with low crystallinity index achieves a

Table 2.1 Chemical composition of some typical lignocellulosic biomass

Classification	LCB type	Cellulose/%	Hemicellulose/%	Lignin/%	References
Softwood	Pine wood	42	14.7	26.3	Du et al. [25]
	Spruce	38.5	4.6	50.6	Cannella et al. [15]
Hardwood	Hybrid poplar	48.95	17.85	23.25	Pan et al. [105]
	Eucalyptus	54.1	18.4	21.5	Isikgor and Becer [53]
	Oak	40.4	35.9	24.1	
	Bamboo	43	23.1	26.2	Tri et al. [134]
Agricultural wastes	Sugarcane bagasse	32–45	20–32	17–22	Arni [5]; Timung et al. [132]
	Rice straw	36.1	24.7	16.8	Zhang et al. [152]
	Rice husks	30.1	13.5	22.4	Ebrahimi et al. [27]
	Corn stover (leaves, stalk, and husk)	35.6–41.2	26.1–37.4	10.7–21	Liu et al. [84] and Reddy and Yang [114]
	Wheat straw	40.5–42.8	25–26	18–19	Tomás-Pejó et al. [133] and Yuan et al. [150]
	Barley straw	38.08	22.63	22.27	Kim et al. [62]
	Oat straw	38.3	22.4	27.8	Dziekońska-Kubczak et al. [26]
	Cotton stalk	31	11	30	Rubio et al. [119]
	Sorghum straw	32.0–35.0	24.0–27.0	15.0–21.0	Isikgor and Becer [53]
Herbaceous crops	Switchgrass (*Panicum virgatum* L.)	31–39	25–33	17–18	Papa et al. [109]
Other wastes	Furfural residues	45.3	0	44.9	Yu et al. [147]
	Newspaper mixture	60.3	16.4	12.4	Lee et al. [70]

higher enzymatic digestibility than the cellulose with high crystallinity index. Specific surface area also has an influence on the cellulose accessibility to enzymes and initial reaction rate [126]. Increasing specific surface areas is usually helpful to improve enzymatic digestibility because most cellulose is exposed to enzyme when the specific surface area increases [89]. The contents and distribution of hemicelluloses and lignin in biomass have a great influence on lignocellulose digestibility because the existence of hemicelluloses and lignin hinders the contact between cellulose and enzyme to a great extent, leading to a low enzymatic hydrolysis efficiency for raw biomass [2, 104]. In addition, nonproductive binding and inactivation of enzymes by lignin are also found to limit cellulose enzymatic hydrolysis [123]. Being unlike the barrier effect from hemicellulose and lignin, nonproductive binding of cellulases by lignin affects the enzymatic digestibility by decreasing the availability of enzymes. Therefore, removing hemicelluloses and lignin can improve enzymatic digestibility significantly by increasing the pore size and accessible specific surface area and reducing the nonproductive binding of cellulase enzymes [100, 126, 128].

Pretreatment is a crucial step for enzymatic hydrolysis of cellulose by cellulase cocktails to produce glucose prior to subsequent fermentation. It also has essential influences on other processes, such as size reduction requirements, enzyme loading, and rates of enzymatic hydrolysis and toxicity of fermentation [143, 151]. An ideal pretreatment process should improve the enzymatic digestibility efficiently with low cost but generate low concentration of degradation products that inhibit enzymatic hydrolysis and fermentation [110]. Recently, a wide range of pretreatment technologies has been developed as reviewed by Kumar and Sharma [68]. All of these pretreatments aim to increase cellulose accessibility by removing hemicelluloses and/or lignin, altering cell structure or modifying lignin chemical structure and surface properties [99]. Unfortunately, nowadays no pretreatment technology could achieve 100% fermentable sugar yield by enzymatic hydrolysis [99]. Some of the traditional methods still only obtain low fermentable sugar yield which is primarily caused by the high residual lignin content in the pretreated solids [161].

2.3 Overview of Organosolv Fractionating Pretreatment

Currently, organosolv fractionating pretreatment (OFP) is a promising method since it is able to remove most of lignin and hemicelluloses and generate a cellulose-rich residue for enzymatic hydrolysis. High-purity lignin could be precipitated by addition of water into the spent liquor, while hemicellulosic sugars and other degradation products (such as furfural, acetic acid) still remain in the liquid [160]. In a biorefinery, this method allows for the integrated utilization of all the major components of LCB to produce high-value- added products with high potential revenue [161]. Since 2010, the interest in OFP has an upward trend as the number of relevant published papers is steadily increasing year by year (Fig. 2.2). Usually, organosolv fractionating pretreatment is performed using a wide range of organic or aqueous-organic solvent systems in presence or absence of catalysts in temperature range from 100 to 250 °C [102]. Mineral acids such as HCl, H_2SO_4, and H_3PO_4 are commonly used to increase delignification and hemicellulose degradation [160]. In addition, some organic acids (e.g., formic acid, acetic acid) can be used as both solvent and acid catalyst during OFP [127]. Addition of alkaline catalysts such as NaOH, KOH, lime, or NH_4OH is helpful to cause the swelling of LCB and accelerate the organosolv delignification [170]. Apart from traditional organosolv pretreatment methods, some novel methods combining with deep eutectic solvent (DES) or ionic liquids have also gained much attention in recent year [55, 56, 86]. According to the solvent used in pretreatment, OFP could be classified into six types, namely, alcohol-based, organic acid-based, ketone-based, combination with ionic liquid, combination with DES, and others (like phenolic-based, ether-based, and amide-based), as summarized in Table 2.2.

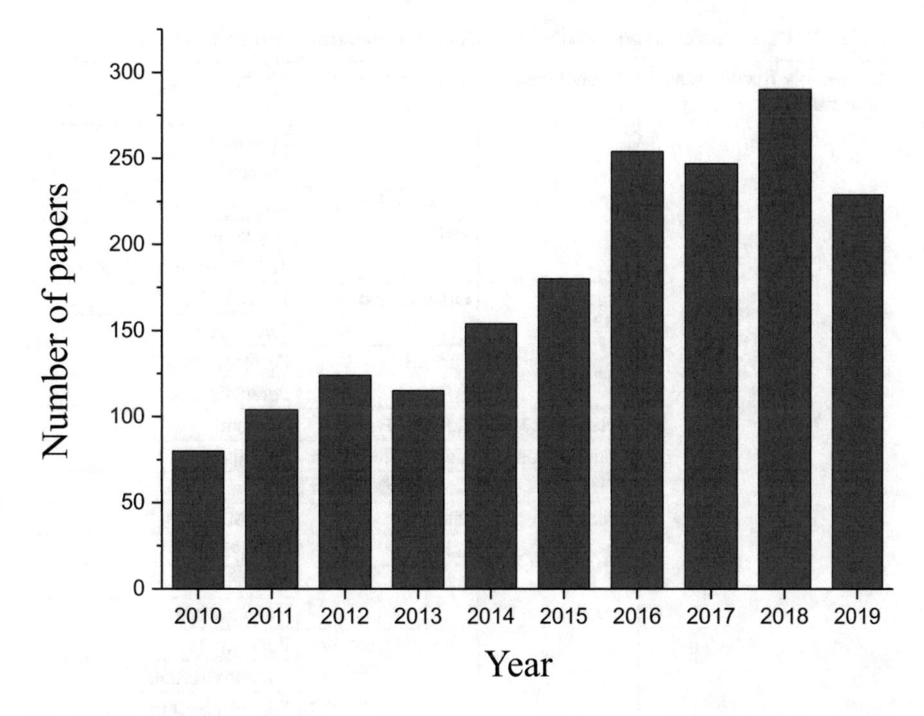

Fig. 2.2 Number of papers on organosolv pretreatment published since 2010 (Results searched by keywords including organosolv and organosolv fractionating pretreatment in Web of Science database)

The typical process for OFP includes unit operations of organosolv cooking, solvent recovery, and lignin recovery [161] (Fig. 2.3). In this chapter, we mainly focus on the recent progress of OFP of LCB for enzymatic hydrolysis of cellulose. Main types of OFP like alcohol-based and organic acid-based are presented in detail, including the process description, mechanisms for improvement of cellulose digestibility, and reaction of lignins. In addition, organosolv-based biorefineries and potential applications of cellulose-rich solid, hemicellulosic sugars, and high-purity lignin are presented. The challenges and perspective of OFP are discussed finally.

2.4 Organosolv Fractionating Pretreatment and Mechanisms

2.4.1 Alcohol-Based Fractionating Pretreatment

2.4.1.1 Process Description

Alcohols are the most frequently used solvents for OFP. Low-boiling-point alcohols (e.g., methanol and ethanol) are more attractive than high-boiling-point alcohols (e.g., ethylene glycol and glycerol) for LCB delignification due to their low cost, easy recovery using simple distillation, and water-miscible property.

Table 2.2 Classification of organosolv fractionating pretreatment based on solvent type

Organosolv fractionating pretreatment	Alcohol-based	Low-boiling-point alcohol	Methanol
			Ethanol
			Propanol
			Butanol
		High-boiling-point alcohol	Ethylene glycol
			Glycerol
			Tetrahydrofurfuryl alcohol
	Organic acid-based	Organic acid	Formic acid
			Acetic acid
		Peracid	Performic acid
			Peracetic acid
	Ketone-based	Simple ketone	Acetone
	Combination	Combination with ionic liquid	
		Combination with DES	
	Others	Phenolic	Phenol
		Ether	Diethyl ether
		Acetate	Methyl acetate
			Ethyl acetate
		Amide	Formamide
			Dimethylformamide
			Dimethylacetamide
		Others	Dioxane, gamma-valerolactone, etc.

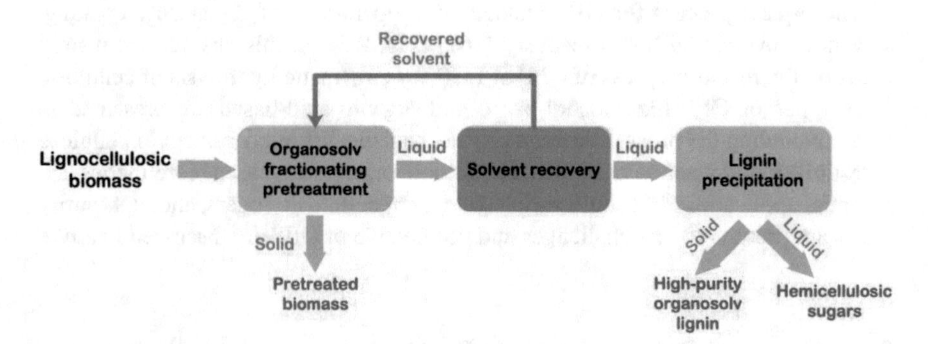

Fig. 2.3 Typical process of organosolv fractionating pretreatment

The typical process scheme of low-boiling-point alcohol organosolv pretreatment is described in Fig. 2.4 [9, 160]. LCB is first cooked with alcohol-water solvent at 100–250 °C. The addition of catalysts could assist pretreatment to be performed under milder conditions compared with that without addition of catalysts. For catalyzed alcohol-based organosolv pretreatment, a variety of chemicals could be used as catalysts for delignification, such as mineral acids [18], organic

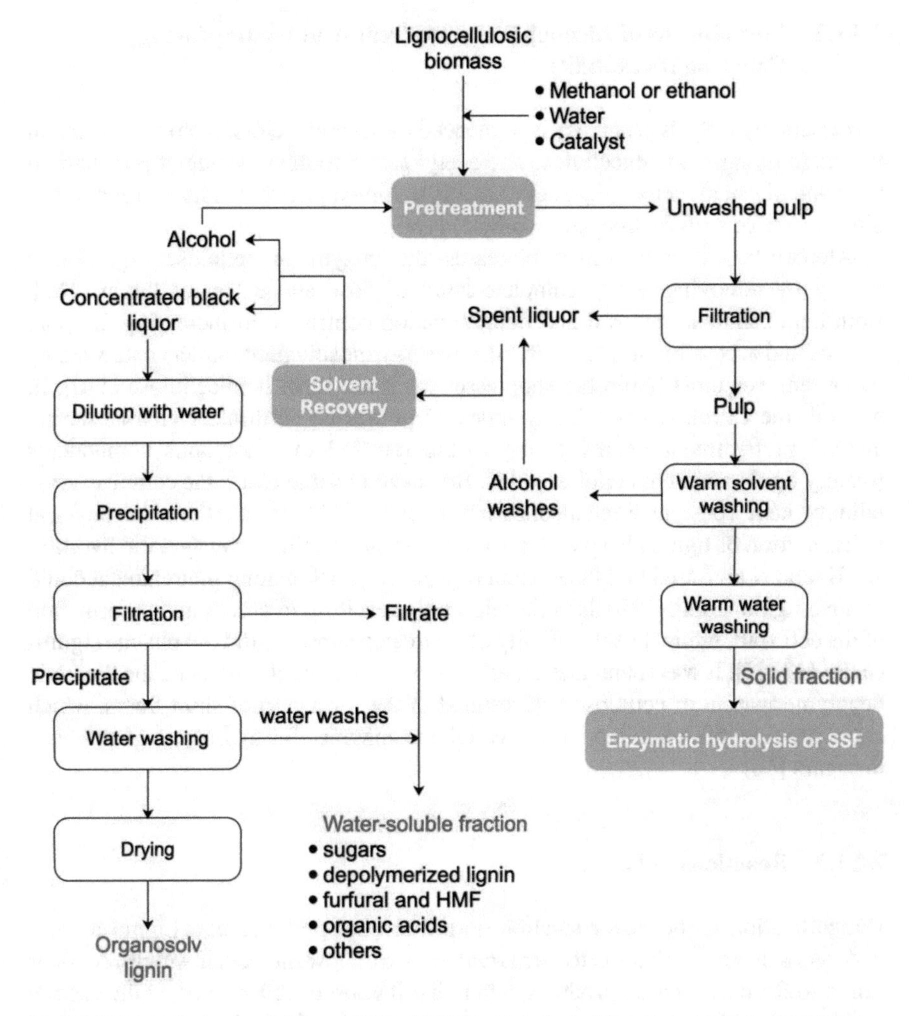

Fig. 2.4 Typical process scheme of low-boiling-point alcohol-based organosolv fractionating pretreatment

acids [1], inorganic bases [2], organic bases [130], and inorganic salts [61]. Being different from acid catalysts, alkaline catalysts could suppress the degradation of carbohydrates (mainly cellulose and hemicelluloses) [2]. After pretreatment, the pretreated biomass is washed with alcohol followed by water. The cellulose-rich solid fraction could be used for enzymatic hydrolysis and fermentation. The spent liquor is discharged from reactor, then evaporated, and condensed. The organic fraction is recycled to pretreatment process. Finally, the condensed spent liquor is diluted with water followed by precipitation and filtration. The precipitated solid mainly consists of lignin, while the filtrate is an aqueous solution composed of water-soluble compounds including sugars, furfural, HMF, and organic acids [9].

2.4.1.2 Mechanisms of Alcohol-Based Pretreatment for Improving Cellulose Digestibility

Enzymatic hydrolysis efficiency is influenced by several factors, including chemical factors (e.g., lignin, hemicellulose and acetyl group content, degree of polymerization) and physical factors (e.g., cellulose crystallinity, physical redistribution of lignin, accessible surface area, pore volume) [160].

Alcohol-based pretreatment increases the enzymatic cellulose digestibility mainly by removing nearly complete hemicellulose and extensive lignin [160]. Both hemicellulose removal and delignification contribute to increasing the pore volume and accessible surface area. However, organosolv pretreatment catalyzed by bases removes most lignin but suppresses the removal of hemicellulose [170]. In general, the cellulose crystallinity index of pretreated substrate increases during alcohol pretreatment mainly owing to the removal of amorphous components (mainly lignin and hemicellulose) [18]. However, in some cases, the cellulose crystallinity kept constant after alcohol pretreatment [16]. In addition, the physical redistribution of lignin also plays an important role in cellulose enzymatic hydrolysis. Hallac et al. found that the ethanol organosolv fractionating pretreatment could remove lignin from middle lamella selectively resulting in cracks and deformation of the cell wall, while the crystallinity of pretreated substrate did not change significantly [43, 44]. It was found that the ethanol pretreatment also induced the dramatic depolymerization of cellulose and resulted in the formation of short fibers, which could provide more binding sites to cellulosic enzymes for hydrolysis of cellulose to sugars [16].

2.4.1.3 Reactions of Lignin

Delignification is the major reaction during alcohol pretreatment. Lignin usually undergoes depolymerization to form fragments with low molecular weight, condensation to form condensed products, and redistribution in cell wall or on the surface of fiber [73, 161]. During alcohol pretreatment, the lignin hydrolysis reactions mainly occurred at the side chains and lignin-hemicellulose bonds. In acidic medium, cleavage of *a*-aryl ether linkage is one of the main reactions, which results in the formation of benzyl carbocation [73]. But β-aryl ether (β-O-4) cleavage is more important than *a*-aryl ether cleavage especially when the pretreatment is under highly serious conditions, because β-aryl ether occupies 40% to 65% of the total linkages in lignins [8]. As found by El Hage et al. [29], cleavage of β-aryl ether linkages leads to an increase of phenolic hydroxyl groups. The cleavage of β-aryl linkage during acid-catalyzed ethanol pretreatment usually takes place in two ways as shown in Fig. 2.5a,b: with elimination of formaldehyde or by formation of Hibbert's ketones [91]. In alkaline medium, β-aryl ether linkage in both phenolic arylpropane and nonphenolic arylpropane units can be cleaved [35–37, 170]. Cleavage of β-aryl ether during alkaline-catalyzed ethanol pretreatment is summarized as shown in Fig. 2.5c,d.

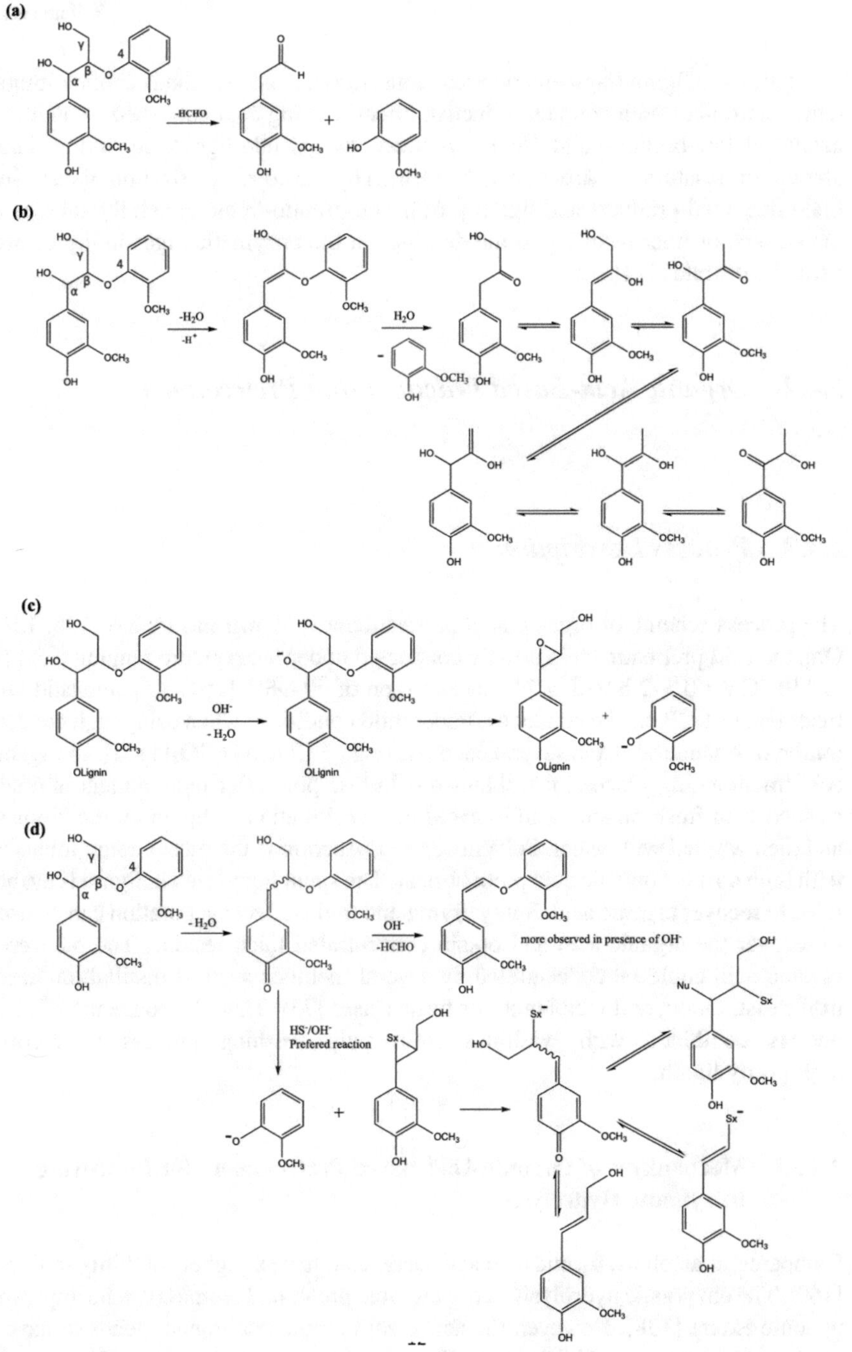

Fig. 2.5 Cleavage of β-aryl ether linkages under acidic and alkaline conditions. (a) Solvolytic cleavage of a β-aryl ether linkage with the elimination of formaldehyde under acidic conditions and (b) solvolytic cleavage of a β-aryl ether linkage to form Hibbert's ketones [91]. (c) Cleavage of nonphenolic β-aryl ether under alkaline conditions and (d) cleavage of phenolic β-aryl ether under alkaline conditions in the presence of HS^{-} [35–37, 170]

Apart from lignin fragmentation reactions, lignin condensation is another important reaction but with counterproductive effect. During acid-catalyzed or alkaline-catalyzed low-boiling-point alcohol process, the pseudo-lignin, containing large amount of unsaturated carbon, may be formed by the copolymerization of carbohydrate degraded products and lignin [161]. The pseudo-lignin generally adheres to the surface of fiber resulting in the decrease in the enzymatic digestibility of pretreated substrates [74].

2.4.2 *Organic Acid-Based Fractionating Pretreatment*

2.4.3 *Process Description*

The process scheme of organic acid pretreatment is shown in Fig. 2.6 [125, 151]. Organic acid pretreatment is usually conducted under temperature ranging from 80 to 110 °C for 0.5–2 h with acid concentration of 60–98% [161]. Organic acid pretreatment of LCB can be conducted under mild conditions when using high concentration of organic acid and several catalysts (e.g., HCl and H_2SO_4) [151]. In organic acid fractionating process, the obtained cellulose pulp after pretreatment is firstly washed with fresh organic acid to avoid the precipitation of lignin on the biomass and then washed with water. Solvent-recovery section is the most energy intensive with high cost for organic acid pretreatment. The spent liquid rich in lignin is evaporated to recover organic acid. Spray drying and multi-effect evaporation can be used to recover the organic acid and obtain concentrated thick residue. The recovered organic acid could be concentrated by several methods such as distillation, azeotropic distillation, and membrane for further uses [73]. Then the concentrated residue is combined with washings from pulp washing process to recover high-purity lignin.

2.4.3.1 Mechanisms of Organic Acid-Based Pretreatment for Improving Enzymatic Hydrolysis

Compared to alcohols, formic acid and acetic acid have a higher solubility to lignin [160]. The enzymatic hydrolysis of organic acid pretreated biomass can be improved to some extent [138]. However, the acetic acid pretreated biomass tends to have a weaker enzymatic digestibility than alcohol pretreated biomass, mainly owing to the cellulose acetylation. It was believed that the acetyl group inhibits the interaction between cellulose and catalytic domain of enzymes [161]. In addition, the acetyl groups might increase the dimensional size of the cellulose chain, which

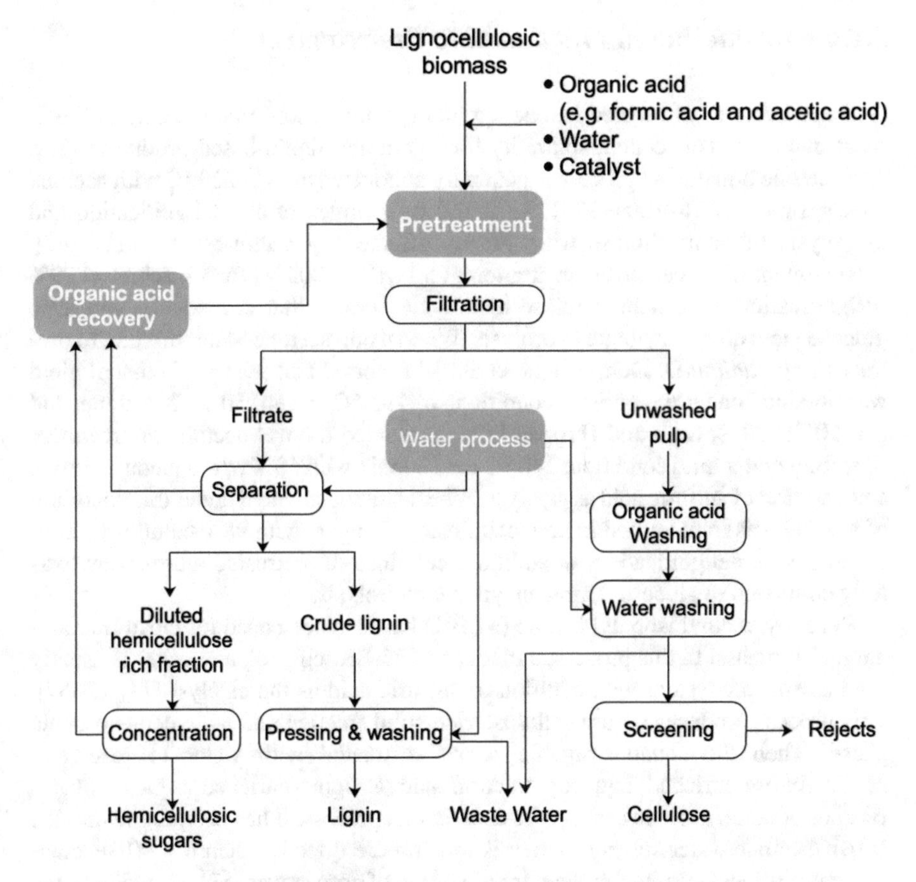

Fig. 2.6 Processing scheme of organic acid pretreatment of lignocellulosic biomass

increases the steric hindrance of enzymes, resulting in the decrease of enzymatic digestibility [166].

2.4.3.2 Reactions of Lignin

During organic acid fractionating pretreatment (e.g., formic acid and acetic acid), the chemical reactions of lignin include β-aryl ether cleavage, lignin condensation, hydrolysis of lignin-hemicellulose bonds, native ester cleavage, and esterification of the hydroxyl groups in lignin [73]. It was found that primary, secondary phenolic hydroxyl groups of lignin model compounds were partially turned into formates when using formic acid for pretreatment [28]. The dissolved lignin from acetic acid pretreatment had more acetyl groups in both Cα and Cγ compared to milled wood lignin, which indicated that acetylation and hydrolysis of native esters occurred simultaneously [73]. In addition, it was confirmed that the hydrolysis of lignin-hemicellulose bonds took place due to the fact that the lignin precipitated from water contained low sugar [139].

2.4.4 Ketone-Based Fractionating Pretreatment

Acetone is the most favored ketone applied for organosolv fractionating pretreatment owing to its excellent solubility for lignin and lignin-based products [124]. The acetone organosolv process is generally conducted at 145–228 °C with acetone concentration of 70–100% [73]. As found by Huijgen et al., delignification and hydrolysis of hemicellulose were greatly affected by acetone-water ratio [47]. Pretreatment with acetone concentration of 50 wt% at 205 °C for 1 h achieved 79% delignification, 82% hemicellulose hydrolysis, 93% cellulose recovery, and 87% glucose yield after enzymatic hydrolysis. When using acetone-water mixture to pretreat *Pinus radiata* D. Don, Araque et al. [4] reported that 99.5% of ethanol yield was obtained under the optimal conditions of 195 °C, 5 min, 50 wt% acetone, and pH 2.0. In 2017, Smit and Huijgen [124] presented a novel acetone pretreatment operating under mild condition: 140 °C for 120 min with 50% w/w aqueous acetone and addition of sulfuric acid as catalyst. Wheat straw pretreated under this condition efficiently was transformed to hemicellulosic C5 sugars with 98% cellulose recovery and 80% delignification. In addition, cellulose of pretreated wheat straw was fully converted to glucose during enzymatic hydrolysis.

Recently, methyl isobutyl ketone (MIBK) has also been used for clean fractionation of biomass. In this process, a mixture of MIBK, ethanol, and water is usually used as solvent system with addition of sulfuric acid as the catalyst [11, 12, 57]. This process produces a pure cellulose-rich solid fraction and an aqueous-organic phase. Then the aqueous-organic phase is treated with water to receive a hemicellulose-enriched aqueous fraction and a lignin-enriched organic phase. Besides ethanol, acetone can also be used in this process. It has been found that the MIBK/acetone/water solvent system is much more effective than the MIBK/ethanol/water solvent system for clean fractionation of corn stover [57]. In addition, the cellulose-rich fraction obtained from clean fractionation should be a great substrate for enzymatic hydrolysis owing to its high glucan content and great cellulose accessibility to enzyme.

2.4.5 Other Organosolv Pretreatment

A wide range of organic solvent can be used for LCB pretreatment. Some other organic solvents such as ethyl acetate [32, 34], acetonitrile [34], ethanolamine, gamma-valerolactone (GVL)[54, 76, 77], and dioxane [33] have been employed. Being different from protic solvents (e.g., alcohols in presence of acids or bases), aprotic solvents such as dimethylformamide mainly result in delignification to form small lignin fragments while protecting carbohydrates from degradation [73]. These organic solvents could be used to increase the enzymatic digestibility of LCB to some extent; however, these processes are still studied in lab scale but lack real industrial application mainly owing to the high cost and difficulty in solvent recovery.

Organosolv also has been combined with ionic liquid to fractionate LCB [85, 87, 88, 153]. Cheng et al. [19] chose a series of alcohol/water solvent (e.g., ethanol/water, glycerol/water, ethylene glycol) for biomass fractionation with the addition of acidic ionic liquid. In this process, acidic ionic liquid can act as catalyst and avoid the equipment corrosion. The participation of acidic ionic liquid increased the lignin yield and lignin extraction rate. Meanwhile, acidic ionic liquid can also promote the degradation of carbohydrates resulting in a lower cellulose-rich material yield and higher weight loss [40]. GVL is a green bio-derived solvent which has been used in organosolv fractionating process of LCB. Though the utilization of GVL showed a great improvement in LCB pretreatment, the pretreatment usually needs mineral acid, and the equipment corrosion, high pressure, and formation of inhibitors during pretreatment are inevitable. Jin et al. [55] investigated the fast dissolution pretreatment of corn stover in GVL with the assistance of ionic liquid. This pretreatment achieved effective delignification, lignin migration, transformation of cellulose I to cellulose II with increased digestibility, and morphological change of fibers.

Deep eutectic solvents (DES) are a new class of solvent with similar characteristics to ionic liquid [59, 112]. DES is usually composed of at least two components (H-bonding acceptor and H-bonding donor) and can increase enzymatic hydrolysis efficiency [168]. DES has already gained great attention mainly because of its low cost, low vapor pressure, low viscosity, high efficiency on lignin depolymerization, and environmental friendliness [122]. Liu et al. [83] investigated the cellulose reservation, xylan loss, and lignin removal using triethylbenzyl ammonium chloride (TEBAC)/lactic acid (LA)-based DES with the TEBAS/lactic acid ratio of 1:5 to 1:13. The highest sugar concentration and theoretical yield were 0.550 g/g wheat straw and 91.27%, respectively. When TEBAS/lactic acid ratio was 1:9, the DES could remove wheat straw lignin selectively to improve enzymatic hydrolysis. Kandanelli et al. [56] combined DES (choline chloride/oxalic acid) and alcohol to achieve an effective delignification. It was found that n-butanol is better than n-propanol and ethyl acetate as cosolvent because of its low miscibility in water and high fractionation of lignin. The combined system using DES and *n*-butanol achieved a high delignification about 50%.

2.5 Comparison of Different Organosolv Fractionating Pretreatments

A comparison of different organic solvent-based pretreatment in terms of enzymatic conversion of cellulose is illustrated in Fig. 2.7. Table 2.3 summarizes some recently published works in which a variety of LCB has been studied. It can be found that the type of LCB also has influence on the choice of organic solvent for pretreatment. In addition, the effects of different types of organosolv pretreatment methods on removal of hemicellulose, delignification, deacetylation, change in lignin structure, increasing accessible area, and decrystallization of cellulose are summarized in Table 2.4.

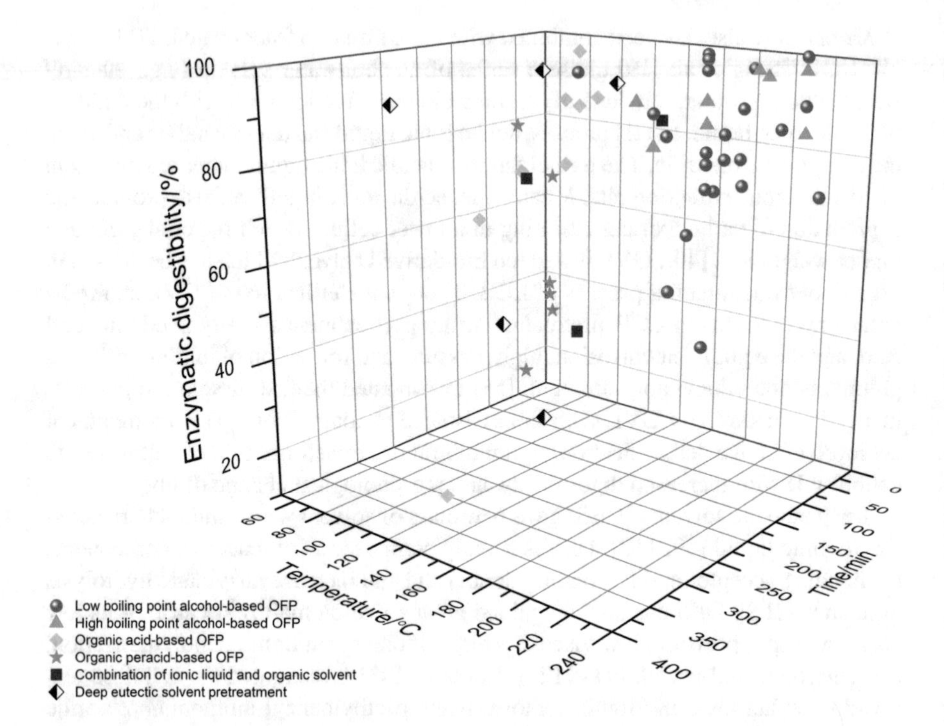

Fig. 2.7 Comparison of different organosolv pretreatment methods in terms of enzymatic digestibility. (Deep eutectic solvent pretreatment data from [41, 42, 75, 122, 131], other pretreatment data from Table 2.3)

2.6 Biorefinery Based on Organosolv Fractionating Pretreatment

2.6.1 Organosolv-Based Biorefinery

Recently, organosolv fractionating processes have been widely employed for biorefining of biomass. Table 2.5 summarizes the existing organosolv-based biorefineries worldwide. It can be known that most of the existing organosolv-based biorefineries are still in pilot or demonstration scale. The Glycell process is the only one industrial-scale organosolv-based process in the world. Compared to other pretreatment processes (e.g., acid hydrolysis and/or steam explosion), the Glycell process could produce more cellulose with less degraded products. Furthermore, it also can improve enzymatic kinetics of pretreated biomass resulting in quick sugar release. Apart from the Glycell process, most of existing organosolv-based biorefineries use alcohol for pretreatment. The Alcell and Organocell processes are two organosolv-based biorefineries that were reported about 30 years ago [7, 66]. However, they are not listed in the table because no up-to-date report or published works can be found to confirm that they are still running.

Table 2.3 A summary of published organosolv fractionating pretreatment for increase of enzymatic hydrolysis of LCB

Lignocellulosic biomass	Solvent	Pretreatment conditions			Enzymatic digestibility	References
		TEMP/°C	t/min	Catalyst		
Alcohols						
Bamboo	60 wt% ethanol	160	60	Absence	60% enzymatic cellulose conversion	Mou and Wu [97]
	75v% ethanol	180	30	2 wt% H_2SO_4	77.1% enzymatic hydrolysis rate	Li et al. [78]
	75 wt% ethanol	180	60	2 wt% H_2SO_4	83.4% enzymatic conversion	Li et al. [79]
	75v% ethanol	120, 140, and 180	30	10 wt% NaOH	Optimal conditions yielded 60.1% cellulose and 45.1% cellulose to glucose	Li et al. [80]
Eucalyptus globulus wood	60% ethanol	3900 H factor		Absence	65% glucose enzymatic hydrolysis	Muñoz et al. [98]
	60% ethanol	12,500 H factor		Absence	77% glucose enzymatic hydrolysis	
	55% glycerol	200	69	Absence	98% @ 20FPU/g glucan	Romaní et al. [116]
Beech	100% glycerol	190	15	–	95%	Hundt et al. [49]
Olive tree	43 wt% ethanol	210	15	Absence	89.6% glucose enzymatic hydrolysis	Diaz et al. [24]
Cottonwood	50% ethanol	170	240	0.2 Mol/L acetic acid	Enzymatically hydrolyzed 2.5 times sugar yield than untreated sample	Neilson and Shafizadeh [101]
Hybrid poplar	25–75v% ethanol	155–205	26–94	0.83–1.67 wt% H_2SO_4	82% hydrolysis efficiency under optimal condition	Pan et al. [105]
Sweetgum	50% ethanol	170	60	1 wt% H_2SO_4	96.6% enzymatic cellulose conversion	Lai et al. [69]
	65% ethanol	170	60	1.1 wt% H_2SO_4	98.9% glucose hydrolysis with 1.19 g/L/h initial rate	Li et al. [72]
Liriodendron tulipifera	50% ethanol	120–140	50	1 wt% H_2SO_4	Up to 77.3% sugar hydrolysis rate	Koo et al. [65]
	50% ethanol	140–160	50	1 wt% NaOH	Increased 65% sugar hydrolysis rate and converted to ethanol at 96% rate	Koo et al. [64]
Willow wood	55–56 wt% ethanol	187	180 including heat up	0.01 M H_2SO_4	87% enzymatic cellulose conversion	Hundt et al. [50]

(continued)

Table 2.3 (continued)

Lignocellulosic biomass	Solvent	Pretreatment conditions			Enzymatic digestibility	References
		TEMP/°C	t/min	Catalyst		
Loblolly pine	65% ethanol	170	60	1.1 wt% H_2SO_4	79% cellulose recovery with 70% enzymatic hydrolysis to glucose	Sannigrahi et al. [121]
	65% ethanol	170	60	1.1 wt% H_2SO_4	76.4% glucose hydrolysis with 1.45 g/L/h initial rate	Li et al. [72]
Lodgepole pine wood	65% ethanol	170	60	1.1% H_2SO_4	~100% enzymatic cellulose conversion	Del Rio et al. [22]
	48–82v% ethanol	153–187	43–77	0.76–1.44 wt% H_2SO_4	97% enzymatic hydrolysis under optimal condition	Pan et al. [108]
	65% ethanol	170	60	1.1% SO_2	71% cellulose digestibility	Del Rio et al. [22]
	65% ethanol	170	60	20% NaOH	35% cellulose digestibility	
	65% butanol	170	60	1.1% H_2SO_4	~90%@20FPU/g cellulose	
	65% butanol	170	60	20% NaOH (based on wood)	~85%@20FPU/g cellulose	
Pitch pine	50v% ethanol	180	0	1% H_2SO_4 (w/v)	56.25% cellulose digestibility	Park et al. [111]
	50v% ethanol	150–210	0–20	1–2% NaOH (w/v)	Highest digestibility of 85.4%	Park et al. [111]
Radiata pine	60v% ethanol	185	18	0.13% H_2SO_4 (w/v)	Fermentation yielded 63.8% by fungus	Monrroy et al. [95]
Norway spruce	63 wt% ethanol	210	15	pH = 3.5	~100% enzymatic cellulose conversion	Agnihotri et al. [1]
	100w% glycerol	230	90	10% KOH	~87%@15FPU/g dry matter	Hundt et al. [50]
	100% glycerol	210	15	Absence	97% cellulose digestibility	Hundt et al. [50]
Sitka spruce sawdust	60v% ethanol	180	60	1 wt% H_2SO_4	~78%@20FPU/g solid	Bouxin et al. [10]
Japanese cypress	50% ethanol	170	45	0.4 wt% HCl	70% enzymatic cellulose conversion	Hideno et al. [45]

Lignocellulosic biomass	Solvent	Pretreatment conditions			Enzymatic digestibility	References
		TEMP/°C	t/min	Catalyst		
Mixed softwood	40–60 wt% ethanol	185–198	30–60	pH 2.0–3.4	Conversion>90% solids to glucose	Pan et al. [103]
Sugarcane bagasse	50v% ethanol	175	60	1.25 wt% H_2SO_4	26.5% enzymatic cellulose digestibility	Mesa et al. [94]
	50v% ethanol	175	60	1.25 wt% H_2SO_4	46.0% cellulose digestibility	Mesa et al. [93]
	50v% ethanol	175	60	1.25 wt% NaOH	38.4% cellulose digestibility	Mesa et al. [94]
	50v% ethanol	140	180	0.8, 1.0 and 1.5 mL GL/g DS	97.7% cellulose digestibility	Yu et al. [148]
	50v% ethanol	80, 100, 140, 160	180	1.5 mL GL/g DS	The highest glucose yield was achieved at a pretreatment temperature of 160 °C	Zhou et al. [171]
	65 wt% ethanol	210	90	pH 3.5	100% enzymatic cellulose digestibility	Agnihotri et al. [1]
	80 wt% glycerol	130	60	1.2% HCl	88%@20FPU/g glucan	Zhang et al. [157]
	90 wt% glycerol	130	30	1.2% H_2SO_4	77%@20FPU/g glucan	Snelders et al. [125]

(continued)

Table 2.3 (continued)

Lignocellulosic biomass	Solvent	Pretreatment conditions			Enzymatic digestibility	References
		TEMP/°C	t/min	Catalyst		
Wheat straw	50 wt% ethanol	210	60	Absence	86% glucose enzymatic hydrolysis	Wildschut et al. [142]
	65 wt% ethanol	220	20	Absence	71% enzymatic cellulose conversion	Chen et al. [18]
	55–56 wt% ethanol	187	180 (including heat up)	0.02 M HCl	99% enzymatic cellulose digestibility	Huijgen et al. [48]
	60 wt% ethanol	190	60	30 mM H_2SO_4	73% enzymatic cellulose digestibility	Chen et al. [18]
	60 wt% ethanol	190	60	30 mM H_2SO_4	89.4% enzymatic cellulose digestibility	Wildschut et al. [142]
Barley straw	50 wt% ethanol	170	60	0.1 wt% H_2SO_4	55% cellulose digestibility	Kim et al. [61]
	50 wt% ethanol	170	60	1 wt% H_2SO_4	85.3% enzymatic cellulose conversion	Kim et al. [60]
Buddleja davidii	50v% ethanol	180	40	1.75 wt% H_2SO_4	98% cellulose digestibility	Hallac et al. [44]
	65v% ethanol	195	60	1.50 wt% H_2SO_4	98% cellulose digestibility	
Empty palm fruit bunch	65 wt% ethanol	160–200	45–90	0.5–2 wt% H_2SO_4	96.3% glucose recovery yielded under optimal conditions	Goh et al. [39]
Rice straw	75v% ethanol	150	60	1 wt% H_2SO_4	46.2% glucose hydrolysis yield	Amiri et al. [3]
	70% glycerol	190	600	Absence	57.5%@17.5FPU/g dry mass	Trinh et al. [135]
Rye straw	50 wt% ethanol	167	35	0.5 M H_2SO_4	~100% enzymatic cellulose digestibility	Ingram et al. [52]
Furfural residues	50v% ethanol	140	60	1.5 mL GL/g DS	85.9% enzymatic cellulose digestibility	Yu et al. [147]
Hemp hurd	45% methanol	165	20	3% H_2SO_4	~60%@20FPU/g substrate	Gandolfi et al. [31]
Organic acids						

Lignocellulosic biomass	Solvent	Pretreatment conditions			Enzymatic digestibility	References
		TEMP/°C	t/min	Catalyst		
Sugarcane bagasse	78 wt% formic acid	107	90	0.1 wt% H_2SO_4	86.9%@20FPU/g solid	Zhao and Liu [163]
	80 wt% acetic acid	110	120	0.3 wt% H_2SO_4	Followed by 4 wt% NaOH, 120 °C, 1 h Achieve 89.24%@20FPU/g solid	Zhao and Liu [162]
Wheat straw	78 wt% formic acid	107	60	Not mentioned	Followed by 2 wt% ca$(OH)_2$, 120 °C, 1 h Achieve 88%@20FPU/g solid	Chen et al. [18]
	90 wt% acetic acid	110	120	0.3 wt% H_2SO_4	Followed by 2 wt% ca$(OH)_2$, 120 °C, 1 h Achieve 75%@20FPU/g solid	
Empty palm fruit bunch	78 wt% formic acid	107	90	Not mentioned	Followed by 2 wt% ca$(OH)_2$, 120 °C, 1 h Achieve 99%@15FPU/g solid	Cui et al. [20]
Corn stover	88 wt% formic acid	80	180	Absence	62.8%@20FPU/g dry substrate	Yu et al. [146]
Douglas fir	95 wt% acetic acid	110	300	Absence	~10%@20FPU/g cellulose	Pan et al. [106]
Miscanthus	40%FA + 40%AA+20% water	107	180	Absence	75.3% cellulose digestibility	Vanderghem et al. [136]
Organic peracids (PAA-peracetic acid)						
Aspen wood	9% PAA (based on solid)	100	120	Absence	44.4%@60FPU/g cellulose	Xu and Tschirner [144]
Sugarcane bagasse	50% PAA (based on solid)	80	120	Absence	82.07%@20FPU/g cellulose	Zhao et al. [167]
	69.1w% H_2O_2-PAA	80	1590	Absence	93.58%@138FPU/g carbohydrate	Tan et al. [129]
	50 wt% PAA (based on solid)	60	60	Absence	~18%@20FPU/g cellulose	Zhao et al. [167]

(continued)

Table 2.3 (continued)

Lignocellulosic biomass	Solvent	Pretreatment conditions			Enzymatic digestibility	References
		TEMP/°C	t/min	Catalyst		
Crofton weed stem	40 wt% PAA (based on solid)	90	90	Absence	38%@20FPU/g cellulose	Zhao et al. [164]
Siam weed stem	50 wt% PAA (based on solid)	90	90	Absence	70%@10FPU/g cellulose	Zhao et al. [165]
Alfalfa stem	9% PAA (based on solid)	100	120	Absence	47.0%@60FPU/g cellulose	Xu and Tschirner [144]
Other solvents						
Wheat straw	50 wt% acetone	205	60	Absence	87%@38FPU/g solid	Huijgen et al. [47]
Waste paper	Ethylene glycol	150	15	2%H_2SO_4	94% glucan digestibility	Lee et al. [70]
Sugarcane bagasse	Glycerol carbonate	90	30	H_2SO_4	80% glucose yield	Zhang et al. [156]
Prairie cordgrass	MIBK	154	39	0.69 wt%	84% glucose yield	Brudecki et al. [13]
Switchgrass	Ethyl acetate/ethanol/water	140	20	0.46% H_2SO_4	84.8% glucose yield	Cybulska et al. [21]
Rice hull	Ionic liquid/glycerol	110	180	–	~74% glucose yields	Lynam and Coronella [87]
Loblolly pine	Ionic liquid/glycerol	140	180	–	42.6%	Lynam and Coronella [88]
Corn stover	Ionic/GVL	140	45	Absence	84% glucose yield	Jin et al. [55]

Table 2.4 Comparison of organosolv pretreatment with some other pretreatment methods regarding chemical compositions and physical and changes of LCB

Pretreatment methods	Removal of hemicellulose	Removal of lignin	Increase accessible surface area	Decrystallization of cellulose	Change of lignin structure	Acetylation of cellulose
Organosolv alcohol and acetone[1]	++	++	++	ND	++	–
Organic acid[a]	++	++	++	ND	++	++/–
Organic peracid[a]	+	++	++	–	++	+/–
PA-acetone[b]	++	++	++	++	ND	–
Dilute acid[c]	++	ND	++	ND	++	ND
Ionic liquid-GVL[d]	ND	++	++	++	++	ND
DES-alcohol[e]	ND	++	ND	ND	ND	ND

++, positive and major effect; +, positive but not major effect; –, negative effect; ND, not determined

[a]Data from Zhao et al. [160]

[b]Data from Zhang et al. [155] PA refers to concentrated phosphoric acid

[c]Data from Mosier et al. [96]

[d]Data from Jin et al. [55]

[e]Data from Kandanelli et al. [56]

Table 2.5 Some reported organosolv-based biorefineries [9]

Process name	Institute	Solvent	Location	Scale
Lignol	Lignol innovations, Inc.	Ethanol	USA	Demonstration
Lignol	Lignol innovations, Inc.		Canada	Pilot
Lignocellulose biorefinery	Fraunhofer CBP		Germany	Pilot
DHR	Dedini Industrias de base	Alcohol	Brazil	Pilot
Glycell	Leaf resources limited	Glycerol	Australia	Industrial
CIMV	Compagnie Industrielle de la Matie're Vegetale (CIMV)	Aqueous mix of acetic acid and formic acid	France	Pilot/demonstration
AST	American science and technology Corp	A new cocktail of organic solvents	USA	Pilot

Though organosolv fractionating pretreatment is a promising pretreatment method to achieve a high cellulose digestibility and high-value-added products, it is still difficult to achieve industrial application mainly because of high operating cost, high energy consumption, and environmental concerns. Separation and recovery of organic solvents are the most expensive units. In addition, evaporation of organic solvents usually has a hazardous influence on the environment and humans. The solid after flash evaporation still needs to be washed by water to remove the residual organic solvents. As a consequence, effluent with high COD would be produced.

2.6.2 Chemicals, Fuels, and Materials Derived from Organosolv Fractionating Pretreatment

Organosolv fractionating pretreatment is a potential method achieving a coproduction of high-purity lignin and hemicellulosic sugars. Compared to the traditional biorefinery, a biorefinery fulfilling a comprehensive utilization of LCB can achieve higher potential revenue. It has been estimated that complete utilization of LCB with coproduction of ethanol, acetic acid, lignin, and other products would result in a revenue three times higher than that of traditional biorefinery only producing ethanol [161]. The typical chemicals, fuels, and materials derived from organosolv fractionating pretreatment are shown in Fig. 2.8.

2.6.2.1 Cellulose-Derived Products

After organosolv fractionating pretreatment, the pretreated solid with high-purity cellulose is usually converted to fermentable sugars or ethanol by biological approaches. Apart from production of sugars and ethanol, cellulose-rich solid can also be used to produce butanol, biohydrogen, dissolving pulp, microcrystalline cellulose, and other products [170].

Butanol is an alternative fuel with an air-to-fuel ratio similar to gasoline. It has a lower vapor pressure and higher energy content than ethanol [113]. Furthermore, it has a good compatibility with current automobile engines and transportation pipelines, which make it attractive as a substitute for gasoline [113]. Farmanbordar et al.

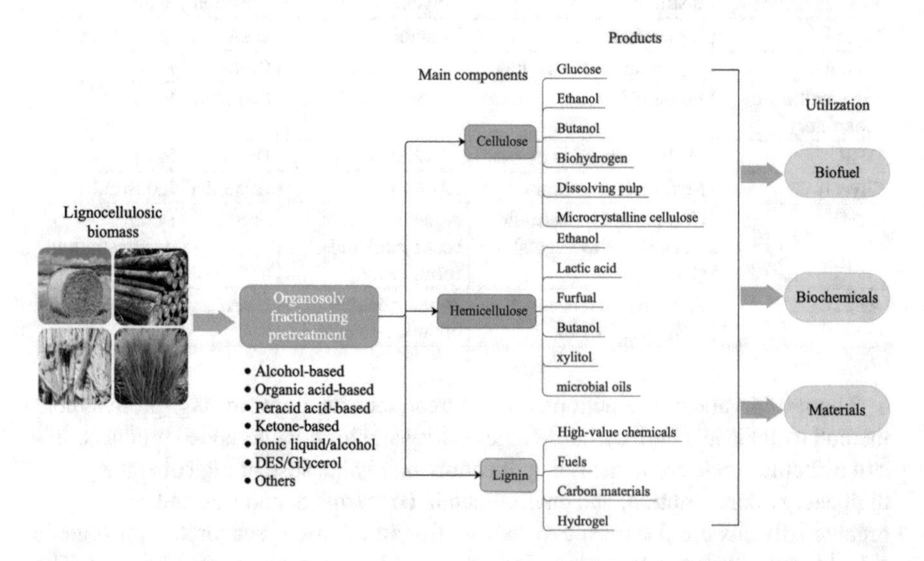

Fig. 2.8 Typical chemicals, fuels, and materials derived from organosolv fractionating pretreatment

[30] have proposed acetone-butanol-ethanol fermentation of ethanol organosolv pretreated municipal solid waste. The highest butanol yield of 8.57 g/L was obtained from the hydrolysate produced from municipal solid waste pretreated at 120 °C for 30 min with 85% ethanol. Biohydrogen produced from LCB is a renewable fuel which can be applied in many industries. Asadi and Zilouei [6] used ethanol organosolv pretreated rice straw for the production of biohydrogen. The highest biohydrogen yield around 19.73 ml/g straw was achieved under pretreatment condition: 180 °C for 30 min with the ethanol concentration of 45 wt%. It is estimated that about 355.8 kilotons of biohydrogen would be produced according to the experimental results. Dissolving pulp is a pulp with high-grade cellulose, which can be used to produce cellulose-derived products, such as regenerated fibers and cellulose esters. Besides sulfite pulping, acid-catalyzed organosolv pulping is also a promising method to make dissolving pulp [90].

2.6.2.2 Hemicellulose-Derived Products

During organosolv fractionating pretreatment (especially autocatalyzed and acid-catalyzed), most of hemicelluloses are removed resulting in organosolv hydrolysate rich in hemicellulosic sugars. Utilization of organosolv hydrolysate is a promising way to produce hemicellulose-based products. Recently, hemicellulosic sugars could be fermented to produce ethanol by several yeasts such as *Pichia* and *Kluyveromyces* [92]. Furthermore, Glaser and Venus [38] have investigated the co-fermentation of main sugars (such as xylose and glucose) from beech wood organosolv hydrolysate by five strains of *Bacillus coagulans* to produce lactic acid, which has been widely used in food industry and a common feedstock for production of biodegradable polymer polylactic acid. Furfural is considered as an important bio-based chemical, which can be used to produce furan, furfuryl alcohol, and other nonfossil-based compounds. Köchermann et al. [63] have studied the conversion of an aqueous hemicellulose and D-xylose into furfural using a continuous tube reactor at temperatures in the range of 160 °C to 200 °C. Except for ethanol, lactic acid, and furfural, hemicellulosic sugars can also be converted to butanol, xylitol, microbial oils, etc., by microbial fermentation.

2.6.2.3 Lignin-Derived Products

Organosolv pretreatment could remove most lignin, while the high-purity organosolv lignin could be recovered by precipitation. There are three potential ways to use such high-purity organosolv lignin to produce high-value-added products, including (1) as a raw material to produce high-value chemicals; (2) direct use or as precursor for production of fuel and materials; and (3) as a feedstock for production of drop-in transportation fuels [115].

The base-catalyzed depolymerization of lignin usually produces monomers as well as other phenolic compounds such as dimeric and oligomeric alkyl-functionalized phenolic compounds, which can be used in many industries and replace phenols from fossil resources [154]. Rößiger et al. [118] have investigated the upscaling of base-catalyzed depolymerization of beech wood organosolv lignin to pilot plant scale. Compared to alkaline lignin, ethanol organosolv lignin from oil palm fronds is found to be better as a renewable phenol substitute mainly owing to its higher amount of lignin and thermal decomposition temperature [51]. Besides being used for producing phenolic compounds, organosolv lignin could also be employed as radical scavenger [107]. High content of hydroxyl groups in organosolv lignin makes it potential for polyurethane synthesis. Sakdaronnarong et al. [120] have synthesized lignin-based polyurethane using sugarcane bagasse organosolv lignin. This polyurethane has better thermal-resistant and mechanical properties than that from Kraft lignin. Xue et al. [145] used ethanol organosolv lignin as a reactive filler to prepare acrylamide-based hydrogels, which had good mechanically elastic and high swelling properties. In addition, lignin has a higher density and lower oxygen content than cellulose and hemicellulose, which makes it ideal to produce drop-in fuels [141]. However, the process to produce fuel from organosolv lignin not only requires depolymerization of lignin but also upgrading of depolymerized products to meet standards for drop-in fuels. As a consequence, application of organosolv lignin for production of fuel is still in development.

2.7 Conclusions

Organosolv fractionating pretreatment is a promising pretreatment approach to achieve fractionation of cellulose, hemicellulose, and lignin for further utilization. After organosolv pretreatment, lignocellulosic cell wall was deconstructed due to removal of hemicelluloses and lignin, modification of lignin structure, redistribution of lignin in cell wall, change in the cellulose crystallinity, and degree of polymerization with associated change in substrate structure. The obtained cellulose-rich fraction can be easily hydrolyzed to produce biofuels. However, nowadays, organosolv fractionating pretreatment is not yet competitive to other industrial pretreatment methods such as steam explosion and dilute acid pretreatment mainly owing to the high operation cost and energy consumption for solvent recovery. Efforts have to be made to reduce the cost and energy consumption by developing novel and low-cost solvent systems that can be used at low temperature and easily recovered. In addition, development of advanced processes for utilization of lignin and hemicellulose to produce high-value-added products should also be considered.

Acknowledgments This work was supported by the National Key R & D Program of China (2018YFA0902200) and National Natural Science Foundation of China (No. 21878176; 21808123).

References

1. Agnihotri, S., Johnsen, I. A., Bøe, M. S., Øyaas, K., & Moe, S. (2015). Ethanol organosolv pretreatment of softwood (Picea abies) and sugarcane bagasse for biofuel and biorefinery applications. *Wood Science and Technology, 49*(5), 881–896.
2. Alvira, P., Tomás-Pejó, E., Ballesteros, M., & Negro, M. (2010). Pretreatment technologies for an efficient bioethanol production process based on enzymatic hydrolysis: A review. *Bioresource Technology, 101*(13), 4851–4861.
3. Amiri, H., Karimi, K., & Zilouei, H. (2014). Organosolv pretreatment of rice straw for efficient acetone, butanol, and ethanol production. *Bioresource Technology, 152*, 450–456.
4. Araque, E., Parra, C., Freer, J., Contreras, D., Rodríguez, J., Mendonça, R., & Baeza, J. (2008). Evaluation of organosolv pretreatment for the conversion of Pinus radiata D. Don to ethanol. *Enzyme and Microbial Technology, 43*(2), 214–219.
5. Arni, S. A. (2018). Extraction and isolation methods for lignin separation from sugarcane bagasse: A review. *Industrial Crops and Products, 115*, 330–339.
6. Asadi, N., & Zilouei, H. (2017). Optimization of organosolv pretreatment of rice straw for enhanced biohydrogen production using Enterobacter aerogenes. *Bioresource Technology, 227*, 335–344.
7. Bajpai, P. (2010). Overview of pulp and papermaking processes. In *Environmentally friendly production of pulp and paper* (pp. 8–45). John Wiley & Sons, Inc., New Jersey.
8. Balakshin, M. Y., Capanema, E. A., & H-m, C. (2007). MWL fraction with a high concentration of lignin-carbohydrate linkages: Isolation and 2D NMR spectroscopic analysis. *Holzforschung, 61*(1), 1–7.
9. Borand, M. N., & Karaosmanoğlu, F. (2018). Effects of organosolv pretreatment conditions for lignocellulosic biomass in biorefinery applications: A review. *Journal of Renewable and Sustainable Energy, 10*(3), 033104.
10. Bouxin, F. P., Jackson, S. D., & Jarvis, M. C. (2014). Organosolv pretreatment of Sitka spruce wood: Conversion of hemicelluloses to ethyl glycosides. *Bioresource Technology, 151*, 441–444.
11. Bozell, J. J., Black, S. K., Myers, M., Cahill, D., Miller, W. P., & Park, S. (2011a). Solvent fractionation of renewable woody feedstocks: Organosolv generation of biorefinery process streams for the production of biobased chemicals. *Biomass and Bioenergy, 35*(10), 4197–4208.
12. Bozell, J. J., O'Lenick, C., & Warwick, S. (2011b). Biomass fractionation for the biorefinery: Heteronuclear multiple quantum coherence–nuclear magnetic resonance investigation of lignin isolated from solvent fractionation of switchgrass. *Journal of Agricultural and Food Chemistry, 59*(17), 9232–9242.
13. Brudecki, G., Cybulska, I., Rosentrater, K., & Julson, J. (2012). Optimization of clean fractionation processing as a pre-treatment technology for prairie cordgrass. *Bioresource Technology, 107*, 494–504.
14. Calvo-Flores, F. G., & Dobado, P. J. A. (2010). Lignin as renewable raw material. *ChemSusChem, 3*(11), 1227–1235.
15. Cannella, D., Sveding, P. V., & Jørgensen, H. (2014). PEI detoxification of pretreated spruce for high solids ethanol fermentation. *Applied Energy, 132*, 394–403.
16. Cateto, C., Hu, G., & Ragauskas, A. (2011). Enzymatic hydrolysis of organosolv Kanlow switchgrass and its impact on cellulose crystallinity and degree of polymerization. *Energy & Environmental Science, 4*(4), 1516–1521.
17. Chen, H. (2015). Lignocellulose biorefinery feedstock engineering. In *Lignocellulose Biorefinery Engineering* (pp. 37–86).
18. Chen, H., Zhao, J., Hu, T., Zhao, X., & Liu, D. (2015). A comparison of several organosolv pretreatments for improving the enzymatic hydrolysis of wheat straw: Substrate digestibility, fermentability and structural features. *Applied Energy, 150*, 224–232.

19. Cheng, F., Zhao, X., & Hu, Y. (2018). Lignocellulosic biomass delignification using aqueous alcohol solutions with the catalysis of acidic ionic liquids: A comparison study of solvents. *Bioresource Technology, 249*, 969–975.
20. Cui, X., Zhao, X., Zeng, J., Loh, S. K., Choo, Y. M., & Liu, D. (2014). Robust enzymatic hydrolysis of Formiline-pretreated oil palm empty fruit bunches (EFB) for efficient conversion of polysaccharide to sugars and ethanol. *Bioresource Technology, 166*, 584–591.
21. Cybulska, I., Brudecki, G. P., Hankerson, B. R., Julson, J. L., & Lei, H. (2013). Catalyzed modified clean fractionation of switchgrass. *Bioresource Technology, 127*, 92–99.
22. Del Rio, L. F., Chandra, R. P., & Saddler, J. N. (2010). The effect of varying organosolv pretreatment chemicals on the physicochemical properties and cellulolytic hydrolysis of mountain pine beetle-killed lodgepole pine. *Applied Biochemistry and Biotechnology, 161*(1–8), 1–21.
23. Deshavath, N. N., Veeranki, V. D., & Goud, V. V. (2019). Lignocellulosic feedstocks for the production of bioethanol: Availability, structure, and composition. In *Sustainable Bioenergy* (pp. 1–19).
24. Diaz, M. J., Huijgen, W. J., van der Laan, R. R., Reith, J. H., Cara, C., & Castro, E. (2011). Organosolv pretreatment of olive tree biomass for fermentable sugars. *Holzforschung, 65*(2), 177–183.
25. Du, X., Lucia, L. A., & Ghiladi, R. A. (2016). Development of a highly efficient pretreatment sequence for the enzymatic saccharification of loblolly pine wood. *ACS Sustainable Chemistry & Engineering, 4*(7), 3669–3678.
26. Dziekońska-Kubczak, U., Berłowska, J., Dziugan, P., Patelski, P., Balcerek, M., Pielech-Przybylska, K., & Robak, K. (2019). Two-stage pretreatment to improve saccharification of oat straw and Jerusalem artichoke biomass. *Energies, 12*(9), 1715.
27. Ebrahimi, M., Caparanga, A. R., Ordono, E. E., Villaflores, O. B., & Pouriman, M. (2017). Effect of ammonium carbonate pretreatment on the enzymatic digestibility, structural characteristics of rice husk and bioethanol production via simultaneous saccharification and fermentation process with Saccharomyces cerevisiae Hansen 2055. *Industrial Crops and Products, 101*, 84–91.
28. Ede, R., Brunow, G., Poppius, K., Sundquist, J., & Hortling, B. (1988). Formic acid/peroxyformic acid pulping. *Nordic Pulp & Paper Research Journal, 3*(3), 119–123.
29. El Hage, R., Brosse, N., Sannigrahi, P., & Ragauskas, A. (2010). Effects of process severity on the chemical structure of Miscanthus ethanol organosolv lignin. *Polymer Degradation and Stability, 95*(6), 997–1003.
30. Farmanbordar, S., Amiri, H., & Karimi, K. (2018). Simultaneous organosolv pretreatment and detoxification of municipal solid waste for efficient biobutanol production. *Bioresource Technology, 270*, 236–244.
31. Gandolfi, S., Ottolina, G., Consonni, R., Riva, S., & Patel, I. (2014). Fractionation of hemp hurds by organosolv pretreatment and its effect on production of lignin and sugars. *ChemSusChem, 7*(7), 1991–1999.
32. Ghosh, A., Bai, X., & Brown, R. C. (2018). Solubilized carbohydrate production by acid-catalyzed depolymerization of cellulose in polar aprotic solvents. *ChemistrySelect, 3*(17), 4777–4785.
33. Ghosh, A., & Brown, R. C. (2019). Factors influencing cellulosic sugar production during acid-catalyzed solvent liquefaction in 1, 4-dioxane. *ACS Sustainable Chemistry & Engineering, 7*(21), 18076–18084.
34. Ghosh, A., Brown, R. C., & Bai, X. (2016). Production of solubilized carbohydrate from cellulose using non-catalytic, supercritical depolymerization in polar aprotic solvents. *Green Chemistry, 18*(4), 1023–1031.
35. Gierer, J. (1980). Chemical aspects of kraft pulping. *Wood Science and Technology, 14*(4), 241–266.
36. Gierer, J. (1982). The chemistry of delignification. A general concept. *Holzforschung, 36*(1), 43–51.

37. Gierer, J. (1985). Chemistry of delignification. *Wood Science and Technology, 19*(4), 289–312.
38. Glaser, R., & Venus, J. (2018). Co-fermentation of the main sugar types from a beechwood organosolv hydrolysate by several strains of Bacillus coagulans results in effective lactic acid production. *Biotechnology Reports, 18*, e00245.
39. Goh, C. S., Tan, H. T., Lee, K. T., & Brosse, N. (2011). Evaluation and optimization of organosolv pretreatment using combined severity factors and response surface methodology. *Biomass and Bioenergy, 35*(9), 4025–4033.
40. Guo, F., Fang, Z., & Zhou, T.-J. (2012). Conversion of fructose and glucose into 5-hydroxymethylfurfural with lignin-derived carbonaceous catalyst under microwave irradiation in dimethyl sulfoxide–ionic liquid mixtures. *Bioresource Technology, 112*, 313–318.
41. Guo, Z., Zhang, Q., You, T., Ji, Z., Zhang, X., Qin, Y., & Xu, F. (2019a). Heteropoly acids enhanced neutral deep eutectic solvent pretreatment for enzymatic hydrolysis and ethanol fermentation of Miscanthus x giganteus under mild conditions. *Bioresource Technology, 293*, 122036.
42. Guo, Z., Zhang, Q., You, T., Zhang, X., Xu, F., & Wu, Y. (2019b). Short-time deep eutectic solvent pretreatment for enhanced enzymatic saccharification and lignin valorization. *Green Chemistry, 21*(11), 3099–3108.
43. Hallac, B. B., Ray, M., Murphy, R. J., & Ragauskas, A. J. (2010a). Correlation between anatomical characteristics of ethanol organosolv pretreated Buddleja davidii and its enzymatic conversion to glucose. *Biotechnology and Bioengineering, 107*(5), 795–801.
44. Hallac, B. B., Sannigrahi, P., Pu, Y., Ray, M., Murphy, R. J., & Ragauskas, A. J. (2010b). Effect of ethanol organosolv pretreatment on enzymatic hydrolysis of Buddleja davidii stem biomass. *Industrial and Engineering Chemistry Research, 49*(4), 1467–1472.
45. Hideno, A., Kawashima, A., Endo, T., Honda, K., & Morita, M. (2013). Ethanol-based organosolv treatment with trace hydrochloric acid improves the enzymatic digestibility of Japanese cypress (Chamaecyparis obtusa) by exposing nanofibers on the surface. *Bioresource Technology, 132*, 64–70.
46. Horn, S. J., Vaaje-Kolstad, G., Br, W., & Eijsink, V. G. (2012). Novel enzymes for the degradation of cellulose. *Biotechnology for Biofuels, 5*(1), 45.
47. Huijgen, W. J., Reith, J. H., & den Uil, H. (2010). Pretreatment and fractionation of wheat straw by an acetone-based organosolv process. *Industrial and Engineering Chemistry Research, 49*(20), 10132–10140.
48. Huijgen, W. J., Smit, A. T., Reith, J. H., & Hd, U. (2011). Catalytic organosolv fractionation of willow wood and wheat straw as pretreatment for enzymatic cellulose hydrolysis. *Journal of Chemical Technology and Biotechnology, 86*(11), 1428–1438.
49. Hundt, M., Schnitzlein, K., & Schnitzlein, M. G. (2013a). Alkaline polyol pulping and enzymatic hydrolysis of hardwood: Effect of pulping severity and pulp composition on cellulase activity and overall sugar yield. *Bioresource Technology, 136*, 672–679.
50. Hundt, M., Schnitzlein, K., & Schnitzlein, M. G. (2013b). Alkaline polyol pulping and enzymatic hydrolysis of softwood: Effect of pulping severity and pulp properties on cellulase activity and overall sugar yield. *Bioresource Technology, 134*, 307–315.
51. Hussin, M. H., Rahim, A. A., Ibrahim, M. N. M., & Brosse, N. (2013). Physicochemical characterization of alkaline and ethanol organosolv lignins from oil palm (Elaeis guineensis) fronds as phenol substitutes for green material applications. *Industrial Crops and Products, 49*, 23–32.
52. Ingram, T., Wörmeyer, K., Lima, J. C. I., Bockemühl, V., Antranikian, G., Brunner, G., & Smirnova, I. (2011). Comparison of different pretreatment methods for lignocellulosic materials. Part I: Conversion of rye straw to valuable products. *Bioresource Technology, 102*(8), 5221–5228.
53. Isikgor, F. H., & Becer, C. R. (2015). Lignocellulosic biomass: A sustainable platform for the production of bio-based chemicals and polymers. *Polymer Chemistry, 6*(25), 4497–4559.

54. Jia, L., Qin, Y., Wen, P., Zhang, T., & Zhang, J. (2019). Alkaline post-incubation improves cellulose hydrolysis after γ-valerolactone/water pretreatment. *Bioresource Technology, 278*, 440–443.
55. Jin, L., Yu, X., Peng, C., Guo, Y., Zhang, L., Xu, Q., Zhao, Z. K., Liu, Y., & Xie, H. (2018). Fast dissolution pretreatment of the corn stover in gamma-valerolactone promoted by ionic liquids: Selective delignification and enhanced enzymatic saccharification. *Bioresource Technology, 270*, 537–544.
56. Kandanelli, R., Thulluri, C., Mangala, R., Rao, P. V. C., Gandham, S., & Velankar, H. R. (2018). A novel ternary combination of deep eutectic solvent-alcohol (DES-OL) system for synergistic and efficient delignification of biomass. *Bioresource Technology, 265*, 573–576.
57. Katahira, R., Mittal, A., McKinney, K., Ciesielski, P. N., Donohoe, B. S., Black, S. K., Johnson, D. K., Biddy, M. J., & Beckham, G. T. (2014). Evaluation of clean fractionation pretreatment for the production of renewable fuels and chemicals from corn stover. *ACS Sustainable Chemistry & Engineering, 2*(6), 1364–1376.
58. Kim, D. (2018). Physico-chemical conversion of lignocellulose: Inhibitor effects and detoxification strategies: A mini review. *Molecules, 23*(2), 309.
59. Kim, K. H., Dutta, T., Sun, J., Simmons, B., & Singh, S. (2018). Biomass pretreatment using deep eutectic solvents from lignin derived phenols. *Green Chemistry, 20*(4), 809–815.
60. Kim, Y., Yu, A., Han, M., G-w, C., & Chung, B. (2011a). Enhanced enzymatic saccharification of barley straw pretreated by ethanosolv technology. *Applied Biochemistry and Biotechnology, 163*(1), 143–152.
61. Kim, Y., Yu, A., Han, M., Choi, G. W., & Chung, B. (2010). Ethanosolv pretreatment of barley straw with iron (III) chloride for enzymatic saccharification. *Journal of Chemical Technology and Biotechnology, 85*(11), 1494–1498.
62. Kim, Y., Yu, A., Han, M., Choi, G. W., & Chung, B. (2011b). Enhanced enzymatic saccharification of barley straw pretreated by ethanosolv technology. *Applied Biochemistry and Biotechnology, 163*(1), 143–152.
63. Köchermann, J., Mühlenberg, J., & Klemm, M. (2018). Kinetics of hydrothermal furfural production from organosolv hemicellulose and d-xylose. *Industrial and Engineering Chemistry Research, 57*(43), 14417–14427.
64. Koo, B.-W., Kim, H.-Y., Park, N., Lee, S.-M., Yeo, H., & Choi, I.-G. (2011). Organosolv pretreatment of Liriodendron tulipifera and simultaneous saccharification and fermentation for bioethanol production. *Biomass and Bioenergy, 35*(5), 1833–1840.
65. Koo, B.-W., Min, B.-C., Gwak, K.-S., Lee, S.-M., Choi, J.-W., Yeo, H., & Choi, I.-G. (2012). Structural changes in lignin during organosolv pretreatment of Liriodendron tulipifera and the effect on enzymatic hydrolysis. *Biomass and Bioenergy, 42*, 24–32.
66. Kozlowski, R., & Helwig, M. (1998). Lignocellulosic polymer composites. In *Science and technology of polymers and advanced materials* (pp. 679–698). Springer, US.
67. Krishania, M., Kumar, V., Vijay, V. K., & Malik, A. (2012). Opportunities for improvement of process technology for biomethanation processes. *Green Processing and Synthesis, 1*(1), 49.
68. Kumar, A. K., & Sharma, S. (2017). Recent updates on different methods of pretreatment of lignocellulosic feedstocks: A review. *Bioresources and Bioprocessing, 4*(1), 7.
69. Lai, C., Tu, M., Li, M., & Yu, S. (2014). Remarkable solvent and extractable lignin effects on enzymatic digestibility of organosolv pretreated hardwood. *Bioresource Technology, 156*, 92–99.
70. Lee, D. H., Cho, E. Y., Kim, C. J., & Kim, S. B. (2010). Pretreatment of waste newspaper using ethylene glycol for bioethanol production. *Biotechnology and Bioprocess Engineering, 15*(6), 1094–1101.
71. Lee, H. V., Hamid, S. B., & Zain, S. K. (2014). Conversion of lignocellulosic biomass to nanocellulose: Structure and chemical process. *ScientificWorldJournal, 2014*, 631013.
72. Li, M., Tu, M., Cao, D., Bass, P., & Adhikari, S. (2013a). Distinct roles of residual xylan and lignin in limiting enzymatic hydrolysis of organosolv pretreated loblolly pine and sweetgum. *Journal of Agricultural and Food Chemistry, 61*(3), 646–654.

73. Li, M. F., Sun, S. N., Xu, F., & Sun, R. C. (2012a). Organosolv fractionation of lignocelluloses for fuels, chemicals and materials: A biorefinery processing perspective. In *Biomass conversion* (pp. 341–379). Berlin: Springer.
74. Li, M. F., Yang, S., & Sun, R. C. (2016a). Recent advances in alcohol and organic acid fractionation of lignocellulosic biomass. *Bioresource Technology, 200*, 971–980.
75. Li, P., Zhang, Q., Zhang, X., Zhang, X., Pan, X., & Xu, F. (2019a). Subcellular dissolution of xylan and lignin for enhancing enzymatic hydrolysis of microwave assisted deep eutectic solvent pretreated Pinus bungeana Zucc. *Bioresource Technology, 288*, 121475.
76. Li, S., Ydna, M. Q.-S., & Jeremy, S. L. (2016b). A mild biomass pretreatment using γ-valerolactone for concentrated sugar production. *Green Chemistry, 18*(4), 937–943.
77. Li, Y.-J., Li, H.-Y., Sun, S.-N., & Sun, R.-C. (2019b). Evaluating the efficiency of γ-valerolactone/water/acid system on Eucalyptus pretreatment by confocal Raman microscopy and enzymatic hydrolysis for bioethanol production. *Renewable Energy, 134*, 228–234.
78. Li, Z., Jiang, Z., Fei, B., Cai, Z., & Pan, X. (2012b). Ethanosolv pretreatment of bamboo with dilute acid for efficient enzymatic saccharification. In *Proceedings of the 55th convention of Society of Wood Science and Technology*, August 27–31, 2012 Beijing China. 9, pp 1–9.
79. Li, Z., Jiang, Z., Fei, B., Pan, X., Cai, Z., & Yu, Y. (2012c). Ethanol organosolv pretreatment of bamboo for efficient enzymatic saccharification. *BioResources, 7*(3), 3452–3462.
80. Li, Z., Jiang, Z., Fei, B., Pan, X., Cai, Z., & Yu, Y. (2013b). Ethanosolv with NaOH pretreatment of moso bamboo for efficient enzymatic saccharification. *BioResources, 8*(3), 4711–4721.
81. Liu, J., Li, R., Shuai, L., You, J., Zhao, Y., Chen, L., Li, M., Chen, L., Huang, L., & Luo, X. (2017). Comparison of liquid hot water (LHW) and high boiling alcohol/water (HBAW) pretreatments for improving enzymatic saccharification of cellulose in bamboo. *Industrial Crops and Products, 107*, 139–148.
82. Liu, Y., Nie, Y., Lu, X., Zhang, X., He, H., Pan, F., Zhou, L., Liu, X., Ji, X., & Zhang, S. (2019a). Cascade utilization of lignocellulosic biomass to high-value products. *Green Chemistry, 21*(13), 3499–3535.
83. Liu, Y., Zheng, J., Xiao, J., He, X., Zhang, K., Yuan, S., Peng, Z., Chen, Z., & Lin, X. (2019b). Enhanced enzymatic hydrolysis and lignin extraction of wheat straw by triethylbenzyl ammonium chloride/lactic acid-based deep eutectic solvent pretreatment. *ACS Omega, 4*(22), 19829–19839.
84. Liu, Z. H., Qin, L., Li, B.-Z., & Yuan, Y.-J. (2015). Physical and chemical characterizations of corn stover from leading pretreatment methods and effects on enzymatic hydrolysis. *ACS Sustainable Chemistry & Engineering, 3*(1), 140–146.
85. Long, J., Li, X., Guo, B., Wang, L., & Zhang, N. (2013). Catalytic delignification of sugarcane bagasse in the presence of acidic ionic liquids. *Catalysis Today, 200*, 99–105.
86. Lynam, J. G., Chow, G. I., Hyland, P. L., & Coronella, C. J. (2016). Corn stover pretreatment by ionic liquid and glycerol mixtures with their density, viscosity, and thermogravimetric properties. *ACS Sustainable Chemistry & Engineering, 4*(7), 3786–3793.
87. Lynam, J. G., & Coronella, C. J. (2014). Glycerol as an ionic liquid co-solvent for pretreatment of rice hulls to enhance glucose and xylose yield. *Bioresource Technology, 166*, 471–478.
88. Lynam, J. G., & Coronella, C. J. (2016). Loblolly pine pretreatment by ionic liquid-glycerol mixtures. *Biomass Conversion and Biorefinery, 6*(3), 247–260.
89. Martin-Sampedro, R., Filpponen, I., Hoeger, I. C., Zhu, J. Y., Laine, J., & Rojas, O. J. (2012). Rapid and complete enzyme hydrolysis of lignocellulosic nanofibrils. *ACS Macro Letters, 1*(11), 1321–1325.
90. Martino, D. C., Colodette, J. L., Chandra, R., & Saddler, J. (2017). Steam explosion pretreatment used to remove hemicellulose to enhance the production of a eucalyptus organosolv dissolving pulp. *Wood Science and Technology, 51*(3), 557–569.
91. McDonough, T. J. (1992). The chemistry of organosolv delignification. *Tappi Journal, 76*, 186–193.

92. Menon, V., Prakash, G., & Rao, M. (2010). Value added products from hemicellulose: Biotechnological perspective. *Global Journal of Biochemistry, 1*(1), 36–67.
93. Mesa, L., González, E., Cara, C., Ruiz, E., Castro, E., & Mussatto, S. I. (2010a). An approach to optimization of enzymatic hydrolysis from sugarcane bagasse based on organosolv pretreatment. *Journal of Chemical Technology and Biotechnology, 85*(8), 1092–1098.
94. Mesa, L., González, E., Ruiz, E., Romero, I., Cara, C., Felissia, F., & Castro, E. (2010b). Preliminary evaluation of organosolv pre-treatment of sugar cane bagasse for glucose production: Application of 23 experimental design. *Applied Energy, 87*(1), 109–114.
95. Monrroy, M., Ibanez, J., Melin, V., Baeza, J., Mendonça, R. T., Contreras, D., & Freer, J. (2010). Bioorganosolv pretreatments of P. radiata by a brown rot fungus (Gloephyllum trabeum) and ethanolysis. *Enzyme and Microbial Technology, 47*(1–2), 11–16.
96. Mosier, N., Wyman, C., Dale, B., Elander, R., Lee, Y., Holtzapple, M., & Ladisch, M. (2005). Features of promising technologies for pretreatment of lignocellulosic biomass. *Bioresource Technology, 96*(6), 673–686.
97. Mou, H., & Wu, S. (2017). Comparison of hydrothermal, hydrotropic and organosolv pretreatment for improving the enzymatic digestibility of bamboo. *Cellulose, 24*(1), 85–94.
98. Muñoz, C., Baeza, J., Freer, J., & Mendonça, R. T. (2011). Bioethanol production from tension and opposite wood of Eucalyptus globulus using organosolv pretreatment and simultaneous saccharification and fermentation. *Journal of Industrial Microbiology & Biotechnology, 38*(11), 1861.
99. Mussatto, S. I. (2016). Biomass pretreatment with acids. In *Biomass fractionation technologies for a lignocellulosic feedstock based biorefinery* (pp. 169–185).
100. Mussatto, S. I., Fernandes, M., Milagres, A. M. F., & Roberto, I. C. (2008). Effect of hemicellulose and lignin on enzymatic hydrolysis of cellulose from brewer's spent grain. *Enzyme and Microbial Technology, 43*(2), 124–129.
101. Neilson, J., & Shafizadeh, F. (1983). Evaluation of organosolv pulp as a suitable substrate for rapid enzymatic hydrolysis. *Biotechnology & Bioengineering (United States), 25*(2), 609.
102. Nitsos, C., Rova, U., & Christakopoulos, P. (2018). Organosolv fractionation of softwood biomass for biofuel and biorefinery applications. *Energies, 11*(1), 50.
103. Pan, X., Arato, C., Gilkes, N., Gregg, D., Mabee, W., Pye, K., Xiao, Z., Zhang, X., & Saddler, J. (2005a). Biorefining of softwoods using ethanol organosolv pulping: Preliminary evaluation of process streams for manufacture of fuel-grade ethanol and co-products. *Biotechnology and Bioengineering, 90*(4), 473–481.
104. Pan, X., Dan, X., Gilkes, N., Gregg, D. J., & Saddler, J. N. (2005b). Strategies to enhance the enzymatic hydrolysis of pretreated softwood with high residual lignin content. *Applied Biochemistry and Biotechnology, 124*(1–3), 1069–1079.
105. Pan, X., Gilkes, N., Kadla, J., Pye, K., Saka, S., Gregg, D., Ehara, K., Xie, D., Lam, D., & Saddler, J. (2006a). Bioconversion of hybrid poplar to ethanol and co-products using an organosolv fractionation process: Optimization of process yields. *Biotechnology and Bioengineering, 94*(5), 851–861.
106. Pan, X., Gilkes, N., & Saddler, J. N. (2006b). Effect of acetyl groups on enzymatic hydrolysis of cellulosic substrates. *Holzforschung, 60*(4), 398–401.
107. Pan, X., Kadla, J. F., Ehara, K., Gilkes, N., & Saddler, J. N. (2006c). Organosolv ethanol lignin from hybrid poplar as a radical scavenger: Relationship between lignin structure, extraction conditions, and antioxidant activity. *Journal of Agricultural and Food Chemistry, 54*(16), 5806–5813.
108. Pan, X., Xie, D., Yu, R. W., Lam, D., & Saddler, J. N. (2007). Pretreatment of lodgepole pine killed by mountain pine beetle using the ethanol organosolv process: Fractionation and process optimization. *Industrial and Engineering Chemistry Research, 46*(8), 2609–2617.
109. Papa, G., Rodriguez, S., George, A., Schievano, A., Orzi, V., Sale, K. L., Singh, S., Adani, F., & Simmons, B. A. (2015). Comparison of different pretreatments for the production of bioethanol and biomethane from corn stover and switchgrass. *Bioresource Technology, 183*, 101–110.

110. Parawira, W., & Tekere, M. (2011). Biotechnological strategies to overcome inhibitors in lignocellulose hydrolysates for ethanol production: Review. *Critical Reviews in Biotechnology, 31*(1), 20–31.
111. Park, N., Kim, H.-Y., Koo, B.-W., Yeo, H., & Choi, I.-G. (2010). Organosolv pretreatment with various catalysts for enhancing enzymatic hydrolysis of pitch pine (Pinus rigida). *Bioresource Technology, 101*(18), 7046–7053.
112. Procentese, A., Raganati, F., Olivieri, G., Russo, M. E., Rehmann, L., & Marzocchella, A. (2018). Deep eutectic solvents pretreatment of agro-industrial food waste. *Biotechnology for Biofuels, 11*, 37.
113. Qureshi, N., Liu, S., Hughes, S., Palmquist, D., Dien, B., & Saha, B. (2016). Cellulosic butanol (ABE) biofuel production from sweet sorghum bagasse (SSB): Impact of hot water pretreatment and solid loadings on fermentation employing Clostridium beijerinckii P260. *Bioenergy Research, 9*(4), 1167–1179.
114. Reddy, N., & Yang, Y. (2005). Biofibers from agricultural byproducts for industrial applications. *Trends in Biotechnology, 23*(1), 22–27.
115. Rinaldi, R., Jastrzebski, R., Clough, M. T., Ralph, J., Kennema, M., Bruijnincx, P. C., & Weckhuysen, B. M. (2016). Paving the way for lignin valorisation: Recent advances in bioengineering, biorefining and catalysis. *Angewandte Chemie (International Ed. in English), 55*(29), 8164–8215.
116. Romaní, A., Ruiz, H. A., Pereira, F. B., Domingues, L., & Teixeira, J. A. (2013). Fractionation of Eucalyptus globulus wood by glycerol–water pretreatment: Optimization and modeling. *Industrial and Engineering Chemistry Research, 52*(40), 14342–14352.
117. Rosales-Calderon, O., & Arantes, V. (2019). A review on commercial-scale high-value products that can be produced alongside cellulosic ethanol. *Biotechnology for Biofuels, 12*, 240.
118. Rößiger, B., Röver, R., Unkelbach, G., & Pufky-Heinrich, D. (2017). Production of biophenols for industrial application: Scale-up of the base-catalyzed depolymerization of lignin. *Green and Sustainable Chemistry, 7*(03), 193.
119. Rubio, M., Tortosa, J. F., Quesada, J., & Gómez, D. (1998). Fractionation of lignocellulosics. Solubilization of corn stalk hemicelluloses by autohydrolysis in aqueous medium. *Biomass and Bioenergy, 15*(6), 483–491.
120. Sakdaronnarong, C., Srimarut, N., & Laosiripojana, N. (2015). Polyurethane synthesis from sugarcane bagasse organosolv and Kraft lignin. In *Key engineering materials* (pp. 527–532). Trans Tech Publ, Switzerland.
121. Sannigrahi, P., Miller, S. J., & Ragauskas, A. J. (2010). Effects of organosolv pretreatment and enzymatic hydrolysis on cellulose structure and crystallinity in Loblolly pine. *Carbohydrate Research, 345*(7), 965–970.
122. Shen, X. J., Wen, J. L., Mei, Q. Q., Chen, X., Sun, D., Yuan, T. Q., & Sun, R. C. (2019). Facile fractionation of lignocelluloses by biomass-derived deep eutectic solvent (DES) pretreatment for cellulose enzymatic hydrolysis and lignin valorization. *Green Chemistry, 21*(2), 275–283.
123. Siqueira, G., Arantes, V., Saddler, J. N., Ferraz, A., & Milagres, A. M. F. (2017). Limitation of cellulose accessibility and unproductive binding of cellulases by pretreated sugarcane bagasse lignin. *Biotechnology for Biofuels, 10*, 176.
124. Smit, A., & Huijgen, W. (2017). Effective fractionation of lignocellulose in herbaceous biomass and hardwood using a mild acetone organosolv process. *Green Chemistry, 19*(22), 5505–5514.
125. Snelders, J., Dornez, E., Benjelloun-Mlayah, B., Huijgen, W. J., de Wild, P. J., Gosselink, R. J., Gerritsma, J., & Courtin, C. M. (2014). Biorefining of wheat straw using an acetic and formic acid based organosolv fractionation process. *Bioresource Technology, 156*, 275–282.
126. Sun, S., Sun, S., Cao, X., & Sun, R. (2016). The role of pretreatment in improving the enzymatic hydrolysis of lignocellulosic materials. *Bioresource Technology, 199*, 49–58.
127. Sun, Y., & Cheng, J. (2003). Hydrolysis of lignocellulosic materials for ethanol production. *Bioresource Technology, 83*(1), 1–11.
128. Taherzadeh, M. J., & Karimi, K. (2008). Pretreatment of lignocellulosic wastes to improve ethanol and biogas production: A review. *International Journal of Molecular Sciences, 9*(9), 1621–1651.

129. Tan, H., Yang, R., Sun, W., & Wang, S. (2009). Peroxide– acetic acid pretreatment to remove bagasse lignin prior to enzymatic hydrolysis. *Industrial and Engineering Chemistry Research, 49*(4), 1473–1479.
130. Tang, C., Shan, J., Chen, Y., Zhong, L., Shen, T., Zhu, C., & Ying, H. (2017). Organic amine catalytic organosolv pretreatment of corn stover for enzymatic saccharification and high-quality lignin. *Bioresource Technology, 232*, 222–228.
131. Thi, S., & Lee, K. M. (2019). Comparison of deep eutectic solvents (DES) on pretreatment of oil palm empty fruit bunch (OPEFB): Cellulose digestibility, structural and morphology changes. *Bioresource Technology, 282*, 525–529.
132. Timung, R., Mohan, M., Chilukoti, B., Sasmal, S., Banerjee, T., & Goud, V. V. (2015). Optimization of dilute acid and hot water pretreatment of different lignocellulosic biomass: A comparative study. *Biomass and Bioenergy, 81*, 9–18.
133. Tomás-Pejó, E., Fermoso, J., Herrador, E., Hernando, H., Jiménez-Sánchez, S., Ballesteros, M., González-Fernández, C., & Serrano, D. P. (2017). Valorization of steam-exploded wheat straw through a biorefinery approach: Bioethanol and bio-oil co-production. *Fuel, 199*, 403–412.
134. Tri, C. L., Khuong, L. D., & Kamei, I. (2018). The improvement of sodium hydroxide pretreatment in bioethanol production from Japanese bamboo Phyllostachys edulis using the white rot fungus Phlebia sp. MG-60. *International Biodeterioration & Biodegradation, 133*, 86–92.
135. Trinh, L. T. P., Lee, J.-W., & Lee, H.-J. (2016). Acidified glycerol pretreatment for enhanced ethanol production from rice straw. *Biomass and Bioenergy, 94*, 39–45.
136. Vanderghem, C., Brostaux, Y., Jacquet, N., Blecker, C., & Paquot, M. (2012). Optimization of formic/acetic acid delignification of Miscanthus× giganteus for enzymatic hydrolysis using response surface methodology. *Industrial Crops and Products, 35*(1), 280–286.
137. Vanneste, J., Ennaert, T., Vanhulsel, A., & Sels, B. (2017). Unconventional pretreatment of lignocellulose with low-temperature plasma. *ChemSusChem, 10*(1), 14–31.
138. Vazquez, G., Antorrena, G., Gonzalez, J., Freire, S., & Crespo, I. (2000). The influence of acetosolv pulping conditions on the enzymatic hydrolysis of Eucalyptus pulps. *Wood Science and Technology, 34*(4), 345–354.
139. Villaverde, J. J., Li, J., Ek, M., Ligero, P., & de Vega, A. (2009). Native lignin structure of Miscanthus x giganteus and its changes during acetic and formic acid fractionation. *Journal of Agricultural and Food Chemistry, 57*(14), 6262–6270.
140. Voelker, S. L., Lachenbruch, B., Meinzer, F. C., & Strauss, S. H. (2011). Reduced wood stiffness and strength, and altered stem form, in young antisense 4CL transgenic poplars with reduced lignin contents. *The New Phytologist, 189*(4), 1096–1109.
141. Wang, X., & Rinaldi, R. (2016). Bifunctional Ni catalysts for the one-pot conversion of Organosolv lignin into cycloalkanes. *Catalysis Today, 269*, 48–55.
142. Wildschut, J., Smit, A. T., Reith, J. H., & Huijgen, W. J. (2013). Ethanol-based organosolv fractionation of wheat straw for the production of lignin and enzymatically digestible cellulose. *Bioresource Technology, 135*, 58–66.
143. Wyman, C. E., Dale, B. E., Elander, R. T., Holtzapple, M., Ladisch, M. R., & Lee, Y. Y. (2005). Coordinated development of leading biomass pretreatment technologies. *Bioresource Technology, 96*(18), 1959–1966.
144. Xu, L., & Tschirner, U. W. (2012). Peracetic acid pretreatment of alfalfa stem and aspen biomass. *BioResources, 7*(1), 0203–0216.
145. Xue, B. L., Wen, J. L., & Sun, R. C. (2015). Ethanol organosolv lignin as a reactive filler for acrylamide-based hydrogels. *Journal of Applied Polymer Science, 132*(40), 42638.
146. Yu, G., Li, B., Liu, C., Zhang, Y., Wang, H., & Mu, X. (2013a). Fractionation of the main components of corn stover by formic acid and enzymatic saccharification of solid residue. *Industrial Crops and Products, 50*, 750–757.
147. Yu, H., Xing, Y., Lei, F., Liu, Z., Liu, Z., & Jiang, J. (2014). Improvement of the enzymatic hydrolysis of furfural residues by pretreatment with combined green liquor and ethanol organosolv. *Bioresource Technology, 167*, 46–52.

148. Yu, H., You, Y., Lei, F., Liu, Z., Zhang, W., & Jiang, J. (2015). Comparative study of alkaline hydrogen peroxide and organosolv pretreatments of sugarcane bagasse to improve the overall sugar yield. *Bioresource Technology, 187*, 161–166.
149. Yu, H. L., Tang, Y., Xing, Y., Zhu, L.-W., & Jiang, J.-X. (2013b). Improvement of the enzymatic hydrolysis of furfural residues by pretreatment with combined green liquor and hydrogen peroxide. *Bioresource Technology, 147*(complete), 29–36.
150. Yuan, Z., Wen, Y., & Li, G. (2018). Production of bioethanol and value added compounds from wheat straw through combined alkaline/alkaline-peroxide pretreatment. *Bioresource Technology, 259*, 228.
151. Zhang, K., Pei, Z., & Wang, D. (2016a). Organic solvent pretreatment of lignocellulosic biomass for biofuels and biochemicals: A review. *Bioresource Technology, 199*, 21–33.
152. Zhang, Q., Huang, H., Han, H., Qiu, Z., & Achal, V. (2017). Stimulatory effect of in-situ detoxification on bioethanol production by rice straw. *Energy, 135*, 32–39.
153. Zhang, X., Zhao, W., Li, Y., Li, C., Yuan, Q., & Cheng, G. (2016b). Synergistic effect of pretreatment with dimethyl sulfoxide and an ionic liquid on enzymatic digestibility of white poplar and pine. *RSC Advances, 6*(67), 62278–62285.
154. Zhang, Y., Ye, Y. Y., Fan, J., & Chang, J. (2013a). Selective production of phenol, guaiacol and 2, 6-dimethoxyphenol by alkaline hydrothermal conversion of lignin. *Journal of Biobased Materials and Bioenergy, 7*(6), 696–701.
155. Zhang, Y. H. P., Ding, S. Y., Mielenz, J. R., Cui, J. B., Elander, R. T., Laser, M., Himmel, M. E., McMillan, J. R., & Lynd, L. R. (2007). Fractionating recalcitrant lignocellulose at modest reaction conditions. *Biotechnology and Bioengineering, 97*(2), 214–223.
156. Zhang, Z., Rackemann, D. W., Doherty, W. O., & O'Hara, I. M. (2013b). Glycerol carbonate as green solvent for pretreatment of sugarcane bagasse. *Biotechnology for Biofuels, 6*(1), 153.
157. Zhang, Z., Wong, H. H., Albertson, P. L., Doherty, W. O., & O'Hara, I. M. (2013c). Laboratory and pilot scale pretreatment of sugarcane bagasse by acidified aqueous glycerol solutions. *Bioresource Technology, 138*, 14–21.
158. Zhao, H., Jones, C. L., Baker, G. A., Xia, S., Olubajo, O., & Person, V. N. (2009a). Regenerating cellulose from ionic liquids for an accelerated enzymatic hydrolysis. *Journal of Biotechnology, 139*(1), 47–54.
159. Zhao, J., & Chen, H. (2013). Correlation of porous structure, mass transfer and enzymatic hydrolysis of steam exploded corn stover. *Chemical Engineering Science, 104*, 1036–1044.
160. Zhao, X., Cheng, K., & Liu, D. (2009b). Organosolv pretreatment of lignocellulosic biomass for enzymatic hydrolysis. *Applied Microbiology and Biotechnology, 82*(5), 815–827.
161. Zhao, X., Li, S., Wu, R., & Liu, D. (2017). Organosolv fractionating pre-treatment of lignocellulosic biomass for efficient enzymatic saccharification: Chemistry, kinetics, and substrate structures. *Biofuels, Bioproducts and Biorefining, 11*(3), 567–590.
162. Zhao, X., & Liu, D. (2011). Fractionating pretreatment of sugarcane bagasse for increasing the enzymatic digestibility of cellulose. *Sheng wu gong cheng xue bao= Chin J Biotechnol, 27*(3), 384–392.
163. Zhao, X., & Liu, D. (2012). Fractionating pretreatment of sugarcane bagasse by aqueous formic acid with direct recycle of spent liquor to increase cellulose digestibility–the Formiline process. *Bioresource Technology, 117*, 25–32.
164. Zhao, X., Zhang, L., & Liu, D. (2008). Comparative study on chemical pretreatment methods for improving enzymatic digestibility of crofton weed stem. *Bioresource Technology, 99*(9), 3729–3736.
165. Zhao, X., Zhang, L., & Liu, D. (2010). Pretreatment of Siam weed stem by several chemical methods for increasing the enzymatic digestibility. *Biotechnology Journal, 5*(5), 493–504.
166. Zhao, X., Zhang, L., & Liu, D. (2012). Biomass recalcitrance. Part I: The chemical compositions and physical structures affecting the enzymatic hydrolysis of lignocellulose. *Biofuels, Bioproducts and Biorefining, 6*(4), 465–482.

167. Zhao, X., Wang, L., & Liu, D. (2007). Effect of several factors on peracetic acid pretreatment of sugarcane bagasse for enzymatic hydrolysis. *Journal of Chemical Technology and Biotechnology, 82*(12), 1115–1121.
168. Zhao, Z., Chen, X., Ali, M. F., Abdeltawab, A. A., Yakout, S. M., & Yu, G. (2018). Pretreatment of wheat straw using basic ethanolamine-based deep eutectic solvents for improving enzymatic hydrolysis. *Bioresource Technology, 263*, 325–333.
169. Zheng, Y., Shi, J., Tu, M., & Cheng, Y.-S. (2017). Principles and development of lignocellulosic biomass pretreatment for biofuels. In *Advances in Bioenergy* (pp. 1–68).
170. Zhou, Z., Lei, F., Li, P., & Jiang, J. (2018). Lignocellulosic biomass to biofuels and biochemicals: A comprehensive review with a focus on ethanol organosolv pretreatment technology. *Biotechnology and Bioengineering, 115*(11), 2683–2702.
171. Zhou, Z., Xue, W., Lei, F., Cheng, Y., Jiang, J., & Sun, D. (2016). Kraft GL-ethanol pretreatment on sugarcane bagasse for effective enzymatic hydrolysis. *Industrial Crops and Products, 90*, 100–109.

Chapter 3
New Developments on Ionic Liquid-Tolerant Microorganisms Leading Toward a More Sustainable Biorefinery

André M. da Costa Lopes, Leonardo da Costa Sousa, Rafał M. Łukasik, and Ana Rita C. Morais

3.1 Introduction

Growing concerns about climate change and global energy security have led to the increasing interest in phasing out the use of fossil fuels and enabling the transition to renewable energy resources, including the use of bioenergy. The most abundant bio-based resources available in the planet are lignocellulosic residues, which can be converted to fermentable sugars by hydrolytic enzymes and used by microorganisms to produce a gamut of biofuels and bio-based chemicals in a biorefinery [1]. Though lignocellulosic residues are abundant in nature, they are highly recalcitrant substrates to enzymes due to their tight and complex ultrastructure formed mostly by cellulose, hemicelluloses, and lignin. In a biorefinery process, biomass recalcitrance is generally overcome by a pretreatment step prior to enzymatic hydrolysis and fermentation. Several types of pretreatment technologies have been developed over the years, all of which are able to disrupt the cell wall atrix at various levels and improve enzyme accessibility to carbohydrates [2–4]. Of those pretreatment technologies available today, ILs have been acknowledged as promising and sustainable due to their superior solvation properties and performance relative to classical organic solvents. In addition, these nonconventional solvents have been acclaimed

A. M. da Costa Lopes · R. M. Łukasik
National Laboratory for Energy and Geology, Bioenergy and Biorefineries Unit, Lisbon, Portugal

L. da Costa Sousa
Independent Scholar, East Lansing, MI, USA

A. R. C. Morais (✉)
National Laboratory for Energy and Geology, Bioenergy and Biorefineries Unit, Lisbon, Portugal

National Bioenergy Center, National Renewable Energy Laboratory, Golden, CO, USA
e-mail: ana.morais@nrel.gov

Z.-H. Liu, A. Ragauskas (eds.), *Emerging Technologies for Biorefineries, Biofuels, and Value-Added Commodities*,
https://doi.org/10.1007/978-3-030-65584-6_3

as one of the few feedstock agnostic technologies capable of efficiently handling various feedstocks, both as single and blends [5], effective at high solid loadings, and operated in a continuous mode [6]. Despite these advantages, some hurdles associated with the use of ILs have been hindering the commercial feasibility of this technology and their use in biorefineries. Firstly, some of the best performing ILs for biomass pretreatment are relatively toxic to microorganisms used in fermentation, requiring several water washing steps to remove residual IL from the pretreated slurry prior to fermentation. The excessive use of water, which introduces a significant wastewater treatment requirement, and the need for IL recycling create significant process engineering and economic challenges to the commercial viability of ILs [7–9]. Another challenge is the pH mismatch between pretreatment and fermentation unit operations, namely, for alkaline ILs (e.g., 1-ethyl-3-methylimidazolium acetate or lysinate or cholinium glycine) [10] that generate highly basic pH conditions and thus are incompatible with the wild-type microbial fermentation hosts that require slightly acid or neutral reaction media [9]. To overcome this compatibility issue, a neutralization step is required prior to fermentation, resulting in the formation of complex salts that turn the recovery and reuse of the ILs even more challenging [9]. In addition, ILs are still one the most expensive solvents currently used in the pretreatment of lignocellulosic biomass. These significant challenges must be tackled to design and develop a cost-effective IL pretreatment technology, compatible with microbial catalysts that can properly function in IL-based reaction media, thus reducing multiple downstream steps for separation and recovery of ILs and desirable end products.

Various studies have been published addressing the effectiveness of ILs and/or aqueous-IL media for biomass pretreatment and microbial conversion to biofuels and other value-added chemicals. The research that has been performed in this field approaches the problem of microbial tolerance to ILs using two major strategies: one strategy focuses on the development of new IL pretreatment systems using more biocompatible ILs, while the other prioritizes the screening and discovery of microbial strains that offer effective tolerance mechanisms to high concentration of ILs in the media. This chapter summarizes and discusses the efforts performed on the study of various microbial strains and their response to the presence of ILs, as well as the attempts to produce more biocompatible IL pretreatment systems that promote biomass deconstruction and its bioconversion to valuable products in a biorefinery concept.

3.2 Growth of Microorganisms in the Presence of Ionic Liquids

The conversion of fermentable sugars to bio-based fuels and chemicals in the presence of ILs requires that microorganisms can grow under such conditions. Several strategies have been implemented to isolate promising microbial strains that can tolerate the presence of inhibitors, such as ILs. A well-established approach is to study cell growth potential in complex lab media with increasing concentrations of the growth inhibitor. This type of study is important to understand the factors that influence cell growth and the development of tolerance to ILs under well-defined growth conditions [11–15]. Another approach to microbial strain isolation involves the selection of naturally occurring microorganisms in lignocellulosic substrates that are further screened for tolerance to the presence of increasing IL concentrations [16–19]. The use of IL-pretreated biomass substrates is used as the ultimate approach to evaluate the growth potential of microbial strains and, eventually, their use in future biorefining applications [12]. This section describes the research that has been developed for the selection of IL-tolerant organisms, including fungi and bacteria, pointing out their main progress, challenges, and limitations. An overview of the advances in this topic is demonstrated in Table 3.1.

3.2.1 Fungi

The fungi is a major kingdom of organisms that have been widely studied and used for biotechnological purposes, with particular attention to the *Saccharomyces* yeast genus. From the practical point of view, it makes sense to evaluate the tolerance of these well-established yeast strains to the presence of ILs, which could rapidly be deployed in industrial applications. On this note, the impact of ILs on *Saccharomyces* yeast growth has been studied in culture experiments performed with complex lab media containing a range of known [emim][OAc] concentrations [12, 13], as it is one of the most widely used ILs for pretreating biomass. Generally, microorganism growth under the presence of [emim][OAc] was measured by spectrophotometry (OD_{600}) relative to a control experiment, where no IL was added to the media. Those studies showed high growth inhibition for several *Saccharomyces* strains when cultured in the presence of up to 5% [emim][OAc] [13]. To better understand this phenomenon, Ouellet et al. studied the independent effects of both [emim][OAc] cation and anion on *Saccharomyces cerevisiae* growth [12]. For that purpose, yeast cell cultures in media containing NaOAc, [emim][Cl], and NaCl at varying concentrations were performed, and cell growth was compared with cultures containing [emim][OAc]. In that work, NaOAc was used to evaluate the effect of IL anion alone, which showed a slightly negative impact on the yeast growth. The IL cation effect was scrutinized by the presence of [emim][Cl] in the culture medium, which demonstrated to inhibit cell growth significantly, while the presence of NaCl showed no inhibitory effect on yeast growth at the examined concentration range. As [emim][OAc] showed greater levels of inhibition relative to [emim][Cl] and NaOAc at equivalent concentrations, the overall study suggests that $[emim]^+$ and $[OAc]^-$ display a negative synergistic effect on *S. cerevisiae* growth. The study showed that *S. cerevisiae* was only tolerant to very low levels of [emim][OAc] (< 5.90 mM), as those conditions did not impact cell growth significantly relative to a control experiment, where no IL was added to the media. In the presence of [emim][OAc]

Table 3.1 Overview of the microorganism growth in the presence of ILs

Microorganism		Ionic liquid		Incubation conditions			Relative growth (%)[a]	Ref.
Class	Species	Type	Concentration	T (°C)	t (h)	Medium		
Fungi	*Fusarium oxysporum* BN	[amim][Cl]	[1%–5%] (w/v)	30	72	LB	[57–0]	[19]
		[bmim][Cl]	[1%–5%] (w/v)	30	72	LB	[63–0]	
		[bmim][FA]	[1%–5%] (w/v)	30	72	LB	[79–48]	
		[emim][DMP]	[1%–5%] (w/v)	30	72	LB	[98–81]	
		[emim][H_2PO_2]	[1%–5%] (w/v)	30	72	LB	[82–60]	
		[emim][OAc]	[1%–5%] (w/v)	30	72	LB	[83–57]	
		[emim][(MeO)HPO_2]	[1%–5%] (w/v)	30	72	LB	[95–58]	
		[emim][$MeSO_4$]	[1%–5%] (w/v)	30	72	LB	[102–71]	
		HEMA	[1%–5%] (w/v)	30	72	LB	[80–55]	
	Saccharomyces cerevisiae	[emim][Cl]	≤ 136.0 mM	30	48	SD	~ 100	[12]
		[emim][OAc]	≤ 5.90 mM	30	48	SD	~ 100	
		[emim][OAc]	[17.6–59.0] mM	30	48	SD	< 5	
	Yarrowia lipolytica	[emim][OAc]	5%	28	48	MpA	~ 70	[20]
		[emim][OAc]	10%	28	48	MpA	~ 40	
		[emim][OAc]	20%	28	48	MpA	~ 15	

(continued)

Table 3.1 (continued)

Microorganism		Ionic liquid		Incubation conditions			Relative growth (%)[a]	Ref.
Class	Species	Type	Concentration	T (°C)	t (h)	Medium		
Bacteria	*Actinobacillus succinogenes*	[amim][Cl]	1%	37	12	LB	~ 95	[21]
		[amim][Cl]	5%	37	12	LB	~ 80	
		[amim][Cl]	≥ 10%	37	12	LB	< 5	
	Bacillus coagulans	[emim][Cl]	236 mM	55	26	MI	~ 70	[17]
		[emim][OAc]	236 mM	55	26	MI	~ 60	
	Clostridium sp.	[emim][DEP]	0.5 g/L	26	140	MS	~ 100	[14]
		[emim][DEP]	[1.0–15] g/L	26	140	MS	~ [85–35]	
		[emim][OAc]	≤ 5 g/L	26	140	MS	~ 140[b]	
			≥ 7.5 g/L	26	140	MS	< 30	
		[mmim][DMP]	[0.4–8] g/L	26	140	MS	~ [100–70]	
		[mmim][DMP]	12 g/L	26	140	MS	~ 0	
		[emim][DEP]	0.5 g/L	26	96	MS	~ 105	[22]
		[emim][DEP]	[1.0–5] g/L	26	96	MS	~ [95–85]	
		[emim][DEP]	10 g/L	26	96	MS	~ 30	
		[emim][OAc]	≤ 2.5 g/L	26	96	MS	~ 150	
		[emim][OAc]	≥ 5 g/L	26	96	MS	< 30	
		[mmim][DMP]	[0.5–10] g/L	26	96	MS	~ [90–75]	
	Escherichia coli	[amim][Cl]	1%	37	12	LB	~ 125	[21]
		[amim][Cl]	5%	37	12	LB	~ 85	
		[amim][Cl]	≥ 10%	37	12	LB	< 25	
	Pseudomonas putida	[emim][OAc]	0.5 g/L	30	72	MS	~ 100	[22]

T - Temperature, t - Time. LB - Luria-Bertani medium, YEPD - Yeast extract peptone Dextrose, SD - Synthetic defined medium, MpA - Not defined, MI - G. Thermoglucosidasius medium, MS - Mineral salts medium

[a]Percentage growth relative to control experiment in absence of IL

[b]It was reported that [emim][OAc] was used as carbon source

concentrations greater than 5.90 mM, *S. cerevisiae* growth inhibition became evident, as depicted by the OD_{600} curves reported by Ouellet et al. in Fig. 3.1 [12]. The fact that *S. cerevisiae* was not inhibited by NaCl and it was highly inhibited by [emim][OAc] means that its tolerance to saline environments cannot be an absolute predictor of IL tolerance. This was also confirmed by testing several halotolerant yeast strains in both NaCl- and [emim][OAc]-containing environments, which showed that highly tolerant strains to the presence of high salt concentrations did not display tolerance to high concentrations of [emim][OAc]. In addition to *S. cerevisiae*, several other yeast strains with good biotechnological potential have been tested for tolerance to increasing [emim][OAc] concentrations. For example, *Kazachstania telluris*, *Wickerhamomyces anomalus*, *Debaryomyces hansenii*, *Galactomyces geotrichum*, and *Clavispora* species have shown different levels of tolerance, up to a maximum of 5% [emim][OAc] in complex lab media [13]. Also, the development of more biocompatible cholinium-based ILs, such as cholinium acetate ([Ch][OAc]), to achieve high microbial tolerance to ILs has been investigated. Similarly to [emim][OAc], cholinium-based ILs can also pretreat lignocellulosic biomass very effectively; however, they have demonstrated to inhibit *S. cerevisiae* growth in a much lower extent relative to [emim][OAc] [23]. Exploring the idea of combining microbial strain selection with the development of more biocompatible ILs for effectively pretreating lignocellulosic biomass is a major stepping-stone that could accelerate the viability of IL-based biorefineries.

In addition to microbial selection strategies using complex lab media, efforts have been made to explore the biodiversity of microbial communities present in the natural environment, notably in lignocellulosic biomass-rich environments, while challenging those organisms to the presence of ILs [19]. For instance, several *Aspergillus* strains were identified to naturally grow in a green waste compost medium containing switchgrass as substrate and relatively high levels of [emim][OAc] [18]. Singer et al. verified that the presence of ILs allows selective growth of *Aspergillus* strains, while a diverse prokaryotic-enriched microbial community (no fungal mat) was grown without [emim][OAc]. It was suggested that the IL inhibited the bacterial community that would normally establish during switchgrass decomposition, but allowed the selective growth of fungi, in particular *Aspergillus* strains [18]. Subsequently, three of those strains were cultivated in agar plates with Luria-Bertani (LB) media containing different concentrations of [emim][OAc]. All strains grew successfully in up to 10% [emim][OAc] concentration. However, when [bmim][Cl] was used in replacement of [emim][OAc], no growth was observed in 5% [bmim][Cl] media. The authors suggested that the $[OAc]^-$ ion from ILs could be assimilated as carbon source by *Aspergillus* fungi. This hypothesis was further tested and confirmed by the addition of NaOAc in media containing [bmim][Cl], which allowed the normal growth of *Aspergillus* strains [18]. This effect could also be observed during *S. cerevisiae* growth in the presence of [emim][OAc], as previously mentioned. Albeit *S. cerevisiae* growth inhibition was detected in the presence of certain concentrations of [emim][OAc], the final OD_{600} values for *S. cerevisiae* cultures containing up to 1% [emim][OAc] were surprisingly higher than that observed for the control experiment in absence of IL. A similar observation was reported by Huang et al. on the growth of *Rhodosporidium toruloides* yeast in the presence of [emim][OAc] as a sole carbon source [11]. In this case, the concentration of $[OAc]^-$ was monitored by HPLC, which confirmed the reduction of $[OAc]^-$ levels in the culture medium during cell growth. Those results suggested that these yeasts were able to assimilate acetate anions as carbon source during their growth process.

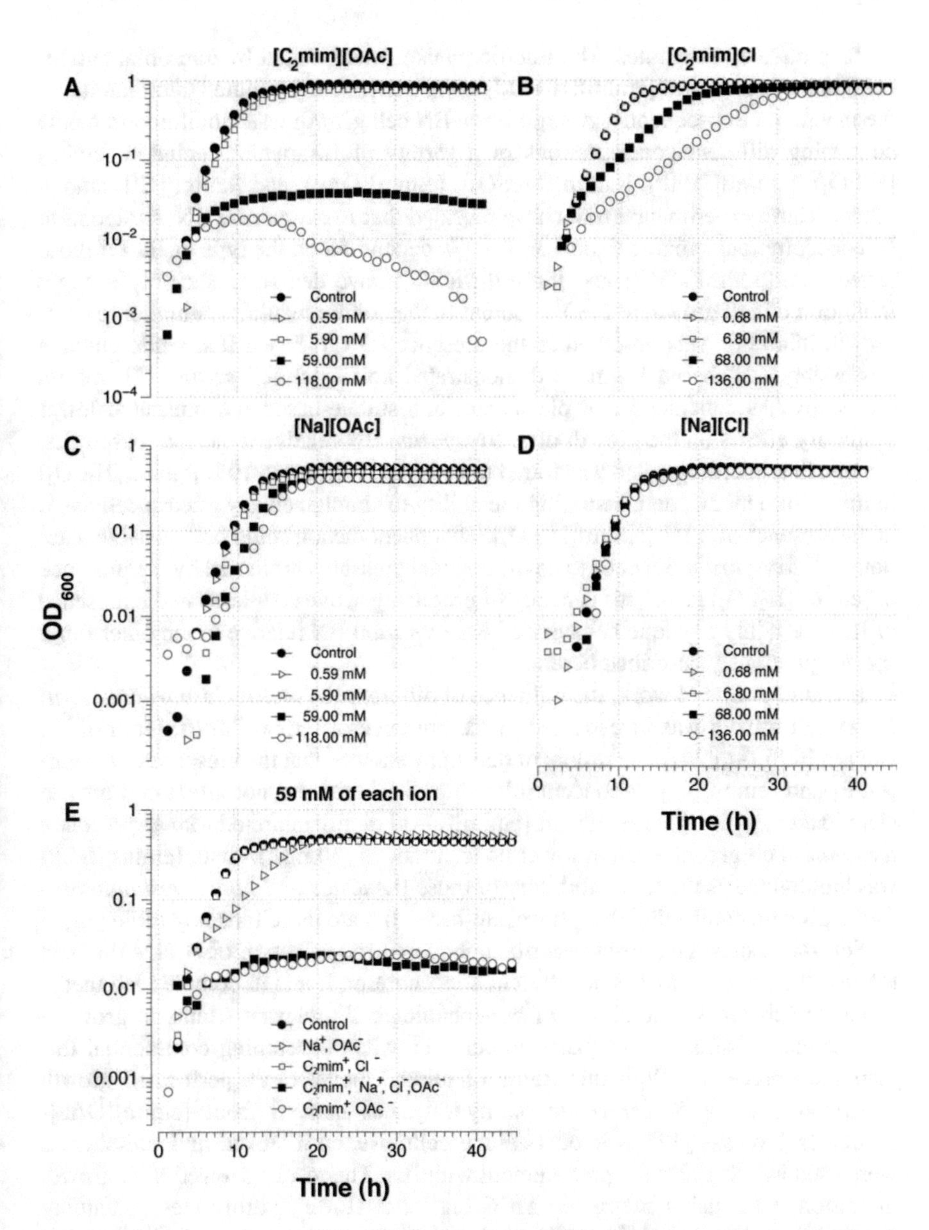

Fig. 3.1 Aerobic growth of *S. cerevisiae* in 20 g/L glucose in the presence of (**a**) [emim][OAc], (**b**) [emim][Cl], (**c**) sodium acetate, (**d**) sodium chloride, and (**e**) in media containing 59 mM of each anion and cation. Screened acetate concentrations using sodium acetate reflect the stoichiometric equivalent of the acetate in [emim][OAc] and are represented using the same symbol. (Reproduced from Ref. [12] with permission from The Royal Society of Chemistry)

Another interesting study showed the potential of the novel IL-tolerant fungal strain *Fusarium oxysporum* BN [19]. Rather than IL tolerance, this microorganism demonstrated to release cellulases in the presence of ILs, which is a relevant advantage if we consider the development of a consolidated bioprocessing (CBP) approach on IL-pretreated substrates. The microorganism was isolated by microbial enrichment in the presence of [emim][H_2PO_2], using a chemically polluted microhabitat as the inoculum. Subsequently, *F. oxysporum* BN cell growth was monitored in media containing different concentrations of a variety of ILs, which included [emim][H_2PO_2], [emim][DMP], [emim][$MeSO_4$], [emim][OAc], and [amim][Cl], among others. These experimental trials have revealed that *F. oxysporum* BN is tolerant to IL concentrations ranging from 1% to 10%, depending on the type of IL. Of those tested ILs, [emim][DMP] and [emim][$MeSO_4$] have demonstrated only a slight inhibition of cell growth at 5% IL content in the culture medium. A higher level of growth inhibition was observed in the presence [OAc]$^-$-based ILs, while cultures possessing [Cl]$^-$-based ILs have demonstrated no fungal cell growth. Therefore, that study has concluded that phosphate- and sulfate-based ILs presented lower inhibitory effects on the growth of *F. oxysporum* BN relative to acetate-based ILs. Finally, that microorganism was found to be highly tolerant to 10% [emim][H_2PO_2] in the culture media, demonstrating the ability to simultaneously release cellulases in the presence of 2.1% [emim][H_2PO_2]. That phenomenon could be a possible reaction of *F. oxysporum* BN cells to environmental pressures promoted by the presence of [emim][H_2PO_2] [19]. This capacity of producing active cellulases in the presence of IL is certainly a unique behavior of *F. oxysporum* BN relative to the other fungi species previously described herein.

In a similar line of work, the influence of different ILs on *Rhodosporidium toruloides* cell growth was investigated in the presence of [emim][DEP], [emim][Cl], and [emim][OAc][24]. The results of that study showed that the presence of [emim][DEP] and [emim][Cl] in concentrations up to 60 mM did not affect cell growth significantly, while the presence of [emim][OAc] highly inhibited *Rhodosporidium toruloides* cell growth even at lower concentrations (30 mM). Also, [emim][DEP] was highlighted as the least inhibitory of those ILs tested on *Rhodosporidium toruloides* growth, confirming that phosphate-based ILs are more biocompatible [24].

Several studies have been described above on the selection of strains that can tolerate the presence of ILs at different concentration levels in complex lab media or native substrates. The ability of biotechnologically relevant strains to grow on commercial substrates is of prime importance when developing commercial fermentation processes. With this frame of mind, Ouellet et al. performed growth experiments using *S. cerevisiae* on hydrolysates derived from [emim][OAc]-pretreated biomass [12]. Microcrystalline cellulose, corn stover, and switchgrass were used as substrates for pretreatment with ILs. The results showed high growth inhibition when using pretreated MCC and corn stover hydrolysates containing residual [emim][OAc] at 33.5 and 52.4 mM concentration, respectively. In other two trials, pretreated corn stover and switchgrass with low levels of IL (< 1 mM concentration of [emim][OAc]) had low impact on *S. cerevisiae* cell growth relative to the control. That study confirmed the difficulties of growing conventional *S. cerevisiae* in the presence of [emim][OAc] for use in IL-based biorefineries, as also ascertained in other research [13].

3.2.2 Bacteria

Bacteria is another important domain of organisms with high biotechnological potential and industrial relevance, which could be developed for future IL pretreatment-based biorefineries. The influence of ILs on the growth of certain genera of bacteria, such as *Clostridium* sp. and *Bacillus* sp., has been tackled in a number of studies [14, 16, 17]. For example, *Clostridium* sp. growth potential was evaluated in the presence of a variety of ILs, notably [emim][OAc], [emim][DEP], and [mmim][DMP] [14]. Generally, at low IL concentrations (<2.5 g/L), *Clostridium* sp. growth curves were comparable to those observed for the control, where no IL was added to the culture media. In contrast, higher IL concentrations (> 2.5 g/L) lead to retardation or complete inhibition of bacterial growth. Particularly, culture media with low concentrations of [emim][OAc] increased the final concentration of bacterial cells to approximately 40% relative to control, despite the inhibitory effect of that IL to cell growth rates [14]. This observation was also reported for yeast growth in the presence of the same IL, as discussed in the previous subsection [11, 13]. Furthermore, Nancharaiah et al. verified that a pH decrease from approximately 7 to 3, which occurred naturally in control experiments, was not observed in the presence of [emim][OAc] [14]. The medium pH was always higher during bacterial growth in the presence of different concentrations of [emim][OAc] than that noticed in control, as depicted in Fig. 3.2. This observation was not recorded for the other growth trials containing [emim][DEP] or [mmim][DMP]. Based on those observations, the authors hypothesized the occurrence of a hormetic effect promoted by the presence of [emim][OAc], mainly due to the buffering action provided by the $[OAc]^-$ ion. At lower IL concentrations, fast and higher growth of *Clostridium* sp. was allowed by the buffering effect of $[OAc]^-$ ions, while bacterial growth inhibition was evident at higher concentrations [14]. The hormetic effect of [emim][OAc] was also observed in studies aimed to evaluate the effect of IL concentration on *Pseudomonas putida* cell growth [22]. In such case, the final concentration of bacterial cells was almost 400% higher than that monitored for the control experiment in absence of IL. Similar pH buffering effect was observed after 40 h of incubation, in which the pH of the culture media was maintained above 6.0. It is worth to point out that culture media influences the hormetic effect caused by [emim][OAc]. For instance, using tryptic soy broth (TSB) or mineral salts (MS) medium supplemented with a non-fermentable carbon source, such as acetate, no hormetic effect was observed during *P. putida* cell growth [22].

Microbial enrichment studies have also been performed using microbial communities from compost, under thermophilic conditions, in the presence of [emim][OAc] and [emim][Cl] [17]. Under such conditions, *Bacillus* bacteria were dominant with a noticeable presence of *Bacillus coagulans*, which revealed high tolerance to both [emim][OAc] and [emim][Cl]. Respiration monitoring during cell growth revealed that these bacteria are tolerant up to 4% IL, as no microbial activity was detected in the culture media at higher IL levels than those. Furthermore, it was discovered that the influence of ILs on *B. coagulans* cell growth is highly dependent on whether the culture was in liquid or solid state. The authors observed higher

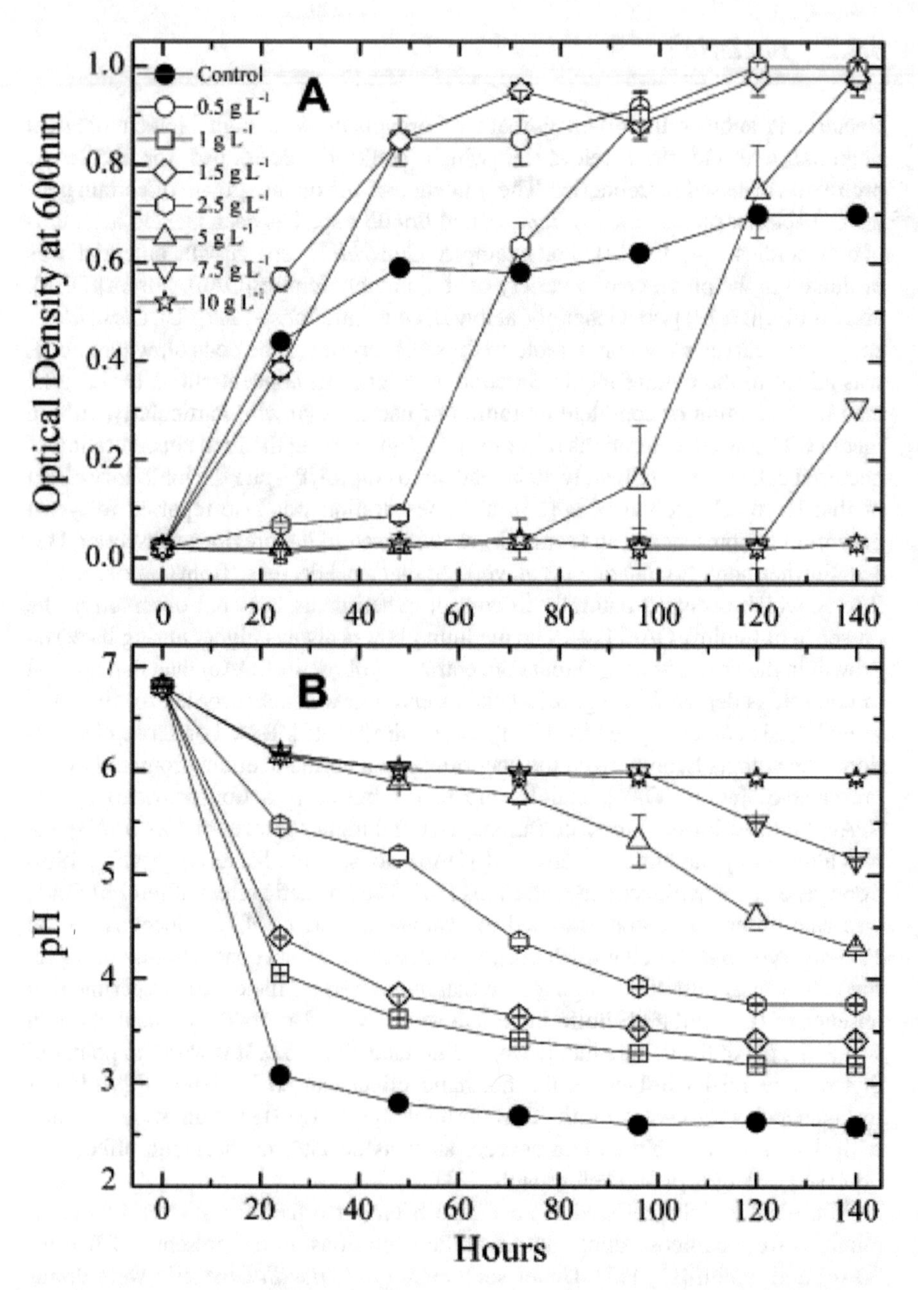

Fig. 3.2 Optical density (**a**) and culture medium pH (**b**) during the growth of *Clostridium* sp. in MS medium (control) and MS medium containing different concentrations of [emim][OAc]. (Reprinted from [14] Bioresource Technology, Vol. 102, Nancharaiah, Y V, Francis, A J, Alkylmethylimidazolium ionic liquids affect the growth and fermentative metabolism of Clostridium sp., Pages No. 6573–6578, Copyright (2011), with permission from Elsevier)

growth rate of bacterial cells in solid-state culture, suggesting that some tolerance mechanisms triggered by *B. coagulans* might be more effective in solid state relative to liquid cultures, in the presence of ILs [17].

3.3 Bioconversion in the Presence of Ionic Liquids

One of the major factors hindering the economic viability of IL-based biorefineries is the low biocompatibility of ILs, which adds expensive water washing requirements to remove residual ILs prior to enzymatic hydrolysis and fermentation. Though microbial cell growth could be largely affected by the presence of ILs, experimental optimizations and microbial enrichment studies allowed the discovery of organisms that can tolerate low concentrations of certain types of ILs. In addition to their IL tolerance, these organisms must be able to produce high titer of the desired products under stress conditions, as commercial biotechnological processes require the efficient production of fermentation-derived products. In this section, examples of microorganism-mediated bioconversion in the presence of ILs for production of ethanol [12, 24], organic acids [21], and lipids [11, 14] are described. Here, the key literature results, challenges, and prospects are given, and an overview of the key data related to bioconversion in ionic liquids is presented in Table 3.2.

3.3.1 Ethanol

One of the first studies testing the ability of microorganisms to produce ethanol in the presence of ILs was performed by Ouellet et al. [12]. That study has shown that *S. cerevisiae* cell growth is highly inhibited by the presence of [emim][OAc] at concentrations greater than 5.90 mM, while ethanol production by the same microbial strain started to decline in the presence of at least 33.5 mM IL concentration (Fig. 3.3). Furthermore, experiments using biomass hydrolysates as carbon source, which derived from IL-pretreated corn stover with a few washing steps (containing 52.4 mM [emim][OAc]), showed low ethanol production due to inhibitory effects attributed to the presence of IL. A more extensive washing procedure performed by the same authors resulted in a lignocellulosic hydrolysate with lower IL concentration (0.79 mM [emim][OAc]), which led to 92% conversion of glucose into ethanol by *S. cerevisiae* [12].

The quantity of residual IL in the pretreated biomass is correlated to the severity of the washing procedure, but it also depends on the type of IL used for pretreating the biomass. For example, it has been shown that pretreatment of biomass with either [emim][OAc] or [Ch][OAc] has resulted in pretreated feedstocks with distinct morphologies, which has highly impacted the performance of subsequent processes, such as saccharification and fermentation [23]. Using [emim][OAc] for biomass

Table 3.2 Overview of bioconversion in ionic liquids

Microorganism		Ionic liquid		Incubation conditions						
Family	Species	Type	Concentration	T (°C)	t (h)	Medium	Substrate	Product	Product concentration / yield	Ref.
Fungi	*Fusarium oxysporum* BN	[emim][H_2PO_2]	–	35	192	–	Rice straw	C_2H_5OH	0.125 g/g biomass (64.2% theoretical)	[19]
	Rhodosporidium toruloides	[emim][Cl]	[30–60] mM	30	100	YEPD	$C_6H_{12}O_6$	Lipids	~ [12–11] g/L	[11]
		[emim][DEP]	[30–60] mM	30	100	YEPD	$C_6H_{12}O_6$	Lipids	~ [10–8] g/L	
		[emim][OAc]	[30–60] mM	30	100	YEPD	$C_6H_{12}O_6$	Lipids	~ [2–1] g/L	
	Saccharomyces cerevisiae	[emim][Cl]	100 mM	30	6	YP	$C_6H_{12}O_6$	C_2H_5OH	~ 18 g/L	[24]
		[emim][DEP]	100 mM	30	6	YP	$C_6H_{12}O_6$	C_2H_5OH	~ 17 g/L	
		[emim][DEP]	200 mM	34	170	YP	Cellulose	C_2H_5OH	~ 1.5 g/L	
		[emim][OAc]	100 mM	30	6	YP	$C_6H_{12}O_6$	C_2H_5OH	~ 17 g/L	
		[emim][OAc]	[5.90–33.5] mM	30	72	SD	$C_6H_{12}O_6$	C_2H_5OH	~ [10–9] g/L	[12]
		[emim][OAc]	[52.4–59.0] mM	30	72	SD	$C_6H_{12}O_6$	C_2H_5OH	~ 2 g/L	
		[emim][DEP]	200 mM	30	96	YP	$C_5H_{10}O_5$	C_2H_5OH	~ 8 g/L	[25]
		[Ch][Lys][a]	10% (w/w)	37	96	–	Corn stover	C_2H_5OH	41.1 g/L (74.8% on glucose basis)	[26]
		[bmpy][Cl]	200 mM	30	72	YPD	Sugarcane bagasse	C_2H_5OH	84.0%	[25]
		[Ch][Lys][a, b]	[1–5] wt %	37	72	YPD	Switchgrass	C_2H_5OH	0.139 g/g switchgrass (83.3% on glucose basis)	[9]
		[Ch][OAc]	[1–5] wt %	30	48	YPD	Sugarcane bagasse	C_2H_5OH	~ [33–63] % theoretical	[23]
		[emim][OAc]	[1–5] wt %	30	48	YPD	Sugarcane bagasse	C_2H_5OH	~ [16–58] % theoretical	
	Wickerhamomyces anomalus	[emim][OAc]	3.2% (w/v)	26	144	YMD	Switchgrass hydrolysate	C_2H_5OH	70% theoretical	[27]

Microorganism		Ionic liquid		Incubation conditions						
Family	Species	Type	Concentration	T (°C)	t (h)	Medium	Substrate	Product	Product concentration / yield	Ref.
	Rhodosporidium toruloides	[Ch][Lys]	10% (w/w)	30	288	YPD	Sorghum hydrolysate	$C_{15}H_{24}$	2.2 g/L	[28]
	Yarrowia lipolytica	[emim][OAc]	10% (v/v)	28	48	MpA	Avicel	$C_5H_6O_5$	Up to 92% of maximum theoretical yield	[20]
Bacteria	*Actinobacillus succinogenes*	[amim][Cl]	0.01%	37	12	LB	$C_6H_{12}O_6$	$C_4H_6O_4$	14.65 g/L	[21]
		[amim][Cl]	0.1%	37	12	LB	$C_6H_{12}O_6$	$C_4H_6O_4$	16.00 g/L	
		[amim][Cl]	1.0%	37	12	LB	$C_6H_{12}O_6$	$C_4H_6O_4$	12.41 g/L	

T - Temperature, t - Time. YEPD - Yeast extract peptone dextrose, YP - Yeast peptone, YMD - Yeast malt dextrose, YPD - Yeast peptone dextrose, SD - Synthetic defined medium, MpA - Not defined, LB - Luria-Bertani

[a]Simultaneous enzymatic hydrolysis and fermentation

[b]In the presence of 1 MPa of CO_2

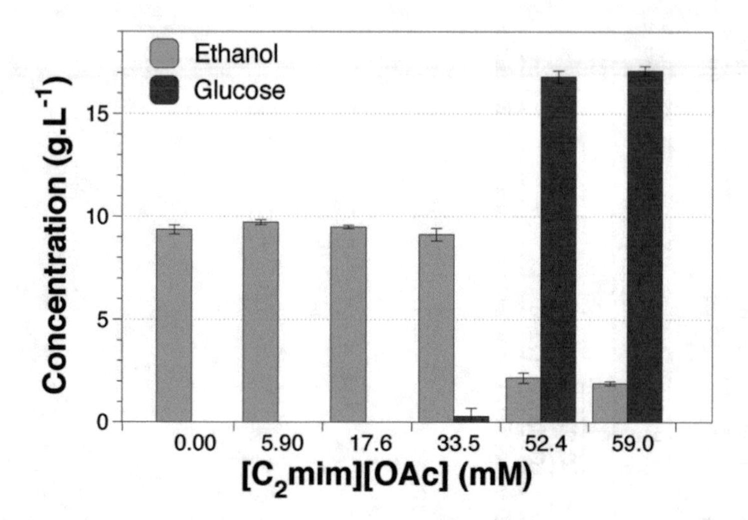

Fig. 3.3 Ethanol and glucose final concentrations after 72 h fermentation in 20 g/L glucose medium culture in the presence of different [emim][OAc] concentrations. (Reproduced from Ref. [12] with permission from The Royal Society of Chemistry)

pretreatment results in a gel-like solid that retains the IL within its structure. This phenomenon was not observed for biomass pretreated with [Ch][OAc]. After water washing, the quantity of IL remaining in both pretreated solids decreased. However, researchers have verified that the [Ch][OAc]-treated biomass contained much lower levels of residual IL than that observed for [emim][OAc]-treated biomass using the same amount of water.

The remarkable ability of [emim][OAc] to swell lignocellulosic biomass and to reduce cellulose crystallinity into a more amorphous structure has led to the deficient dispersion of IL in the antisolvent and/or in the solvent used for the subsequent washing steps. After enzymatic saccharification of the pretreated biomass, the effect of the number of washing times on *S. cerevisiae* fermentative ability was further examined in the presence of residual ILs [23]. Ethanol production by *S. cerevisiae* increased with the number of washing steps, and the sugar conversion to ethanol was greater in the presence of [Ch][OAc] relative to [emim][OAc], for most of the experimental conditions tested in that study. Such results suggested that pretreating the biomass with [Ch][OAc] could reduce the amount of solvent used for washing while removing the great majority of the residual IL after biomass pretreatment, thus leading to improved fermentation performances relative to [emim][OAc]. However, one can note that a maximum of 60% ethanol yield was achieved in those experiments. Though the usage of [Ch][OAc] allowed great improvements in ethanol yields relative to [emim][OAc], it is still evident that *S. cerevisiae* fermentation performance in the presence of [Ch][OAc] needs to be improved in order to become a commercial process in the future [23]. More advances in this field have been reported in the last years [27, 29].

A more recent study by Xu et al. shows the impact of a newly developed IL pretreatment system using the bio-based [Ch][Lys] on *S. cerevisiae* BY4741 growth and ethanol production [26]. The authors achieved 74.8% conversion of glucose to ethanol with a final ethanol concentration of 41.1 g/L, in a media containing 10 wt% IL in water. This "one pot" approach enables the possibility of recovering the IL after bioconversion to fuels and chemicals with no apparent impacts on microbial activity, even without the extensive water washing requirements. The same approach was later adopted by Amoha et al. using genetically engineered *S. cerevisiae* strains XR-XDH and XI, to consume xylose as carbon source [25]. Those authors tested five different ILs, of which [bmpy][Cl] showed the greatest biocompatibility, allowing the *S. cerevisiae* XI strain to convert 84% of the total fermentable sugar available in the sugarcane bagasse hydrolysate to ethanol, in a media containing 200 mM of IL in water. Though the conversion of sugars to ethanol was quite promising, the enzymatic hydrolysis was performed at low solid loading, which in turn resulted in concentrations of ethanol around 1.6 g/L after 24 h fermentation. More work in this area is required to evaluate the potential use of [bmpy][Cl] in biomass pretreatment along with lignocellulose bioconversion by *S. cerevisiae* XI at industrially relevant conditions.

As described in the previous section, *F. oxysporum* BN fungus was discovered by microbial enrichment technique to grow in the presence of [emim][H_2PO_2] IL [19]. This microorganism can release cellulases that hydrolyze cellulose into glucose, which can be used by the organism to produce ethanol. Albeit *F. oxysporum* BN has demonstrated to grow in the presence of [emim][H_2PO_2], cellulases' activity and ethanol fermentation is highly inhibited by it. This suggests that the isolation of organisms that can tolerate ILs to a certain extent, like *F. oxysporum* BN, may not have the in-built ability to effectively grow and produce value-added products, notably ethanol, in the presence of ILs [19]. The improvement of these strains is possible with proper understanding of inhibition mechanisms and evolving the strains by adaptation or genetic engineering.

A unique conversion process was developed in a more recent study, which consisted in IL pretreatment of cellulose, followed by simultaneous saccharification and fermentation to produce ethanol in "one pot" (Fig. 3.4) [24]. Functional ("arming") yeasts containing three main cellulases immobilized on its surface, namely, endoglucanase II (EG) and cellobiohydrolase II (CBH) from *Trichoderma reesei* and β-glucosidase I (BGL) from *Aspergillus aculeatus*, were used in this process [24, 30]. That study also focused on the influence of [emim][DEP], [emim][Cl], and [emim][OAc] for pretreating cellulose and the subsequent saccharification and fermentation. All three ILs tested in that study allowed total dissolution of cellulose with further reduction of its crystallinity, which improved the enzymatic activity on the pretreated cellulose relative to the native counterpart. Furthermore, high ethanol production from IL-treated cellulose was observed in this study, contrasting with the control experiment where untreated cellulose could not be effectively converted to ethanol. The maximum concentration of ILs in the final solution without affecting ethanol fermentation was determined to be 200 mM. However, yeast cellular

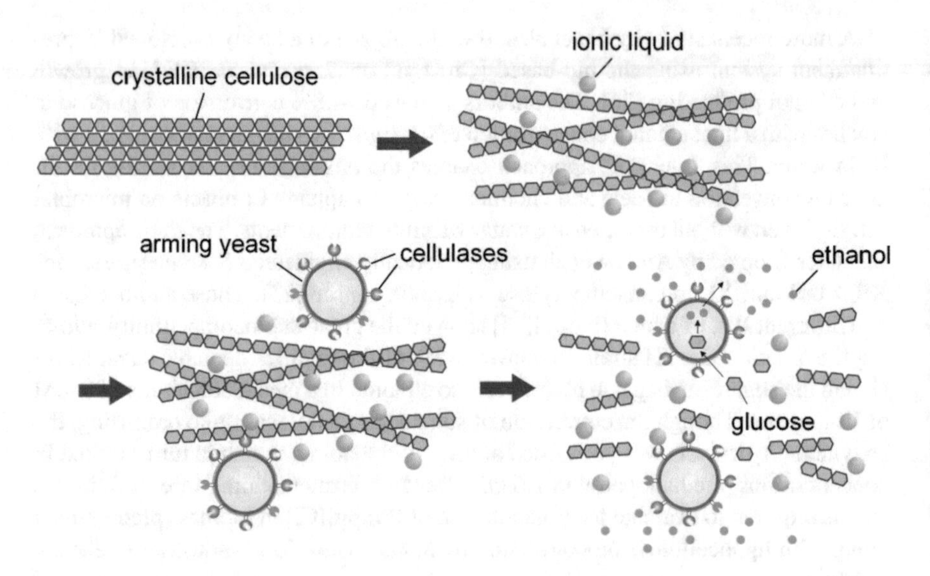

Fig. 3.4 Cellulose pretreatment with IL and simultaneous saccharification and fermentation process to produce ethanol by functional yeasts. (Reproduced from Ref. [24] with permission from the Royal Society of Chemistry)

viability after fermentation largely depended on the IL used. For example, [emim][OAc] was the only IL tested in that study that negatively affected the viability of the yeasts after fermentation. Such observation was probably due to an increase of pH in the culture medium promoted by the presence of high concentration of $[OAc]^-$ anions. The most efficient production of ethanol was achieved with [emim][DEP], which revealed to be less toxic to enzymes and yeasts. However, after 7 days of incubation, approximately 1.6 g/L ethanol was obtained, which is far from ethanol titers attained with conventional industrial fermentations [24].

3.3.2 *Organic Acids*

Acetate, butyrate, and lactate are organic acids with industrial relevance, which are commonly produced by *Clostridium* sp. fermentation. As lignocellulosic biorefineries become a reality, the commercial production of these organic acids will derive from lignocellulosic sugars. Thus, in the context of an IL pretreatment-based biorefinery, *Clostridium* sp. fermentation performance could be highly impacted by the presence of residual IL in the pretreated biomass, especially if the biorefinery viability demands minimal washing of the pretreated biomass. To evaluate the impact of IL in the production of acetate, butyrate, and lactate, studies were performed in mineral salts (MS) medium with glucose, in the presence of various levels of [emim][OAc], [emim][DEP], and [mmim][DMP] [14]. Based on those studies, the

Clostridium sp. tolerance to the presence of ILs highly depended on the type of IL used. For example, cell growth was inhibited at concentrations of [emim][OAc] greater than 2.5 g/L, while it could tolerate concentrations up to 4 g/L for the remaining ILs tested. Interestingly, a significant hormetic effect was observed in the presence of [emim][OAc] at concentrations lower or equal to 2.5 g/L, where the total organic acid production improved between 65% and 45% relative to the control (no IL present). However, a significant drop in total organic acid production relative to the control was observed for [emim][OAc] concentrations greater or equal to 5 g/L. A milder hormetic effect was observed in the presence of [mmim][DMP], where organic acid production was slightly improved (between 9% and 7.5% relative to control) by IL concentrations lower than 2 g/L. In the presence of [emim][DEP], the organic acid production consistently dropped with the presence of IL concentrations greater than 0.5 g/L. Though these experiments were important to understand the effects of various ILs during organic acid production by *Clostridium* sp., the organic acid concentration levels obtained in these experiments with MS medium were quite low for an industrial process, with a maximum of 63.5 g/L [14]. Considering that lignocellulosic hydrolysates offer additional challenges to fermentative organisms due to the presence of inhibitory compounds naturally occurring in the plants, it is critical that *Clostridium* sp. strains are developed to perform well in IL-containing lignocellulosic hydrolysates.

Another organic acid with great industrial relevance is succinic acid, which can be used as a platform chemical to produce various precursors for the polymer and chemical industries [31, 32]. Succinic acid can be produced via fermentation by several bacterial species, including *Actinobacillus succinogenes*, which has been used to produce succinic acid in LB media and lignocellulosic hydrolysates derived from [amim][Cl]-pretreated corn stover and pinewood [21]. That study by Wang et al. revealed that *A. succinogenes* 130Z cell growth is inhibited by the presence of [amim][Cl] at concentrations greater than 5% (v/v) in LB media, while succinic acid production is highly reduced even in the presence of 0.01% (v/v) [amim][Cl]. Based on these results, a maximum IL content of 0.01% (v/v) was allowed to be present in the lignocellulosic hydrolysate derived from [amim][Cl]-treated pinewood and corn stover prior to fermentation. Such conditions allowed a succinic acid yield of 0.65 g/g consumed sugar, which is comparable or better succinic acid production relative to other pretreatment technologies, such as dilute acid and steam explosion. However, the extensive water washing requirements to remove residual IL levels down to 0.01% could undermine the economic viability of the biorefinery [21]. Therefore, new bacterial strains should be developed in the future to tolerate concentrations of IL comparable to those obtained after performing minimal-to-no washing steps.

Another study has revealed an effective process to produce α-ketoglutaric acid, mediated by *Yarrowia lipolytica* yeast in the presence of [emim][OAc] [20]. For this purpose, simultaneous saccharification and fermentation was conducted on 10 g/L Avicel PH-101 in defined MpA medium, containing 10% (v/v) IL concentration, at 28 °C and pH 6.3 for 72 h. Surprisingly, under these conditions *Y. lipolytica* produced 92% of the maximum theoretical yield of α-ketoglutaric acid. In comparison to other organisms, *Y. lipolytica* yeast was able to produce value-added products with very high efficiency in the presence of relatively high IL concentrations. It was suggested by the authors that *Y. lipolytica* increases the composition of cyclopropane fatty acids in the cell membrane under stress conditions, offering high IL tolerance to the yeast [20]. The knowledge and understanding of the mechanisms that provide *Y. lipolytica* tolerance to high concentrations of IL could benefit the development of other strains and their effective usage in the production of other value-added chemicals in a future IL pretreatment-based lignocellulosic biorefinery.

3.3.3 Lipids

Lipids are also very important platform chemicals used in lubricants, coolants, polymers, adhesives, detergents, and biofuels, among other commodities. In recent years, there has been increasing interest in production of lipids by oleaginous yeasts using sugars as carbon source [11]. As lignocellulosic sugars become less expensive and more readily available, the production of lipids could derive from that carbon source in a future IL pretreatment-based biorefinery. As mentioned in the previous sections, the viability of an IL pretreatment-based biorefinery hinges on several factors, including the ability of the microorganisms to grow and convert the carbon source effectively in the presence of ILs, in order to avoid expensive biomass washing steps after pretreatment. In this perspective, the development of oleaginous yeasts tolerant to the presence of ILs in the culture media is of prime importance to avoid expensive washing steps. Based on this requirement, the production of lipids by the yeast *R. toruloides* was studied in YEPD medium, in the presence of different types of ILs, with varying IL concentrations [11]. The results showed that lipid production was inhibited by the ILs in the following order: [emim][OAc] > [emim][DEP] > [emim][Cl], as depicted in Fig. 3.5.

Based on those results, it was possible to obtain lipid yields in media containing 60 mM [emim][Cl] comparable to those of the control (no IL added), while only up to 30 mM [emim][DEP] allowed the same level of lipid production. In contrast, [emim][OAc] acted as a strong inhibitor for the range of IL concentrations tested, leading to very low lipid yields produced by *R. toruloides*. Normally, the culture medium pH dropped from 5.0 to 3.4 over a 24 h culture, which has been observed for the control and all experiments with [emim][DEP] and [emim][Cl]. However, the culture medium pH increased to 8.0 for cultures in the presence of [emim][OAc]. Such high pH is not favorable for the yeast to grow and produce lipids, as it interferes with the biosynthesis of lipids by impeding elongation and fatty acid desaturation [11]. Though these results are very promising by revealing the high potential of *R. toruloides* for lipid production in an IL pretreatment-based biorefinery, it is worth to mention that the high IL tolerance was demonstrated by this yeast while growing in YEPD media. It is important to note that lignocellulosic substrates

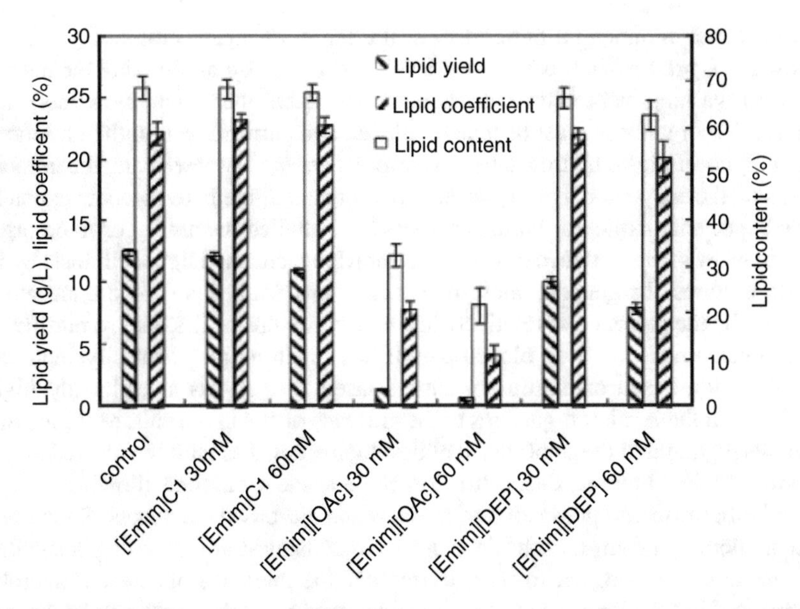

Fig. 3.5 Lipid production by *R. toruloides* in the presence of [emim][Cl] and [emim][OAc], [emim][DEP] at 30 mM and 60 mM concentration. (Reprinted from [11] Bioresource Technology, Vol. 130, Qitian Huang, Qian Wang, Zhiwei Gong, Guojie Jin, Hongwei Shen, Shan Xiao, Haibo Xie, Shuhong Ye, Jihui Wang, Zongbao K. Zhao, Effects of selected ionic liquids on lipid production by the oleaginous, Pages No. 339–344, Copyright (2013), with permission from Elsevier)

are often more inhibitory to microorganisms and may not carry the nutrient balance required to achieve high tolerance to environmental stress and high yield of lipids. Therefore, more work in this line of research is required to fully develop oleaginous yeast strains that can perform well on lignocellulosic hydrolysates in the presence of high concentration of ILs.

3.4 Conclusions

ILs are unique solvents that enable the fractionation of biomass to its components, exposing carbohydrates to enzymatic action like very few other pretreatment technologies today. Though ILs are very promising solvents for plant cell wall deconstruction, their strong interactions with the pretreated biomass make their recovery and reuse more challenging. Also, ineffective removal of ILs from the pretreated biomass can negatively impact the biorefining process, as ILs inhibit most enzymes and microorganisms if they are present at certain concentrations, depending on the organism. Up until recently, extensive biomass washing steps have been commonly used to remove residual IL present in the biomass after pretreatment, thus enabling its efficient conversion to bio-based products like ethanol, organic acids, and lipids, among others. As these extensive biomass washing steps are expensive and not practical at a commercial biorefinery scale, the most recent biorefinery concepts based on IL pretreatment pay a great level of attention on minimizing the need for biomass washing steps after pretreatment. To enable these new biorefinery concepts, microorganisms must be able to tolerate and perform well in the presence of ILs in lignocellulosic hydrolysates. Therefore, over the last few years, the influence of ILs on the enzymatic hydrolysis and microorganism-mediated bioconversion has been expectedly explored. Various authors have studied the ability of microorganisms to grow in the presence of ILs, both using lab media and lignocellulosic hydrolysate cultures. In general, most microorganisms studied so far are negatively affected by the presence of relatively low concentrations of ILs, like [emim][OAc]. The development of more biocompatible ILs, such as phosphate, lysinate, and sulfate anion-based or cholinium cation-based ILs, allows significantly higher tolerance of those microorganisms to the presence of ILs in the culture media. Such findings highlighted the importance of developing new ILs with the desired properties for efficient biomass deconstruction while at the same time allowing effective microbial growth and production of biofuels and bio-based chemicals. By the same token, microorganisms must be developed, either using strain screening, adaptation, or genetic engineering techniques, to improve the tolerance of relevant microbial strains to ILs. To achieve such goal, the understanding of the genetic traits that concede strain tolerance to ILs is quite important. For example, *Y. lipolytica* can grow in lab culture media containing up to 10% IL and produce 92% of the maximum theoretical yield of α-ketoglutaric acid. The mechanisms by which this and other strains adapted to the presence of ILs could be investigated and potentially implemented in organisms of high biotechnological potential, such as *S. cerevisiae*. These and other advancements, notably those related to IL recovery and recycling and reduction of IL manufacturing costs, are acknowledged key factors for the economic viability and success of the future IL pretreatment-based biorefinery.

Nomenclature of Ionic Liquids

Cations

[amim] – 1-allyl-3-methylimidazolium
[bmim] – 1-butyl-3-methylimidazolium
[bmpy] – 1-butyl-3-methylpyridinium
[Ch] – cholinium
[emim] – 1-ethyl-3-methylimidazolium
[mmim] – 1,3-dimethylimidazolium

Anions

[Cl] – chloride
[DEP] – diethylphosphate
[DMP] – dimethylphosphate
[FA] – formate
[H_2PO_2] – phosphinate
[Lys] – lysinate
[(MeO)HPO_2] – dimethylphosphate
[$MeSO_4$] – methylsulphate

[OAc] – acetate

Other
HEMA – tris(2-hydroxyethyl)-methylammonium methylsulfate

Acknowledgments This research was supported by the Fundação para a Ciência e a Tecnologia (FCT, Portugal) through grant IF/00471/2015 (RML) and SFRH/BD/94297/2013 (ARCM). This work was authored in part by the National Renewable Energy Laboratory, operated by Alliance for Sustainable Energy, LLC, for the US Department of Energy (DOE) under Contract No. DE-AC36-08GO28308. Funding provided by the US Department of Energy, Office of Energy Efficiency and Renewable Energy, Bioenergy Technologies Office. The views expressed in the article do not necessarily represent the views of the DOE or the US Government. The US Government retains, and the publisher, by accepting the article for publication, acknowledges that the US Government retains a nonexclusive, paid-up, irrevocable, worldwide license to publish or reproduce the published form of this work, or allow others to do so, for US Government purposes.

References

1. Clark, J. H., Deswarte, F. E. I., & Farmer, T. J. (2009). The integration of green chemistry into future biorefineries. *Biofuels, Bioproducts and Biorefining*, *3*, 72–90. https://doi.org/10.1002/bbb.119.
2. Morais, A. R. C., da Costa Lopes, A. M., & Bogel-Łukasik, R. (2015). Carbon dioxide in biomass processing: Contributions to the green biorefinery concept. *Chemical Reviews, 115*, 3–27. https://doi.org/10.1021/cr500330z.
3. Morais, A. R. C., Pinto, J. V., Nunes, D., et al. (2016). Imidazole: Prospect solvent for Lignocellulosic biomass fractionation and delignification. *ACS Sustainable Chemistry & Engineering, 4*, 1643–1652. https://doi.org/10.1021/acssuschemeng.5b01600.
4. Silveira, M. H. L., Morais, A. R. C., da Costa Lopes, A. M., et al. (2015). Current pretreatment technologies for the development of cellulosic ethanol and biorefineries. *ChemSusChem, 8*, 3366–3390.
5. Shi, J., Thompson, V. S., Yancey, N. A., et al. (2013). Impact of mixed feedstocks and feedstock densification on ionic liquid pretreatment efficiency. *Biofuels, 4*, 63–72. https://doi.org/10.4155/bfs.12.82.
6. Da Silva, A. S. A., Teixeira, R. S. S., Endo, T., et al. (2013). Continuous pretreatment of sugarcane bagasse at high loading in an ionic liquid using a twin-screw extruder. *Green Chemistry, 15*, 1991–2001. https://doi.org/10.1039/c3gc40352a.
7. Klein-Marcuschamer, D., Simmons, B. A., & Blanch, H. W. (2011). Techno-economic analysis of a lignocellulosic ethanol biorefinery with ionic liquid pre-treatment. *Biofuels, Bioproducts and Biorefining, 5*, 562–569. https://doi.org/10.1002/bbb.303.
8. Shi, J., Gladden, J. M., Sathitsuksanoh, N., et al. (2013). One-pot ionic liquid pretreatment and saccharification of switchgrass. *Green Chemistry, 15*, 2579–2589. https://doi.org/10.1039/c3gc40545a.
9. Sun, J., Murthy Konda, N. V. S. N., Shi, J., et al. (2016). CO2 enabled process integration for the production of cellulosic ethanol using bionic liquids. *Energy & Environmental Science, 9*, 2822–2834. https://doi.org/10.1039/c6ee00913a.
10. Sun, J., Konda, N. V. S. N. M., Parthasarathi, R., et al. (2017). One-pot integrated biofuel production using low-cost biocompatible protic ionic liquids. *Green Chemistry, 19*, 3152–3163. https://doi.org/10.1039/c7gc01179b.

11. Huang, Q., Wang, Q., Gong, Z., et al. (2013). Effects of selected ionic liquids on lipid production by the oleaginous yeast Rhodosporidium toruloides. *Bioresource Technology, 130*, 339–344. https://doi.org/10.1016/j.biortech.2012.12.022.
12. Ouellet, M., Datta, S., Dibble, D. C., et al. (2011). Impact of ionic liquid pretreated plant biomass on Saccharomyces cerevisiae growth and biofuel production. *Green Chemistry, 13*, 2743–2749. https://doi.org/10.1039/c1gc15327g.
13. Sitepu, I. R., Shi, S., Simmons, B. A., et al. (2014). Yeast tolerance to the ionic liquid 1-ethyl-3-methylimidazolium acetate. *FEMS Yeast Research, 14*, 1286–1294. https://doi.org/10.1111/1567-1364.12224.
14. Nancharaiah, V. Y., & Francis, A. J. (2011). Alkyl-methylimidazolium ionic liquids affect the growth and fermentative metabolism of Clostridium sp. *Bioresource Technology, 102*, 6573–6578. https://doi.org/10.1016/j.biortech.2011.03.042.
15. Yang, Z. H., Zeng, R., Wang, Y., et al. (2009). Tolerance of immobilized yeast cells in imidazolium-based ionic liquids. *Food Technology and Biotechnology, 47*, 62–66.
16. Reddy, A. P., Simmons, C. W., Claypool, J., et al. (2012). Thermophilic enrichment of microbial communities in the presence of the ionic liquid 1-ethyl-3-methylimidazolium acetate. *Journal of Applied Microbiology, 113*, 1362–1370. https://doi.org/10.1111/jam.12002.
17. Simmons, C. W., Reddy, A. P., Vandergheynst, J. S., et al. (2014). Bacillus coagulans tolerance to 1-ethyl-3-methylimidazolium-based ionic liquids in aqueous and solid-state thermophilic culture. *Biotechnology Progress, 30*, 311–316. https://doi.org/10.1002/btpr.1859.
18. Singer, S. W., Reddy, A. P., Gladden, J. M., et al. (2011). Enrichment, isolation and characterization of fungi tolerant to 1-ethyl-3-methylimidazolium acetate. *Journal of Applied Microbiology, 110*, 1023–1031. https://doi.org/10.1111/j.1365-2672.2011.04959.x.
19. Xu, J., Wang, X., Hu, L., et al. (2015). A novel ionic liquid-tolerant Fusarium oxysporum BN secreting ionic liquid-stable cellulase: Consolidated bioprocessing of pretreated lignocellulose containing residual ionic liquid. *Bioresource Technology, 181*, 18–25. https://doi.org/10.1016/j.biortech.2014.12.080.
20. Ryu, S., Labbé, N., & Trinh, C. T. (2015). Simultaneous saccharification and fermentation of cellulose in ionic liquid for efficient production of α-ketoglutaric acid by Yarrowia lipolytica. *Applied Microbiology and Biotechnology, 99*, 4237–4244. https://doi.org/10.1007/s00253-015-6521-5.
21. Wang, C., Yan, D., Li, Q., et al. (2014). Ionic liquid pretreatment to increase succinic acid production from lignocellulosic biomass. *Bioresource Technology, 172*, 283–289. https://doi.org/10.1016/j.biortech.2014.09.045.
22. Nancharaiah, Y. V., & Francis, A. J. (2015). Hormetic effect of ionic liquid 1-ethyl-3-methylimidazolium acetate on bacteria. *Chemosphere, 128*, 178–183. https://doi.org/10.1016/j.chemosphere.2015.01.032.
23. Ninomiya, K., Omote, S., Ogino, C., et al. (2015). Saccharification and ethanol fermentation from cholinium ionic liquid-pretreated bagasse with a different number of post-pretreatment washings. *Bioresource Technology, 189*, 203–209. https://doi.org/10.1016/j.biortech.2015.04.022.
24. Nakashima, K., Yamaguchi, K., Taniguchi, N., et al. (2011). Direct bioethanol production from cellulose by the combination of cellulase-displaying yeast and ionic liquid pretreatment. *Green Chemistry, 13*, 2948–2953. https://doi.org/10.1039/c1gc15688h.
25. Amoah, J., Ogura, K., Schmetz, Q., et al. (2019). Co-fermentation of xylose and glucose from ionic liquid pretreated sugar cane bagasse for bioethanol production using engineered xylose assimilating yeast. *Biomass and Bioenergy, 128*, 15283. https://doi.org/10.1016/j.biombioe.2019.105283.
26. Xu, F., Sun, J., Konda, N. V. S. N. M., et al. (2016). Transforming biomass conversion with ionic liquids: Process intensification and the development of a high-gravity, one-pot process for the production of cellulosic ethanol. *Energy & Environmental Science, 9*, 1042–1049. https://doi.org/10.1039/c5ee02940f.
27. Sitepu, I. R., Enriquez, L. L., Nguyen, V., et al. (2019). Ethanol production in switchgrass hydrolysate by ionic liquid-tolerant yeasts. *Bioresource Technology Reports, 7*. https://doi.org/10.1016/j.biteb.2019.100275.

28. Sundstrom, E., Yaegashi, J., Yan, J., et al. (2018). Demonstrating a separation-free process coupling ionic liquid pretreatment, saccharification, and fermentation with: Rhodosporidium toruloides to produce advanced biofuels. *Green Chemistry, 20*, 2870–2879. https://doi.org/10.1039/c8gc00518d.
29. Egorova, K. S., & Ananikov, V. P. (2018). Ionic liquids in whole-cell biocatalysis: A compromise between toxicity and efficiency. *Biophysical Reviews, 10*, 881–900. https://doi.org/10.1007/s12551-017-0389-9.
30. Fujita, Y., Ito, J., Ueda, M., et al. (2004). Synergistic Saccharification, and direct fermentation to ethanol, of amorphous cellulose by use of an engineered yeast strain codisplaying three types of cellulolytic enzyme. *Applied and Environmental Microbiology, 70*, 1207–1212. https://doi.org/10.1128/AEM.70.2.1207-1212.2004.
31. Bechthold, I., Bretz, K., Kabasci, S., et al. (2008). Succinic acid: A new platform chemical for biobased polymers from renewable resources. *Chemical Engineering and Technology, 31*, 647–654.
32. Takkellapati, S., Li, T., & Gonzalez, M. A. (2018). An overview of biorefinery-derived platform chemicals from a cellulose and hemicellulose biorefinery. *Clean Technologies and Environmental Policy, 20*, 1615–1630. https://doi.org/10.1007/s10098-018-1568-5.

Chapter 4
Liquid Hot Water Pretreatment for Lignocellulosic Biomass Biorefinery

Xinshu Zhuang, Wen Wang, Bing Song, and Qiang Yu

4.1 Introduction

Biorefinery refers to pretreatment and bioconversion steps to produce biofuels, bioproducts, and other high-value-added products from biomass. Plant biomass or lignocellulose which has dominant amount in the world is mainly composed of cellulose, hemicellulose, and lignin. These three components connect with each other via covalent/non-covalent bonds to form compact structure to endow the recalcitrance of plant biomass for microbial or enzymatic attack [1]. Pretreatment which can break up the close connection among components and make the structure loose is a prerequisite step for the following effective bioconversion process of lignocellulosic materials [2, 3]. Liquid hot water (LHW), also called as hot-compressed water, is one of pretreatment technologies to be conducted at elevated temperatures (160–240 °C) and above saturated vapor pressure with liquid water as the sole solvent and catalyst [4, 5]. At the first stage of LHW process, water would be autoionized to generate hydronium ions to cleave O-acetyl and uronic acid substitutions from hemicellulose [6, 7]. Acetic and other organic acids are subsequently formed and would further result in hydronium ion generation via autoionization. The hydronium ions originated from these organic acids are more significant than those from water to alter physicochemical features of biomass, especially depolymerize hemicellulose [6, 7]. Due to no chemicals added, LHW pretreatment has advantages of low corrosion to the equipment and little pollution to the environment [8, 9].

X. Zhuang (✉) · W. Wang · Q. Yu
Guangzhou Institute of Energy Conversion, Chinese Academy of Sciences,
Guangzhou, China
e-mail: zhuangxs@ms.giec.ac.cn

B. Song
Scion, Te Papa Tipu Innovation Park, Rotorua, New Zealand

Z.-H. Liu, A. Ragauskas (eds.), *Emerging Technologies for Biorefineries, Biofuels, and Value-Added Commodities*,
https://doi.org/10.1007/978-3-030-65584-6_4

LHW pretreatment can be widely used for biorefinery of herbaceous and woody biomass and microalgae. Biomass treated with LHW would generate liquid and solid fractions. Hemicellulose is the main component to be degraded in the liquid stream, while cellulose and lignin are chiefly remained in the solid fraction [5, 10]. Xylooligosaccharide (XOS) which has healthcare functions can be obtained from the liquid part [11]. The utilization of cellulose is promoted due to the physicochemical changes like hemicellulose removal, loose surface structure, and so on after LHW pretreatment. Lignin in LHW-treated biomass has different physicochemical properties from its original [12] and is the major component to affect enzymatic hydrolysis of cellulose via hindering the access of enzyme to cellulose or unproductively adsorbing enzyme [13]. The degradation degree of hemicellulose and the physicochemical changes of LHW-treated solid residue are closely affected by pretreatment temperature and time which can be expressed as severity factor (R_0, Eq. 4.1) [7, 14]:

$$R_0 = t \cdot \exp\left(\frac{T - 100}{14.75}\right) \tag{4.1}$$

where t is pretreatment time (min) and T is pretreatment temperature (°C). The log (R_0) is usually presented as a parameter to investigate the combined effect of temperature and time on lignocellulose suffering LHW pretreatment [15, 16]. Furthermore, the flow-through LHW technology is developed, and the water flow is introduced as another parameter influencing the effectiveness of LHW pretreatment [17–19]. This chapter focuses on the principle and application of LHW technology for biomass biorefinery.

4.2 Physicochemical Changes of Feedstocks After LHW Pretreatment

Various lignocelluloses treated with LHW will have changes on chemical components and structural characteristics and recover 40–75% solid residues. Table 4.1 shows the compositional changes of herbaceous and woody biomass after LHW pretreatment. It indicates that LHW pretreatment can largely degrade hemicellulose and partially remove lignin, which subsequently alter the physicochemical properties of feedstocks.

4.2.1 Hemicellulose Depolymerization

Hemicellulose is a heterogeneous and highly branched polysaccharide, which may consist of glycosyl units from pentoses (β-D-xylose, α-L-arabinose), hexoses (β-D-glucose, β-D-mannose, α-D-galactose), and/or uronic acid residues (α-D-glucuronic,

Table 4.1 Components of untreated and LHW-treated lignocelluloses at various conditions

Lignocelluloses	Solid recovery (%)	Cellulose (%)	Hemicellulose (%)	Lignin (%)	Ref.
Sugarcane bagasse					[20]
Untreated		33.1	26.3	18.2	
S:L=1:10, 170 °C, 15 min	63.6	43.6	21.8	27.0	
S:L=1:10, 195 °C, 10 min	50.8	58.8	7.5	28.9	
S:L=1:10, 220 °C, 5 min	45.9	56.7	2.0	37.0	
S:L=1:10, 220 °C, 15 min	43.3	54.7	1.8	36.9	
Corn stover					[21]
Untreated		30.9	29.9	26.8	
S:L=1:10, 170 °C, 40 min	68.5	40.1	29.4	33.5	
S:L=1:10, 180 °C, 40 min	58.6	48.7	13.1	37.0	
S:L=1:10, 190 °C, 40 min	53.2	51.0	9.2	38.5	
Wheat straw					[22]
Untreated		28.2	13.1	19.1	
S:L=1:4, 170 °C, 10 min	74.0	33.9	17.6	25.1	
S:L=1:4, 170 °C, 40 min	61.6	42.4	13.0	20.1	
S:L=1:4, 180 °C, 10 min	69.8	35.6	16.3	28.5	
S:L=1:4, 180 °C, 40 min	53.0	48.3	4.0	23.0	
S:L=1:4, 190 °C, 10 min	55.8	43.0	10.8	28.0	
S:L=1:4, 200 °C, 10 min	46.6	50.9	4.3	24.9	
Safflower straw					[16]
Untreated		35.1	19.6	22.0	
S:L=1:11, 150 °C, 1 h	≈65.0	40.4	18.1	21.5	
S:L=1:11, 150 °C, 2 h	≈62.0	42.5	17.8	24.7	
S:L=1:11, 180 °C, 1 h	≈55.0	48.9	7.1	27.1	
S:L=1:11, 180 °C, 2 h	≈50.0	55.9	3.3	33.1	

(continued)

Table 4.1 (continued)

Lignocelluloses	Solid recovery (%)	Cellulose (%)	Hemicellulose (%)	Lignin (%)	Ref.
Beechwood					[23]
Untreated		42.1	25.4	26.1	
S:L=1:15, 190 °C, 15 min	70.5	54.4	7.6	28.8	
S:L=1:15, 220 °C, 15 min	61.0	60.2	0.7	37.7	
Eucalyptus					[24]
Untreated		44.4	21.8	27.7	
S:L=1:8, 195 °C, 7 min	74.7	57.9	4.5	34.4	
S:L=1:8, 205 °C, 7 min	71.4	59.1	3.6	31.1	
S:L=1:8, 220 °C, 7 min	70.3	64.9	1.3	31.4	
S:L=1:8, 230 °C, 7 min	69.9	62.6	1.0	35.1	

$$\begin{array}{l}\text{Fast-Xylan (XF)} \overset{k1f}{\rightarrow} \\ \qquad\qquad \text{Xylo-oligomers (XO)} \overset{k2}{\rightarrow} \text{Xylose (Xyl)} \overset{k3}{\rightarrow} \text{Degradation products (DP)} \\ \text{Slow-Xylan (XS)} \overset{k1s}{\rightarrow}\end{array}$$

Fig. 4.1 Reaction pathway of xylan hydrolysis

α-D-4-O-methylgalacturonic, and α-D-galacturonic acids) [25]. Different biological origins have different types of hemicellulose. Hardwoods mainly have xylan-type hemicellulose like glucuronoxylans (O-acetyl-4-O-methyglucuronoxylans) and arabinoglucuronoxylans (arabino-4-O-methylglucuronoxylans) for herbaceous biomass, while softwoods mainly possess mannan-type hemicellulose such as galactoglucomannans (O-acetyl-galactoglucomannans) [25].

According to the fundamental of LHW pretreatment, the depolymerization of hemicellulose in the process belongs to acid-catalyzed reaction. Over 80% hemicellulose would be degraded into saccharides, furfural, acetic acid, etc. [8, 26]. Generally, xylan is the main polymer chain of hemicellulose in various lignocelluloses. It has biphasic pattern in hydrolytic process comprising fast-rate and slow-rate hydrolysis portions. A comprehensive model of hemicellulose hydrolysis has been proposed on the basis of pseudo-homogeneous assumption following a first-order reaction (Fig. 4.1) [26–28]. The kinetic model can be developed described as Eqs. 4.1–4.6. C_X, C_{XF}, C_{XS}, C_{XO}, C_{Xyl}, and C_{DP} are the concentrations of xylan, fast-xylan, slow-xylan, xylooligomers, xylose, and degradation products, respectively. k_{1f}, k_{1s}, k_2, and k_3 are the specific reaction rates varied with temperature. The relation between k value and temperature can be determined by the Arrhenius equation (Eq. 4.7), where A is a pre-exponential factor depending on acid concentration, Ea is activation energy (J/mol), R is the gas constant, and T is kelvin temperature (K).

$$C_X = C_{XF} + C_{XS} \tag{4.1}$$

$$\frac{\mathrm{d}C_{XF}}{\mathrm{d}t} = \; k_{1f}\mathrm{C}_{\mathrm{XF}} \tag{4.2}$$

$$\frac{\mathrm{d}C_{XS}}{\mathrm{d}t} = \; k_{1s}\mathrm{C}_{\mathrm{XS}} \tag{4.3}$$

$$\frac{\mathrm{d}C_{XO}}{\mathrm{d}t} = k_{1f}\mathrm{C}_{\mathrm{XF}} + k_{1s}\mathrm{C}_{\mathrm{XS}} \quad k_2\mathrm{C}_{\mathrm{XO}} \tag{4.4}$$

$$\frac{\mathrm{d}C_{Xyl}}{\mathrm{d}t} = k_2\mathrm{C}_{\mathrm{XO}} \quad k_3\mathrm{C}_{\mathrm{Xyl}} \tag{4.5}$$

$$\frac{\mathrm{d}C_{DP}}{\mathrm{d}t} = k_3\mathrm{C}_{\mathrm{Xyl}} \tag{4.6}$$

$$k = Ae^{\frac{-E_a}{RT}} \tag{4.7}$$

The fast- or slow-xylan is not a real representation of the intrinsic characteristic of xylan but only a conception for developing a kinetic model. Several studies neglected the biphasic style of hemicellulose and developed kinetic models initially from single xylan without fast/slow separation, which also fitted the experimental data well [29, 30]. The monophasic/biphasic reaction is related to the autohydrolysis condition [30]. Different feedstocks have different activation energies of LHW hydrolysis, which finally affect the specific reaction rate. For example, the activation energies of pure xylan, rice straw, and palm shell are 65 580, 68760, and 95 190 J/mol, respectively [31]. Yu et al. compared LHW hydrolysis of hemicellulose from sweet sorghum bagasse and *Eucalyptus grandis* chips in batch and flow-through reactors and claimed that reactor type contributed more on hemicellulose hydrolysis kinetics than feedstock category [29]. Pronyk et al. also pointed out that the kinetic rate constants are not only affected by temperature and acid concentration but also related to fluid velocity in the reactor [30].

4.2.2 Lignin Change

Besides hemicellulose degradation, lignin will be departed into two parts after LHW pretreatment: one is dissolved in the liquid, and the other is retained in the solid residue. Lignin is a complex heteropolymer composed of three aromatic phenylpropane units, namely, guaiacyl (G), syringyl (S), and *p*-hydroxyphenyl (H) moieties, derived from coniferyl, sinapyl, and *p*-coumaryl alcohols, respectively [32]. Syringyl units of lignin have been thought as the most susceptible fragments to hydrothermal treatment [7]. Lignin dissolution comprises reactions of lignin-carbohydrate bond cleavage and lignin depolymerization [7]. The content of lignin degradation products in LHW pretreatment liquid will be increased at the elevated temperatures, of which a partial will recondense on the surface of LHW-treated solid residue to form droplets, spheres, or layers with LHW pretreatment liquid cooling down (Fig. 4.2) [23, 33]. In addition, pseudo-lignin originating from lignocellulosic carbohydrates could be also in the form of spherical droplets. The mechanism of pseudo-lignin formation has been proposed that hydroxymethylfurfural (HMF) or furfural derived from six-carbon or five-carbon sugars can produce aromatic compounds which can be further synthesized to polyphenolic structures via polymerization reactions [34].

Most lignin is remained in the solid residues of lignocellulose after LHW pretreatment. It has been reported that lignin has glass transition behavior which is described as the transformation from a hard or glassy status into a viscous or rubbery status when being heated [35]. The glass transition temperature (T_g) is a sensitive indicator of structural changes of lignin [12]. Various lignocelluloses have different lignin constituents. For example, the lignin of wheat straw contains 5% H unit, 49% G unit, and 46% S unit, while that of rice straw has 15% H unit, 45% G unit, and 40% S unit [32]. Lignin originating from different lignocelluloses or prepared from the same lignocellulose with various methods has obviously different T_g values. Ko et al. reported that hardwood lignins obtained at elevated LHW pretreatment severities had increased T_g values [12]. In addition, the distribution of lignin in the cell wall would be changed after LHW pretreatment. Yu et al. adopted SEM-EDXA method to detect $KMnO_4$ staining lignin in the cell wall of untreated and LHW-treated sugarcane bagasse and found that lignin can redistribute among different layers of the cell wall in the LHW process [10]. Lignin distributed in different layers of the cell wall has different dissolution ability during LHW pretreatment. Ma et al. used confocal Raman microscopy to observe that lignin in the middle layer of the secondary cell wall (S2 layer) was dissolved more than that in compound middle lamella (CML) [36].

The chemical groups, chemical bonds, and characteristic monomeric units of lignin are also changed after LHW pretreatment, and their changes can be detected

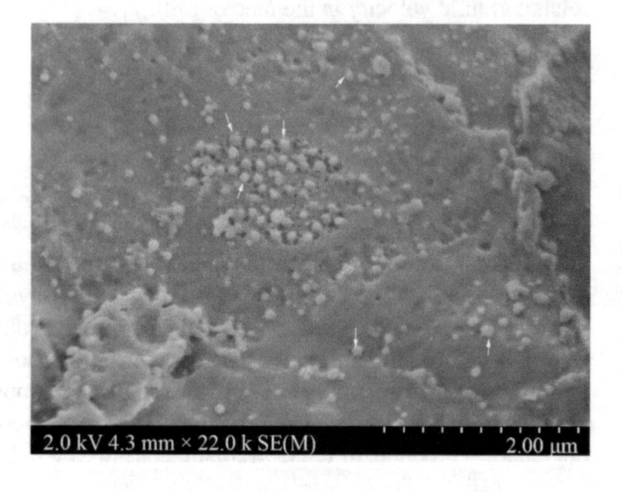

Fig. 4.2 Droplets formed on the surface of LHW-treated sugarcane bagasse (pointed by white arrows)

by Fourier transform infrared (FTIR) spectroscopy, nuclear magnetic resonance (NMR) spectroscopy, pyrolysis gas chromatography/mass spectrometer (Py-GC/MS), and so on. FTIR detection can monitor the changes of functional chemical groups like carbonyl, aldehyde group, carboxyl, aromatic group, conjugate group, chemical bonds, etc. Generally, the FTIR characteristic peak intensities of lignin after LHW pretreatment are changed, which indicates that the amount of chemical groups in lignin is increased or decreased [23, 37–39]. The NMR data can give information of lignin linkage and monolignol variation [40]. Two-dimensional (2D) heteronuclear single-quantum coherence (HSQC) and solid-state cross-polarization/magic angle spinning (CP/MAS) NMR spectroscopy are commonly adopted to measure the organic solvent dissolved and undissolved lignin, respectively [38, 40]. The Py-GC/MS method can reveal the structural information via products generated during the pyrolysis process [40–42].

4.2.3 *Microstructural Change*

After LHW pretreatment, the microstructure of lignocellulose including surface morphology, cell wall, micropores, and crystallinity can be changed.

The compactness of surface morphology of lignocellulose becomes loose after LHW pretreatment, and the longitudinal cracks appear among the fascicular structure (Fig. 4.3) [39, 43]. Spherical droplets are found to deposit on the surface of LHW-treated lignocellulose, which are ascribed to lignin or pseudo-lignin [34, 44]. Compared with the untreated lignocellulose, the specific surface area of LHW-treated lignocellulose is usually increased to one- to threefold [33, 45], and the pore diameter of LHW-treated lignocellulosic surface is enlarged more obviously with severity factor of pretreatment intensified [45]. The cell wall of raw lignocellulose has apparently various layers, namely, CML, primary cell wall (PCW), outer layer of secondary cell wall (SCW, S1), middle layer of secondary cell wall (S2), and inner layer of secondary cell wall (S3) [10], among which the boundaries become

Fig. 4.3 SEM observation of untreated (**a**) and LHW-treated (**b**) sweet sorghum bagasse

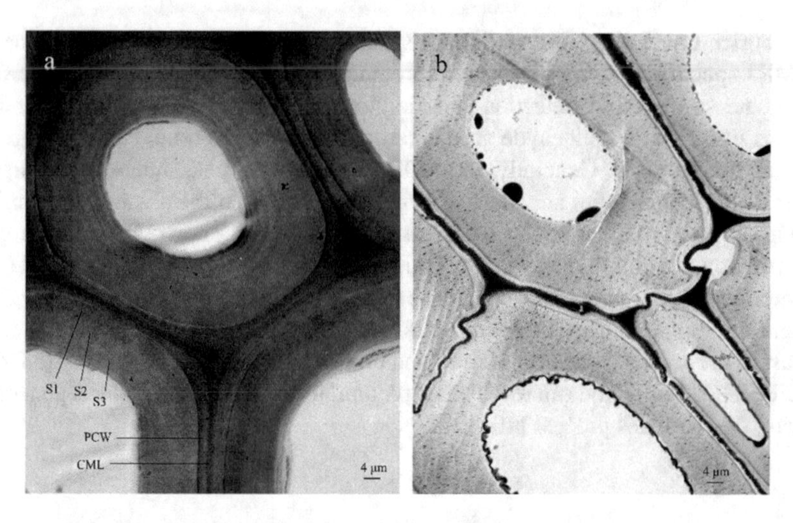

Fig. 4.4 TEM observation of untreated (**a**) and LHW-treated (**b**) sugarcane bagasse

blurry after LHW pretreatment (Fig. 4.4) [46, 47]. The cell wall is destroyed due to the removal of hemicellulose and lignin [46, 47]. The crystallinity of lignocellulose indicated with crystallization index (CrI) is usually increased after LHW pretreatment due to the removal of amorphous components and the retention of crystalline cellulose. The CrI value is usually calculated with XRD data (Eq. 4.8) and sometimes with FTIR (Eq. 4.9) and NMR data (Eq. 4.10). The type of cellulose cannot be transformed after LHW pretreatment.

$$\mathrm{CrI} = \frac{I_{002} - I_{am}}{I_{002}} \times 100 \tag{4.8}$$

where I_{002} is the diffraction intensity of 002 lattice plane at around 2θ=22.5° and I_{am} is the baseline intensity at about 2θ=18.4° [48]:

$$\mathrm{CrI} = \frac{I_{WN1}}{I_{WN2}} \times 100 \tag{4.9}$$

where I_{WN1} is band intensity at 1430 cm^{-1} only for cellulose I or 1371 cm^{-1} for cellulose I and II and I_{WN2} is band intensity at 889 cm^{-1} only for cellulose I or 665 cm^{-1} for cellulose I and II [49]:

$$\mathrm{CrI} = \frac{S_{\delta 86.4-93.0}}{S_{\delta 81.0-93.0}} \times 100 \tag{4.10}$$

where $S_{\delta 86.4\text{-}93.0}$ and $S_{\delta 81.0\text{-}93.0}$ are the integral areas of bands at the chemical shift from 86.4 to 93.0 ppm and 81.0 to 93.0, respectively [50].

4.3 Technology Development of LHW Pretreatment

The last decades have witnessed the development of techniques in LHW pretreatment of biomass on various aspects for different purposes. In this section, the developments in reactors, catalysts, and the use of cosolvents are mainly discussed to update the progress of LHW pretreatment techniques. Integration of LHW treatment and other techniques for biomass pretreatment is also considered as a development in LHW techniques and is introduced in this section.

4.3.1 Development in Reactors

In general, the reactors used for LHW pretreatment of biomass are in three categories, i.e., batch, semicontinuous, and continuous reactors, based on their flow types. Each of these types has its advantages and is used for different purposes [51]. In brief, using batch reactors minimizes the use of water or solvents, thus reducing the cost of solvents, catalysts, and the energy consumption for heating up the system. However, pretreatment with batch reactors has longer residence time for both dissolved and solid residues compared with treatments under semicontinuous and continuous conditions. The long residence time can cause further degradation of products such as xylose and glucose to various compounds, some of which (e.g., furans, acids, and furfurals) are toxic to enzymes and microorganisms [52]. By using semicontinuous reactors, it is achievable to separate water or solvent-soluble compounds from solids by their phases and to minimize the further degradation of soluble compounds [53]. Interestingly, by taking advantage of the distinct decomposition temperatures of hemicelluloses (160–210 °C), cellulose (above 240 °C), and lignin (insoluble at 160–210 °C), near-complete fractionation and recovery of hemicellulose can be achieved by LHW pretreatment under moderate conditions [54–56]. Continuous reactors are mainly used to investigate the mechanisms of biomass decomposition under hydrothermal conditions, because these reactors can minimize the retention time of whole feedstocks; thus, the further degradation of both soluble and solid products can be minimized. However, a further extraction or fractionation process is needed to achieve the separation of different compounds after continuous treatment.

The development in reactors can be evidenced by the improvement in heating types. Conventional reactors generally use furnaces or bathe such as oil bath (less than 300 °C) and sand bath (less than 600 °C) as heating sources [57–61]. In recent years, microwave has been widely applied for biomass pretreatment. Compared with conventional ways, microwave-assisted heating can achieve uniform heating of biomass/water mixtures more rapidly with high efficient energy transmission [62]. Moreover, compared with conventional heating methods, microwave-assisted pretreatment has shown significant enhancements of the accessibility of cellulose to enzymes and promoted the biodegradability of lignocellulosic biomass [63, 64].

More recently, solar-assisted heating is also applied for biomass LHW pretreatment [65, 66]. Compared with other heating methods, solar-assisted heating has obvious advantages because it eliminates the cost for heating supply, thus significantly reduces the overall cost of biorefinery. More interestingly, unlike LHW pretreatment with conventional or microwave-assisted heating, solar-assisted pretreatment has shown to be more effective for the delignification of biomass and maintain most of the cellulose at the same time [65]. However, the use of solar energy is limited by the availability and intensity of sunlight in different areas. In general, the new heating types are highlighted as these methods can drastically save the process cost and promote the scale-up of biorefinery.

4.3.2 Development in Catalysts

Catalysts are commonly used in LHW pretreatment of lignocellulosic biomass to optimize the biodegradability of cellulose. Diluted sulfuric acid (e.g., ~2%) is one of the most commonly used catalysts for biomass pretreatment which can not only enhance the recovery of hemicellulose but also promote the rearrangement of lignin, thus enhancing the biodigestibility of cellulose [67, 68]. However, undesirable degradation of cellulose can also happen during LHW/sulfuric acid pretreatment, for instance, the pretreatment of switchgrass with 0.5% of sulfuric acid (140 °C for 10 min) can cause more than 20% of glucan loss [69]. Thus, weaker acids such as acetic acid and aqueous CO_2 are also used to reduce the loss of glucan [70, 71]. Moreover, pretreatment of lignocellulosic biomass with acetic acid and sodium chlorite at moderate conditions (~105 °C) can selectively remove lignin, maintain the majority of hemicellulose and cellulose, and reduce the polymerization of cellulose [72–74]. Alkaline reagents such as sodium chloride and aqueous ammonia are also widely used as additives for the delignification of biomass during LHW pretreatment. The addition of alkalines enhances the dissolution of lignin during LHW pretreatment. Also, hemicellulose can be removed by setting proper temperatures. Compared with acid-catalyzed LHW pretreatment, alkali-catalyzed ways are more effective in delignification [75].

A major issue of acid- or alkaline-based pretreatment is the cost of chemicals and the extra process cost for the neutralization or recovery of catalysts. To overcome this shortcoming, heterogeneous catalysts have been developed as recyclable alternatives to replace the use of homogeneous catalysts [76]. LHW pretreatment with solid or polymer-supported acids as catalysts can not only achieve the rapid decomposition of hemicellulose but also stabilize acids on solids to avoid the waste or pollution caused by catalysts [77, 78]. The development from homogeneous to heterogeneous catalysts is plausible, but effective recovery methods to separate catalysts from the pretreated biomass residues are needed to achieve practical reuse of catalysts.

4.3.3 Development in Solvent/Water Systems

Modifying the solvent systems of LHW is another key technique applied to lignocellulosic biomass. Efforts have been taken mainly to achieve the dissolution of lignin. In general, various cosolvents including aprotic solvents, protic solvents, ionic liquids (ILs), and deep eutectic solvents (DESs) have been used in LHW pretreatment. Some of the main developments in these solvents are highlighted and discussed as follows.

Protic Solvent/Water Ethanol is the most commonly used protic solvent for the hydrothermal pretreatment of lignocellulosic biomass. Some early studies (years 2007 to 2009) primarily developed the method of using ethanol/water mixtures as solvents for the hydrothermal pretreatment of lignocellulosic biomass [79–81]. After that, following studies combined ethanol/water pretreatment with the use of various catalysts to further enhance the performances of this treatment method and applied it onto various feedstocks [82–86]. Despite the dissolution of lignin, the use of ethanol also activates the etherification between lignin function groups and ethanol, in which reaction can enhance the enzymatic hydrolysis of pretreated biomass by reducing the affinity of enzymes onto lignin [87]. An advantage of using ethanol as a cosolvent for biomass pretreatment is that ethanol is widely produced from lignocellulose biomass via hydrothermal pretreatment with subsequent enzymatic hydrolysis and fermentation. Therefore, using ethanol as a cosolvent for biomass pretreatment is considered as a sustainable way of ethanol production and application [88].

Aprotic Solvent/Water Aprotic solvents such as acetone, tetrahydrofuran (THF), methyl isobutyl ketone (MIBK), gamma-valerolactone (GVL), and imidazole have also attracted wide interests as cosolvents for the pretreatment of lignocellulosic biomass [15, 61, 89–91]. Among these solvents, GVL is highlighted as a promising cosolvent because it is a biorefinery-derived chemical and its superior performances for lignocellulosic biomass pretreatment. Some of the recent studies have comprehensively studied the produce of GVL from biomass-derived chemicals [92], the use of GVL/water for biomass pretreatment with subsequent bioconversion [61, 93], the solvent effect of GVL on biomass decomposition mechanisms and kinetics [94–96], and the recovery of products and solvents from GVL/water/biomass pretreated mixtures [61, 97]. It is plausible that GVL-based biorefinery has broadened the future of biomass valorization.

ILs/Water ILs were first introduced as solvents or cosolvents for biomass pretreatment in 2007, when 1-n-butyl-3-methylimidazolium chloride and 1-allyl-3-methylimidazolium chloride were used for the dissolution of lignin from lignocellulosic biomass, and the pretreatment with ILs showed significant enhancement to the biodegradability of biomass [98, 99]. Unlike organic solvents, the solubilities of ILs can be tuned by using different ion donators and/or accepters or adjusting the ratio of water in ILs/water mixtures. In other words, ILs can be used to selectively dissolute lignin, hemicellulose, or cellulose for corresponding purposes [100–105]. Despite the promising future of ILs/water in biomass pretreatment, most ILs are toxic to enzymes or microorganisms, and ILs are generally expensive [106, 107]. The use of "greener" and cost-effective ILs is encouraged for biomass pretreatment to achieve scale-up applications [105].

DESs/Water Similar to ILs, DESs (prepared with various hydrogen donators and accepters) were also developed to achieve biomass fractionation by taking advantage of their different solubilities on lignin, cellulose, and hemicellulose. DESs/water was first studied as a solvent for biomass pretreatment in 2012 [108]. After that, an increasing number of studies have been conducted to replace ILs with DESs for biomass pretreatment [109–113]. Compared with ILs, DESs have less toxicity because the hydrogen donator can be various biomass-derived chemicals such as sugar monomers and biomass-derived acids. Moreover, the chemicals used for DES preparation are much cheaper than those used for ILs, and using DESs is more cost-effective [114, 115]. However, the use of DESs/water on biorefinery is still at its infantry level and needs to be further studied.

4.3.4 LHW Pretreatment Combined with Other Pretreatments

Properly designed integrated pretreatment with LHW and other techniques can also enhance the sugar yields with minimized consumption. For instance, the integration of mechanical size reduction (e.g., ball milling) and hydrothermal pretreatment was studied to enhance the removal of hemicellulose and lignin in lignocellulosic biomass, and the enzymatic hydrolysis efficiency of pretreated biomass was significantly enhanced [116, 117]. It should be noted that mechanical pretreatment can also enhance the unwanted loss of glucan during LHW pretreatment [118, 119]. Therefore, an assessment on the overall glucose recovery is required, and it has been suggested that the combination of hydrothermal pretreatment with subsequent size reduction can achieve higher glucan bioconversion than the other way around [120]. Moreover, steam explosion has been proved to be applicable to increase the porosity of biomass, thus enhancing the pretreatment performances of LHW or LHW/solvents [121, 122]. In general, compared with the developments in reactors, catalysts, and cosolvents, less progress in integrated pretreatment techniques has been achieved.

4.4 Factors Influencing Lignocellulosic Biomass Bioconversion Based on LHW Pretreatment

The bioconversion or degradation (enzymatic hydrolysis, anaerobic digestion, or fermentation) of lignocellulosic biomass mainly uses the cellulose or glucan of biomass. Whether a pretreatment process is successful or not should be assessed based on the recovery of glucan after pretreatment and the bioconversion performances of pretreated biomass, as well as the cost consumed for the overall treatment. Therefore, to generate minimal yields of unwanted products and maximal yields of target products in cost-effective ways, it is of significance to understand the factors influencing the bioconversion process of biomass pretreatment. This section mainly discusses soluble degradation products, lignin, structural features, and solid loading, which are LHW pretreatment correlated key factors that influence the lignocellulose bioconversion.

4.4.1 Soluble Degradation Products

There are multiple products during the hydrothermal treatment of lignocellulosic biomass. Some of these products have been identified as inhibitors to the bioconversion of pretreated biomass and need to be minimized during or after pretreatment, while some other products are wanted. These products are summarized as follows:

1. The hydrothermal decomposition of hemicellulose generates corresponding sugar oligomers and monomers such as xylooligosaccharide, xylose, arabinan, and arabinose. These sugars can suppress the enzymatic hydrolysis performance of cellulase [123]. For instance, even a concentration of xylose at 1 g/L significantly decreases the enzyme efficiency of cellulase [124].
2. LHW with high treatment severities can easily cause the further decomposition of hemicellulose and cellulose-derived sugars that can generate chemicals such as furans, furfurals, and organic acids, which are toxic to enzymes and microorganisms [52, 124, 125].
3. Part of lignin also decomposes and dissolves in LHW as phenolics. These chemicals are toxic to enzymes and can significantly inhibit the bioconversion of cellulose, and even a reduction of cellulose hydrolysis rate of 92% can be induced by the presence of phenolics at 1 g/L [124, 126].
4. The other products, especially cellulose-derived monomers and oligomers, are preferred for the subsequent bioconversion such as anaerobic digestion and fermentation. For instance, LHW pretreatment of biomass at moderate temperature

can cause the release of some soluble compounds and the overall solubility of biomass, which are preferred to enhance the anaerobic digestion of biomass [127, 128].

4.4.2 Lignin

Lignin has been long considered as the dominant factor that suppresses the bioconversion of lignocellulosic biomass [129]. In general, it has been demonstrated that the biodegradability of cellulose decreases with the increase of lignin content in pretreated biomass [130–132]. The detailed effect of lignin on the bioconversion of cellulose is concluded as follows:

1. Lignin naturally seals in the cell wall structure of lignocellulosic biomass and performances as a cover for cellulose. The covering effect of lignin can block the accessibility of cellulose to cellulase. Therefore, part of the cellulose in lignocellulosic biomass cannot be hydrolyzed. In other words, the enzymatic hydrolysis efficiency of cellulose is strongly suppressed for raw biomass because of the presence of lignin [133].
2. Even for the exposed cellulose, the loading of cellulase is also influenced by lignin because the functional groups of lignin are more attractive for cellulase compared with that of cellulose, which results in the less distribution of cellulase on cellulose but accumulation of cellulase on lignin [87].
3. As mentioned above, part of lignin decomposes under LHW process and generates phenolics.

As abovementioned, the modification, rearrangement, and removal of lignin via LHW pretreatment are all effective ways to minimize the inhibition effect of lignin and enhance the biodegradability of biomass.

4.4.3 Structural Features

During the LHW pretreatment of lignocellulosic biomass, modifications or changes of the structural features also happen, except for the decomposition of different fractions. Herein, the changes in biomass particulate properties (e.g., size and porosity) and cellulose crystals are summarized.

LHW pretreatment can cause not only the partial release of compounds but also the collapses of cell structures of biomass, both of which can cause particle size reductions [76, 134]. Moreover, the porosity of particles is also enhanced after pretreatment [135]. These changes can result in the increase of specific surface area of feedstocks, while higher specific surface areas are preferred for the enzymatic bioconversion of cellulose.

Changes also happen to the crystalline structure of cellulose. The structural changes of cellulose crystals can be characterized by the degree of polymerizations (DPs) and crystalline indexes (CrIs). The DPs of cellulose are generally decreased after LHW pretreatment due to the cleavage of glycosidic bonds [136]. However, changes in crystalline indexes vary with different pretreatment methods. In some of the previous studies with only LHW or LHW/acids for biomass pretreatment, the CrIs of cellulose were slightly increased [135, 137], while for other treatments using ILs/water or DESs/water, the CrIs of cellulose were decreased [138, 139]. Herein, it is arbitrary to make a statement that LHW pretreatment can ensure the increase or decrease of CrIs. However, compared with the dominant effect of lignin, the changes in CrIs and DPs are less effective to the digestibility of the biomass. In fact, increases in bioconversion of pretreated feedstocks were achieved in all referred studies [134–138]. Rather, it is plausible that the decrease of CrIs can enhance the initial hydrolysis rate during the enzymatic hydrolysis of cellulose [140, 141].

4.4.4 Solid Loading Ratio (%, W/W)

The solid loading ratio of biomass varies with different bioconversion techniques and biomass feedstocks. For LHW-pretreated biomass, most of the previous bioconversion techniques are either anaerobic digestion (AD) for biogas production or enzymatic hydrolysis (EH) for saccharification. Herein, the influences of LHW pretreatment with different solid loading for subsequent AD and EH are discussed, respectively.

Typical solid contents or solid loading of lignocellulose in AD treatment ranges from 10% to 50% [142]. Previous studies have classified the anaerobic digestion systems into solid-state anaerobic digestion (SS-AD, solid content >15%) and liquid-state anaerobic digestion (LS-AD, solid content < 15%) based on the content of total solids [143]. Obviously, higher solid loading ratio is preferred to enhance the capacity of treatment systems and reduce the energy cost for temperature maintenance. However, with the increase of solid loading, the concentrations of intermediate compounds, especially acids, can be over-generated, resulting in the failure of the AD [144]. Optimization of the loading rate is required for different cases. As to raw lignocellulosic biomass (mainly grasses, energy corps, or agriculture wastes), a loading ratio at 15% to 30% is likely optimal for their AD conversion, while the value is expected to reduce after LHW because of the increase of digestible fractions [145].

The enzymatic hydrolysis efficiency of biomass is strongly correlated with the treatment time, solid loading, and enzyme loading. In general, higher solid loading ratio is also preferred to maximize the capacity of each treatment. However, increasing solid content can result in the increase of sugar concentrations in the solution and cause end-product inhibition [146]. Moreover, with the increase of solid content, longer reaction time or higher enzyme loading (determined as FPU per gram biomass dry matter, or FPU per gram glucan in the dry matter) is required to achieve the maximum sugar recovery of the feedstocks, and reducing the use of enzyme with optimal sugar recovery has been more discussed than shortening the reaction time due to the high cost of enzymes [146–148]. In general, a solid loading ratio should be determined based on an overall assessment of treatment capacity, sugar recovery, and enzyme consumption. Based on the above discussion, an adequate LHW pretreatment can enhance the glucan content in pretreated biomass and modify or reduce the amount of lignin. Therefore, lower enzyme loading rate is required to achieve a similar sugar recovery from raw biomass, while the solid loading ratio is expected to be reduced for LHW-pretreated biomass to eliminate the end-product inhibition.

4.5 Bioproduct Production Based on LHW Pretreatment

After LHW pretreatment, hemicellulose in lignocellulose would be degraded into oligosaccharides and monosaccharides in the liquid portion, from which xylooligosaccharide can be extracted. Cellulose and lignin are mainly retained in the solid part, which would be directly used as feedstock for producing biogas or biocarbon or further converted into glucose for biochemicals like ethanol, butanol, lactic acid, etc., and the resulting lignin-rich solid residue can be used as fungi medium or for productions of lignin derivatives. The conversion route of lignocellulose based on LHW pretreatment is depicted as in Fig. 4.5.

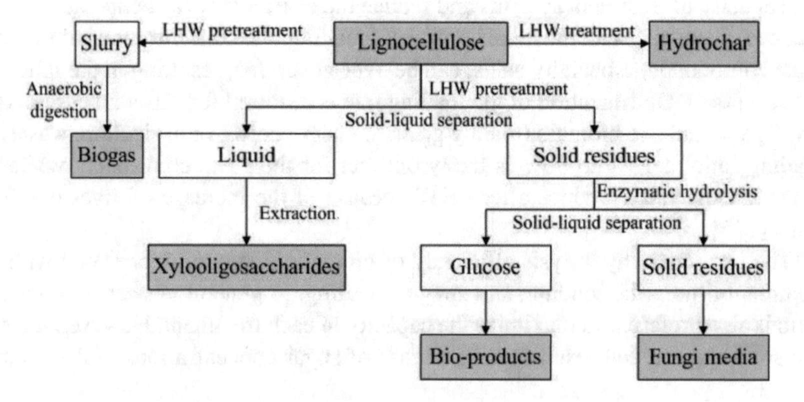

Fig. 4.5 Products converted from lignocellulose based on LHW technology

4.5.1 Xylooligosaccharides

The degradation products of hemicellulose that mainly consist of xylan with degree of polymerization (DP) less than 20 can be defined as xylooligosaccharides (XOS), among which those possessing 2–7 xylose units with β-1,4 glycoside bond linkage have been considered as prebiotics [149, 150]. XOS cannot be digested in human body due to the lack of enzymes and thus are suitable for diabetics as dietary sweeteners [150]. XOS have multiple health benefits like adjusting proliferative activities of intestinal beneficial flora especially *Bifidobacteria*, improving immunity of human body, reducing cholesterol content in serum, lowering down the risk of colon cancer, and so on [151, 152].

In the LHW process, XOS with different DP are distributed in the pretreatment liquid. The DP of XOS are affected by the pretreatment condition concerning temperature and retention time. The yield of high-DP XOS is more than that of low-DP XOS at mild pretreatment condition, while less at intense condition [153, 154]. Wang et al. [153] detected XOS from rapeseed straw with LHW pretreatment at elevated temperatures from 145 to 205 °C for remaining from 15 to 120 min and pointed out high content and well-distributed XOS would be obtained at a proper high temperature for a short time or low temperature for a long time. Generally, high-DP (DP>6) XOS are the predominant components in the pretreatment liquid [153, 155]. Due to the existence of degradation products from lignin and monosaccharides, it needs to extract XOS from the pretreatment liquid. Several methods like activated carbon adsorption [156], resin adsorption [11, 157], and so on are used to extract and purify XOS. The recovery of XOS is around 50% [156, 157], and the purity of XOS can reach to higher than 90% [11, 157].

4.5.2 Ethanol

LHW-treated lignocellulose is rich in cellulose and lignin. The bioconversion of cellulose into glucose is the essential pathway for various bioproduct productions. Glucose is a universal and high-quality carbon source for microbial metabolism. It can be converted into different bioproducts via inoculation of different microbial strains, such as ethanol [158, 159], butanol [160, 161], fatty acid [162], and so on.

Ethanol is the main target product from the fermentation of glucose originated from the enzymatic hydrolysis of LHW-treated lignocelluloses like agricultural crops [11, 22, 45, 163], hardwood [164], energy crops [158, 159], bamboo [165], palm-oil residues [166], and so on. The yeast strain *Saccharomyces cerevisiae* is largely used to ferment glucose into ethanol under anaerobic condition through the Embden-Meyerhof-Parnas (EMP) pathway involving 11 reaction steps [158, 167]. The bacterial strain *Zymomonas mobilis* is occasionally applied for ethanol fermentation via the Entner-Doudoroff (ED) pathway comprising 5 reaction steps [2, 168]. Other strains like engineering *Escherichia coli* are also employed for converting glucose into ethanol [169]. After fermentation, 1 mol glucose can be fermented into 2 mol ethanol (Eq. 4.11), in which the theoretic conversion efficiency can be calculated as 0.51:

$$C_6H_{12}O_6 = 2CH_3CH_2OH + 2CO_2 \tag{4.11}$$

The separate hydrolysis and fermentation (SHF) process and simultaneous saccharification and fermentation (SSF) process are two main ways for ethanol production from glucose [170]. The SHF process performs enzymatic hydrolysis and fermentation at their optimum temperatures (generally, 50 °C for enzymatic hydrolysis and 30 °C for fermentation) to obtain high yield of glucose, which has the disadvantage that glucose will accumulate to a certain amount as the enzymatic hydrolysis proceeds, posing feedback inhibition on enzyme activity. The SSF process conducts enzymatic hydrolysis and fermentation at the same temperature to timely eliminate feedback inhibition from glucose accumulation, but the temperature for enzymatic hydrolysis is usually lower than its optimum temperature due to the vitality protection of fermenting microorganism, which means that the enzymatic hydrolysis cannot reach its maximum efficiency. The fermenting microorganism which can endure higher temperature is more suitable for SSF process to achieve higher ethanol production. *Kluyveromyces* strains which can grow at temperature as high as 46 °C are usually used for high-temperature SSF process [171].

4.5.3 Medium

After enzymatic hydrolysis and fermentation of LHW-treated lignocellulose, the solid residues containing fermented lignocellulose and cells of yeast strain are good materials to prepare edible fungi medium. Wang et al. reported that the enzymatic hydrolyzed LHW-treated sugarcane bagasse can be mixed with wheat bran to prepare medium for mushroom growth [172]. The hydrolyzed LHW-treated sugarcane bagasse still contained 49.1% of cellulose, 7.1% of hemicellulose, and 37.0% of Klason lignin, which could provide carbon source, and the wheat bran could provide nitrogen source [172]. The carbon-to-nitrogen (C/N) ratio is the key nutrient factor for culturing edible fungi. Normally, the proper C/N ratio in medium for mycelium growth is (15~20):1, and that of (20~35):1 is suitable for fruiting [173]. Under proper condition control, the edible fungi can grow on the medium made of enzymatic hydrolyzed or fermented lignocellulose [172]. It is reported that the mushroom cultured in the medium composed of hydrolyzed sugarcane bagasse and wheat bran had superior nutritive value to the market mushroom [172].

4.5.4 Biogas

Biogas is a kind of renewable energies which is usually produced from wastewater of food industry, livestock and poultry manure, and other organic matters through anaerobic digestion involving microbial community [174, 175]. These organic materials are easily digested by microorganisms. Recently, lignocellulose has been developed as a feedstock for biogas production. Due to the complex and compact structure, the microorganisms cannot efficiently assimilate lignocellulose, which leads to the introduction of pretreatment for improving biogas production from lignocellulose [176, 177]. LHW pretreatment has gained more attention due to no addition of chemical reagents except water, lower production of inhibitors, and less corrosion to equipment [16, 178–180].

The pretreatment liquid and solid residue obtained after LHW pretreatment can be directly used for biogas production [16]. Hashemi et al. found that the pretreatment liquid obtained at higher temperature could give higher biogas production, while the solid residue acquired at higher temperature had lower biogas yield [16]. When the pretreatment temperature is equal to or lower than 150 °C, the biogas production from LHW-treated lignocellulose will increase with increasing pretreatment time [16]. It means that lignocellulose treated at temperatures no higher than 150 °C can produce more biogas as the severity factor increases. Furthermore, Hashemi et al. also reported that safflower straw treated at 120 °C could yield more biogas than those treated at 150 and 180 °C [16]. When the pretreatment temperature surpasses 170 °C, higher severity factor of LHW pretreatment would not result in higher biogas production. Jiang et al. reported that the giant reed treated at 170 °C can produce higher biogas amount than those treated at 190, 210, and 230 °C [181]. It indicates that LHW pretreatment at lower temperature is better for biogas production from lignocellulose.

4.5.5 Biomaterials

LHW pretreatment can be directly used for producing biomaterials like microcrystalline cellulose and hydrochar.

Microcrystalline cellulose is a kind of high-value functional cellulosic product that can be used for pharmaceutical industry, light industry, and so on. Yue et al. prepared microcrystalline cellulose from bleached softwood sulfate pulp with transition metal ion-enhanced LHW pretreatment [182]. It can effectively remove residual hemicellulose to obtain purified microcrystalline cellulose which has similar quality as the commercial product [182]. LHW pretreatment enhanced by transition metal ions is a promising method for preparing microcrystalline cellulose from lignin-free lignocellulose.

Hydrochar is originated from hydrothermal carbonization of biowastes at medium temperatures (180–260 °C) and its vapor pressure [183, 184], which is different from the pyrolysis biochar. Hydrothermal carbonization is suitable for carbonizing wet feedstocks to avoid drying process and thus reduces energy input [185]. It is particularly suitable to upgrade biomass with high inorganic elements to improve its energy content and creates hydrochar presenting coal-like features [186]. The hydrochar can be used as substance in combustion system or applied for contaminant remediation, soil amelioration, etc. [185–187].

4.6 Prospective

LHW treatment is a green and versatile technology for biomass biorefinery. It can effectively separate hemicellulose from lignocellulose due to the chemical bonds in hemicellulose that are vulnerable to the attack of hydrogen cation. The LHW-treated lignocellulosic solid residues have good property for being converted to biochemical products or biomaterials. However, high energy input and high water consumption are the bottlenecks for its industrial application. To add the indigenous small molecular organics from lignocellulose into LHW system as catalysts or solvent is one way to reduce the consumption of energy and water, which can keep LHW as a green process simultaneously. Applying LHW treatment for the production of high-value-added products is another way to overcome the two bottlenecks. Additionally, combined with lignin-removal technologies, LHW treatment is an effective method to separate hemicellulose, lignin, and cellulose. In a word, LHW treatment is a promising technology in lignocellulose biorefinery system.

References

1. Himmel, M. E., et al. (2007). Biomass recalcitrance: Engineering plants and enzymes for biofuels production. *Science, 315*(5813), 804–807.
2. Liu, C. G., et al. (2019). Cellulosic ethanol production: Progress, challenges and strategies for solutions. *Biotechnology Advances, 37*(3), 491–504.
3. Kumari, D., & Singh, R. (2018). Pretreatment of lignocellulosic wastes for biofuel production: A critical review. *Renewable & Sustainable Energy Reviews, 90*, 877–891.
4. Zhuang, X. S., et al. (2016). Liquid hot water pretreatment of lignocellulosic biomass for bioethanol production accompanying with high valuable products. *Bioresource Technology, 199*, 68–75.
5. Li, M., et al. (2017). The effect of liquid hot water pretreatment on the chemical-structural alteration and the reduced recalcitrance in poplar. *Biotechnology for Biofuels, 10*, 237.
6. Akhtar, N., et al. (2016). Recent advances in pretreatment technologies for efficient hydrolysis of lignocellulosic biomass. *Environmental Progress & Sustainable Energy, 35*(2), 489–511.
7. Garrote, G., Domínguez, H., & Parajó, J. C. (1999). Hydrothermal processing of lignocellulosic materials. *Holz als Roh- und Werkstoff, 57*, 191–202.
8. Alvira, P., et al. (2010). Pretreatment technologies for an efficient bioethanol production process based on enzymatic hydrolysis: A review. *Bioresource Technology, 101*(13), 4851–4861.

9. Silveira, M. H. L., et al. (2015). Current pretreatment technologies for the development of cellulosic ethanol and biorefineries. *ChemSusChem, 8*(20), 3366–3390.
10. Yu, Q., et al. (2013). Liquid hot water pretreatment of sugarcane bagasse and its comparison with chemical pretreatment methods for the sugar recovery and structural changes. *Bioresource Technology, 129*, 592–598.
11. Yu, Q., et al. (2015). Xylo-oligosaccharides and ethanol production from liquid hot water hydrolysate of sugarcane bagasse. *Bioresources, 10*(1), 30–40.
12. Ko, J. K., et al. (2015). Effect of liquid hot water pretreatment severity on properties of hardwood lignin and enzymatic hydrolysis of cellulose. *Biotechnology and Bioengineering, 112*(2), 252–262.
13. Li, X., & Zheng, Y. (2017). Lignin-enzyme interaction: Mechanism, mitigation approach, modeling, and research prospects. *Biotechnology Advances, 35*(4), 466–489.
14. Overend, R. P., & Chornet, E. (1987). Fractionation of lignocellulosics by steam-aqueous pretreatments. *Philosophical Transactions of the Royal Society London A, 321*, 523–536.
15. Toscan, A., et al. (2019). New two-stage pretreatment for the fractionation of lignocellulosic components using hydrothermal pretreatment followed by imidazole delignification: Focus on the polysaccharide valorization. *Bioresource Technology, 285*, 121346.
16. Hashemi, S. S., Karimi, K., & Mirmohamadsadeghi, S. (2019). Hydrothermal pretreatment of safflower straw to enhance biogas production. *Energy, 172*, 545–554.
17. Reddy, P., et al. (2015). Structural characterisation of pretreated solids from flow-through liquid hot water treatment of sugarcane bagasse in a fixed-bed reactor. *Bioresource Technology, 183*, 259–261.
18. Archambault-Léger, V., Losordo, Z., & Lynd, L. R. (2015). Energy, sugar dilution, and economic analysis of hot water flow-through pre-treatment for producing biofuel from sugarcane residues. *Biofuels, Bioproducts and Biorefining, 9*(1), 95–108.
19. Yu, Q., et al. (2011). The effect of metal salts on the decomposition of sweet sorghum bagasse in flow-through liquid hot water. *Bioresource Technology, 102*(3), 3445–3450.
20. Batista, G., et al. (2019). Effect of severity factor on the hydrothermal pretreatment of sugarcane straw. *Bioresource Technology, 275*, 321–327.
21. Zhou, W., et al. (2014). Changes in plant cell-wall structure of corn stover due to hot compressed water pretreatment and enhanced enzymatic hydrolysis. *World Journal of Microbiology and Biotechnology, 30*(8), 2325–2333.
22. Han, Q., et al. (2015). Autohydrolysis pretreatment of waste wheat straw for cellulosic ethanol production in a co-located straw pulp mill. *Applied Biochemistry and Biotechnology, 175*(2), 1193–1210.
23. Nitsos, C. K., et al. (2019). Enhancing lignocellulosic biomass hydrolysis by hydrothermal pretreatment, extraction of surface lignin, wet milling and production of cellulolytic enzymes. *ChemSusChem, 12*(6), 1179–1195.
24. Romaní, A., et al. (2010). Experimental assessment on the enzymatic hydrolysis of hydrothermally pretreated eucalyptus globulus wood. *Industrial and Engineering Chemistry Research, 49*, 4653–4663.
25. Girio, F. M., et al. (2010). Hemicelluloses for fuel ethanol: A review. *Bioresource Technology, 101*(13), 4775–4800.
26. Shi, S., et al. (2017). Reaction kinetic model of dilute acid-catalyzed hemicellulose hydrolysis of corn stover under high-solid conditions. *Industrial & Engineering Chemistry Research, 56*(39), 10990–10997.
27. Conner, A. H. (1984). Kinetic modeling of hardwood prehydrolysis. Part 1. Xylan removal by water prehydrolysis. *Wood and Fiber Science, 16*(2), 268–277.
28. Nabarlatz, D., Farriol, X., & Montané, D. (2004). Kinetic modeling of the autohydrolysis of lignocellulosic biomass for the production of hemicellulose-derived oligosaccharides. *Industrial and Engineering Chemistry Research, 43*, 4124–4131.
29. Yu, Q., et al. (2012). Hydrolysis of sweet sorghum bagasse and eucalyptus wood chips with liquid hot water. *Bioresource Technology, 116*, 220–225.

30. Pronyk, C., & Mazza, G. (2010). Kinetic modeling of hemicellulose hydrolysis from triticale straw in a pressurized low polarity water flow-through reactor. *Industrial and Engineering Chemistry Research, 49*, 6367–6375.
31. Zhuang, X., et al. (2009). Kinetic study of hydrolysis of xylan and agricultural wastes with hot liquid water. *Biotechnology Advances, 27*(5), 578–582.
32. Ralph, J., Lapierre, C., & Boerjan, W. (2019). Lignin structure and its engineering. *Current Opinion in Biotechnology, 56*, 240–249.
33. Nitsos, C. K., Matis, K. A., & Triantafyllidis, K. S. (2013). Optimization of hydrothermal pretreatment of lignocellulosic biomass in the bioethanol production process. *ChemSusChem, 6*(1), 110–122.
34. Meng, X., & Ragauskas, A. J. (2017). Pseudo-lignin formation during dilute acid pretreatment for cellulosic ethanol. *Recent Advance in Petrochem Science, 1*(1), 555551.
35. Hatakeyama, H., et al. (2010). Thermal decomposition and glass transition of industrial hydrolysis lignin. *Journal of Thermal Analysis and Calorimetry, 101*(1), 289–295.
36. Ma, J., et al. (2014). Revealing the changes in topochemical characteristics of poplar cell wall during hydrothermal pretreatment. *Bioenergy Research, 7*(4), 1358–1368.
37. Zhang, H. D., & Wu, S. B. (2014). Impact of liquid hot water pretreatment on the structural changes of sugarcane bagasse biomass for sugar production. *Mechanical Science and Engineering Iv, 472*, 774–779.
38. Xiao, L. P., et al. (2011). Impact of hot compressed water pretreatment on the structural changes of woody biomass for bioethanol production. *Bioresources, 6*(2), 1576–1598.
39. Wang, W., et al. (2012). Effect of structural changes on enzymatic hydrolysis of eucalyptus, sweet sorghum bagasse, and sugarcane bagasse after liquid hot water pretreatment. *Bioresources, 7*(2), 2469–2482.
40. Zhang, L., et al. (2015). Characterization of lignin derived from water-only and dilute acid flowthrough pretreatment of poplar wood at elevated temperatures. *Biotechnology for Biofuels, 8*, 203.
41. Moghaddam, L., et al. (2017). Structural characteristics of bagasse furfural residue and its lignin component. An NMR, Py-GC/MS, and FTIR study. *ACS Sustainable Chemistry & Engineering, 5*(6), 4846–4855.
42. Qin, Z., et al. (2018). Structural elucidation of lignin-carbohydrate complexes (LCCs) from Chinese quince (Chaenomeles sinensis) fruit. *International Journal of Biological Macromolecules, 116*, 1240–1249.
43. Zeng, M. J., et al. (2012). Tissue-specific biomass recalcitrance in corn stover pretreated with liquid hot-water: SEM imaging (part 2). *Biotechnology and Bioengineering, 109*(2), 398–404.
44. Wang, W., et al. (2015). Investigation of the pellets produced from sugarcane bagasse during liquid hot water pretreatment and their impact on the enzymatic hydrolysis. *Bioresource Technology, 190*, 7–12.
45. Li, X. Z., et al. (2014). Characteristics of corn stover pretreated with liquid hot water and fed-batch semi-simultaneous saccharification and fermentation for bioethanol production. *PLoS One, 9*(4).
46. Shi, J., et al. (2018). Effect of thermal treatment with water, H2SO4 and NaOH aqueous solution on color, cell wall and chemical structure of poplar wood. *Scientific Reports, 8*(1), 17735.
47. Qiang, Y., et al. (2014). Change of ultrastructure and composition of sugarcane bagasse in liquid hot water. *CIESC Journal, 65*(12), 5010–5016.
48. L, S., et al. (1959). An empirical method for estimating the degree of crystallinity of native cellulose using the X-ray diffractometer. *Textile Research Journal, 29*, 786–794.
49. He, J., Cui, S., & Wang, S.-y. (2008). Preparation and crystalline analysis of high-grade bamboo dissolving pulp for cellulose acetate. *Journal of Applied Polymer Science, 107*, 1029–1038.
50. Ali, M., et al. (1996). A solid - state NMR study of cellulose degradation. *Cellulose, 3*, 77–90.
51. Yu, Y., & Wu, H. (2010). Understanding the primary liquid products of cellulose hydrolysis in hot-compressed water at various reaction temperatures. *Energy & Fuels, 24*(3), 1963–1971.

52. Dogaris, I., et al. (2009). Hydrothermal processing and enzymatic hydrolysis of sorghum bagasse for fermentable carbohydrates production. *Bioresource Technology, 100*(24), 6543–6549.
53. Gallina, G., et al. (2016). Optimal conditions for hemicelluloses extraction from Eucalyptus globulus wood: Hydrothermal treatment in a semi-continuous reactor. *Fuel Processing Technology, 148*, 350–360.
54. Cocero, M. J., et al. (2018). Understanding biomass fractionation in subcritical & supercritical water. *The Journal of Supercritical Fluids, 133*, 550–565.
55. Liaw, S. B., Yu, Y., & Wu, H. (2016). Association of inorganic species release with sugar recovery during wood hydrothermal processing. *Fuel, 166*, 581–584.
56. Makishima, S., et al. (2009). Development of continuous flow type hydrothermal reactor for hemicellulose fraction recovery from corncob. *Bioresource Technology, 100*(11), 2842–2848.
57. Yu, Y. (2009). Formation and characteristics of glucose oligomers during the hydrolysis of cellulose in hot-compressed water. Curtin University.
58. Duan, P., & Savage, P. E. (2011). Hydrothermal liquefaction of a microalga with heterogeneous catalysts. *Industrial & Engineering Chemistry Research, 50*(1), 52–61.
59. Gallina, G., et al. (2018). Hydrothermal extraction of hemicellulose: From lab to pilot scale. *Bioresource Technology, 247*, 980–991.
60. Tolonen, L. K., et al. (2011). Structural changes in microcrystalline cellulose in subcritical water treatment. *Biomacromolecules, 12*(7), 2544–2551.
61. Luterbacher, J. S., et al. (2014). Nonenzymatic sugar production from biomass using biomass-derived gamma-valerolactone. *Science, 343*(6168), 277–280.
62. Chen, W.-H., Tu, Y.-J., & Sheen, H.-K. (2011). Disruption of sugarcane bagasse lignocellulosic structure by means of dilute sulfuric acid pretreatment with microwave-assisted heating. *Applied Energy, 88*(8), 2726–2734.
63. Hu, Z., & Wen, Z. (2008). Enhancing enzymatic digestibility of switchgrass by microwave-assisted alkali pretreatment. *Biochemical Engineering Journal, 38*(3), 369–378.
64. Ooshima, H., et al. (1984). Microwave treatment of cellulosic materials for their enzymatic hydrolysis. *Biotechnology Letters, 6*(5), 289–294.
65. Gabhane, J., et al. (2015). Solar assisted alkali pretreatment of garden biomass: Effects on lignocellulose degradation, enzymatic hydrolysis, crystallinity and ultra-structural changes in lignocellulose. *Waste Management, 40*, 92–99.
66. Xiao, C., et al. (2019). A solar-driven continuous hydrothermal pretreatment system for biomethane production from microalgae biomass. *Applied Energy, 236*, 1011–1018.
67. Lloyd, T. A., & Wyman, C. E. (2005). Combined sugar yields for dilute sulfuric acid pretreatment of corn stover followed by enzymatic hydrolysis of the remaining solids. *Bioresource Technology, 96*(18), 1967–1977.
68. Wyman, C. E., et al. (2009). Comparative sugar recovery and fermentation data following pretreatment of poplar wood by leading technologies. *Biotechnology Progress, 25*(2), 333–339.
69. Djioleu, A., & Carrier, D. J. (2016). Effects of dilute acid pretreatment parameters on sugar production during biochemical conversion of switchgrass using a full factorial design. *ACS Sustainable Chemistry & Engineering, 4*(8), 4124–4130.
70. Matsushita, Y., et al. (2010). Enzymatic saccharification of Eucalyptus bark using hydrothermal pre-treatment with carbon dioxide. *Bioresource Technology, 101*(13), 4936–4939.
71. Petrik, S., Kádár, Z., & Márová, I. (2013). Utilization of hydrothermally pretreated wheat straw for production of bioethanol and carotene-enriched biomass. *Bioresource Technology, 133*, 370–377.
72. Song, B., et al. (2019). Importance of lignin removal in enhancing biomass hydrolysis in hot-compressed water. *Bioresource Technology, 288*, 121522.
73. Hubbell, C. A., & Ragauskas, A. J. (2010). Effect of acid-chlorite delignification on cellulose degree of polymerization. *Bioresource Technology, 101*(19), 7410–7415.
74. Siqueira, G., et al. (2013). Enhancement of cellulose hydrolysis in sugarcane bagasse by the selective removal of lignin with sodium chlorite. *Applied Energy, 102*, 399–402.

75. Kim, J. S., Lee, Y., & Kim, T. H. (2016). A review on alkaline pretreatment technology for bioconversion of lignocellulosic biomass. *Bioresource Technology, 199*, 42–48.
76. Mosier, N., et al. (2005). Features of promising technologies for pretreatment of lignocellulosic biomass. *Bioresource Technology, 96*(6), 673–686.
77. Vu, A., Wickramasinghe, S. R., & Qian, X. (2018). Polymeric solid acid catalysts for lignocellulosic biomass fractionation. *Industrial & Engineering Chemistry Research, 57*(13), 4514–4525.
78. Qi, W., et al. (2018). Carbon-based solid acid pretreatment in corncob saccharification: Specific xylose production and efficient enzymatic hydrolysis. *ACS Sustainable Chemistry & Engineering, 6*(3), 3640–3648.
79. Pan, X., et al. (2007). Pretreatment of lodgepole pine killed by mountain pine beetle using the ethanol organosolv process: Fractionation and process optimization. *Industrial & Engineering Chemistry Research, 46*(8), 2609–2617.
80. Pan, X., et al. (2008). The bioconversion of mountain pine beetle-killed lodgepole pine to fuel ethanol using the organosolv process. *Biotechnology and Bioengineering, 101*(1), 39–48.
81. Brosse, N., Sannigrahi, P., & Ragauskas, A. (2009). Pretreatment of Miscanthus x giganteus using the ethanol organosolv process for ethanol production. *Industrial & Engineering Chemistry Research, 48*(18), 8328–8334.
82. Huijgen, W. J., et al. (2011). Catalytic organosolv fractionation of willow wood and wheat straw as pretreatment for enzymatic cellulose hydrolysis. *Journal of Chemical Technology & Biotechnology, 86*(11), 1428–1438.
83. Mesa, L., et al. (2011). The effect of organosolv pretreatment variables on enzymatic hydrolysis of sugarcane bagasse. *Chemical Engineering Journal, 168*(3), 1157–1162.
84. Wildschut, J., et al. (2013). Ethanol-based organosolv fractionation of wheat straw for the production of lignin and enzymatically digestible cellulose. *Bioresource Technology, 135*, 58–66.
85. Zhang, H., & Wu, S. (2014). Efficient sugar release by acetic acid ethanol-based organosolv pretreatment and enzymatic saccharification. *Journal of Agricultural and Food Chemistry, 62*(48), 11681–11687.
86. Tang, C., et al. (2017). Organic amine catalytic organosolv pretreatment of corn stover for enzymatic saccharification and high-quality lignin. *Bioresource Technology, 232*, 222–228.
87. Lai, C., et al. (2017). Lignin alkylation enhances enzymatic hydrolysis of lignocellulosic biomass. *Energy & Fuels, 31*(11), 12317–12326.
88. Lynd, L. R., et al. (1991). Fuel ethanol from cellulosic biomass. *Science, 251*(4999), 1318–1323.
89. Huijgen, W. J., Reith, J. H., & den Uil, H. (2010). Pretreatment and fractionation of wheat straw by an acetone-based organosolv process. *Industrial & Engineering Chemistry Research, 49*(20), 10132–10140.
90. Mostofian, B., et al. (2016). Local phase separation of co-solvents enhances pretreatment of biomass for bioenergy applications. *Journal of the American Chemical Society, 138*(34), 10869–10878.
91. Teng, J., et al. (2016). Catalytic fractionation of raw biomass to biochemicals and organosolv lignin in a methyl isobutyl ketone/H2O biphasic system. *ACS Sustainable Chemistry & Engineering, 4*(4), 2020–2026.
92. Wettstein, S. G., et al. (2012). Production of levulinic acid and gamma-valerolactone (GVL) from cellulose using GVL as a solvent in biphasic systems. *Energy & Environmental Science, 5*(8), 8199–8203.
93. Shuai, L., Questell-Santiago, Y. M., & Luterbacher, J. S. (2016). A mild biomass pretreatment using γ-valerolactone for concentrated sugar production. *Green Chemistry, 18*(4), 937–943.
94. Mellmer, M. A., et al. (2014). Solvent effects in acid-catalyzed biomass conversion reactions. *Angewandte Chemie International Edition, 53*(44), 11872–11875.
95. Song, B., Yu, Y., & Wu, H. (2018). Solvent effect of gamma-valerolactone (GVL) on cellulose and biomass hydrolysis in hot-compressed GVL/water mixtures. *Fuel, 232*, 317–322.

96. Song, B., Yu, Y., & Wu, H. (2017). Insights into hydrothermal decomposition of cellobiose in gamma-valerolactone/water mixtures. *Industrial & Engineering Chemistry Research, 56*(28), 7957–7963.
97. Lê, H. Q., et al. (2018). Chemical recovery of γ-valerolactone/water biorefinery. *Industrial & Engineering Chemistry Research, 57*(44), 15147–15158.
98. Fort, D. A., et al. (2007). Can ionic liquids dissolve wood? Processing and analysis of lignocellulosic materials with 1-n-butyl-3-methylimidazolium chloride. *Green Chemistry, 9*(1), 63–69.
99. Kilpeläinen, I., et al. (2007). Dissolution of wood in ionic liquids. *Journal of Agricultural and Food Chemistry, 55*(22), 9142–9148.
100. Brandt, A., et al. (2010). The effect of the ionic liquid anion in the pretreatment of pine wood chips. *Green Chemistry, 12*(4), 672–679.
101. Brandt, A., et al. (2013). Deconstruction of lignocellulosic biomass with ionic liquids. *Green Chemistry, 15*(3), 550–583.
102. Brandt, A., et al. (2011). Ionic liquid pretreatment of lignocellulosic biomass with ionic liquid–water mixtures. *Green Chemistry, 13*(9), 2489–2499.
103. Zhang, Y., et al. (2010). Ionic liquid– water mixtures: Enhanced K w for efficient cellulosic biomass conversion. *Energy & Fuels, 24*(4), 2410–2417.
104. Dibble, D. C., et al. (2011). A facile method for the recovery of ionic liquid and lignin from biomass pretreatment. *Green Chemistry, 13*(11), 3255–3264.
105. Brandt-Talbot, A., et al. (2017). An economically viable ionic liquid for the fractionation of lignocellulosic biomass. *Green Chemistry, 19*(13), 3078–3102.
106. Lê, H. Q., Sixta, H., & Hummel, M. (2018). Ionic liquids and gamma-valerolactone as case studies for green solvents in the deconstruction and refining of biomass. *Current Opinion in Green and Sustainable Chemistry, 18*, 20–24.
107. Wahlström, R., & Suurnäkki, A. (2015). Enzymatic hydrolysis of lignocellulosic polysaccharides in the presence of ionic liquids. *Green Chemistry, 17*(2), 694–714.
108. Francisco, M., van den Bruinhorst, A., & Kroon, M. C. (2012). New natural and renewable low transition temperature mixtures (LTTMs): Screening as solvents for lignocellulosic biomass processing. *Green Chemistry, 14*(8), 2153–2157.
109. Kumar, A. K., Parikh, B. S., & Pravakar, M. (2016). Natural deep eutectic solvent mediated pretreatment of rice straw: Bioanalytical characterization of lignin extract and enzymatic hydrolysis of pretreated biomass residue. *Environmental Science and Pollution Research, 23*(10), 9265–9275.
110. Kim, K. H., et al. (2018). Biomass pretreatment using deep eutectic solvents from lignin derived phenols. *Green Chemistry, 20*(4), 809–815.
111. Satlewal, A., et al. (2018). Assessing the facile pretreatments of bagasse for efficient enzymatic conversion and their impacts on structural and chemical properties. *ACS Sustainable Chemistry & Engineering, 7*(1), 1095–1104.
112. Chen, L., et al. (2019). A novel deep eutectic solvent from lignin-derived acids for improving the enzymatic digestibility of herbal residues from cellulose. *Cellulose, 26*(3), 1947–1959.
113. Yu, Q., et al. (2018). Deep eutectic solvents from hemicellulose-derived acids for the cellulosic ethanol refining of Akebia'herbal residues. *Bioresource Technology, 247*, 705–710.
114. Yu, Q., et al. (2019). In situ deep eutectic solvent pretreatment to improve lignin removal from garden wastes and enhance production of bio-methane and microbial lipids. *Bioresource Technology, 271*, 210–217.
115. Paiva, A., et al. (2014). Natural deep eutectic solvents–solvents for the 21st century. *ACS Sustainable Chemistry & Engineering, 2*(5), 1063–1071.
116. Yuan, Z., et al. (2015). Process intensification effect of ball milling on the hydrothermal pretreatment for corn straw enzymolysis. *Energy Conversion and Management, 101*, 481–488.
117. Deng, A., et al. (2016). Production of xylo-sugars from corncob by oxalic acid-assisted ball milling and microwave-induced hydrothermal treatments. *Industrial Crops and Products, 79*, 137–145.

118. Yu, Y., & Wu, H. (2011). Effect of ball milling on the hydrolysis of microcrystalline cellulose in hot-compressed water. *AIChE Journal, 57*(3), 793–800.
119. Ruiz, H. A., et al. (2011). Evaluation of a hydrothermal process for pretreatment of wheat straw—effect of particle size and process conditions. *Journal of Chemical Technology & Biotechnology, 86*(1), 88–94.
120. Kim, S. M., Dien, B. S., & Singh, V. (2016). Promise of combined hydrothermal/chemical and mechanical refining for pretreatment of woody and herbaceous biomass. *Biotechnology for Biofuels, 9*(1), 97.
121. Liu, C.-G., et al. (2015). Combination of ionic liquid and instant catapult steam explosion pretreatments for enhanced enzymatic digestibility of rice straw. *ACS Sustainable Chemistry & Engineering, 4*(2), 577–582.
122. Wojtasz-Mucha, J., Hasani, M., & Theliander, H. (2017). Hydrothermal pretreatment of wood by mild steam explosion and hot water extraction. *Bioresource Technology, 241*, 120–126.
123. Qing, Q., Yang, B., & Wyman, C. E. (2010). Xylooligomers are strong inhibitors of cellulose hydrolysis by enzymes. *Bioresource Technology, 101*(24), 9624–9630.
124. Rajan, K., & Carrier, D. J. (2016). Insights into exo-cellulase inhibition by the hot water hydrolyzates of rice straw. *ACS Sustainable Chemistry & Engineering, 4*(7), 3627–3633.
125. Yu, Q., et al. (2010). Two-step liquid hot water pretreatment of Eucalyptus grandis to enhance sugar recovery and enzymatic digestibility of cellulose. *Bioresource Technology, 101*(13), 4895–4899.
126. Lin, R., et al. (2015). Inhibitory effects of furan derivatives and phenolic compounds on dark hydrogen fermentation. *Bioresource Technology, 196*, 250–255.
127. Wang, D., et al. (2018). Can hydrothermal pretreatment improve anaerobic digestion for biogas from lignocellulosic biomass? *Bioresource Technology, 249*, 117–124.
128. Lin, R., et al. (2019). Improving gaseous biofuel production from seaweed Saccharina latissima: The effect of hydrothermal pretreatment on energy efficiency. *Energy Conversion and Management, 196*, 1385–1394.
129. Ragauskas, A. J., et al. (2014). Lignin Valorization: Improving Lignin Processing in the Biorefinery. *Science, 344*, 1246843.
130. Xu, J., et al. (2016). Correlation between physicochemical properties and enzymatic digestibility of rice straw pretreated with cholinium ionic liquids. *ACS Sustainable Chemistry & Engineering, 4*(8), 4340–4345.
131. Mou, H., & Wu, S. (2017). Comparison of hydrothermal, hydrotropic and organosolv pretreatment for improving the enzymatic digestibility of bamboo. *Cellulose, 24*(1), 85–94.
132. Kang, X., et al. (2018). Improving methane production from anaerobic digestion of Pennisetum hybrid by alkaline pretreatment. *Bioresource Technology, 255*, 205–212.
133. Yang, B., & Wyman, C. E. (2004). Effect of xylan and lignin removal by batch and flowthrough pretreatment on the enzymatic digestibility of corn stover cellulose. *Biotechnology and Bioengineering, 86*(1), 88–98.
134. Li, M.-F., Chen, C.-Z., & Sun, R.-C. (2014). Effect of pretreatment severity on the enzymatic hydrolysis of bamboo in hydrothermal deconstruction. *Cellulose, 21*(6), 4105–4117.
135. Nitsos, C. K., et al. (2016). Optimization of hydrothermal pretreatment of hardwood and softwood lignocellulosic residues for selective hemicellulose recovery and improved cellulose enzymatic hydrolysis. *ACS Sustainable Chemistry & Engineering, 4*(9), 4529–4544.
136. Martínez, J. M., et al. (1997). Hydrolytic pretreatment of softwood and almond shells. Degree of polymerization and enzymatic digestibility of the cellulose fraction. *Industrial & Engineering Chemistry Research, 36*(3), 688–696.
137. Xiao, X., et al. (2014). Enhanced enzymatic hydrolysis of bamboo (Dendrocalamus giganteus Munro) culm by hydrothermal pretreatment. *Bioresource Technology, 159*, 41–47.
138. Zhang, J., et al. (2014). Understanding changes in cellulose crystalline structure of lignocellulosic biomass during ionic liquid pretreatment by XRD. *Bioresource Technology, 151*, 402–405.

139. Fu, D., & Mazza, G. (2011). Aqueous ionic liquid pretreatment of straw. *Bioresource Technology, 102*(13), 7008–7011.
140. Li, C., et al. (2010). Comparison of dilute acid and ionic liquid pretreatment of switchgrass: Biomass recalcitrance, delignification and enzymatic saccharification. *Bioresource Technology, 101*(13), 4900–4906.
141. Laureano-Perez, L., et al. (2005). Understanding factors that limit enzymatic hydrolysis of biomass. *Applied Biochemistry and Biotechnology, 124*(1-3), 1081–1099.
142. Lehtomäki, A., et al. (2008). Anaerobic digestion of grass silage in batch leach bed processes for methane production. *Bioresource Technology, 99*(8), 3267–3278.
143. Xu, F., & Li, Y. (2012). Solid-state co-digestion of expired dog food and corn stover for methane production. *Bioresource Technology, 118*, 219–226.
144. Li, L., et al. (2010). Effect of temperature and solid concentration on anaerobic digestion of rice straw in South China. *International Journal of Hydrogen Energy, 35*(13), 7261–7266.
145. Sawatdeenarunat, C., et al. (2015). Anaerobic digestion of lignocellulosic biomass: Challenges and opportunities. *Bioresource Technology, 178*, 178–186.
146. Wang, L., Templer, R., & Murphy, R. J. (2012). High-solids loading enzymatic hydrolysis of waste papers for biofuel production. *Applied Energy, 99*, 23–31.
147. Gao, Y., et al. (2014). Optimization of fed-batch enzymatic hydrolysis from alkali-pretreated sugarcane bagasse for high-concentration sugar production. *Bioresource Technology, 167*, 41–45.
148. Larnaudie, V., Ferrari, M. D., & Lareo, C. (2019). Enzymatic hydrolysis of liquid hot water-pretreated switchgrass at high solid content. *Energy & Fuels, 33*(5), 4361–4368.
149. Xiao, L., Ning, J., & Xu, G. H. (2012). Application of Xylo-oligosaccharide in modifying human intestinal function. *African Journal of Microbiology Research, 6*(9), 2116–2119.
150. Carvalho, A. F. A., et al. (2013). Xylo-oligosaccharides from lignocellulosic materials: Chemical structure, health benefits and production by chemical and enzymatic hydrolysis. *Food Research International, 51*(1), 75–85.
151. Chung, Y. C., et al. (2007). Dietary intake of xylooligosaccharides improves the intestinal microbiota, fecal moisture, and pH value in the elderly. *Nutrition Research, 27*(12), 756–761.
152. Lecerf, J. M., et al. (2012). Xylo-oligosaccharide (XOS) in combination with inulin modulates both the intestinal environment and immune status in healthy subjects, while XOS alone only shows prebiotic properties. *British Journal of Nutrition, 108*(10), 1847–1858.
153. Wang, Z. W., et al. (2016). Comprehensive evaluation of the liquid fraction during the hydrothermal treatment of rapeseed straw. *Biotechnology for Biofuels, 9*, 142.
154. Griebl, A., et al. (2006). Xylo-oligosaccharide (XOS) formation through hydrothermolysis of xylan derived from viscose process. *Macromolecular Symposia, 232*, 107–120.
155. Surek, E., & Buyukkileci, A. O. (2017). Production of xylooligosaccharides by autohydrolysis of hazelnut (Corylus avellana L.) shell. *Carbohydrate Polymers, 174*, 565–571.
156. Chen, M.-H., et al. (2014). Autohydrolysis of Miscanthus x giganteus for the production of xylooligosaccharides (XOS): Kinetics, characterization and recovery. *Bioresource Technology, 155*, 359–365.
157. Chen, M. H., et al. (2016). Miscanthus x giganteus xylooligosaccharides: Purification and fermentation. *Carbohydrate Polymers, 140*, 96–103.
158. Alam, A., et al. (2019). A finalized determinant for complete lignocellulose enzymatic saccharification potential to maximize bioethanol production in bioenergy Miscanthus. *Biotechnology for Biofuels, 12*, 99.
159. Yu, Q., et al. (2016). Hemicellulose and lignin removal to improve the enzymatic digestibility and ethanol production. *Biomass & Bioenergy, 94*, 105–109.
160. Qureshi, N., et al. (2018). Butanol production from sweet sorghum bagasse with high solids content: Part I-Comparison of liquid hot water pretreatment with dilute sulfuric acid. *Biotechnology Progress, 34*(4), 960–966.

161. Su, H. F., et al. (2015). A biorefining process: Sequential, combinational lignocellulose pretreatment procedure for improving biobutanol production from sugarcane bagasse. *Bioresource Technology, 187*, 149–160.
162. Shang, C., et al. (2017). Oil production from enzymatic hydrolysate of sugarcane bagasse by Trichosporon cutaneum. *Journal of the Chinese Cereals and Oils Association, 32*(7), 74–78.
163. Wang, W., et al. (2012). High consistency enzymatic saccharification of sweet sorghum bagasse pretreated with liquid hot water. *Bioresource Technology, 108*, 252–257.
164. Tian, D., et al. (2019). Liquid hot water extraction followed by mechanical extrusion as a chemical-free pretreatment approach for cellulosic ethanol production from rigid hardwood. *Fuel, 252*, 589–597.
165. Yang, H., et al. (2019). Bioethanol production from bamboo with alkali-catalyzed liquid hot water pretreatment. *Bioresource Technology, 274*, 261–266.
166. Cardona, E., et al. (2018). Liquid-hot-water pretreatment of palm-oil residues for ethanol production: An economic approach to the selection of the processing conditions. *Energy, 160*, 441–451.
167. Cuevas, M., et al. (2015). Enhanced ethanol production by simultaneous saccharification and fermentation of pretreated olive stones. *Renewable Energy, 74*, 839–847.
168. Klinke, H. B., Thomsen, A. B., & Ahring, B. K. (2004). Inhibition of ethanol-producing yeast and bacteria by degradation products produced during pre-treatment of biomass. *Applied Microbiology and Biotechnology, 66*(1), 10–26.
169. Brandon, S. K., et al. (2011). Ethanol and co-product generation from pressurized batch hot water pretreated T85 bermudagrass and Merkeron napiergrass using recombinant Escherichia coli as biocatalyst. *Biomass & Bioenergy, 35*(8), 3667–3673.
170. Wang, W., et al. (2016). Highly efficient conversion of sugarcane bagasse pretreated with liquid hot water into ethanol at high solid loading. *International Journal of Green Energy, 13*(3), 298–304.
171. Kádár, Z., Szengyel, Z., & Réczey, K. (2004). Simultaneous saccharification and fermentation (SSF) of industrial wastes for the production of ethanol. *Industrial Crops and Products, 20*, 103–110.
172. Wang, W., et al. (2013). Reuse of enzymatic hydrolyzed residues from sugarcane bagasse to cultivate lentinula edodes. *Bioresources, 8*(2), 3017–3026.
173. X, W. H., & Fungi, E. (2004). *China Agricultural*. Beijing: University Press.
174. Molino, A., et al. (2013). Biomethane production by anaerobic digestion of organic waste. *Fuel, 103*, 1003–1009.
175. Mao, C., et al. (2015). Review on research achievements of biogas from anaerobic digestion. *Renewable and Sustainable Energy Reviews, 45*, 540–555.
176. Yu, Q., et al. (2019). A review of crop straw pretreatment methods for biogas production by anaerobic digestion in China. *Renewable and Sustainable Energy Reviews, 107*, 51–58.
177. Taherzadeh, M. J., & Karimi, K. (2008). Pretreatment of lignocellulosic wastes to improve ethanol and biogas production: A review. *International Journal of Molecular Sciences, 9*(9), 1621–1651.
178. Gaworski, M., et al. (2017). Enhancing biogas plant production using pig manure and corn silage by adding wheat straw processed with liquid hot water and steam explosion. *Biotechnology for Biofuels, 10*, 259.
179. Calicioglu, O., Richard, T. L., & Brennan, R. A. (2019). Anaerobic bioprocessing of wastewater-derived duckweed: Maximizing product yields in a biorefinery value cascade. *Bioresource Technology, 289*, 121716.
180. Ahmad, F., Silva, E. L., & Varesche, M. B. A. (2018). Hydrothermal processing of biomass for anaerobic digestion – A review. *Renewable and Sustainable Energy Reviews, 98*, 108–124.
181. Jiang, D., et al. (2016). Comparison of liquid hot water and alkaline pretreatments of giant reed for improved enzymatic digestibility and biogas energy production. *Bioresource Technology, 216*, 60–68.

182. Yue, X., et al. (2019). A novel method for preparing microcrystalline cellulose from bleached chemical pulp using transition metal ions enhanced high temperature liquid water process. *Carbohydrate Polymers, 208*, 115–123.
183. Zhao, P., et al. (2014). Clean solid biofuel production from high moisture content waste biomass employing hydrothermal treatment. *Applied Energy, 131*, 345–367.
184. Gupta, D., Mahajani, S. M., & Garg, A. (2019). Effect of hydrothermal carbonization as pretreatment on energy recovery from food and paper wastes. *Bioresource Technology, 285*, 121329.
185. Fang, J., et al. (2015). Hydrochars derived from plant biomass under various conditions: Characterization and potential applications and impacts. *Chemical Engineering Journal, 267*, 253–259.
186. Volpe, M., Goldfarb, J. L., & Fiori, L. (2018). Hydrothermal carbonization of Opuntia ficus-indica cladodes: Role of process parameters on hydrochar properties. *Bioresource Technology, 247*, 310–318.
187. Gao, P., et al. (2016). Preparation and characterization of hydrochar from waste eucalyptus bark by hydrothermal carbonization. *Energy, 97*, 238–245.

Chapter 5
Multiproduct Biorefining from Lignocellulosic Biomass Using Steam Explosion Technology

Zhi-Hua Liu

5.1 Introduction

The global increase in energy consumption and demands, the depletion of fossil fuel reserves, and the concerns of security and environment have led to the urgent development of more sustainable energy systems [1–5]. Biorefining of lignocellulosic biomass (LCB) to produce renewable biofuels, chemicals, and materials provides a sustainable alternative to fossil energy systems due to the large availability, renewability, and the reduction of CO_2 emissions [3, 5–7]. LCB mainly includes agricultural residues (e.g., corn stover), forestry residues (e.g., poplar trees), and energy crops (e.g., sweet sorghum), and it is a low-cost and promising raw material on Earth compared with other ones based on energy content, such as crude oil, natural gas, and various fermentation products (sugars, organic acids, drink softeners, etc.) [8–10]. Valorization of LCB to bioproducts presents a viable option for improving energy security, increasing economic development, and reducing greenhouse emissions [11–14]. Despite several years of research, more effort is still needed to improve the profitability of biorefinery and make it for the industrial implementation.

The bioprocessing strategy of LCB is generally including five major sections: biomass preparation, pretreatment, hydrolysis, fermentation, and product separation [15–18]. LCB generally consists of cellulose, hemicellulose, lignin, proteins, pectins, ash, and other components, which have very different chemical structures. The compositions in common LCB are shown in Table 5.1 [16, 19–25]. These components are intertwined with each other in a hetero-matrix to different degrees

Z.-H. Liu (✉)
School of Chemical Engineering and Technology, Tianjin University, Tianjin, China

Department of Plant Pathology and Microbiology, Texas A&M University, College Station, Texas, USA
e-mail: zhliu@tju.edu.cn

Z.-H. Liu, A. Ragauskas (eds.), *Emerging Technologies for Biorefineries, Biofuels, and Value-Added Commodities*,
https://doi.org/10.1007/978-3-030-65584-6_5

Table 5.1 Cellulose, hemicellulose, and lignin contents in lignocellulosic biomass (LCB) [16] [19–25]

Lignocellulosic biomass	Cellulose	Hemicellulose						
	Glucan	Xylan	Araban	Mannan	Lignin	Acetyl	Extractives	Ash
Agricultural biomass								
Corn Stover	31.7	17.1	2.6		12.6	2.9	22.5	4.3
Corn cobs	40	41.4			5.8			
Wheat straw	38.2	21.2	2.5	0.3	23.4			
Rice straw	36.1	27.2			19.7			12.1
Energy crops								
Switchgrass	43.8	28.8			9.2			
Sugarcane bagasse	40.1	26	1.7		19.1	3.4	9.2	1.0
Sugarcane leaf	35.3	23	3.4		19.6		7.5	7.8
Sweet sorghum bagasse	27.3	13.1	1.4		14.3		32.3	
Forestry biomass								
Poplar	39.2	18.8			29.6			1.5
Spruce	43.3	4.9	1.1	12	28.1			
Willow	43	24.9	1.2	3.2	24.2			
Pine	44.9	6.2	1.9	11.5	26.2		3.4	

and varying relative composition depending on the type, species, and source of LCB [1]. Most importantly, in plant cell wall, cellulose has a crystalline structure, and hemicellulose and lignin are connected by hydrogen bond, while hemicellulose combines lignin with covalent bond to form a lignin-carbohydrate complex (LCC), contributing to the biomass recalcitrance [1, 26]. Owing to biomass recalcitrance, pretreatment is a crucial step to disrupt the LCC structure and increase the conversion efficiency of the compositions, which determines the process cost of biorefinery [16, 18, 27]. Specifically, the aims of pretreatment are to alter the content and structures of lignin and/or the hemicellulose, simultaneously disrupt the architecture of plant cell wall, and hence increase the accessible surface area of cellulose to enzymes [28, 29]. Various pretreatments have been developed to convert LCB to biobased products, among which steam explosion (SE) has been generally recognized as one of the most effective pretreatments for decreasing the biomass recalcitrance due to lower environmental impact, less hazardous process chemicals, and greater potential for energy efficiency compared with other methods. Steam explosion has been extensively investigated to produce biofuels from a wide range of LCB feedstocks such as corn stover, poplar, wheat straw, sugarcane straw and bagasse, and so on [16, 19, 30–32].

Although pretreatments such as steam explosion show the capacity to deconstruct the LCC and overcome the biomass recalcitrance, each pretreatment has its own unique disadvantages, and it is unlikely that a single pretreatment can become the universal method for the full utilization of multiple compositions [4, 11, 33]. Because of the intrinsic characteristics of LCB including multi-compositions, macromolecules, and recalcitrance, an integrated process may be feasible for the full conversion of LCB to multiproducts in the large-scale application of biorefinery [16, 18].

The purpose of this chapter is to summarize the production of multiproducts from LCB using the integrated process of steam explosion. First, stream explosion pretreatment technology and its impaction on overcoming biomass recalcitrance were introduced systematically. Second, the production of biobased products especially biofuels based on the steam explosion technology was summarized. After that, to make a full utilization of LCB in a biorefinery, the advanced and integrated process of stream explosion for the multiproducts was talked about across the diversity of raw materials. Finally, several pilot- and demonstration-scale operations of the biorefinery for the multiproduct production had been described.

5.2 Steam Explosion Pretreatment Overcoming Biomass Recalcitrance

Owing to structural recalcitrance of LCB, pretreatment is an essential step for breaking down the LCC structure to facilitate the enzymatic hydrolysis of carbohydrates [16, 27, 34]. Generally, an effective pretreatment approach can be accessed by several criteria, such as the need for size reduction of biomass, the hemicellulose sugar recovery, the degradation product formation, the energy demands, and process cost. Besides these, an effective pretreatment should also employ the mild operation conditions, such as low pretreatment severity and low catalyst cost, and need to produce higher-value lignin as feedstock for coproducts. Steam explosion has been considered as one of the most promising pretreatments used for the deconstruction of LCB in biorefinery. As a physicochemical treatment technology, steam explosion was first used to defibrate wood into fiber for board production [35, 36]. Steam explosion possesses less hazardous process chemicals and environmentally benign relative to other technologies. Besides these, it requires low capital investment and low severity for the pretreatment of most LCB feedstocks to fractionate biopolymer constituents. Overall, steam explosion technology has potential as cost-effective pretreatment of LCB for biorefinery.

Steam explosion refers to a pretreatment technique for treating LCB resources by employing a combination of chemical and mechanical action [37, 38]. The simple process of steam explosion pretreatment has been shown in Fig. 5.1. The conditions employed in steam explosion pretreatment and its influence on the structure of LCB feedstocks have been shown in Table 5.2 [37–45]. In general, it includes two stages: auto-hydrolysis and explosive depressurization. During the steam explosion, LCB feedstock is loaded into the reactor chamber and then rapidly heated by high-pressure steam without addition of any chemicals as catalyst. The reactor system is maintained under a certain temperature (160 ~ 240 °C) and pressure for a period of time (several seconds to minutes) to promote the cleavage of acetyl group, which form acetic acid to in turn further catalyze the hydrolysis of the hemicellulose. This

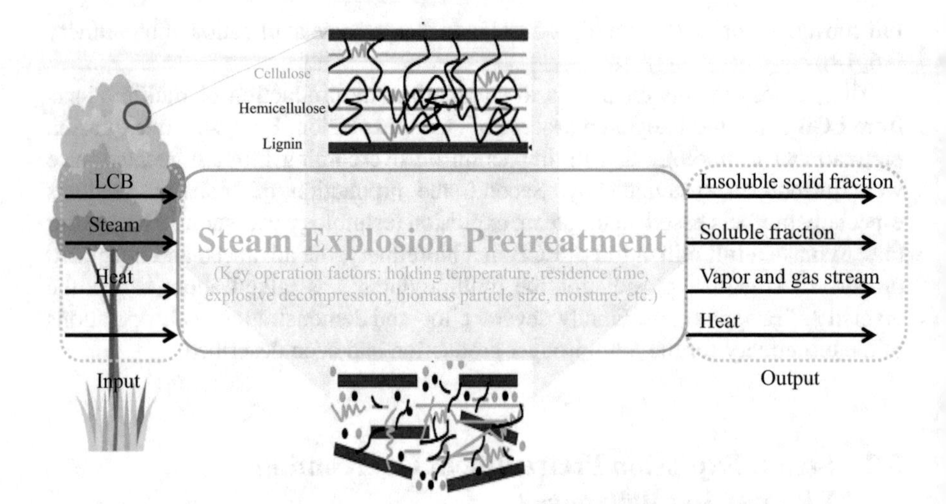

Fig. 5.1 Schematic of steam explosion pretreatment process to fractionate the lignocellulosic biomass in a biorefinery concept

Table 5.2 Steam explosion pretreatment conditions and the key changes in the structure of lignocellulosic biomass feedstocks [37–43]

Lignocellulosic biomass	The conditions of steam explosion pretreatment				The changes in the structures
	Holding temperature	Residence time	Particle size	Pressure	
Agricultural biomass	150–200 °C	3–10 min		<1.5 MPa	(i) The removal of hemicellulose sugars. (ii) The cleavage of glycosidic links. (iii) The cleavage of lignin-carbohydrate complex bonds. (iv) The cleavage of β-ether linkages of lignin. (v) Lignin melt, modification, and reorganization. (vi) Some degradation products. (vii) Decreased cellulose crystallinity. (viii) Defibrillization. (ix) Increased porosity of LCB. (x) Increased accessible surface area.
Energy crops	180–210 °C	5–15 min		<1.6 MPa	
Forestry biomass	190–230 °C	10–20 min	<5 cm	>1.6 MPa	

stage is named auto-hydrolysis process, during which the structure of LCB was modified as follows: (1) the cleavage of accessible glycosidic links, (2) the cleavage of lignin-carbohydrate complex bonds, and (3) the cleavage of β-ether linkages of lignin [16, 37, 46, 47]. For the lignin, steam explosion will modify the lignin structure at the high temperature by inducing the homolytic cleavage of the β-O-4 ether and other acid labile linkages. This process may produce a series of cinnamyl alcohol derivatives and other condensed molecules [47–49]. After maintaining for a short period of residence time, the pretreatment of LCB was terminated by a sudden explosive decompression, which will rapidly reduce the temperature and quench the reaction. This process is named explosive depressurization stage, which are featured by the deconstruction of biomass particulate structure but enhancement of the cellulose hydrolysis. The pretreated solid is only weakly correlated with this physical effect resulting in further substantial breakdown of the LCB structure and depolymerization of the lignin components [37]. Besides these, due to the instantaneous decompression during escape of high-pressure steam, the "steam explosion circles" may be formed, and the defibrillization may occur as confirmed by the irregular and fuzzy boundaries of the fibers [37]. Steam explosion pretreated biomass solid also exhibits the low crystallinity and high porosity to facilitate the bioconversion [37].

As a result, after steam explosion pretreatment, cellulose, lignin, and partial hemicellulose are retained in insoluble solid fraction, while partial hemicellulose sugars and extractives can be extracted into a liquid fraction. Steam explosion with two stages of auto-hydrolysis and explosive depressurization produces partial hemicellulose solubilization and lignin reorganization, which increase the cellulose accessibility to enzymatic attack by exposing the cellulose surface. It also produces partial sugars and lignin degradation to generate soluble compounds that are inhibitory or toxics for the enzymes and fermenting microorganisms in subsequent steps. Overall, steam explosion shows the effective capacity to deconstruct the LCC and fractionate components from LCB for overcoming the biomass recalcitrance.

5.3 Steam Explosion Pretreatment for Biorefinery

The most important variables determining the effectiveness of steam explosion are holding temperature, residence time, explosive decompression, and biomass particle size. Both holding temperature and residence time are essential to the steam explosion of LCB, and their performance can be described as Log R_0, which is generally used for assessing pretreatment severity: [50]

$$\log R_0 = t \times \exp\left[\left(T - T_b\right) / \omega\right]$$

where t is the residence time, min; T is the holding temperature, °C; T_b is the base temperature, 100 °C; and ω is the fitted value based on the activation energy, 14.75.

The combined severity factor log R_0" is employed to assess the pretreatment severity when acid catalyst is added in the steam explosion pretreatment: [51]

$$\log R_0'' = \log R_0 + |\text{pH-7}|$$

The development and optimization of process is very crucial to improve the pretreatment efficiency. A range of steam explosion pretreatment conditions have been evaluated based on sugar recovery, degradation product generation, and hydrolysis yields at high sold loadings. Steam explosion at 200 °C for 10 min produced higher enzymatic hydrolysis yield (91.7%), corresponding to a total glucose yield of 35.4 g glucose/100 g wheat straw [52]. However, under the same conditions, it also produced more degradation compounds. The characteristics of the pretreated wheat straw with a lower severity of log $R_0 = 3.65$, which correspond to steam explosion at 190 °C for 10 min, minimized the degradation of sugars and the formation of degradation compounds, which indicated a great potential for maximizing total sugar yield by optimizing enzymatic hydrolysis strategies [52]. The effects of the pretreatment severity of steam explosion on the bioconversion of corn stover biomass to ethanol had been investigated [53]. Results indicated that steam explosion with higher severity increased the specific surface area, the swollen volume, and the water holding capacity of pretreated corn stover and hence facilitated the efficiency of hydrolysis and fermentation. Even steam explosion with shorter residence time (6 min) recovered more glucan and xylan; the enzymatic hydrolysis and fermentation performance was obviously improved with longer residence time (9 min). Under the optimal conditions of 1.5 MPa and 9 min, glucan conversion was 87.2%, while the concentration and yield of ethanol reached 45.0 g/L and 85.6%, respectively [53]. Uncatalyzed steam explosion was conducted as a pretreatment to improve the hydrolysis performance of rapeseed straw [54]. Experimental statistical design and response surface methodology were employed to evaluate the effects of the holding temperature of 185 to 215 °C and the residence time of 2.5 to 7.5 min on the deconstruction efficiency of rapeseed straw [54]. Results showed that steam explosion at 215 °C and 7.5 min was the optimal ones to maximize the sugar yield of 72.3%, equivalent to 81% of potential glucose in enzymatic hydrolysis. Under the optimal conditions, the concentrations of ethanol in fermentation reached 43.6 g/L using 20% (w/v) solid loading, which is equivalent to 12.4 g ethanol/100 g biomass [54]. These studies showed that the steam explosion is an effective approach for biorefinery and it can enhance the bioconversion efficiency of LCB under a high severity, but it may degrade hemicellulose sugar especially xylan to generate various by-products as inhibitors for hydrolysis and fermentation. As the ultimate goal of biorefinery is "zero waste" by converting all sugars or carbons efficiently, the operation modes and conditions of steam explosion should be further optimized to improve the sustainability of biorefinery.

A novel steam explosion with low holding temperature and high exploding pressure had been exploited to increase the hydrolysis and sugar release from corn stover biomass [55]. By evaluating the operation mode, glucan and xylan recovery decreased with the increase of holding temperature and residence time, respectively, while glucan and xylan conversion in enzymatic hydrolysis showed an opposite trend. Under the optimal mode employing 160 °C and 48 min, glucose, xylose, and total sugar yield reached to 77.3%, 62.8%, and 72.3%, respectively, in a high solid process. Most importantly, the yield of hydroxymethyl furfural, furfural, and lignin-derived products was 6.3×10^{-2}, 7.5×10^{-2}, and less than 3.7×10^{-2} g/100 g feedstock, respectively, which was significantly lower than that from pretreatment with high severity [55]. This novel operation mode of steam explosion process improved the biomass conversion efficiency by increasing sugar recovery and yield, reducing degradation products, and enhancing hydrolysis efficiency [55]. All these aforementioned results suggested that the severity factors reflected the impactions of holding temperature and residence time on the pretreatment performance. Previous study also reported the effect of the explosive decompression of steam explosion on the enzymatic cellulose hydrolysis of pretreated spruce wood chips [56]. Results showed that the explosive depressurization stage had a significant influence on the accessibility of pretreated solid, improving the hydrolysis performance by up to 90% as compared with only steam pretreatment. Two key factors, pretreatment severity and explosion pressure, had been identified that were responsible for the effect of the explosive depressurization on the enzymatic hydrolysis [56]. Steam explosion with a higher severity can soften up and weaken the spruce wood structure; explosive depressurization stage can then better break up the spruce wood chip and decrease its particle size. Most importantly, increasing the pressure difference of the explosion will result in serious defibration, produce smaller particles, and thus facilitate the enzymatic hydrolysis of pretreated biomass. The most feature of the biomass modification impacting the accessibility was the size reduction of the macroscopic particles by the explosion. Thus, steam explosion with a high severity and a high-pressure difference of the explosive depressurization can lead to a comparatively high enzymatic hydrolysis performance of LCB [56].

Besides these, other important factors that have an influence on stream explosion performance are the biomass particles employed in the biorefinery [16, 57]. To make an effective LCB utilization, a size reduction step is deemed necessary for pretreatment and enzymatic hydrolysis to proceed effectively. Size reduction of LCB is energy-intensive and expensive in biorefinery. Most importantly, biomass particles not only influence the design of handling, transportation, and conversion facilities but also determine the efficiency of pretreatment as particle size has implication in mass and heat transfer processes [19, 58]. Heat transfer limitation should be a key issue for steam explosion pretreatment, which depends on the biomass particles used. It may result in overcooking the surface part of the larger biomass particles and incomplete pretreatment of the interior part, while it may be prone to degrade hemicellulose in smaller biomass particles into by-products due to the intense degree of heat [19]. To achieve a high sugar conversion, the impactions of particle size on pretreatment and their correlations with enzymatic hydrolysis performance were investigated [19]. The conversion efficiency under five biomass particle sizes of 2.5, 2.0, 1.5, 1.0, and 0.5 cm was systematically investigated and compared [19]. Results showed that the highest sugar recovery had been reached with the smaller biomass particle size of 1.0 and 0.5 cm, while the highest glucan and xylan conversion in enzymatic hydrolysis had been observed with the larger one of 2.5 cm. Interestingly, enzymatic hydrolysis rate and conversion of pretreated solid were also higher when biomass particle size was used. By characterizing the structure modification of pretreated solid, the specific surface area of pretreated solid apparently increased with the increase of biomass particle size, while the crystallinity index decreased [19]. Based on these results, the larger biomass particles within the range of 0.5–2.5 cm had been confirmed as a suitable feedstock to achieve the high stream explosion pretreatment performance. The proposed mechanism for these phenomena was that the porosity of the biomass pile was higher for larger particles than that for smaller ones because of the lower bulk density of larger particles. High-pressure steam can penetrate easier the interior of the biomass pile of larger particles to enhance the efficiency of auto-hydrolysis and explosive depressurization. Additionally, the larger particles submerged in condensate water were less than smaller ones, which resulted in the high efficiency of explosive depressurization on the biomass solids. As a result, steam explosion produced the pretreated larger solid particles with higher specific surface area and lower crystallinity, which improved the sugar conversions and yields.

The effects of the most important variables, including temperature, residence time, and chip size, on various physical/chemical parameters of pine biomass in steam explosion had been investigated [59]. Some parameters were employed as an alternative tool to evaluate the effect of steam explosion on lignocellulosic materials [59]. They found that the center of bigger chips remained "uncooked" after the pretreatment at low holding temperature for short residence time because steam has not reached that part of the chip likely due to transport phenomena effects. A more dark and brownish appearance in pretreated solid was observed after steam explosion at a high severity or with smaller chips likely due to the chemical breakdown of lignin and wood extractives. Results showed that more condensed lignin-derived substances may appear as holding temperature of steam explosion increases from 190 °C to 210 °C. The degradation of hemicellulose would be implied in the generation of these condensed substances. Smaller chips were prone to obtain higher content of acid-insoluble substances, which may be related to the formation of pseudo-lignin [48, 60, 61]. Overall, with the harshness of pretreatment and smaller chips employed, more condensed substances would be generated [59]. As woody biomass is highly recalcitrant to sugar release, severe pretreatment and size reduction are required to achieve economically viable sugar yields. To understand the effects of biomass particles on the effectiveness of steam explosion and subsequent hydrolysis, a novel downscaled analysis and high-throughput pretreatment and hydrolysis were employed to evaluate the composition and digestibility of pretreated wood chip [62]. Woody biomass after the pretreatment at 180 °C for 8 min or longer produced reasonably uniform enzymatic sugar yields across their entire thickness. Various technologies, including heat transfer modeling, Simons' stain testing, magnetic resonance imaging, and scanning electron microscopy, had been used to investigate the effects of pretreatment on the modification of woody biomass for revealing the potential causes of variation. The heat transfer modeling suggested that high sugar yields will be obtained uniformly if the residence time was sufficient to allow the target holding temperature to be achieved throughout the entire thickness of the biomass chip. Magnetic resonance imaging and Simons' stain testing confirmed that the enzymatic hydrolysis of the pretreated biomass chip was dependent on the pore size distribution. By applying these techniques, it was demonstrated that the effectiveness of steam explosion pretreatment varied substantially with the thickness of biomass chip at short residence times, which significantly influenced overall lower sugar yields in the subsequent conversion process [62]. These results suggested that steam explosion with lower holding temperature was well suited for larger wood chips possibly because of the nonuniformity in holding temperature and digestibility profiles that can result from high holding temperature and short residence time [62]. To enhance the digestion of grass biomass to produce biomethane, steam explosion pretreatment of wheat straw with different particle sizes had been evaluated [63]. The maximum biomethane yield was obtained under 200 °C for 5 min, which showed a 27% increase in biomethane productivity as compared with non-treated wheat straw. The performance of biomethane production was better for wheat straw with higher particle size (3–5 cm) as compared with smaller ones (<1 mm). However, the economic impact of LCB milling on the pretreatment would be absolutely negative by increasing the energy input cost for size reduction [63].

Overall, the choice of biomass particle size would determine the steam explosion pretreatment and thus the sugar release in the following hydrolysis. The optimal particles will be helpful to facilitate the pretreatment performance and make a sustainable biorefinery.

5.4 Steam Explosion Pretreatment Impacts Up- and Downstreams of Biorefinery

The pretreatment and its up- and downstream units will interact with each other and thus determine the biorefining profitability [64–67]. Storage is the upstream unit for LCB bioconversion in biorefinery, and storage characteristics will affect the following conversion performance. On the one hand, efficient storage will significantly improve the efficiency of pretreatment and hydrolysis and hence facilitate the profitability of the biorefinery. On the other hand, the pretreatment option will impact the storage methods employed [64, 66].

To fundamentally and systematically understand the effects of storages on the bioconversion of LCB, the storage factors such as moisture, time, and particles had been evaluated, and the effects of storage strategies on the steam explosion pretreatment and enzymatic hydrolysis had been investigated [68]. Wet-stored biomass produced higher sugar conversion and yield as compared with dry-stored one after steam explosion and enzymatic hydrolysis. Even shredded biomass after storage reduced sugar conversion, it increased sugar yield obviously. After 3-month storage, non-shredded wet-stored biomass led to higher glucan conversion and glucose yield (91.5% and 87.6%, respectively) than other approaches. By characterizing the microstructure and crystallinity of corn stover biomass, it is shown that corn stover biomass maintained the flexible and porous structure after wet storage and possessed the high permeability, which facilitate the enhancement of the steam explosion and hydrolysis efficiency. The mechanism for the effects of storages on the bioconversion had been proposed based on the analysis of physicochemical properties of corn stover biomass [68]. During shredded dry storage, water-extractable sugars were consumed likely due to that the efficient aeration promoted microbial metabolism in the shredded corn stover biomass. Shredded biomass with dry storage will also allow the inter-fibril water to escape more rapidly, leading to the cross-linking of microfibrils to shrink due to the surface tension forces and thus form new hydrogen bonds between adjacent microfibrils. Most importantly, rapid evaporation could result in the pore collapse by reducing a number of larger pores and decreasing the average pore volume and pore size and thus lead to the fiber aggregation. As a result, the fiber hornification could be formed through covalent lactone bridges establishing between part of the existent carboxylic acid and hydroxyl groups in neighboring polymeric chains. Taken these together, the shrinkage of the cross-linking of microfibrils, the collapse of pore structure, the reorganization of hydrogen bonds in cellulose, and the fiber hornification resulted in the decrease of specific surface area and the increase of crystallinity. Dried stored biomass with these properties prevented the high-pressure steam from permeating efficiently into the internal plant cell walls, leading to the poor efficiency of auto-hydrolysis and explosion. Compared with dry storage, plant respiration continued as well as the microbial degradation due to the presence of air at the beginning of wet storage. After the oxygen in interstitial voids of biomass was exhausted, anaerobic fermentation began. Various microorganisms naturally grew on corn stover biomass and typically attacked the cell lumens or middle lamella regions of plant cell walls due to easier physical accessibility of nutrients. This process could spoil the structures, part of which was likely to be extensively loosened and became more flexible during wet storage. Meanwhile, microorganisms converted water-extractable sugars into organic acids (e.g., lactic acid and acetic acid) and created a low pH environment to prevent corn stover biomass from further biological degradation. Additionally, the inherent capillaries of biomass were filled with water, while the fibers remain swollen and porous, leading to the high permeability of biomass. Consequently, high-pressure steam easily permeated into the plant cell walls, and the organic acids produced by microbes also helped auto-hydrolysis effect. During the depressurization stage, the rapid pressure release resulted in the evaporation and explosion of water inside the biomass, causing the convective transport of extractable components toward the cell lumen and cell corners through the middle lamella. The rigid and highly ordered fibrils were loosened and disrupted, and the porosity of biomass greatly increased. These changes of microstructures led to the high conversion efficiency of corn stover biomass [68]. Therefore, wet storage methods would be desirable for the conversion of corn stover biomass to fermentable sugars with the employment of steam explosion and enzymatic hydrolysis [68].

The bioconversion of LCB to bioethanol needs to integrate several steps, mainly including pretreatment, hydrolysis, glucose fermentation, and xylose fermentation. Simultaneous saccharification and fermentation (SSF) is one of promising processes for the bioconversion of LCB to ethanol. The SSF of steam-exploded corn stover for ethanol production was investigated at a high glucan loading and temperature. The SSF conditions were systematically evaluated, and the optimized ones were inoculation optical density OD_{600} 4.0, initial pH 4.8, 50% nutrients added, 36 hours pre-hydrolysis time, 39 °C, and 12% glucan loading (20% solid loading). With the optimal conditions, a glucan conversion of 78.6%, an ethanol yield of 77.2%, and a final ethanol concentration of 59.8 g/L have been reached, respectively. By employing steam explosion pretreatment and thermal-tolerant yeast, the inhibitory effect from degradation products on the hydrolysis and fermentation was not obvious even the SSF was conducted at high glucan loading. As compared with separate hydrolysis and fermentation (SHF), glucan conversion and final ethanol concentration increased by 13.6% and 18.7% in SSF, respectively. Therefore, the SSF performance was significantly improved by employing high effective steam explosion strategy using a novel thermal-tolerant strain. Simultaneous saccharification and co-fermentation (SScF) referred to the process that enzymatic hydrolysis can be conducted simultaneously with the fermentation. SScF offers several advantages such as reduced inhibition effects of enzymes, short processing time, and low contamination risk as compared with separate SHF and separate hydrolysis and co-fermentation (SHcF). It is also superior to SSF because of the potential high productivity and yield of ethanol by co-utilizing of glucose and xylose. To improve xylose utilization, ethanol yield, and ethanol concentration in SScF, the SScF using steam-exploded corn stover (SECS) had been further investigated at high solid loadings compared with other conversion processes. The concentration, yield, and productivity of ethanol reached 34.3 g/L, 90.0%, and 2.61 g/L/h, respectively, by the coculture of 60 g/L glucose and 10 g/L xylose using *Saccharomyces cerevisiae* IPE003. Glucan and xylan conversion in enzymatic hydrolysis was 82.0% and 82.1% in SScF at a high solid loading of 20% (w/w), respectively. As a result, the concentration, yield, and productivity of ethanol reached 60.8 g/L, 75.3%, and 0.63 g/L/h, respectively, by SScF. Therefore, SScF of pretreated corn stover biomass by steam explosion enhanced the bioconversion performance of LCB to bioethanol by increasing sugar utilization and ethanol productivity [69].

Overall, the key factors determining the steam explosion performance and its interdependent and synergistic relationships among up- and downstreams need to be systematically investigated to improve the steam explosion efficiency for making a sustainable biorefinery.

5.5 Integrated Technologies for Promoting Multiproducts in Biorefinery

In biorefinery, the multilevel fractionation and biological conversion processes could be required to improve the profitability of LCB biorefinery [70–72]. Through the biological conversion process, the production of multiproducts would be more economically feasible based on partial or complete utilization of main compositions (cellulose, hemicellulose, and lignin) of LCB. The integrated technologies should be very vital to achieve this goal and thus the success and implementation of biorefineries because economic analysis based on product revenues also suggested that the full component utilization of LCB will increase the net margin of biorefinery [8].

To overcome the inhomogeneity and improve the utilization performance of wheat straw biomass, a novel integrated process, multilevel composition fractionation, and integrated conversion process have been designed and evaluated (Fig. 5.2) [73]. Based on this process, wheat straw biomass was initially pretreated by steam explosion, and the liquid stream was separated to fractionate hemicellulose sugars. The delignification of steam-exploded wheat straw followed by green solvent extraction and gradient acid precipitation was then conducted to separate and purify the lignin fraction. The cellulose fraction in wheat straw was sieved into

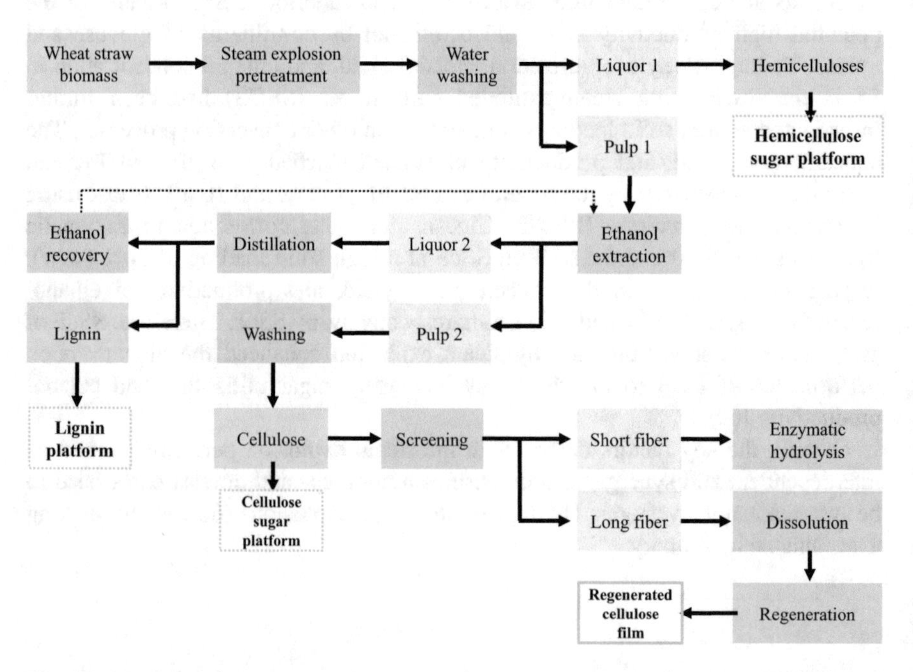

Fig. 5.2 Flow diagrams of the production of multiproducts in a biorefinery by the integrated processes of steam explosion: the multilevel composition fractionation process of the preparation of regenerated cellulose film and bioethanol

long fibers and short fibers. Long fibers were used as a feedstock for the preparation of regenerated cellulose film (RCF), while short fibers were hydrolyzed for the production of ethanol. As a result, by this integrated process, the hemicellulose fractionation yield reached 73% with steam explosion at 1.6 MPa for 5.2 min. The lignin yield reached 90% from the delignification with the extraction conditions of 160 °C, 2 h, and 60% ethanol (v/v). After steam explosion and extraction, the cellulose recovery was 93%. After being screened under 40 mesh, the short fibers were hydrolyzed, and the glucan conversion reached 90% in 9.0 h hydrolysis. Long fibers accounted for 90% of the total cellulose fibers. Regenerated cellulose film was prepared using long fibers by [bmim]Cl, of which tensile strength and breaking elongation were 120 MPa and 4.8%, respectively. The cross section of regenerated cellulose film displayed homogeneous structure, indicating a better mechanical performance. In this multilevel composition fractionation and conversion process, wheat straw biomass could be fractionated into different polymeric fractions with high yield. The platforms of cellulose, sugar, and lignin were established to facilitate the production of high-value multiproducts. Therefore, the integrated process employing steam explosion, ethanol extraction, and ionic liquid technology was an effective strategy to fractionate multi-components for the production of multiproducts [73].

The potential process of coproducing two different biofuels (bioethanol and bio-oil) from wheat straw in a biorefinery concept had also been investigated (Fig. 5.3) [74]. Wheat straw was pretreated in a steam explosion pilot plant with 200 °C for 10 min. To improve the coproduction efficiency of biorefinery, the SSF of the hemicellulose and cellulose fractions was evaluated to maximize the production of bioethanol [74]. Using washed water-insoluble solid fraction from the pretreated wheat straw, higher ethanol productivities at 24 h of SSF were obtained when the inoculum size increased from 1 to 3 g/L. The residual lignin fraction was valorized

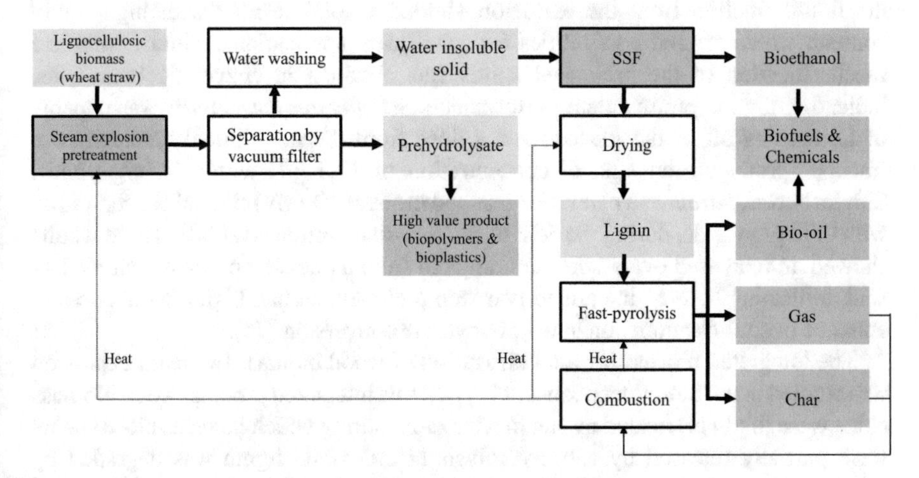

Fig. 5.3 The integration of biological and thermal processing of lignocellulosic biomass for transport fuels and chemicals by employing steam explosion technology in a biorefinery concept

by thermal fast pyrolysis into bio-oil, which can be further converted into other biofuels or biochemicals. By using the residual lignin, the yield of bio-oil was 31.9 wt% (water-free basis), which confirmed that it is mostly composed of oxygenated aromatics generated from the lignin monomers. Furthermore, catalytic fast pyrolysis of the residual lignin was conducted by employing HZSM-5 zeolite to promote decarbonylation and cracking of the primary vapors. By coupling both processes, the production of liquid products significantly enhanced, improving the utilization efficiency of the wheat straw biomass. Compared with a simple process of bioethanol production, the process that developed by intergrading steam explosion, SSF, and lignin valorization led to remarkable improvements in terms of mass and chemical energy yields, which increased 1.9- and 1.7-fold, respectively [74]. Overall, the results confirmed that the coproduction of bioethanol and bio-oil was an advantageous option and approach for the valorization of LCB in a biorefinery concept.

The integration of steam explosion followed by alkaline delignification was explored to treat sugarcane bagasse [75]. In the process, around 82.6% of pentosan and most acetyl group fractions were dissolved in the liquid stream in stream explosion pretreatment, while 90.2% of cellulose and 87.0% lignin were recovered in the solid fraction. After the pretreatment, the delignification solubilized approximately 91% of the lignin and 72.5% of the pentosans in the steam-exploded solids, leading to a pulp with almost 90% of cellulose. The acidification of the liquid stream from delignification allowed recovery of 48.3% of the lignin based on the weight in the feedstock. Around 14% of lignin, 22% of cellulose, and 26% of pentosans were lost during the integrated process [75].

A two-step pretreatment process employing steam explosion and extrusion technologies had been evaluated for the optimal fractionation of LCB [76]. By this integrated two-step pretreatment, overall glucan, hemicellulose, and lignin recovery yields of barley straw reached to 84%, 91%, and 87%, respectively. Precipitation of the liquid fraction from the extrusion yielded a solid residue with high lignin content, which offered possibilities for subsequent applications. Almost complete saccharification of the pretreated solids was obtained in enzymatic hydrolysis, indicating a good pretreatment performance. *Scheffersomyces stipitis* was capable of fermenting all of the glucose and xylose from the non-diluted hemicellulose fraction, producing an ethanol concentration of 17.5 g/L with 0.34 g/g yields. Similarly, *Saccharomyces cerevisiae* resulted in about 4% (v/v) ethanol concentration with 0.40 g/g yields, during the SSF of the two-step pretreated solids. These results showed an increased overall conversion yield from a one-step pretreatment by 1.4-fold, indicating that the integrated two-step pretreatment had high effectiveness to enhance overall fractionation and carbohydrate conversion [76].

The integrated process for total utilization of wood biomass by steam explosion was studied in the biorefinery concept [77]. In this integrated process, wood biomass chips were firstly pretreated by steam explosion, during which hemicellulose sugars were partially released by auto-hydrolysis effect, while lignin was degraded by extensive cleavage of α- and β-aryl ether linkages. The three main components of steam-exploded woods (cellulose, hemicellulose, and lignin) were then fractionated by successive extraction with water and 90% dioxane. The water extractives were decolored and purified by chromatography on synthetic adsorbents and ion-exchange resins to yield a mixture of xylooligosaccharides and xylose, which can be used as a sweetener or food additive. The lignin fraction obtained from extraction can be used to prepare thermoplastic materials, lignin-pitch, by phenolysis followed by heat treatment. The lignin-pitch was well spun into thermoplastic in the temperature range 150–190 °C at a speed of 500–1000 m/min using the melt-spinning method. The filaments were carbonized by heating from room temperature up to 1000 °C in a stream of nitrogen. The carbon fiber was fabricated with a more than 40% yield. The properties of the carbon fiber were equivalent to a commercial one made from petroleum pitch. The lignin fraction separated from the dioxane extraction can also be used to produce adhesives. The cellulose after extraction was enzymatically hydrolyzed with a cellulase preparation, and the hydrolyzate can be fermented to single-cell protein and/or ethanol. The mass balance was conducted by assuming a plant processing 30,000 t of birch wood biomass chips per year (100 t per day) based on dry weight. As a result, the annual production of the xylooligosaccharides, carbon fiber, and ethanol can reach 4500 t, 1542 t, and 6658 kl, respectively [77].

5.6 Pilot-Scale Platforms of Biorefinery with Steam Explosion Technology

The bioconversion of LCB to ethanol via steam explosion and enzymatic hydrolysis had been studied, and the economy of a biorefinery had been evaluated in pilot-scale and demonstration-scale [8, 43, 78–83]. Industrial steam explosion was applied to pretreat sugarcane straw for the production of ethanol in a pilot-scale [84]. Steam explosion at 180, 190, and 200 °C for 15 min was carried out for pretreating sugarcane straw in an industrial sugar/ethanol reactor (2.5 m^3) with a mass rate of 53 kg/m^3 [84]. The stream explosion led to the remarkable solubilization of hemicellulose, with a maximum value of 92.7% under the condition of 200 °C. The pretreated straw was delignificated by sodium hydroxide, leading to lignin solubilization of 86.7% at 180 °C and 81.3% at 200 °C. Steam explosion improved the enzymatic hydrolysis of cellulose in pretreated solid, with a maximum conversion of 80.0% achieved at 200 °C. The delignification process increased the enzymatic hydrolysis of cellulose from 58.8% to 85.1% of the biomass pretreated at 180 °C. Interestingly, for the biomass pretreated at 190 °C, the increase of sugar conversion in enzymatic hydrolysis was less remarkable, while that at 200 °C the sugar conversion decreased after the delignification, which could be attributed by rearrangements in cellulose structure or cellulignin recovery by pseudo-lignin [84]. Besides these, the use of grinded straw made viable the large-scale pretreatment process [84].

A demonstration-scale project with annual straw ethanol outputs of 3000 t had been reported to be successfully established [85]. Besides steam explosion, this demonstration-scale project integrated several other technologies, including gas double dynamic solid-state fermentation, solid-state hydrolysis coupled with ethanol fermentation and gas stripping, and organic fertilizer production. The major equipment employed in this demonstration project includes a 5 m^3 steam explosion tank, two 100 m^3 fermenters, a 110 m^3 reactor for solid-state simultaneous hydrolysis, fermentation and ethanol separation, and four ethanol distillation towers. In this process, straw biomass was steam-exploded, and the liquid steam was then obtained to produce the hemicellulose sugars. After the discoloration and ion exchange process, xylooligosaccharide was produced. A small part of steam-exploded straw was used for the production of cellulase for enzymatic hydrolysis. The hydrolysate from steam-exploded straw was utilized in fermentation to produce ethanol, while the fermentation residues were used for the production of organic fertilizers. The process operating results showed that the ethanol yield was more than 0.15 g/g dry straw, and the final ethanol concentration desorbed from activated carbon was higher than 69.8%. The overall production cost was around 5900 RMB/ton ethanol with an annual production of 3000 t bioethanol and coproduction of 200 t xylooligosaccharides and 3000 t organic fertilizers [85].

An industrial system for bioethanol production from corn stover biomass had been reported by integrating feedstock handling, steam explosion, and simultaneous saccharification and co-fermentation [86]. The major equipment of this industrial system includes six 50 m^3 steam explosion reactors, sixteen 400 m^3 fermenters, one ethanol distillation tower, and six tanks of natural gas fermentation. Upon this industrial system, the feedstock cost employed the handling strategy was reduced by 19.4% compared with traditional conversion process. The enzyme loading used in enzymatic hydrolysis was reduced by 25.0% using synergistic cellulase systems as compared to that with the addition of cellulase only. The pre-hydrolysis and SScF without detoxification operation had been developed using inhibitor-tolerant *Saccharomyces cerevisiae*. More than 4% (w/w) ethanol concentration was produced, which corresponded to 72.3% of the ethanol theoretical yield. The improvement in the conversion efficiency of pentose resulted in 41% increase of ethanol yield. Lignin plastic composite material and compress natural gas were coproduced in this industrial system, which proportioned the capital cost of process and hence facilitated the economic feasibility of process. As a result, the corresponding ethanol total cost was reduced to 5571.6 RMB/t. In this industrial system, the low cost of feedstock, integrated conversion process, and the coproduction of multiproducts contributed to the profitability of biorefinery. Overall, steam explosion represents a prospective pretreatment technology for the multiproducts from LCB using integrated technologies in a biorefinery (Fig. 5.4).

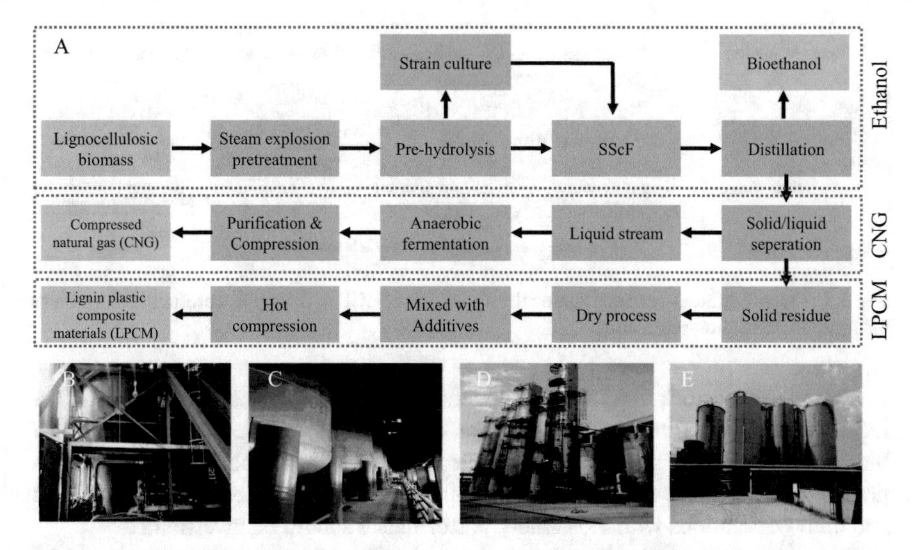

Fig. 5.4 An industrial system for the production of bioethanol and multiproducts from corn stover biomass established by integrating feedstock handling, steam explosion, and simultaneous saccharification and co-fermentation (SScF) (**a**) and the major equipment of the industrial system: (**b**) 50 m^3 tank of steam explosion; (**c**) 400 m^3 fermenters of SScF; (**d**) distillation tower of ethanol; (**e**) methane fermenters. CNG represents compressed natural gas; LPCM represents lignin plastic composite material

5.7 Direction of Future Work of Biorefinery with Stream Explosion

Even if steam explosion shows the potential to enhance the bioconversion of LCB and improve the profitability of biorefinery, several issues still exist in processing of LCB, including the insufficient understanding of the intrinsic characteristics of LCB, unsatisfactory deconstruction and fractionation performance, and the high cost for the technology of product production in biorefinery. To solve these problems, the development and design of the integrated systems for biorefinery will be crucial for coproducing multiproducts, including chemicals, materials, fuels, and energy, to make a sustainable biorefinery. Based on a biorefinery concept, overall strategy employing steam explosion as a core technology has been proposed to improve the profitability of biorefinery from a totally new perspective (Fig. 5.5).

Multiproduct biorefinery needs to be developed by taking the advanced and integrated process and technology based on the total understanding of intrinsic properties of LCB. Through facing to the aspects of feedstock, process, and product, an environmentally friendly and ecological efficient biorefinery strategy would be developed. Therefore, in the general principles of multiproduct biorefinery from LCB, these aspects should be focused on and investigated to make a profitable biorefinery.

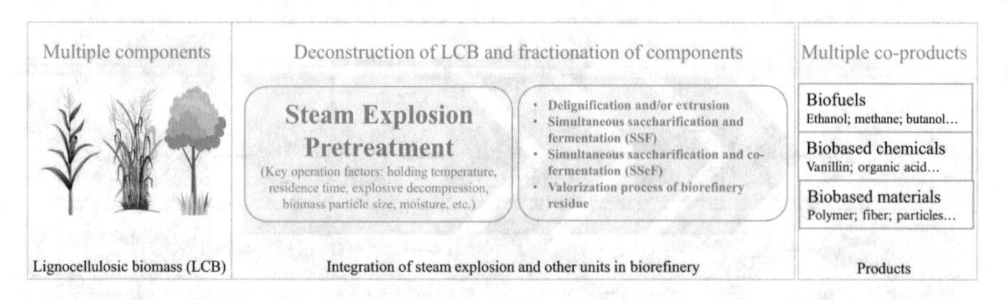

Fig. 5.5 Overall strategy for maximizing the utilization of LCB and making a profitable biorefinery using steam explosion pretreatment

Facing to feedstock, the scientific understanding of the intrinsic characteristics of LCB needs to be further evaluated. For example, the molecular mechanism of biomass recalcitrance and its impact on the pretreatment and subsequent conversion process should be revealed. The multi-compositions and macromolecules of LCB and their correlations with biorefinery performance should be investigated.

Facing to process, the integration principles for biorefinery system should be proposed to guide the development of process and the improvement of biorefinery performance. The goals of the integrated system are to fractionate LCB into cellulose, hemicellulose, lignin, and other intermediates effectively. This integrated system should first retain the original features of macromolecules for atom economy. Second, it could activate the conversion performance of components which should be suitable for enzymatic hydrolysis or subsequent conversion. Third, it could maximize the value of intermediate products as far as possible. Therefore, based on these principles, the development of thermal, chemical, and physical processes of LCB should be further carried out. The principle of directional multilevel conversion and selective structural fractionation should be feasible and conform to the multiproduct biorefinery concept.

Facing to products, the coproduction of multiproducts such as biofuels, materials, chemicals, and energy by integrated systems should be necessary to make a profitable biorefinery. Therefore, with the integrated systems facing to feedstock, process, and product, the maximum functionality of LCB could be achieved in a biorefinery.

References

1. Agbor, V. B., Cicek, N., Sparling, R., Berlin, A., & Levin, D. B. (2011). Biomass pretreatment: Fundamentals toward application. *Biotechnology Advances, 29*, 675–685.
2. Himmel, M. E., et al. (2007). Biomass recalcitrance: Engineering plants and enzymes for biofuels production. *Science, 315*, 804–807.
3. Liu, Z. H., et al. (2019). Identifying and creating pathways to improve biological lignin valorization. *Renewable and Sustainable Energy Reviews, 105*, 349–362.
4. Rinaldi, R., et al. (2016). Paving the way for lignin valorisation: Recent advances in bioengineering, biorefining and catalysis. *Angewandte Chemie. International Edition, 55*, 8164–8215.

5. Ding, S. Y., et al. (2012). How does plant cell wall nanoscale architecture correlate with enzymatic digestibility? *Science, 338*, 1055–1060.
6. Barros-Rios, J., Romani, A., Garrote, G., & Ordas, B. (2015). Biomass, sugar, and bioethanol potential of sweet corn. *GCB Bioenergy, 7*, 153–160.
7. Liu, Z. H., et al. (2019). Defining lignin nanoparticle properties through tailored lignin reactivity by sequential organosolv fragmentation approach (SOFA). *Green Chemistry, 21*, 245–260.
8. Zhang, Y. H. P. (2008). Reviving the carbohydrate economy via multi-product lignocellulose biorefineries. *Journal of Industrial Microbiology & Biotechnology, 35*, 367–375.
9. Liu, Z. H., et al. (2019). Codesign of Combinatorial Organosolv Pretreatment (COP) and Lignin Nanoparticles (LNPs) in biorefineries. *ACS Sustainable Chemistry & Engineering, 7*, 2634–2647.
10. Reddy, N., & Yang, Y. (2005). Biofibers from agricultural byproducts for industrial applications. *Trends in Biotechnology, 23*, 22–27.
11. Ozdenkci, K., et al. (2017). A novel biorefinery integration concept for lignocellulosic biomass. *Energy Conversion and Management, 149*, 974–987.
12. Liu, Z. H., et al. (2017). Synergistic maximization of the carbohydrate output and lignin processability by combinatorial pretreatment. *Green Chemistry, 19*, 4939–4955.
13. Liu, Z. H., Xie, S. X., Lin, F. R., Jin, M. J., & Yuan, J. S. (2018). Combinatorial pretreatment and fermentation optimization enabled a record yield on lignin bioconversion. *Biotechnology for Biofuels, 11(1)*, 1–20.
14. Cherian, B. M., et al. (2010). Isolation of nanocellulose from pineapple leaf fibres by steam explosion. *Carbohydrate polymers, 81*, 720–725.
15. Yang, H. B., et al. (2019). Overcoming cellulose recalcitrance in woody biomass for the lignin-first biorefinery. *Biotechnology for Biofuels, 12*, 1–18.
16. Chen, H. Z., & Liu, Z. H. (2015). Steam explosion and its combinatorial pretreatment refining technology of plant biomass to bio-based products. *Biotechnology Journal, 10*, 866–885.
17. Kim, K. H., et al. (2019). Integration of renewable deep eutectic solvents with engineered biomass to achieve a closed-loop biorefinery. *Proceedings of the National Academy of Sciences of the United States of America, 116*, 13816–13824.
18. Liu, Z. H., et al. (2019). Cooperative valorization of lignin and residual sugar to polyhydroxyalkanoate (PHA) for enhanced yield and carbon utilization in biorefineries. *Sustainable Energy & Fuels, 3*, 2024–2037.
19. Liu, Z. H., et al. (2013). Effects of biomass particle size on steam explosion pretreatment performance for improving the enzyme digestibility of corn stover. *Industrial Crops and Products, 44*, 176–184.
20. Krishnan, C., et al. (2010). Alkali-based AFEX pretreatment for the conversion of sugarcane bagasse and cane leaf residues to ethanol. *Biotechnology and Bioengineering, 107*, 441–450.
21. Ewanick, S. M., Bura, R., & Saddler, J. N. (2007). Acid-catalyzed steam pretreatment of lodgepole pine and subsequent enzymatic hydrolysis and fermentation to ethanol. *Biotechnology and Bioengineering, 98*, 737–746.
22. Jorgensen, H., Kristensen, J. B., & Felby, C. (2007). Enzymatic conversion of lignocellulose into fermentable sugars: Challenges and opportunities. *Biofuels Bioproducts and Biorefining, 1*, 119–134.
23. Carvalheiro, F., Duarte, L. C., & Girio, F. M. (2008). Hemicellulose biorefineries: A review on biomass pretreatments. *Journal of Scientific and Industrial Research India, 67*, 849–864.
24. Sorek, N., Yeats, T. H., Szemenyei, H., Youngs, H., & Somerville, C. R. (2014). The implications of lignocellulosic biomass chemical composition for the production of advanced biofuels. *Bioscience, 64*, 192–201.
25. dos Santos, A. C., Ximenes, E., Kim, Y., & Ladisch, M. R. (2019). Lignin-enzyme interactions in the hydrolysis of lignocellulosic biomass. *Trends in Biotechnology, 37*, 518–531.
26. Balat, M. (2011). Production of bioethanol from lignocellulosic materials via the biochemical pathway: A review. *Energy Conversion and Management, 52*, 858–875.

27. Jonsson, L. J., & Martin, C. (2016). Pretreatment of lignocellulose: Formation of inhibitory by-products and strategies for minimizing their effects. *Bioresource Technology, 199*, 103–112.
28. Zhao, X. B., Li, S. M., Wu, R. C., & Liu, D. H. (2017). Organosolv fractionating pre-treatment of lignocellulosic biomass for efficient enzymatic saccharification: Chemistry, kinetics, and substrate structures. *Biofuels, Bioproducts and Biorefining, 11*, 567–590.
29. Zhao, X. B., Zhang, L. H., & Liu, D. H. (2012). Biomass recalcitrance. Part I: The chemical compositions and physical structures affecting the enzymatic hydrolysis of lignocellulose. *Biofuels, Bioproducts and Biorefining, 6*, 465–482.
30. Chen, H. Z., Liu, Z. H., & Dai, S. H. (2014). A novel solid state fermentation coupled with gas stripping enhancing the sweet sorghum stalk conversion performance for bioethanol. *Biotechnology for Biofuels, 7*, 1–13.
31. Di Risio, S., Hu, C. S., Saville, B. A., Liao, D., & Lortie, J. (2011). Large-scale, high-solids enzymatic hydrolysis of steam-exploded poplar. *Biofuels, Bioproducts and Biorefining, 5*, 609–620.
32. Liu, Z. H., & Chen, H. Z. (2016). Mechanical property of different corn stover morphological fractions and its correlations with high solids enzymatic hydrolysis by periodic peristalsis. *Bioresource Technology, 214*, 292–302.
33. Melati, R. B., et al. (2019). Key factors affecting the recalcitrance and conversion process of biomass. *Bioenergy Research, 12*, 1–20.
34. Sindhu, R., Binod, P., & Pandey, A. (2016). Biological pretreatment of lignocellulosic biomass – An overview. *Bioresource Technology, 199*, 76–82.
35. Vignon, M. R., GarciaJaldon, C., & Dupeyre, D. (1995). Steam explosion of woody hemp chenevotte. *International Journal of Biological Macromolecules, 17*, 395–404.
36. He, Q., et al. (2020). Lignin-first integrated steam explosion process for green wood adhesive application. *ACS Sustainable Chemistry & Engineering, 8*, 5380–5392.
37. Liu, Z. H., Qin, L., Li, B. Z., & Yuan, Y. J. (2015). Physical and chemical characterizations of corn stover from leading pretreatment methods and effects on enzymatic hydrolysis. *ACS Sustainable Chemistry & Engineering, 3*, 140–146.
38. Rocha, G. J. M., Goncalves, A. R., Nakanishi, S. C., Nascimento, V. M., & Silva, V. F. N. (2015). Pilot scale steam explosion and diluted sulfuric acid pretreatments: Comparative study aiming the sugarcane bagasse saccharification. *Industrial Crops and Products, 74*, 810–816.
39. Liu, Z. H., & Chen, H. Z. (2016). Biomass-water interaction and its correlations with enzymatic hydrolysis of steam-exploded corn stover. *ACS Sustainable Chemistry & Engineering, 4*, 1274–1285.
40. Auxenfans, T., Cronier, D., Chabbert, B., & Paes, G. (2017). Understanding the structural and chemical changes of plant biomass following steam explosion pretreatment. *Biotechnology for Biofuels, 10*(36), 1-16
41. Liu, Z. H., & Chen, H. Z. (2016). Periodic peristalsis enhancing the high solids enzymatic hydrolysis performance of steam exploded corn stover biomass. *Biomass and Bioenergy, 93*, 13–24.
42. Semwal, S., et al. (2019). Process optimization and mass balance studies of pilot scale steam explosion pretreatment of rice straw for higher sugar release. *Biomass and Bioenergy, 130(105390)*, 1–9.
43. Bhatia, R., et al. (2020). Pilot-scale production of xylo-oligosaccharides and fermentable sugars from Miscanthus using steam explosion pretreatment. *Bioresource Technology, 296(122285)*, 1–9.
44. Datar, R., et al. (2007). Hydrogen production from the fermentation of corn stover biomass pretreated with a steam-explosion process. *International Journal of Hydrogen Energy, 32*, 932–939.
45. Zhang, X., Yuan, Q. P., & Cheng, G. (2017). Deconstruction of corncob by steam explosion pretreatment: Correlations between sugar conversion and recalcitrant structures. *Carbohydrate Polymer, 156*, 351–356.

46. Deepa, B., et al. (2011). Structure, morphology and thermal characteristics of banana nano fibers obtained by steam explosion. *Bioresource Technology, 102*, 1988–1997.
47. Li, J. B., Henriksson, G., & Gellerstedt, G. (2007). Lignin depolymerization/repolymerization and its critical role for delignification of aspen wood by steam explosion. *Bioresource Technology, 98*, 3061–3068.
48. Sannigrahi, P., Kim, D. H., Jung, S., & Ragauskas, A. (2011). Pseudo-lignin and pretreatment chemistry. *Energy & Environmental Science, 4*, 1306–1310.
49. Pandey, M. P., & Kim, C. S. (2011). Lignin depolymerization and conversion: A review of thermochemical methods. *Chemical Engineering and Technology, 34*, 29–41.
50. Overend, R. P., & Chornet, E. (1987). Fractionation of lignocellulosics by steam-aqueous pretreatments. *Philos T R Soc A, 321*, 523–536.
51. Pedersen, M., & Meyer, A. S. (2010). Lignocellulose pretreatment severity – Relating pH to biomatrix opening. *New Biotechnology, 27*, 739–750.
52. Alvira, P., Negro María, J., Ballesteros, I., González, A., & Ballesteros, M. (2016). Bioethanol (Vol. 2).
53. Liu, Z. H., & Chen, H. Z. (2017). Two-step size reduction and post-washing of steam exploded corn stover improving simultaneous saccharification and fermentation for ethanol production. *Bioresource Technology, 223*, 47–58.
54. Lopez-Linares, J. C., et al. (2015). Optimization of uncatalyzed steam explosion pretreatment of rapeseed straw for biofuel production. *Bioresource Technology, 190*, 97–105.
55. Liu, Z. H., & Chen, H. Z. (2015). Xylose production from corn stover biomass by steam explosion combined with enzymatic digestibility. *Bioresource Technology, 193*, 345–356.
56. Pielhop, T., Amgarten, J., von Rohr, P. R., & Studer, M. H. (2016). Steam explosion pretreatment of softwood: The effect of the explosive decompression on enzymatic digestibility. *Biotechnology for Biofuels, 9*(152), 1–13.
57. Baral, N. R., & Shah, A. (2017). Comparative techno-economic analysis of steam explosion, dilute sulfuric acid, ammonia fiber explosion and biological pretreatments of corn stover. *Bioresource Technology, 232*, 331–343.
58. Vidal, B. C., Dien, B. S., Ting, K. C., & Singh, V. (2011). Influence of feedstock particle size on lignocellulose conversion – A review. *Applied Biochemistry and Biotechnology, 164*, 1405–1421.
59. Negro, M. J., Manzanares, P., Oliva, J. M., Ballesteros, I., & Ballesteros, M. (2003). Changes in various physical/chemical parameters of Pinus pinaster wood after steam explosion pretreatment. *Biomass & Bioenergy, 25*, 301–308.
60. Hu, F., Jung, S., & Ragauskas, A. (2012). Pseudo-lignin formation and its impact on enzymatic hydrolysis. *Bioresource Technology, 117*, 7–12.
61. Shinde, S. D., Meng, X. Z., Kumar, R., & Ragauskas, A. J. (2018). Recent advances in understanding the pseudo-lignin formation in a lignocellulosic biorefinery. *Green Chemistry, 20*, 2192–2205.
62. DeMartini, J. D., et al. (2015). How chip size impacts steam pretreatment effectiveness for biological conversion of poplar wood into fermentable sugars. *Biotechnology for Biofuels, 8(209)*, 1–16.
63. Ferreira, L. C., Nilsen, P. J., Fdz-Polanco, F., & Perez-Elvira, S. I. (2014). Biomethane potential of wheat straw: Influence of particle size, water impregnation and thermal hydrolysis. *Chemical Engineering Journal, 242*, 254–259.
64. Sahoo, K., & Mani, S. (2017). Techno-economic assessment of biomass bales storage systems for a large-scale biorefinery. *Biofuels Bioproducts and Biorefining, 11*, 417–429.
65. Budzianowski, W. M., & Postawa, K. (2016). Total chain integration of sustainable biorefinery systems. *Applied Energy, 184*, 1432–1446.
66. Audsley, E., & Annetts, J. E. (2003). Modelling the value of a rural biorefinery - Part I: The model description. *Agricultural Systems, 76*, 39–59.

67. Kataria, R., Mol, A., Schulten, E., Happel, A., & Mussatto, S. I. (2017). Bench scale steam explosion pretreatment of acid impregnated elephant grass biomass and its impacts on biomass composition, structure and hydrolysis. *Industrial Crops and Products, 106*, 48–58.
68. Liu, Z. H., et al. (2013). Evaluation of storage methods for the conversion of corn stover biomass to sugars based on steam explosion pretreatment. *Bioresource Technology, 132*, 5–15.
69. Liu, Z. H., & Chen, H. Z. (2016). Simultaneous saccharification and co-fermentation for improving the xylose utilization of steam exploded corn stover at high solid loading. *Bioresource Technology, 201*, 15–26.
70. Yang, B., & Wyman, C. E. (2008). Pretreatment: The key to unlocking low-cost cellulosic ethanol. *Biofuels, Bioproducts and Biorefining, 2*, 26–40.
71. Mokomele, T., et al. (2018). Ethanol production potential from AFEX (TM) and steam-exploded sugarcane residues for sugarcane biorefineries. *Biotechnology for Biofuels, 11*(127), 1–21.
72. Liu, Z. H., Qin, L., Zhu, J. Q., Li, B. Z., & Yuan, Y. J. (2014). Simultaneous saccharification and fermentation of steam-exploded corn stover at high glucan loading and high temperature. *Biotechnology for Biofuels, 7(167)*, 1–16.
73. Chen, H. Z., & Liu, Z. H. (2014). Multilevel composition fractionation process for high-value utilization of wheat straw cellulose. *Biotechnology for Biofuels, 7(137)*, 1–12.
74. Tomas-Pejo, E., et al. (2017). Valorization of steam-exploded wheat straw through a biorefinery approach: Bioethanol and bio-oil co-production. *Fuel, 199*, 403–412.
75. Rocha, G. J. M., Martin, C., da Silva, V. F. N., Gomez, E. O., & Goncalves, A. R. (2012). Mass balance of pilot-scale pretreatment of sugarcane bagasse by steam explosion followed by alkaline delignification. *Bioresource Technology, 111*, 447–452.
76. Oliva, J.M. et al. (2017). A sequential steam explosion and reactive extrusion pretreatment for lignocellulosic biomass conversion within a fermentation-based biorefinery perspective. *Fermentation, 3(15)*, 1–15.
77. Shimizu, K., et al. (1998). Integrated process for total utilization of wood components by steam-explosion pretreatment. *Biomass and Bioenergy, 14*, 195–203.
78. Kheshgi, H. S., Prince, R. C., & Marland, G. (2000). The potential of biomass fuels in the context of global climate change: Focus on transportation fuels. *Annual Review of Energy and the Environment, 25*, 199–244.
79. Pielhop, T., Amgarten, J., Studer, M. H., & von Rohr, P. R. (2017). Pilot-scale steam explosion pretreatment with 2-naphthol to overcome high softwood recalcitrance. *Biotechnology for Biofuels, 10(130)*, 1–13.
80. Pal, S., et al. (2017). Pilot-scale pretreatments of sugarcane bagasse with steam explosion and mineral acid, organic acid, and mixed acids: Synergies, enzymatic hydrolysis efficiencies, and structure-morphology correlations. *Biomass Conversion and Biorefinery, 7*, 179–189.
81. De Bari, I., et al. (2002). Ethanol production at flask and pilot scale from concentrated slurries of steam-exploded aspen. *Industrial and Engineering Chemistry Research, 41*, 1745–1753.
82. Liu, Z. H., & Chen, H. Z. (2016). Periodic peristalsis releasing constrained water in high solids enzymatic hydrolysis of steam exploded corn stover. *Bioresource Technology, 205*, 142–152.
83. Bonfiglio, F., et al. (2019). Pretreatment of switchgrass by steam explosion in a semi-continuous pre-pilot reactor. *Biomass & Bioenergy, 121*, 41–47.
84. Oliveira, F. M. V., et al. (2013). Industrial-scale steam explosion pretreatment of sugarcane straw for enzymatic hydrolysis of cellulose for production of second generation ethanol and value-added products. *Bioresource Technology, 130*, 168–173.
85. Chen, H. Z., & Qiu, W. H. (2010). Key technologies for bioethanol production from lignocellulose. *Biotechnology Advances, 28*, 556–562.
86. Chen, H. Z., & Fu, X. G. (2016). Industrial technologies for bioethanol production from lignocellulosic biomass. *Renewable and Sustainable Energy Reviews, 57*, 468–478.

Chapter 6
Fundamentals of Lignin-Carbohydrate Complexes and Its Effect on Biomass Utilization

Usama Shakeel, Saif Ur Rehman Muhammad, Yong Zhao, Hongqiang Li, Xia Xu, Yong Sun, and Jian Xu

6.1 Introduction

Due to excessive usage of fossil fuel by ever-increasing population, fossil fuel reserves are depleting continuously and causing environment challenges in our society. Therefore, researchers are trying to find some sustainable carbon-neutral alternatives to fossil fuel with low environmental threats [57]. Biomass has shown significant potential and considered as an attractive candidate to replace the fossil fuel as it is composed of nature's most abundant biopolymers, cellulose, hemicellulose (carbohydrates), and lignin [49, 76]. These non-edible biopolymers have great potential toward biorefinery, paper and pharmaceutical industries, ethanol production, and various valuable chemicals [24, 54, 68, 69, 91, 110]. Despite its unmatchable potential in energy industry, efficient fractionation of biomass into its constitutional components limits the direct use of biomass into industry [5, 40]. Lignin and carbohydrates are chemically linked, and they are locked into biomass hierarchical complexity and form so-called lignin-carbohydrate complex (LCC) [7, 18, 23]. It is suggested that this complex plays a very crucial role in the recalcitrance of biomass during the fractionation process as it is reported that all and 47–66% lignin moieties are chemically bounded to carbohydrates in the softwood and

U. Shakeel · S. U. R. Muhammad · Y. Zhao · X. Xu · J. Xu (✉)
Biochemical Engineering Research Center, Anhui University of Technology, Anhui, China

School of Chemistry and Chemical Engineering, Anhui University of Technology, Anhui, China

H. Li
Beijing Jiansheng Pharmaceuticals, Ltd, Beijing, China

Y. Sun
School of Engineering, Edith Cowan University, Joondalup, WA, Australia

Z.-H. Liu, A. Ragauskas (eds.), *Emerging Technologies for Biorefineries, Biofuels, and Value-Added Commodities*,
https://doi.org/10.1007/978-3-030-65584-6_6

hardwood, respectively [31, 32, 48, 51]. Lignin linked to carbohydrates through five different types of bondings such as benzyl ether (BE), benzyl ester, phenyl glycosidic (PG), ferulate/coumarate esters (FE/CE), and hemiacetal/acetal linkages [10, 38, 39, 41, 43, 45, 55, 64, 70, 77, 93, 101, 102]. Benzyl ester bond present between the uronic acid of carbohydrates and hydroxyl group of lignin and its concentration varies as species, while benzyl ether and phenyl glycosidic bond links glycosyl of carbohydrates with phenolic or hydroxyl group of lignin and prevalent in softwood and hardwood LCCs, respectively [3, 27, 42, 100, 108]. Acetal bond combines the moieties of carbonyl group to the hydroxyl group of carbohydrates and ferulate or deferulate bonding majorly present in grass or non-woody plants [28, 109]. Benzyl ether, benzyl ester, and PG bonding are abundantly present in biomass and benzyl ether, and benzyl ether with hydroxyl group can easily be cleaved under acidic condition while both are alkaline stable and alkali labile, respectively [3, 6, 12, 88]. Similarly, it is reported that benzyl ester linkages are alkaline sensitive, while phenyl glycoside bond is alkaline stable [12, 42, 86]. Hardwood contains high amount of phenyl glycoside bond and showed lowest yield of Kraft pulping; another study showed only 4% PhyGlc hydrolyzed under neutral condition while 96% cleaved in acidic condition, which can elucidate alkaline-stable nature of PG [42, 88]. Ferulate ester is highly alkali sensitive and can be easily hydrolyzed under alkaline condition at ambient temperature [42]. Although much information has been revealed until now, the detailed insight of hierarchically complex structure of biomass regarding its recalcitrance is essential for biorefinery concept to address current environmental issues. This review will cover the basic essential aspects of biomass such as LCC extractions, presence of different bonds among different parts of biomass, direct and indirect analysis of LCCs, and future outlook.

6.2 Extraction of LCCs

To unveil the biomass structure, recalcitrance, and the challenges associated with its direct use of biomass into biorefinery, extraction of LCC is crucially important and unavoidable. For a better understanding of biomass recalcitrance, extensively detailed investigation of LCCs demands the extraction of LCCs by keeping its native structure intact. Although the milder LCCs fractionation protocols are preferable to minimize the structural alteration in native biomass, the adoption of harsh conditions under the compulsion of recalcitrance is unavoidable. Many researchers stated the existence of LCC for altered materials, and some of them tried different protocols for the extraction or fractionation of LCCs. This section will briefly cover the various LCCs isolation protocols.

6.2.1 Ball Milling Prior to Solvent Extraction

LCB is very tough and compact causing the structural components difficult to be dissolved in general neutral solvent. The existing LCC extraction methods generally include a milling step, in particular, a vibratory milling process. Bjorkman [8] initially used a vibrating ball mill to crush the raw material and investigated the effect of dispersants, temperature, and milling time on the distribution and extraction rate of LCCs. A standardized pretreatment procedure was finally proposed: raw material pre-sieved through a 20-mesh sieve was extracted with organic solvent to remove its nonstructural constituents. The mixture was pulverized twice with a Lampen mill for 48 h followed by a vibratory ball milling for 48 h. Toluene was used as the dispersing solvent medium in the crushing process.

Koshijima et al. [44] used an improved method: 200–230 g wood flour (40–60 mesh) was ball milled on a vibrating ball mill for 120 h. The vibration amplitude was set to 9 mm throughout the experiment. Nitrogen was reinjected to replace the gas in the tank every 2 days. This method was used to ensure that the material temperature did not exceed 30 °C during the milling process.

Ever since, the researchers basically have used a similar approach to process LCB with minor changes. For example, Yelle et al. [105] first dried the sample and then ball milled using zirconia balls. The parameters of ball milling operation were set with eight balls (10 mm) plus three balls (20 mm), 300 rpm, 20 min followed by a 10 min pause. The 20 mm balls were then replaced with two 10 mm balls, and the speed increased to 600 rpm. The milling was continued for another 12 h (20 min running plus 10 min pause).

6.2.2 Water or Water Solution Extraction

Traynard et al. [90] developed a method to extract water-soluble LCCs from plant materials at 140 °C with the yield of 16–18%. The composition and structural characteristics showed that water-soluble LCCs were rich in carbohydrates. Elevated extraction temperature probably caused LC bond cleavage resulting in low LCC yield. However, the LCCs isolated with this method were partially representative of the original LCCs.

In order to improve the purity of LCCs and to extend its applicability, Watanabe et al. [94] developed a method for extracting water-soluble LCC from Korean pine (*Pinus koraiensis*) using cold and hot water (20 °C, 80 °C). This method was a simple separation procedure for grasses with higher content of acetylated glucomannan. The LCCs obtained were more representative of original LCCs than those prepared by Traynard et al. [90]. However, this method is not applicable to the separation of the water-soluble LCCs in the hardwood with low content of acetylated glucomannan.

Lawoko et al. [50] firstly proposed a quantitative LCC separation method with three successive steps: ball milling, endoglucan hydrolysis, and alkali dissolution. All lignins in softwood can be recovered in the format of LCCs. However, this method includes several steps and is only applicable to softwood.

6.2.3 Organic Solvent Extraction

In order to find out the original LCC fraction, Bjorkman [9] used organic solvent to separate and extract lignin and LCCs from spruce. In this method, the residue obtained from the milled wood lignin (MWL) preparation process was further extracted with dimethyl sulfoxide (DMSO) or dimethylformamide (DMF) to obtain a crude LCC fraction. It was then eluted with 50% acetic acid and 1,2-dichloroethane/ethanol solution (volume ratio of 2:1) to obtain original LCC fraction, known as Björkman LCCs. The results showed that this method was suitable for the separation of needle-hardwood and non-wood raw material LCCs. LCC yield and relative molecular mass of LCCs extracted by DMSO (Mn > 15,000) were higher than that of LCCs extracted by DMF (Mn < 4000–5000) under the same condition. The composition of LCCs and ^{13}C-NMR structure analysis showed that Björkman LCCs can represent the original LCCs in raw material. Although the purity of LCCs prepared was high, the yield was relatively low.

Koshijima et al. [44] improved the extraction conditions of Björkman method. The yield of LCCs increased with an increase in milling time. However, the polymer chains in LCCs were broken down at milling time > 48 h. It was found that DMF could only dissolve low molecular weight components (4000–5000) compared with DMSO (10,000–15,000).

Balakshin et al. [5] isolated and purified LCCs from crude mulberry lignin (MWL) employing ethanol extraction during the separation and purification of Korean pine mulberry lignin (MWL). This method comprises the following steps (Fig. 6.1): (1) dissolving crude MWL fraction prepared with Björkman method in 90% ethanol solution, (2) precipitating MWL by adding water, (3) centrifuging and removing ethanol, and (4) evaporating and vacuum-drying to obtain LCC fraction. This method yielded LCCs only about 3%. Two-dimensional NMR analysis of LCCs showed that lignin and carbohydrate contents in LCCs were almost equal and rich in arabinose and galactose. Ester bond between uronic acid and the C-position of the side chain of lignin connection was observed.

Li et al. [53] proposed a simpler method for quantitative LCC isolation based on ball milling and dissolution in DMSO and TBAH. The milling was carried out for 12 h and lignin structure was well retained. Biomass whether natural or treated was first completely dissolved in the solvent so that all LCCs were guaranteed to be in

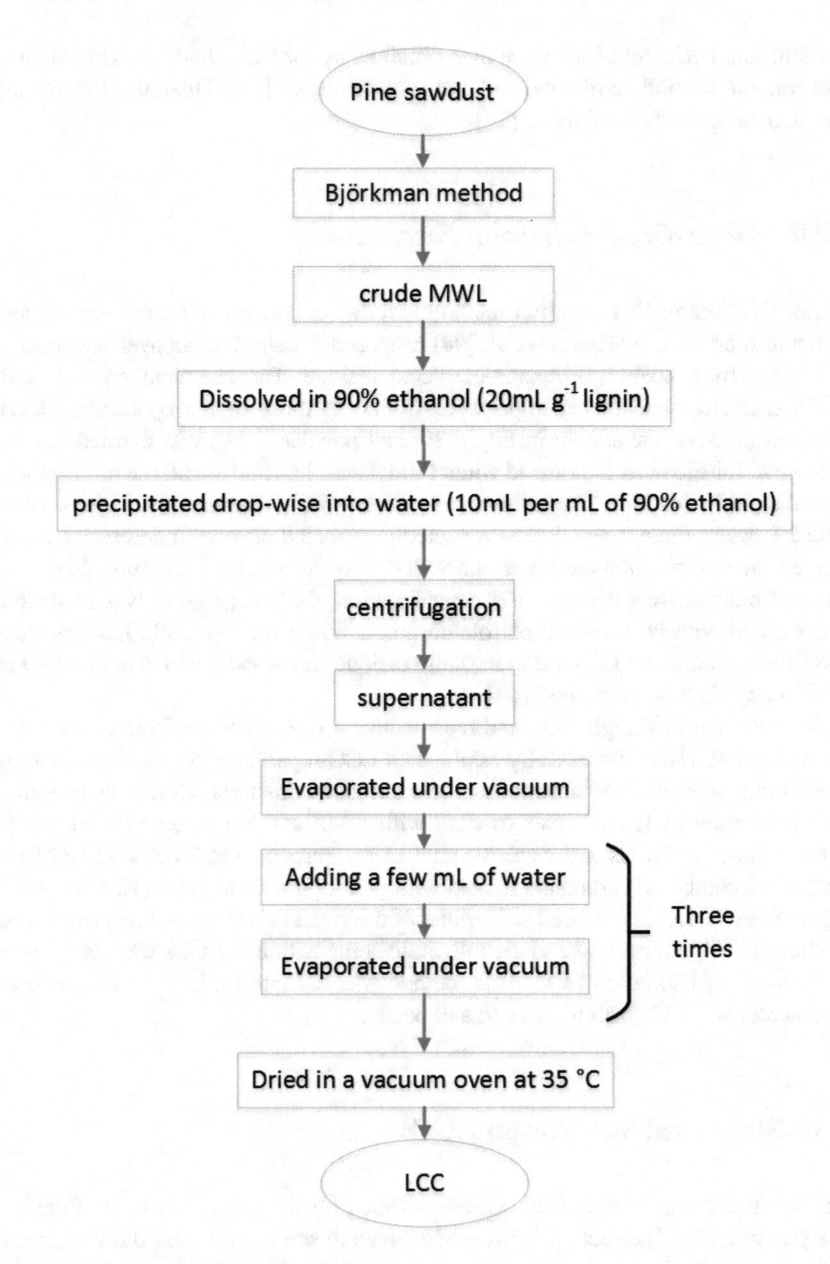

Fig. 6.1 LCC isolation and purification

the solution. Different LCCs were then obtained by further extraction. This method was suitable for both hardwood and non-wood samples [14]. The modified method can also be applied to softwood [17].

6.2.4 Water-Organic Solvent Extraction

Although Björkman's extraction method is reliable, it is still time-consuming and difficult to proceed. Watanabe et al. [94] proposed a method to recover water-soluble LCCs from 80% 1,4-dioxane-extracted residue. The chemical properties of LCCs extracted with this method were similar to those of the Björkman LCCs. Detailed methods are shown in Fig. 6.2: plant powder (1 kg) was extracted twice with 80% 1,4-dioxane. Deionized water (10 L) was then added and the mixture was stirred at 25 °C for 24 h. The solution was filtered with Whatman No. 2 and washed with 1 L water three times. It was repeatedly extracted once with deionized water. The extracts were combined and concentrated under reduced pressure. Five volumes of ethanol were added. After centrifugation, the precipitation was recovered and washed with ethanol and petroleum ether. White water-soluble LCC powder was finally obtained. LCCs left in the solid residue can be extracted using DMSO or DMF using Koshijima method [44].

In order to obtain high yield and representative LCC fractions from raw materials, Yaku et al. [101] successfully established LCC separation from Korean pine by combining water with organic solvent extraction. In this method, the residue after the preparation of lignin was extracted with DMF and hot water (70–80 °C) to obtain Björkman LCCs and water-soluble LCCs, respectively. The two LCC fractions were combined and characterized with better purity and yield (12.8%), much higher than the LCCs prepared with either of the water or organic solvent extraction method. The lignin content and carbohydrate composition in LCCs were very similar to those of Björkman LCC. This method was suitable for hardwood to extract representative LCCs with high yield and purity.

6.3 Structural Analysis on LCCs

Various direct and indirect methods are outlined for the confirmation or structural analysis of LCCs. Indirect methods are involved in some acidic and basic degradation of specific LC bond, methylation, oxidation, and enzymatic hydrolysis. Generally, these approaches selectively degrade, substitute, or functionalize the specific analysis sites. Although the selectivity of wet chemistry analysis approaches is questionable, it is still debatable. Among various direct analysis methods, NMR has emerged as a powerful analytical tool for probing the LCCs. Various NMR techniques such as 1D (1H and 13C), 2D (HSQC, HMQC, HMBC, TOCSY) and the 3D (TOCSY-HSQC) have employed to get satisfying evidence of LCCs [36].

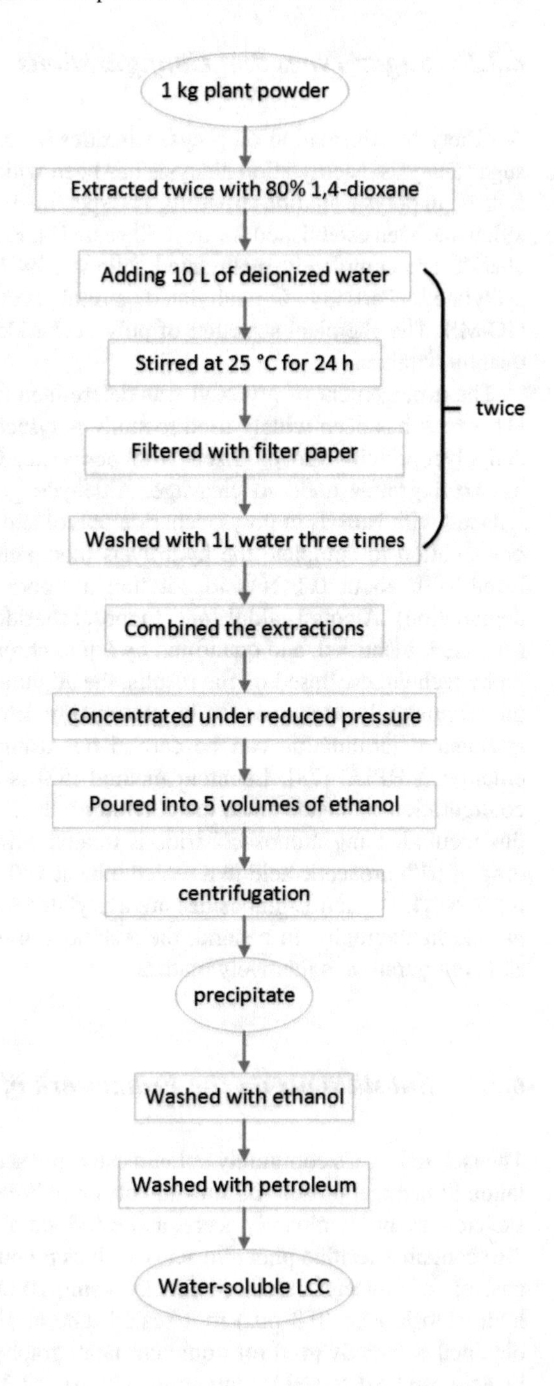

Fig. 6.2 Extraction process of water-soluble LCC

6.3.1 Sugar Types and Configurations

A variety of information on polysaccharides in LCC structure can be obtained by sugar analysis. Methylation analysis has been widely used to determine glycosidic linkage in polysaccharide structure. The specific methylation analysis on the branch xylan has been established for over 50 years [1]. Free hydroxyl groups in a polysaccharide are completely methylated followed by being hydrolyzed, reduced, and acetylated. Partially O-methylated alditol acetates are finally analyzed by GC-MS. The chemical structure of polysaccharide glycosidic bond type is consequently obtained.

The arrangement of glycosyl was determined using Smith degradation method [1], which has been widely used to study polysaccharide structure. When glycans and glycoproteins are oxidized with periodate, C-C bonds between the linked hydroxyl groups undergo cleavage. Aldehyde group of the reaction product is reduced with $NaBH_4$ to polysaccharide polyol and heated with 0.5–1 N acid aqueous solution to complete the hydrolysis (complete Smith degradation) or hydrolyzed with about 0.1 N acid solution at room temperature (controlled Smith degradation). Alcohol, aldol, and monosaccharide formed by the hydrolysis are separated, identified, and quantified by liquid chromatography and gas chromatography techniques. Based on the results, the alignment of saccharide composition in the original polysaccharide can be presumably determined.

Sugar determination can be carried out using a two-step acidolysis method employing HPLC [78]. Leontein method [52] is recommended to determine the configuration of sugars since the derived products obtained are very stable. Using this method, 1 mg monosaccharide is treated with 0.5 mL (+)-2-octanol and one drop of trifluoroacetic acid in a sealed tube at 130 °C overnight. After evaporation, (+)-2-octyl, D-, and L-glucosides are acetylated and quantitatively analyzed using gas chromatography. In general, the technique used to analyze sugar composition and configuration is relatively mature.

6.3.2 Substituents on the Framework of Polysaccharides

The isolated LCCs commonly contain esterified acetic acid due to the partial acetylation of hemicellulose. The substituents on polysaccharide (mainly hemicellulose) skeleton are predominantly acetyl and esterified phenolic acids. Herbaceous plants also contain esterified phenolic acids such as p-coumaric acid and ferulic acid. The analysis of O-acetyl can be performed using 10 mg sample hydrolyzed with 1 M hydrochloric acid (0.3 mL) in a sealed tube at 100 °C for 2 h. The hydrolysate obtained is directly used for liquid chromatography analysis. The esterified phenolic acid can be detected by saponification using 1 M NaOH and then analyzed with HPLC [30].

The etherified phenolic acids can be released with acid hydrolysis including cleavage of arylglycerol-p-aryl ether linkages and benzyl aryl ether linkages. The saponified residue (20 mg) is dissolved in 10 mL solution of 1,4-dioxane and 2 M hydrochloric acid (9: 1, v/v) and then refluxed for 2 h. The solution is diluted with 100 mL distilled water. After pH is adjusted to 4 using saturated $NaHCO_3$ aqueous solution, dioxane is then removed by distillation. Subsequently, phenolic acid is extracted using diethyl ether and analyzed with liquid chromatography [30]. Additionally, there is 4-O-methylglucuronic acid, which could be determined by the methods of sugar composition and configuration determination.

6.3.3 Lignin Structural Monomers

The structure of lignin monomers can be determined using traditional alkaline nitrobenzene oxidation method developed by Roadhouse and MacDougall [71]. Due to its complicated operational procedure, Sudo et al. [81] developed another approach without making the derivatives of original compound. LCCs containing 40 mg lignin are heated in 4.0 mL of 2 M NaOH aqueous solution containing 0.24 mL nitrobenzene at 170 °C. After 2 h, the reaction mixture is cooled and washed with distilled water. The excess nitrobenzene is extracted with ether. After acidification using 2 M HCl solution, the oxidized products are extracted using diethyl ether and analyzed quantitatively using HPLC. Almost all of the structural units of lignin can be detected with this method such as hardwood lignin [104].

The reactivity of lignin is determined by the monomer composition, functional groups, and the linkages between the structural units. Characterization of monomer composition, functional groups, and linkages in lignin structure requires more than one analytical technique. Common NMR techniques such as ^{1}H-NMR, ^{13}C-NMR, 2D HSQC-NMR, and ^{31}P-NMR are the most powerful tools applied for lignin structure analysis. In the future, it is believed that lignin structure will be fully characterized as more efficient, and accurate analysis technologies with higher signal/noise ratio are being developed.

6.3.4 Analysis of Linkages Between Lignin and Carbohydrates

There is a great deal of evidence that lignin and polysaccharides are covalently linked, although the linkage bonds are relatively small in numbers. There are three types of LCC linkages: benzyl ether, phenylglycoside, and esters (Fig. 6.3). The occurrence of stable lignin-carbohydrate bonds generates significant problems in selective separation and isolation of lignin and hemicellulose from lignocellulose. Therefore, LCC structure analysis is fundamental and of practical importance.

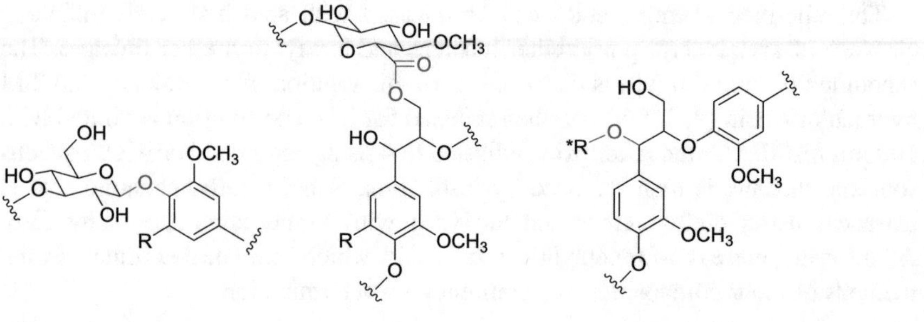

Phenyl glycoside Ester Benzyl ether

R=H or OCH3

BE1: R*=C6 in Glc, Man, Gal, C5 in Ara
BE2: R*=C2 or C3 in Xyl, Glc, Man, Gal, Ara

Fig. 6.3 Chemical bonds within LCC

6.3.4.1 Wet Chemistry Method

Oikawa et al. [66] reported selective cleavage of 4-methoxybenzyl esters and 3,4-dimethoxybenzyl esters using 2,3-dichloro-5,6-dicyanobenzoquinone (DDQ) at room temperature to produce alcohols. DDQ is able to release 91–93% glucose from isosporine or its 4-O-methoxy derivative in 50% 1, 4-dioxane aqueous solution for 1 h at 50 °C. However, DDQ is not active for the acetyl and glycosyl linkages. If the acetyl group is introduced into the phenolic hydroxyl group on the 4-hydroxy-3-methoxybenzyl ether, DDQ can not disrupt the benzyl ether bond due to the pull-electron effect of acetyl. Based on the above results, Koshijima et al. [43] and Watanabe et al. [95] analyzed the links between lignin and carbohydrates using DDQ treatment.

The following procedure is recommended for the detection of benzyl ether linkages and conjugated ether linkages in LCCs (Fig. 6.4) [93]: LCC fragment is first methylated with diazomethane to block the phenolic hydroxyl group and then O-acetylated using pyridine-acetic anhydride solution. The methylated and acetylated LCCs (20 mg) are dissolved in a mixture of dichloromethane and water (18:1, v/v) and treated with 50 mg of DDQ at 50 °C for 2 h. This solution is loaded onto a silica gel column with the adsorption of 50 mg of 1,2,3,4,6-penta-O-acetyl glucose/g Silica Gel 60. The unreacted DDQ and the corresponding hydroquinone are thoroughly washed off with deionized water. Afterward, the silica gel column is eluted using methanol, which is evaporated to dryness and methylated using diazomethane-boron trifluoride diethyl ether solution or methyl trifluoromethane sulfonate, followed by hydrolysis using 90% formic acid solution and 0.25 M sulfuric acid solution. Partially methylated monosaccharides are converted to sugar alcohol acetates and analyzed using GC and GC-MS. The position of the methylation is the site to which the lignin is attached [4].

6.3.4.2 NMR

The use of spectroscopic methods for structural analysis of LCCs might be difficult mainly due to overlapping signals from the functional groups in LCCs. Infrared

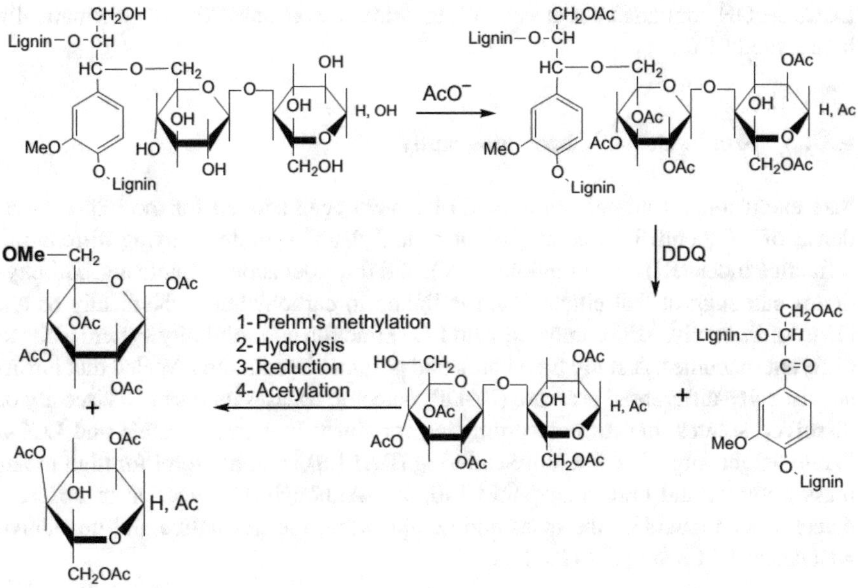

Fig. 6.4 Analysis of LCC linkages with the DDQ method

spectroscopy (IR) is only useful in the study of ester bonds within LCCs [86]. Although 13C NMR can provide partial structural information on carbohydrates and lignin in LCCs, it does not provide information about chemical bonds between LCCs as these signals have a significant overlap with lignin or carbohydrate signals [6]. Xie et al. [99] developed a method to label the specific positions of lignin side chains with ^{13}C. The labeled lignin was then analyzed using ^{13}C NMR. However, subsequent studies showed that the results obtained in this way were not reliable, mainly because 1D ^{13}C NMR was not a reliable technical method to study the chemical bonds within LCCs. 2D NMR can be used to overcome these obstacles. The structures of phenyl glycosides and benzyl ethers in LCCs were first directly detected by high-resolution HSQC 2D NMR [5]. Nowadays, the problem of too low levels of lignin carbohydrates linkages (LCLs) in LCCs can be overcome with newly developed high-sensitivity NMR techniques. With NMR cryogenic probe, the sensitivity can be increased by three to four times. In addition, HSQC and heteronuclear multiple bond correlation (HMBC) techniques are available for efficient analysis of various low-level LCLs in LCCs.

Another way to improve LCC analysis sensitivity is to separate samples containing high levels of LCLs. When LCC samples with high lignin content were degraded by enzymatic and/or ball milling techniques, it was possible to know the specific carbohydrate units linked to lignin by analyzing the constituent units of carbohydrate residues in LCCs. For instance, a sample of LCCs prepared by Balakshin et al. [5] contained higher amounts of arabinose and galactose. The sugar composition of enzymatically hydrolyzed LCCs derived from pine was significantly different from that of LCCs extracted with acetic acid (LCCs-AcOH) from the same sample. Birch

LCCs-AcOH contained about 90% xylan, while it was only 70% in enzymatically hydrolyzed LCCs.

6.3.4.3 Size Exclusion Chromatography

Size exclusion chromatography (SEC) has also been studied for the indirect evidence of LC bond. It is composed of a dual detector system having differential refractive index (RI) and ultraviolet (UV), and the co-elusion of lignin and carbohydrates can suggest that either lignin is linked to carbohydrates chemically or not [19]. Additionally, SEC combined with LC extraction and solubility system leads to different outcomes. A study based on wood polymer interaction revealed that intrinsic viscosity-differential pressure (IV-DP) detector, focuses on intrinsic viscosity of dissolved solutes, hence more promising for linear-free carbohydrate and LCCs. While, Right Angle Laser Light Scattering (RALLS) is more useful for high molar mass polymer and branching points [50, 96]. Although the selection of different detectors can elucidate the mass and composition, the qualitative and quantative analysis of LCCs are yet to be done.

6.3.4.4 FTIR

Infrared spectroscopy is also often used for qualitative analysis on chemical bonds within LCCs. The absorption peaks of acetyl group are shown at 1250 and 1733 cm^{-1} (Fig. 6.5). The major carbohydrate absorption regions are 990, 1030–1040, 1070–1090, and 1150–1180 cm^{-1}, typical of xylan and/or dextran absorption peaks [33]. The disappearance of the absorption peak near 1730 cm^{-1} in Fig. 9 shows that

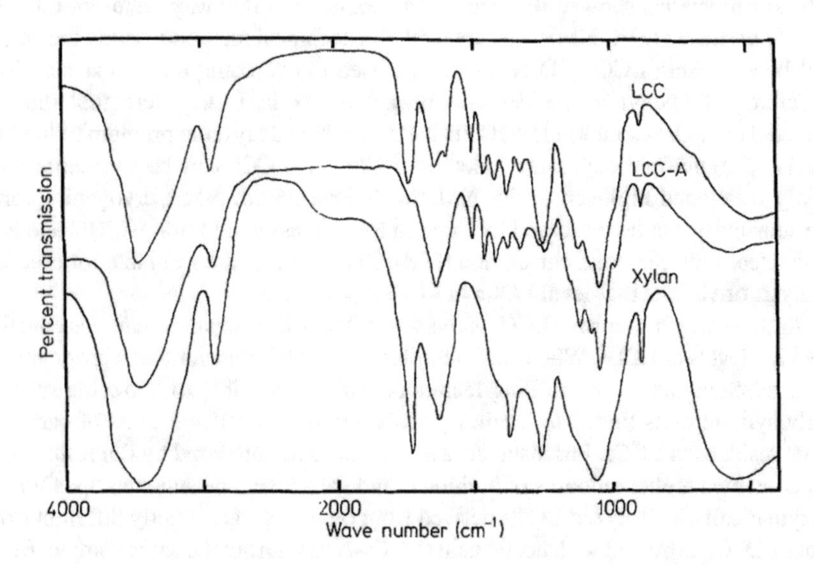

Fig. 6.5 Infrared spectra of LCCs and Xylan

the esterified carbonyl group disappears because the alkali treatment destroys the ester bond within LCCs.

6.4 Bonds Between Lignin and Carbohydrates and Their Breakage

The chemical linkage between lignin and carbohydrates is generated during the lignin biosynthesis and the concentration and type of bond directly dependent on a species. Various types of LC bonds have been studied and reported such as benzyl ether, benzyl ester, and phenyl glycoside bond which defines the nature and properties of LCC fragment. This part will critically focus on the presence of different bonds among different parts of biomass.

6.4.1 Type of Chemical Bonds

6.4.1.1 Benzyl Ether/Ester Bonds

Choi et al. [15] characterized LCCs isolated from unbleached Kraft pulps of spruce and beech wood employing GC and GC-MS techniques. It was found that benzyl ether established a lignin-carbohydrate linkage that structurally varies with biomass. Structurally, galactose either combines with mannose (spruce wood) or xylose (beech wood) to form this linkage. C-6 (in hexose) and C-2/C-3 (in xylose) were found as the principal bonding sites to develop LCCs [56].

Ether bonds are alkali stable which leave residual lignin attached with carbohydrates in the pulping process [26]. Using model compounds, few researchers have found the possibility of LC ether bond formation during Kraft pulping [26, 35]. It was postulated that harsh reaction conditions might cause artificial LC bonding formation (Lawoko 2005). However, Minor [60] argued that only LC bond enrichment occurred during Kraft pulping than bond formation.

Several studies have reported the existence of LC ester bonds in LCB based on their evidences, such as the presence of carboxyls in LC binding [10], the degree of esterification [92], alkali pretreatment disassociating lignin and carbohydrate linkage [21, 101], and 2D NMR observations [6, 99]. LC ester bonds are considered to be alkali labile because they can be easily cleaved by alkalis [5, 25].

6.4.1.2 Phenyl Glycoside (PhGly) Bonds

The mechanism of phenyl glycosides is not yet fully understood. However, it seems that acetylation at C-2 on xylan does not permit the coupling of lignin at the adjacent C-1 hydroxyl, thereby inhibiting the formation of phenyl glycosides. Another opinion assumes that phenyl glycoside formation is an enzymatic process where enzyme specificity causes acetylation to inhibit the coupling of lignin with xylan.

6.4.1.3 Acetal-Type Bonds

Acetal formation is favored in a mild acidic environment (pH ~ 5), and this linkage is present between the lignin structural fragment phenylpropane and the hydroxyl group of carbohydrates. This bond formed when two –OH groups of polysaccharide linked to lignin through acetal, and it is regarded as the rare one [28].

6.4.1.4 Ferulate Ester Bonds

Herbaceous plant, grasses, and non-wood materials contain high concentration of ferulate acid, which is attached to carbohydrates. Ferulic acid showed the tendency to link with the lignin oxidatively, while presence of carboxylic acid group at the terminal end of propenyl group can make ester linkage with carbohydrates as well [89]. As a result, it forms lignin-ferulate-polysaccharide (LFP) complex in which ferulate esters of carbohydrates bonded to lignin oxidatively [16, 106]. The alkali-labile nature of ferulate ester can easily be cleaved under mild alkaline condition and yields lignin and ferulic acid molecules [13].

6.4.2 Breakage of Different Bonds

6.4.2.1 Ester Bond Breakage

Noori and Karimi [62] reported the cleavage of LC ester bonds in pine wood employing NaOH pretreatment. It has been shown that an alkaline process effectively cleaves LC ester bonds between lignin and carbohydrates without disrupting hemicellulose under mild temperature conditions [25]. Alkaline reaction rate is proportional to the concentration of hydroxide ions [40, 61]. After being treated with NaOH, the enzymatic hydrolysis was found to be improved due to the increased pretreated biomass porosity caused by the cleavage of ester bonds between xylan and lignin. However, severe alkali treatment (at elevated temperature and for prolonged time) decomposes hemicellulose into xylan and mannan, and it also results in alkali-stable linkages due to peeling off reactions [74]. Min et al. [59] pretreated corn cob with mild alkaline solution (1 N NaOH) at room temperature that reduced the amount of LCCs from 22% to 2.5% on C9 basis because NaOH cleaved LC ester bonds in LCC.

Santos et al. [75] studied the alkali extraction of wood lignin employing NaOH and found that alkali-extracted lignin and milled wood lignin contained low amount of carbohydrates (2–3%) due to LC ester bond cleavage. In native wood LCCs, ester bonds linked with 4-O-methylglucuronic acid units of xylan with lignin [21] are labile to alkaline pretreatment [21, 63]. Yan et al. [103] studied the alkaline pretreatment of sweet sorghum bagasse employing NaOH and Ca(OH)2 and confirmed that both alkaline treatments cleaved LC ester bonds between hydroxycinnamic acids and lignin [103].

Martínez et al. [58] investigated alkaline deconstruction of sugarcane bagasse using mild (4%) and severe (9%) alkali concentrations. Mild alkaline treatment cleaved ester bonds between ferulic acid (FA) and xylan and between lignin (H unit) and xylan. At 4% NaOH treatment, cleavage of LC ester bonds between lignin and xylan occurred without affecting lignin removal. However, an increase in the severity of alkaline treatment (from 4% to 9%) caused ester bond cleavage as well as lignin disintegration [58]. Xiao et al. [98] studied NaOH treatment of dewaxed maize stems, rye straw, and rice straw. It was found that alkali broke down LC ester bonds between lignin and hemicellulose and FA [80]. Sun et al. [84] isolated different types of lignin from *Tamarix* spp. employing ball milling, alkaline organosolv process, and alkaline treatment. The results revealed that the residual carbohydrates attached to extracted lignins followed the given order: milled wood lignin (MWL), 3.24% > alkaline lignin (AL), 1.86% > organosolv lignin (OL), 0.79%. Alkaline and organosolv processes cleaved LC ether and ester bonds in LCCs and removed maximum amount of carbohydrates [84].

Huynh and Arioka [34] isolated GE from *Neurospora crassa* and confirmed its function to cleave ester link between xylan and lignin using a model compound (3-(4-methoxyphenyl) propyl methyl 4-O-methyl-a-D-glucopyranosiduronate). Kinetic studies of this cleavage reaction revealed that GEs could differentiate alkyl and aryl alkyl esters of 4-O-D-glucuronic acid [79, 97]. Moreover, GEs preferably cleaved 4-O-methyl-D-glucuronic acid esters rather than D-glucuronic acid esters [20]. It was observed that the cleavage mechanism of ester bonds in the synthetic LCC was independent of carbohydrate type during the hydrolysis [34, 79].

Organosolv pretreatment has been evolved by using ethanol, acetone, formic acid, and acetic acid for the depolymerization of hemicellulose and lignin to produce bioethanol or paper products. The obtained products required extensive washing step to purify the product followed by further treatment with enzymes to avoid contamination. The addition of acids accelerates the delignification of LCB and improves xylose yield [85].

6.4.2.2 Ether Bond Breakage

Sun et al. [83] employed alkaline pretreatment to the dewaxed and partially delignified poplar wood and observed that the recovered hemicellulose fragments were linked to low quantity of lignin (<5%). Alkali pretreatment significantly cleaved LC ether bonds between lignin and carbohydrates. It was also noticed that increase in the concentration of alkali (1.5–8.5%) caused extensive bond cleavage and, hence, resulted in reduced lignin content (4.9–2.6%) linked to hemicellulose. In a follow-up study, Sun et al. [82] compared the alkaline process with alkaline peroxide process (NaOH and H2O2) and revealed that alkaline peroxide process hydrolyzed ether linkage better than the alkaline process. Alkaline peroxide process (2 M NaOH and 2% H_2O_2) extracted hemicellulose linked to lower quantity of lignin (<2%). Hydroperoxyl and hydroxyl radicals were found in the suspension after alkaline peroxide process [22]. These radicals break some LC ether bonds causing fragmentation of lignin-hemicellulose complex [22, 67].

Jiang et al. [37] used Na_2CO_3-H_2O-tetrahydrofuran (THF) reaction system to cleave ether bonds between cellulose and lignin. Small amount of unconverted lignin revealed that Na_2CO_3-H_2O-tetrahydrofuran (THF) reaction system could not cleave some of the ether linkages between cellulose-lignin complex. Initial pH value of this reaction system was found as the main driver to cleave LC ether bond [37].

Dilute acid pretreatment at moderate temperature (140–190 °C) can efficiently remove hemicellulose and be recovered in the form of dissolved sugar [47]. Dilute acid pretreatment enhances the cellulose digestibility in the enzymatic hydrolysis process by removing hemicellulose. Benzyl ether bond is reported as acid sensitive; however, benzyl ether linkages with phenolic hydroxyl groups are proposed to be alkali labile. This acid hydrolysis is also used for the ether bond investigation carried out by simultaneous reduction of LCC with sodium borohydride followed by acid hydrolysis of reduced LCC. This process hydrolyzes the ether linkage of LCC and produces new phenolic and benzyl alcohol hydroxyl group.

6.5 Effect of LCC on Biomass Utilization

The biorefinery concept is developing continuously because of increasing environmental concern over greenhouse emission and fossil fuel sustainability issues. At this point, the investigation and in-depth understanding of LCCs is really an essential step toward biomass utilization as it is hypothesized that recalcitrance of biomass directly belongs to LCCs. Although the emerging LCC fractionation process is obtaining the fundamental polymers with high purity, the progress in this field is really slow because of heterogeneous nature of biomass, characterization of low-frequency LCC bonds, and operational cost [46].

6.5.1 Bioactivity of LCC

Sakagami et al. [73] reviewed the potential of LCC in cosmetic and anti-UV applications. LCC showed two times higher anti-UV activity than lower MW polyphenols and hot-water extracts of Kampo medicines and tea leaves. LCC and vitamin C displayed compound effect on anti-UV activity. Furthermore, an LCC-rich drug (alkaline extract of *Sasa senanensis* Rehder (SE) leaves) also exhibited strong antiviral and radical scavenging activity when used along with vitamin C [73].

Researchers studied antimicrobial activity of pine-extracted LCC in mice. It was observed that LCC was effective against several pathogens such as *Staphylococcus aureus* SH10, *Escherichia coli* GN2411, *Pseudomonas aeruginosa* H7, *Klebsiella pneumoniae* ST101, and *Candida albicans* YA2 [29, 65]. It was also noticed that carbohydrate removal from LCC significantly decreased the anti-pathogenic activity [72]. Abe et al. [2] also reported that sugar conjugated lignin (LCC) effectively protected infant mice from Cestoda infection compared with the carbohydrate-free LCC.

6.5.2 Effect of LCC on Separation of Different Components

Strong oxidative agents such as sequential alkaline peroxide process can fragment relatively pure hemicellulose (0.5–5.1% bonded lignin) from LCC. This process produces hydroperoxyl and hydroxyl radicals which significantly cleave the LC linkages between lignin and hemicelluloses in the cell wall of bagasse [82]. Zhang et al. [107] studied the extraction of xylan from sweet gum employing glycerol thermal process (GTP) followed by alkaline extraction. GTP caused lignin fragmentation, and resultant lignin units were loosely bound to xylan via weak intermolecular forces forming lignin-xylan complexes. Follow-up of alkaline process co-extracted the lignin units along with xylan from the lignin-xylan complexes.

6.6 Prospects

Different biorefinery designs are possible as well as viable. Each biorefinery has to focus on one primary objective such as lignin, hemicellulose, or cellulose extraction, and rest of the macromolecules should be secondary objectives. For example, if hemicellulosic extraction (for furfural and HMF) is the prime objective, then the acidic pretreatment could be used [87], and the resultant lignin will have modified structure (more C-C bonds). Accordingly, biorefinery route to use such modified lignin needs to be developed. On the other hand, if one biorefinery targets to extract pure lignin for further applications, then such pretreatment should be selected that specifically extracts the native lignin. The residual LCB after lignin extraction can be used for carbohydrate extraction. For example, Brandt et al. [11] proposed an ionic liquid (IL)-based pretreatment (ionoSolv) that can fit well to lignin-focused biorefinery concept. There are a number of ways to employ ionoSolv process in a biorefinery setting. It can be used to cleave intra-unit lignin bonds (ether and ester linkages) to depolymerize lignin to yield low MW aromatics. Contrary to this pathway, ionoSolv can be manipulated to induce lignin condensation reactions to produce high MW and high phenolic hydroxyl fractions which can be used in the production of fuel blends (resins/additives).

Limited pretreatment processes can isolate lignin and cellulose from LCBs. Lignin is recovered via dissolution, whereas cellulose is isolated as solid residue.

Lignin and carbohydrates are cross-linked within LCC and arise as complex irregular structures due to tight physical and chemical bonding. Probably, single pretreatment method is not sufficient to overcome LCC recalcitrance, and it would result in low yield and impure products. There is a need to search for such optimum pretreatments (single/integrated) which may co-valorize lignin and carbohydrates simultaneously.

Acknowledgment This work was supported by the Major Research Plan of the National Natural Science Foundation of China (91534107), the National Natural Science Foundation of China (21576266), Wanjiang Scholar Program, and Start Fund for Biochemical Engineering Research Center from Anhui University of Technology.

References

1. Abdel-Akher, M., Hamilton, J. K., Montgomery, R., & Smith, F. (1952). A new procedure for the determination of the fine structure of polysaccharides. *Journal of the American Chemical Society, 74*(19), 4970–4971.
2. Abe, M., Okamoto, K., Konno, K., & Sakagami, H. (1989). Induction of antiparasite activity by pine cone lignin-related substances. *In Vivo, 3*, 359–362.
3. Albersheim, P., Darvill, A., Roberts, K., Sederoff, R. & Staehelin, A. (2010). *Plant cell walls, Garland Science.*
4. Azuma, J. I. (1989). Analysis of lignin-carbohydrate complexes of plant cell walls. In H. F. Linskens & J. F. Jackson (Eds.), *Plant fibers* (pp. 100–126). Springer.
5. Balakshin, M. Y., Capanema, E. A., & Chang, H. M. (2007). MWL fraction with a high concentration of lignin-carbohydrate linkages: Isolation and 2D NMR spectroscopic analysis. *Holzforschung, 61*(1), 1–7.
6. Balakshin, M., Capanema, E., Gracz, H., Chang, H. M., & Jameel, H. (2011). Quantification of lignin–carbohydrate linkages with high-resolution NMR spectroscopy. *Planta, 233*(6), 1097–1110.
7. Balakshin, M., Capanema, E., & Berlin, A. (2014). Isolation and analysis of lignin–carbohydrate complexes preparations with traditional and advanced methods: A review. *Studies in Natural Products Chemistry, Elsevier., 42*, 83–115.
8. Björkman, A. (1956). Studies on finely divided wood. Part 1. Extraction of lignin with neutral solvents. *Svensk Papperstidning, 59*(13), 477–485.
9. Björkman, A. (1957). Studies on finely divided wood (part 3)-extraction of lignin-carbohydrate complexes with neutral solvent. *Svensk Papperstding, 60*, 243–251.
10. Bolker, H. I. (1963). A lignin carbohydrate bond as revealed by infra-red spectroscopy. *Nature, 197*(4866), 489–490.
11. Brandt, A., Chen, L., van Dongen, B. E., Welton, T., & Hallett, J. P. (2015). Structural changes in lignins isolated using an acidic ionic liquid water mixture. *Green Chemistry, 17*(11), 5019–5034.
12. Brunow, G., & Lundquist, K. (2010). Functional groups and bonding patterns in lignin (including the lignin-carbohydrate complexes). *Lignin and Lignans: Advances in Chemistry*, 267–299.
13. Buranov, A. U., & Mazza, G. (2008). Lignin in straw of herbaceous crops. *Industrial Crops and Products, 28*(3), 237–259.
14. Cadena, E. M., Du, X., Gellerstedt, G., Li, J., Fillat, A., García-Ubasart, J., Vidal, T., & Colom, J. F. (2011). On hexenuronic acid (HexA) removal and mediator coupling to pulp fiber in the laccase/mediator treatment. *Bioresource Technology, 102*(4), 3911–3917.

15. Choi, J. W., Choi, D.-H., & Faix, O. (2007). Characterization of lignin-carbohydrate linkages in the residual lignins isolated from chemical pulps of spruce (Picea abies) and beech wood (Fagus sylvatica). *Journal of Wood Science, 53*(4), 309–313.
16. de Oliveira, D. M., Finger-Teixeira, A., Rodrigues Mota, T., Salvador, V. H., Moreira-Vilar, F. C., Correa Molinari, H. B., Craig Mitchell, R. A., Marchiosi, R., Ferrarese-Filho, O., & Dantas dos Santos, W. (2015). Ferulic acid: A key component in grass lignocellulose recalcitrance to hydrolysis. *Plant Biotechnology Journal, 13*(9), 1224–1232.
17. Du, X., Li, J., & Gellerstedt, G. (2013). Universal fractionation of lignin–carbohydrate complexes (LCCs) from lignocellulosic biomass: An example using spruce wood. *The Plant Journal, 74*, 328–338.
18. Du, X., Pérez-Boada, M., Fernández, C., Rencoret, J., José, C., Jiménez-Barbero, J., Li, J., Gutiérrez, A., & Martínez, A. T. (2014). Analysis of lignin–carbohydrate and lignin–lignin linkages after hydrolase treatment of xylan–lignin, glucomannan–lignin and glucan–lignin complexes from spruce wood. *Planta, 239*(5), 1079–1090.
19. Dubuis, A., Le Masle, A., Chahen, L., Destandau, E., & Charon, N. (2019). Off-line comprehensive size exclusion chromatography× reversed-phase liquid chromatography coupled to high resolution mass spectrometry for the analysis of lignocellulosic biomass products. *Journal of Chromatography A*, 460505.
20. Duranová, M., Hirsch, J., Kolenová, K., & Biely, P. (2009). Fungal glucuronoyl esterases and substrate uronic acid recognition. *Bioscience Biotechnology, and Biochemistry, 73*, 2483–2487.
21. Eriksson, Ö., Goring, D. A. I., & Lindgren, B. O. (1980). Structural studies on the chemical bonds between lignin and carbohydrates in spruce. *Wood Science and Technology, 14*, 267–279.
22. Fang, J. M., Sun, R. C., Salisbury, D., Fowler, P., & Tomkinson, J. (1999). Comparative study of hemicelluloses from wheat straw by alkali and hydrogen peroxide extractions. *Polymer Degradation and Stability, 66*, 423–432.
23. Feng, N., Ren, L., Wu, H., Wu, Q., & Xie, Y. (2019). New insights on structure of lignin-carbohydrate complex from hot water pretreatment liquor. *Carbohydrate Polymers, 224*, 115130.
24. García, J. C., Zamudio, M. A., Pérez, A., Feria, M. J., Gomide, J. L., Colodette, J. L., & López, F. (2011). Soda-AQ pulping of Paulownia wood after hydrolysis treatment. *BioResources, 6*(2), 971–986.
25. Gaspar, M., Kalman, G., & Reczey, K. (2007). Corn fibre as a raw material for hemicellulose and ethanol production. *Process Biochemistry, 42*, 1135–1139.
26. Gierer, J., & Wännström, S. (1986). Formation of ether bonds between lignins and carbohydrates during alkaline pulping processes. *Holzforschung, 40*(6), 347–352.
27. Giummarella, N., Zhang, L., Henriksson, G., & Lawoko, M. (2016). Structural features of mildly fractionated lignin carbohydrate complexes (LCC) from spruce. *RSC Advances, 6*(48), 42120–42131.
28. Grushnikov, O. P., & Shorygina, N. N. (1970). The present state of the problem of lignin–carbohydrate bonds in plant tissues. *Russian Chemical Reviews, 39*(8), 684.
29. Harada, H., Sakagami, H., Konno, K., Sato, T., Osawa, N., Fujimaki, M., & Komatsu, N. (1988). Induction of antimicrobial activity by antitumor substances from pine cone extract of Pinus parviflora Sieb. et Zucc. *Anticancer Research, 8*, 581–587.
30. Hatfield, R. D., Ralph, J., & Grabber, J. H. (1999). Cell wall cross-linking by ferulates and diferulates in grasses. *Journal of the Science of Food and Agriculture, 79*(3), 403–407.
31. Henriksson, G., et al. (2007a). Lignin-carbohydrate network in wood and pulps: A determinant for reactivity. *Holzforschung, 61*(6), 668–674.
32. Henriksson, G., Lawoko, M., Martin, M. E. E., & Gellerstedt, G. (2007b). Lignin-carbohydrate network in wood and pulps: A determinant for reactivity. *Holzforschung, 61*(6), 668–674.
33. Hong, P., Luo, Q., Ruan, R., Zhang, J., & Liu, Y. (2014). Structural features of lignin and lignin-carbohydrate complexes from bamboo (Phyllostachys pubescens mazel). *BioResources, 9*(1), 1276–1289.

34. Huynh, H. H., & Arioka, M. (2016). Functional expression and characterization of a glucuronoyl esterase from the fungus Neurospora crassa: Identification of novel consensus sequences containing the catalytic triad. *The Journal of General and Applied Microbiology, 62*(5), 217–224.
35. Iversen, T., & Wannström, S. (1986). Lignin-carbohydrate bonds in a residual lignin isolated from pine kraft pulp. *Holzforschung, 40*, 19–22.
36. Jiang, B., Zhang, Y., Zhao, H., Guo, T., Wu, W., & Jin, Y. (2019). Dataset on structure-antioxidant activity relationship of active oxygen catalytic lignin and lignin-carbohydrate complex. *Data in Brief, 25*, 104413.
37. Jiang, Z., Zhang, H., He, T., Lv, X., Yi, J., Li, J., & Hu, C. (2016). Understanding the cleavage of inter-and intramolecular linkages in corncob residue for utilization of lignin to produce monophenols. *Green Chemistry, 18*, 4109–4115.
38. Joseleau, J.-P., & Kesraoui, R. (1986). Glycosidic bonds between lignin and carbohydrates. *Holzforschung-International Journal of the Biology, Chemistry, Physics and Technology of Wood, 40*(3), 163–168.
39. Kato, Y., & Nevins, D. J. (1985). Isolation and identification of O-(5-O-feruloyl-α-L--arabinofuranosyl)-1 (→ 3)-O-β-D-xylopyranosyl-(1→ 4)-D-xylopyranose as a component of Zea shoot cell-walls. *Carbohydrate Research, 137*, 139–150.
40. Kim, J. S., Lee, Y., & Kim, T. H. (2016). A review on alkaline pretreatment technology for bioconversion of lignocellulosic biomass. *Bioresource Technology, 199*, 42–48.
41. Kondo, R., Sako, T., Iimori, T., & Imamura, H. (1990). Formation of glycosidic lignin-carbohydrate complex in the enzymatic dehydrogenative polymerization of coniferyl alcohol. *Mokuzai Gakkaishi= Journal of the Japan Wood Research Society, 36*(4), 332–338.
42. Koshijima, T., & Watanabe, T. (2013). *Association between lignin and carbohydrates in wood and other plant tissues*. Springer.
43. Koshijima, T., Watanabe, T., & Azuma, J.-i. (1984). Existence of benzylated carbohydrate moiety in lignin–carbohydrate complex from pine wood. *Chemistry Letters, 13*(10), 1737–1740.
44. Koshijima, T., Taniguchi, T., & Tanaka, R. (1972). Lignin carbohydrate complex. Pt. I. the influences of milling of wood upon the Björkman LCC. *Holzforschung, 26*(6), 211–217.
45. Košíková, B., Joniak, D., & Kosakova, L. (1979). On the properties of benzyl ether bonds in the lignin-saccharidic complex isolated from spruce. *Holzforschung-International Journal of the Biology, Chemistry, Physics and Technology of Wood, 33*(1), 11–14.
46. Kumar, B., Bhardwaj, N., Agrawal, K., Chaturvedi, V., & Verma, P. (2020). Current perspective on pretreatment technologies using lignocellulosic biomass: An emerging biorefinery concept. *Fuel Processing Technology, 199*, 106244.
47. Kumari, D., & Singh, R. (2018). Pretreatment of lignocellulosic wastes for biofuel production: A critical review. *Renewable and Sustainable Energy Reviews, 90*, 877–891.
48. Lawoko, M., Henriksson, G., & Gellerstedt, G. (2005a). Structural differences between the lignin carbohydrate complexes in wood and in chemical pulps. *Biomacromolecules, 6*, 3467–3473.
49. Lawoko, M. (2013). Unveiling the structure and ultrastructure of lignin carbohydrate complexes in softwoods. *International Journal of Biological Macromolecules, 62*, 705–713.
50. Lawoko, M., Henriksson, G., & Gellerstedt, G. (2003). New method for quantitative preparation of lignin-carbohydrate complex from unbleached softwood kraft pulp: Lignin-polysaccharide networks I. *Holzforschung, 57*(1), 69–74.
51. Lawoko, M., Henriksson, G., & Gellerstedt, G. (2005b). Structural differences between the lignin– carbohydrate complexes present in wood and in chemical pulps. *Biomacromolecules, 6*(6), 3467–3473.
52. Leontein, K., Lindberg, B., & Lõnngren, J. (1978). Assignment of absolute configuration of sugars by glc of their acetylated glycosides formed from chiral alcohols. *Carbohydrate Research, 62*(2), 359–362.

53. Li, J., Martin-Sampedro, R., Pedrazzi, C., & Gellerstedt, G. (2011). Fractionation and characterization of lignin-carbohydrate complexes (LCCs) from eucalyptus fibers. *Holzforschung, 65*, 43–50.
54. Lora, J. H., & Glasser, W. G. (2002). Recent industrial applications of lignin: A sustainable alternative to nonrenewable materials. *Journal of Polymers and the Environment, 10*(1–2), 39–48.
55. Lundquist, K., Simonson, R., & Tingsvik, K. (1983). Lignin carbohydrate linkages in milled wood lignin preparations from spruce wood. *Svensk Papperstidning, 86*(6), 44–47.
56. Lupoi, J. S., Singh, S., Parthasarathi, R., Simmons, B. A., & Henry, R. J. (2015). Recent innovations in analytical methods for the qualitative and quantitative assessment of lignin. *Renewable and Sustainable Energy Reviews, 49*, 871–906.
57. Mahmood, H., Moniruzzaman, M., Iqbal, T., & Khan, M. J. (2019). Recent advances in the pretreatment of lignocellulosic biomass for biofuels and value-added products. *Current Opinion in Green and Sustainable Chemistry*.
58. Martínez, P. M., Punt, A. M., Kabel, M. A., & Gruppen, H. (2016). Deconstruction of lignin linked p-coumarates, ferulates and xylan by NaOH enhances the enzymatic conversion of glucan. *Bioresource Technology, 216*, 44–51.
59. Min, D. Y., Jameel, H., Chang, H. M., Lucia, L., Wang, Z. G., & Jin, Y. C. (2014). The structural changes of lignin and lignin–carbohydrate complexes in corn stover induced by mild sodium hydroxide treatment. *RSC Advances, 4*(21), 10845–10850.
60. Minor, J. L. (1986). Chemical linkage of polysaccharides to residual lignin in loblolly pine kraft pulps. *Journal of Wood Chemistry and Technology, 6*, 185–201.
61. Modenbach, A. (2013). *Sodium hydroxide pretreatment of corn Stover and subsequent enzymatic hydrolysis: An investigation of yields, kinetic modeling and glucose recovery*, PhD dissertation at the University of Kentucky.
62. Noori, M. S., & Karimi, K. (2016). Chemical and structural analysis of alkali pretreated pinewood for efficient ethanol production. *RSC Advances, 6*(70), 65683–65690.
63. Obst, J. (1982). Frequency and alkali resistance of wood lignin-carbohydrate bonds in wood. *Tappi/April, 65*(4), 109–112.
64. Eriksson, O., & Lindgren, B. (1977). Linkage in wood between lignin and hemicellulose. Abstracts of papers of the American chemical society, 173, 17–17.
65. Oh-Hara, T., Sakagami, H., Kawazoe, Y., Kaiya, T., Komatsu, N., Ohsawa, N., Fujimaki, M., Tanuma, S., & Konno, K. (1990). Antimicrobial spectrum of lignin-related pine cone extracts of Pinus parviflora Sieb. et Zucc. *In Vivo, 4*, 7–12.
66. Oikawa, Y., Yoshioka, T., & Yonemitsu, O. (1982). Specific removal of o-methoxybenzyl protection by DDQ oxidation. *Tetrahedron Letters, 23*(8), 885–888.
67. Pan, G. X., Bolton, J. L., & Leary, G. J. (1998). Determination of ferulic and p-coumaric acids in wheat straw and the amounts released by mild acid and alkaline peroxide treatment. *Journal of Agriculture and Food Chemistry, 46*, 5283–5288.
68. Pothiraj, C., Kanmani, P., & Balaji, P. (2006). Bioconversion of lignocellulose materials. *Mycobiology, 34*(4), 159–165.
69. Radhika, K., Ravinder, R., & Ravindra, P. (2011). Bioconversion of pentose sugars into ethanol: A review and future directions. *Biotechnology and Molecular Biology Reviews, 6*(1), 8–20.
70. Ralph, J., Lundquist, K., Brunow, G., Lu, F., Kim, H., Schatz, P. F., Marita, J. M., Hatfield, R. D., Ralph, S. A., & Christensen, J. H. (2004). Lignins: Natural polymers from oxidative coupling of 4-hydroxyphenyl-propanoids. *Phytochemistry Reviews, 3*(1–2), 29–60.
71. Roadhouse, F. E., & MacDougall, D. (1956). A study of the nature of plant lignin by means of alkaline nitrobenzene oxidation. *The Biochemical Journal, 63*, 33.
72. Sakagami, H., Kushida, T., Oizumi, T., Nakashima, H., & Makino, T. (2010). Distribution of lignin–carbohydrate complex in plant kingdom and its functionality as alternative medicine. *Pharmacology & Therapeutics, 128*, 91–105.

73. Sakagami, H., Sheng, H., Okudaira, N., Yasui, T., Wakabayashi, H., Jia, J., Natori, T., Suguro-Kitajima, M., Oizumi, H., & Oizumi, T. (2016). Prominent anti-UV activity and possible cosmetic potential of lignin-carbohydrate complex. *In Vivo (Brooklyn), 30*, 331–339.
74. Salehian, P., Karimi, K., Zilouei, H., & Jeihanipour, A. (2013). Improvement of biogas production from pine wood by alkali pretreatment. *Fuel, 106*, 484–489.
75. Santos, R. B., Capanema, E. A., Balakshin, M. Y., Chang, H. M., & Jameel, H. (2012). Lignin structural variation in hardwood species. *Journal of Agricultural and Food Chemistry, 60*(19), 4923–4930.
76. Santos, R. B., Hart, P., Jameel, H., & Chang, H. M. (2013). Wood based lignin reactions important to the biorefinery and pulp and paper industries. *BioResources, 8*(1), 1456–1477.
77. Smelstorius, J. (1974). Chemical composition of wood of Australian-grown *Pinus radiata* D. don. III. lignin-polysaccharide complexes. *Holzforschung-International Journal of the Biology, Chemistry, Physics and Technology of Wood, 28*(3), 99–101.
78. Sone, Y., Kakuta, M., & Misaki, A. (1978). Isolation and characterization of polysaccharides of "Kikurage", fruit body of Auricularia auricula-judae. *Agricultural and Biological Chemistry, 42*(2), 417–425.
79. Spániková, S., Poláková, M., Joniak, D., Hirsch, J., & Biely, P. (2007). Synthetic esters recognized by glucuronoyl esterase from Schizophyllum commune. *Archives of Microbiology, 188*, 185–189.
80. Spencer, R. R., & Akin, D. E. (1980). Rumen microbial degradation of potassium hydroxide-treated coastal bermuda grass leaf blades examined by electron microscopy. *Journal of Animal Science, 51*(5), 1189–1196.
81. Sudo, K., Shimizu, K., & Sakurai, K. (1985). Characterization of steamed wood lignin from beech wood. *Holzforschung, 39*(5), 281–288.
82. Sun, J. X., Sun, X. F., Sun, R. C., & Su, Y. Q. (2004). Fractional extraction and structural characterization of sugarcane bagasse hemicelluloses. *Carbohydrate Polymers, 56*(2), 195–204.
83. Sun, R. C., Fang, J. M., Tomkinson, J., Geng, Z. C., & Liu, J. C. (2001). Fractional isolation, physico-chemical characterization and homogeneous esterification of hemicelluloses from fast-growing poplar wood. *Carbohydrate Polymers, 44*, 29–39.
84. Sun, Y. C., Wang, M., & Sun, R. C. (2015). Toward an understanding of inhomogeneities in structure of lignin in green solvents biorefinery. Part 1: Fractionation and characterization of lignin. *ACS Sustainable Chemistry & Engineering, 3*(10), 2443–2451.
85. Sun, Y., & Cheng, J. (2002). Hydrolysis of lignocellulosic materials for ethanol production: A review. *Bioresource Technology, 83*, 1–11.
86. Takahashi, N., & Koshijima, T. (1988). Ester linkages between lignin and glucuronoxylan in a lignin-carbohydrate complex from beech (Fagus crenata) wood. *Wood Science and Technology, 22*(3), 231–241.
87. Tan, M., Ma, L., Rehman, M. S. U., Ahmed, M. A., Sajid, M., Xu, X., Sun, Y., Cui, P., & Xu, J. (2019). Screening of acidic and alkaline pretreatments for walnut shell and corn Stover biorefining using two way heterogeneity evaluation. *Renewable Energy, 132*, 950–958.
88. Tarasov, D., Leitch, M., & Fatehi, P. (2018). Lignin–carbohydrate complexes: Properties, applications, analyses, and methods of extraction: A review. *Biotechnology for Biofuels, 11*(1), 269.
89. Terrett, O. M., & Dupree, P. (2019). Covalent interactions between lignin and hemicelluloses in plant secondary cell walls. *Current Opinion in Biotechnology, 56*, 97–104.
90. Traynard, P., Ayroud, A., & Eymery, A. (1953). Existence d'une liaison lignine-hydrates de carbone dans le bois. *Assoc Tech Ind Papetiere Bull, 2*, 45–52.
91. Vishtal, A. G., & Kraslawski, A. (2011). Challenges in industrial applications of technical lignins. *BioResources, 6*(3), 3547–3568.
92. Wang, P. Y., Bolker, H. I., & Purves, C. B. (1967). Uronic acid ester groups in some softwoods and hardwoods. *Tappi, 3*, 123–124.
93. Watanabe, T., & Koshijima, T. (1988). Evidence for an ester linkage between lignin and glucuronic acid in lignin–carbohydrate complexes by DDQ-oxidation. *Agricultural and Biological Chemistry, 52*(11), 2953–2955.

94. Watanabe, T., Azuma, J., & Koshijima, T. (1987). *A convenient method for preparing lignin-carbohydrate complex from Pinus densiflora wood*. Soc: Journal of the Japan Wood Research.
95. Watanabe, T., Kaizu, S., & Koshijima, T. (1986). Binding sites of carbohydrate moieties toward lignin in "lignin–carbohydrate complex" from *Pinus densiflora* wood. *Chemistry Letters, 15*(11), 1871–1874.
96. Westermark, U., & Gustafsson, K. (1994). Molecular size distribution of wood polymers in birch kraft pulps. *Holzforschung-International Journal of the Biology, Chemistry, Physics and Technology of Wood, 48*(s1), 146–150.
97. Wong, D. W. S., Chan, V. J., McCormack, A. a., Hirsch, J., & Biely, P. (2012). Functional cloning and expression of the Schizophyllum commune glucuronoyl esterase gene and characterization of the recombinant enzyme. *Biotechnology Research International, 2012*, 951267.
98. Xiao, B., Sun, X., & Sun, R. (2001). Chemical, structural, and thermal characterizations of alkali-soluble lignins and hemicelluloses, and cellulose from maize stems, rye straw, and rice straw. *Polymer Degradation and Stability, 74*(2), 307–319.
99. Xie, Y., Yasuda, S., Wu, H., & Liu, H. (2000). Analysis of the structure of lignin-carbohydrate complexes by the specific 13C tracer method. *Journal of Wood Science, 46*(2), 130–136.
100. Xu, G., Shi, Z., Zhao, Y., Deng, J., Dong, M., Liu, C., Murugadoss, V., Mai, X., & Guo, Z. (2019). Structural characterization of lignin and its carbohydrate complexes isolated from bamboo (Dendrocalamus sinicus). *International Journal of Biological Macromolecules, 126*, 376–384.
101. Yaku, F., Yamada, Y., & Koshijima, T. (1976). Lignin carbohydrate complex Pt. II. Enzymic degradation of acidic polysaccharide in Björkman LCC. *Holzforschung-International Journal of the Biology, Chemistry, Physics and Technology of Wood, 30*(5), 148–156.
102. Yamamoto, E., Bokelman, G. H., & Lewis, N. G. (1989). *Phenylpropanoid metabolism in cell walls: An overview*. ACS Publications. Washington DC.
103. Yan, Z., Li, J., Chang, S., Cui, T., Jiang, Y., Yu, M., & Li, S. (2015). Lignin relocation contributed to the alkaline pretreatment efficiency of sweet sorghum bagasse. *Fuel, 158*, 152–158.
104. Yang, G., Jahan, M. S., & Ni, Y. (2013). Structural characterization of pre-hydrolysis liquor lignin and its comparison with other technical lignins. *Current Organic Chemistry, 17*(15), 1589–1595.
105. Yelle, D. J., Kaparaju, P., Hunt, C. G., Hirth, K., Kim, H., Ralph, J., & Felby, C. (2013). Two-dimensional NMR evidence for cleavage of lignin and Xylan substituents in wheat straw through hydrothermal pretreatment and enzymatic hydrolysis. *Bio Energy Research, 6*(1), 211–221.
106. You, T. T., Zhang, L. M., Zhou, S. K., & Xu, F. (2015). Structural elucidation of lignin–carbohydrate complex (LCC) preparations and lignin from Arundo donax Linn. *Industrial Crops and Products, 71*, 65–74.
107. Zhang, Z., Harrison, M. D., Rackemann, D. W., Doherty, W. O., & O'Hara, I. M. (2016). Organosolv pretreatment of plant biomass for enhanced enzymatic saccharification. *Green Chemistry, 18*(2), 360–381.
108. Zhao, B. C., Chen, B. Y., Yang, S., Yuan, T. Q., Charlton, A., & Sun, R. C. (2016). Structural variation of lignin and lignin–carbohydrate complex in Eucalyptus grandis × E. urophylla during its growth process. *ACS Sustainable Chemistry & Engineering, 5*(1), 1113–1122.
109. Zhao, X., Qi, F., & Liu, D. (2017). Hierarchy nano-and ultrastructure of lignocellulose and its impact on the bioconversion of cellulose. *Nanotechnology for bioenergy and biofuel production* (pp. 117–151). Springer.
110. Zheng, Y., Pan, Z., & Zhang, R. (2009). Overview of biomass pretreatment for cellulosic ethanol production. *International Journal of Agricultural and Biological Engineering, 2*(3), 51–68.

Chapter 7
Systematic Metabolic Engineering of *Saccharomyces cerevisiae* for Efficient Utilization of Xylose

Jing Han, Guoli Gong, Xia Wu, and Jian Zha

Abbreviations

PPP	Pentose phosphate pathway
TCA	Tricarboxylic acid
XDH	Xylitol dehydrogenase
XI	Xylose isomerase
XR	Xylose reductase

7.1 Introduction

Biomass is a sustainable source of fermentation feedstock for the generation of value-added chemicals such as fuels and attracts much attention with the rising concern of rapidly decreasing reserve of fossil fuels and insufficient supply of cereals as staple food [1, 2]. The decomposition product of biomass, also termed lignocellulosic hydrolysate, consists of hexoses (mainly glucose) and pentoses (mainly xylose) [3]. Glucose can be readily converted to ethanol, organic acids, and other bulk chemicals by *Saccharomyces cerevisiae*, one of the most commonly used industrial fermentation microorganisms with robust resistance to environmental stresses and phage infection. However, wild-type *S. cerevisiae* strains cannot metabolize xylose, thus leading to inefficient utilization of lignocellulosic hydrolysates. To solve such a problem, extensive efforts of metabolic engineering have been made in the past two decades toward breeding of an efficient xylose-utilizing *S. cerevisiae* [4].

J. Han · G. Gong · X. Wu · J. Zha (✉)
School of Food and Biological Engineering, Shaanxi University of Science and Technology, Xi'an, Shaanxi, China
e-mail: gongguoli@sust.edu.cn; wuxia@sust.edu.cn; zhajian1985@sust.edu.cn

Z.-H. Liu, A. Ragauskas (eds.), *Emerging Technologies for Biorefineries, Biofuels, and Value-Added Commodities*,
https://doi.org/10.1007/978-3-030-65584-6_7

The key to the development of xylose-fermenting *S. cerevisiae* is the reconstruction of a xylose catabolic pathway in the host strain. The earliest explored pathway derived from natural xylose-utilizing yeasts, such as *Scheffersomyces stipitis*, converts xylose to xylulose in two steps. Xylose is reduced to xylitol by xylose reductase (XR) and then oxidized to xylulose by xylitol dehydrogenase (XDH) (Fig. 7.1). A more direct pathway is the xylose isomerase (XI) pathway whereby xylose is converted into xylulose by XI in a single step (Fig. 7.1). Xylulose is then channeled into glycolysis via phosphorylation and multiple biochemical reactions in the non-oxidative pentose phosphate pathway (PPP). Both xylose catabolic pathways have been functionally constructed in *S. cerevisiae*, and many attempts have been made to enhance xylose utilization for the production of ethanol, the major end product in the waves of biofuels. In recent years, the end products have been expanded to other chemicals such as lactate and isobutanol [5]. However, the introduced pathway often causes inefficient energy supply and other metabolic mismatch with native metabolic network, especially under anaerobic conditions [6]. Despite some great success on engineering xylose metabolism in recombinant yeasts, multiple key problems are still present in terms of xylose consumption rate, the generation of by-products, etc.

In this chapter, we will summarize the metabolic engineering of *S. cerevisiae* to achieve the goal of efficient xylose utilization (Fig. 7.2) with a focus on the strategies to regulate redox imbalance inherited with the XR-XDH pathway, to improve XI activity in the XI pathway, and to solve other problems existing in both pathways, such as insufficient metabolic flux of the non-oxidative PPP and glucose-mediated inhibition of xylose transportation. We will also discuss the future directions on *S. cerevisiae* engineering for highly efficient xylose metabolism.

7.2 Redox Imbalance

In the XR-XDH pathway, a prominent problem is redox imbalance and subsequent formation of the by-product xylitol. The redox imbalance is caused by poor recycling of redox cofactors in the initial oxidoreductive steps catalyzed by NADPH-preferring XR (encoded by *Xyl1*) and strictly NAD^+-dependent XDH (encoded by *Xyl2*) [7, 8]. Thus, many rational approaches to balancing cofactor recycling have been attempted. First, protein engineering of XR or XDH has been performed to alter the cofactor dependence for cofactor matching. For example, a mutant XR (R276H) based on the XR from *S. stipitis* prefers NADH to NADPH, and its incorporation into the xylose-assimilating pathway leads to better cofactor recycling, less xylitol production, and higher ethanol production [9]. In another case, a $NADP^+$-specific XDH mutant (D207A/I208R/F209S/N211R) has been constructed and its expression with NADPH-preferred wild-type XR also demonstrates the feasibility of protein engineering [10]. Second, regulation of cellular cofactor metabolism has been carried out to balance cofactor recycling for decreased xylitol formation and improved ethanol production [11, 12]. A water-forming NADH oxidase (NoxE)

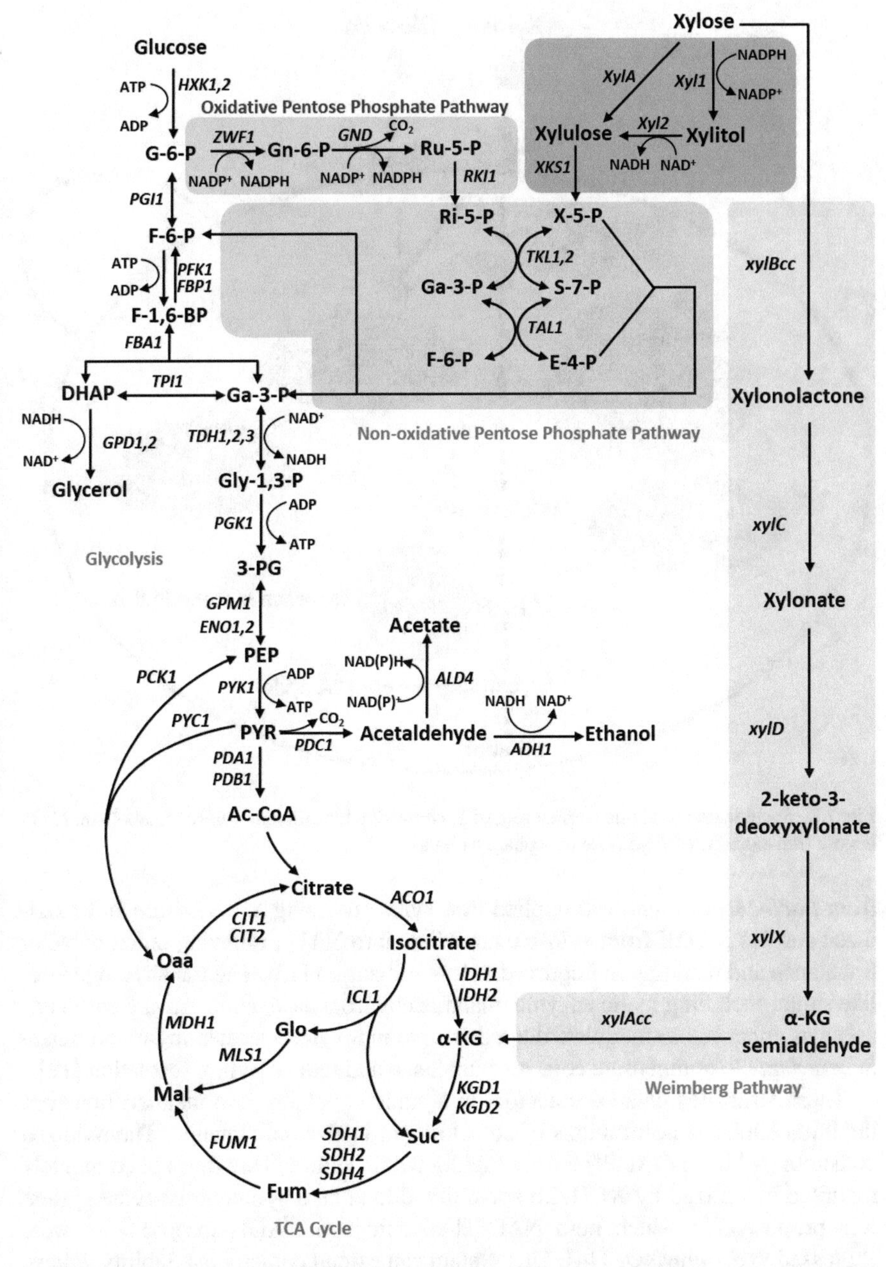

Fig. 7.1 The metabolic pathways for xylose metabolism in *S. cerevisiae*. Symbols: *X-5-P* xylulose-5-phosphate, *Ri-5-P* ribose-5-phosphate, *Ru-5-P* ribulose-5-phosphate, *Ga-3-P* glyceraldehyde-3-phosphate, *S-7-P* sedoheptulose-7-phosphate, *F-6-P* fructose-6-phosphate; *E-4-P* erythrose-4-phosphate, *G-6-P* glucose-6-phosphate, *Gn-6-P* 6-phosphogluconate, *F-1,6-BP* fructose-1,6-bisphosphate, *DHAP* dihydroxyacetonephosphate, *Gly-1,3-P* 1,3-bisphosphoglycerate, *3-PG* 3-phosphoglycerate, *PEP* phosphoenolpyruvate, *PYR* pyruvate, *Ac-CoA* acetyl-CoA, *Oaa* oxaloacetate, *Mal* malate, *Fum* fumarate, *Suc* succinate, *α-KG* α-ketoglutarate

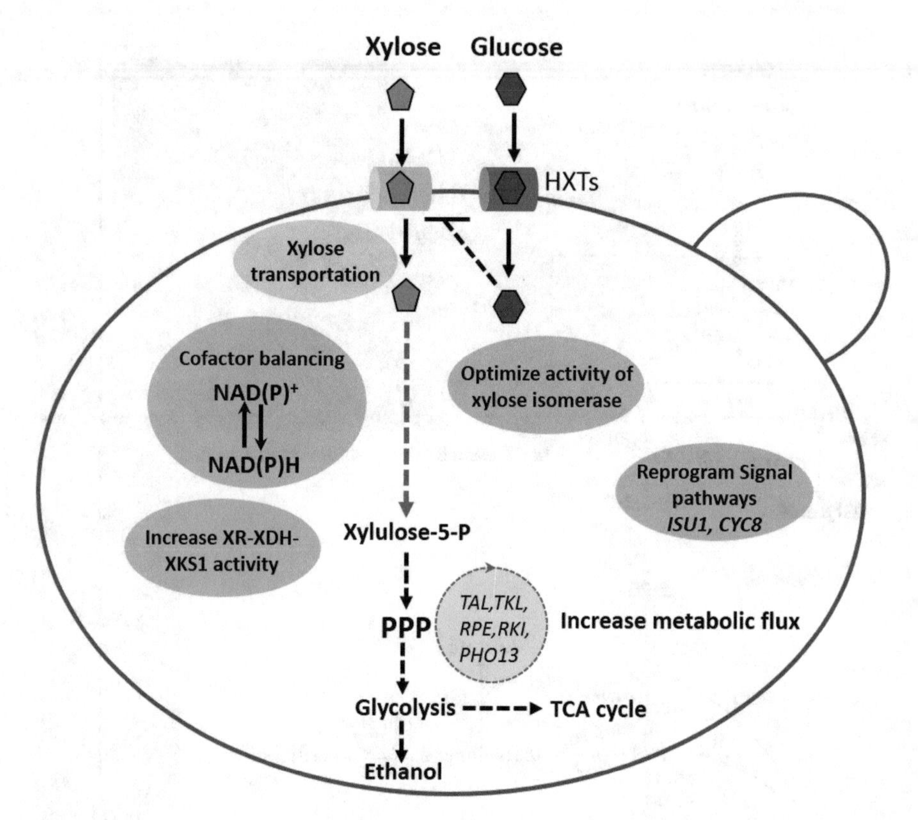

Fig. 7.2 Systematic metabolic engineering of *S. cerevisiae* for efficient xylose metabolism. *HXTs* hexose transporters, *PPP* pentose phosphate pathway

from *Lactococcus lactis* was applied in a xylose-utilizing *S. cerevisiae* and reoxidized surplus NADH from xylose metabolism into NAD^+, relieving redox cofactor imbalance and resulting in improved xylose utilization [13]. The transhydrogenase-like shunt, including malic enzyme, malate dehydrogenase, and pyruvate carboxylase, was enhanced to modulate the redox state in a xylose-fermenting recombinant *S. cerevisiae* to compensate cofactor imbalance and reduce xylitol formation [12].

These strategies balance cofactors in a static stoichiometric manner; however, the intracellular concentrations of cofactors are in dynamic change. The oxidized cofactors (NAD^+ or $NADP^+$) diffuse rapidly within cells and are thus not completely recruited or utilized by XDH. To solve this dilemma, a synthetic isozyme system was proposed, in which both NADPH-specific and NADH-specific XRs were expressed in *S. cerevisiae* [14]. This system can extend cofactor availability, relieve cofactor imbalance, and improve xylose metabolism.

Besides direct regulation of redox imbalance to reduce xylitol formation, an indirect way of regulating xylitol formation is regulation of xylitol secretion. In *S. cerevisiae*, aquaglyceroporin Fps1p is suggested to contribute to xylitol export, and its deletion can reduce xylitol production and increase ethanol yields [15].

7.3 Activity of Enzymes in Xylose Assimilation Pathway

The activities of enzymes in the initial metabolic pathway are important for efficient xylose metabolism in both the XR-XDH and the XI pathways. The XI pathway will be discussed in next section given the relative complexity of improving XI activity. For the XR-XDH pathway, the activities of XR and XDH determine the rate of xylose consumption [16]. A higher-level expression of *Xyl1* can lead to more active XR, providing stronger driving force for xylose metabolism, whereas XDH with a higher activity can convert xylitol into xylulose more efficiently, leading to less production of xylitol. The importance of XR and XDH activities has been confirmed in a couple of studies by direct modifications on *Xyl1* and *Xyl2* [17, 18], and comparative analysis of evolved strains and parental strains also suggests the role of high-level expression of *Xyl1* and *Xyl2* [19]. In addition to the activities of XR and XDH, XKS1 activity is also crucial to xylose metabolism as native XKS1 activity is not strong enough to phosphorylate xylulose into xylulose-5-phosphate efficiently [20]. The significance of XKS1 activity is evidenced by the fact that elevated expression of *XKS1* is often observed in evolved strains [21]. To increase the overall activities of intracellular XR, XDH, and XKS1, multiple integration of the xylose assimilation pathway is commonly performed through δ-integration or GIN11/FRT-based multigene integration system, and strains with improved xylose metabolism are selected based on colony size on plates with xylose as the sole carbon source [22].

High-level expression of *Xyl1*, *Xyl2*, and *XKS1* does not guarantee efficient xylose metabolism; these genes need to be expressed in a balanced manner. Pathway optimization methods, such as DNA-assembler-based combination and golden-gate-based pathway assembly, have been applied to optimize the expression of *Xyl1*, *Xyl2*, and *XKS1* [23, 24], which help to enhance xylose metabolism and reduce xylitol production.

7.4 Improving the Activity of Xylose Isomerase

The major problem with the XI pathway is the low XI activity when expressed in *S. cerevisiae*. *XylA* from *Piromyces* is the first reported XI gene that is functional in *S. cerevisiae* [25], and *XylA* from *Clostridium phytofermentans* is the first reported prokaryotic XI gene functional in *S. cerevisiae* [26]. Several other *XylA* genes have been reported to encode active XI in *S. cerevisiae* after deep mining of thousands of genome sequences (Table 7.1). However, XI activity in these systems is still low, hindering efficient xylose utilization.

XI is partially inhibited by xylitol generated from xylose by the endogenous aldose reductase GRE3 [25]; hence the *GRE3* gene is commonly knocked out when expressing *XylA* in *S. cerevisiae*. Nonetheless, XI activity is still not high enough

Table 7.1 Properties of various xylose isomerases functionally expressed in *S. cerevisiae*

Source of gene	Km(xylose) (mM)	Ki(xylitiol) (mM)	Vmax (μmol/ mg protein/min)	Classification	References
Piromyces sp. *E2*	49.85 ± 2.82	4.6 ± 1.78	0.0538	Fungal	[25]
C. phytofermentans	66.01 ± 1.00	14.5 ± 1.08	0.0344	Bacterial	[27]
Bacteroides stercoris	54.03 ± 1.24	5.1 ± 1.15	–	Bacterial	[27]
Burkholderia cenocepacia	–	–	–	Bacterial	[28]
Orpinomyces sp.	–	–	1.91	Fungal	[29]
Prevotella ruminicola TC2–24	40	–	0.28	Bacterial	[30]
Termite hindgut	10.52 ± 1.10	–	0.0074	Protists	[31]
Lachnoclostridium phytofermentans	41.8 ± 0.8		0.064 ± 0.002	Bacterial	[32]
Ruminococcus flavefaciens	117	–	0.29	Bacterial	[33]

due to the poor expression of this enzyme. To overcome this obstacle, XI from *Propionibacterium acidipropionici* and the molecular chaperon GroEL-GroES from *Escherichia coli* were co-expressed in *S. cerevisiae* for better enzyme folding, resulting in functional XI despite slow xylose consumption [34]. Additionally, error-prone PCR-based random mutagenesis of *XylA* from *Clostridium phytofermentans* led to the identification of two mutants with improved xylose isomerization activity both in vivo and in vitro [32].

During adaptation of recombinant *S. cerevisiae* expressing *XylA*, multiple genetic changes occur to *XylA* as well as other genes, leading to improved xylose isomerization. In one study, insertion of *XylA* close to an ARS (autonomously replicating sequence) resulted in ~ninefold amplification of its copy number in the evolved strain, which conferred higher *XylA* expression and elevated XI activity [35]. Interestingly, reduced *XylA* expression was observed but with higher XI activity in other studies [21, 36]. Other mutations that contribute to elevated activity of XI have been reported. In one evolution, mutations (including ASK10M475R and ASK10 deletion) on ASK10, the gene encoding a stress response regulator, were identified, which upregulated the expression levels of molecular chaperone-encoding genes (HSP26, SSA1, and HSP104) and subsequently improved the folding and activity of XI [37]. A mutation was identified in *PMR1*, which encodes a high-affinity Ca^{2+}/Mn^{2+} ATPase. In *pmr1* strains, intracellular concentration of Mn^{2+}, the essential metal for XI activity, was much higher than that in the parental strain, causing elevated XI activity and accelerated xylose consumption [38–40].

7.5 The Low Metabolic Flux of the Non-oxidative Pentose Phosphate Pathway

The non-oxidative PPP is the channel to direct xylulose-5-phosphate to glycolysis pathway through the XR-XDH pathway or the XI pathway (Fig. 7.1). In natural *S. cerevisiae*, PPP acts as the pool to provide C-5 sugars for the biosynthesis of nucleotides, aromatic amino acids, and so on, and its metabolic flux is naturally lower than that of glycolysis or the tricarboxylic acid (TCA) cycle. In recombinant yeast without any modification on PPP, intermediates such as xylulose-5-phosphate and sedoheptulose-7-phosphate accumulate during xylose fermentation, indicating that the native PPP lacks sufficient metabolic activity in one or more steps in xylose catabolism to allow for significant cell growth and xylose fermentation either aerobically or anaerobically [20]. Metabolomic analysis and transcriptional analysis also support this inference. Thus, overexpression of genes in the non-oxidative PPP for increased xylose consumption is favored in the construction of xylose-utilizing *S. cerevisiae*. A golden-gate-based combinatorial optimization method was applied to optimize the PPP, with five promoters incorporated into the whole pathway [24]. The improved strain harbored *RPE* under the control of the strong promoter *TDH3* and other pathway genes (*RKI*, *TKL*, and *TAL*) under the control of moderate promoters. In another combinatorial test, overexpression of *RKI1* and *TAL1* derived from *Kluyveromyces marxianus* with *S. cerevisiae TKL1* increased xylose consumption rate by nearly twofold relative to the parental strain [41]. In addition to overexpression of pathway genes, inactivation of *PHO13* was also reported to be effective in enhancing the metabolic flux of PPP. *PHO13* can repress TAL1 activity, and the limited transaldolase activity can cause dephosphorylation of sedoheptulose-7-phosphate, leading to carbon loss and inefficient xylose metabolism [42].

7.6 Xylose Transportation

S. cerevisiae lacks a xylose-specific transport system, and the transport of xylose into its cytoplasm is mediated by hexose transporters which show much higher affinity to glucose than to xylose [8]. Such discrimination on affinity causes differential uptake and utilization in the presence of multiple sugars, with glucose having the highest priority and metabolized first, thus leading to "glucose repression" [43]. Such sequential utilization of xylose after glucose exhaustion severely decreases xylose consumption rate. Thus, identification and development of xylose-preferring transporters are crucial for efficient xylose fermentation.

Since native hexose transporters recognize both glucose and xylose, it is useful to develop hexose transporter-null strains, such as EBY.VW4000 with more than 20 hexose transporter genes knocked out, for screening and characterization of new transporters by functional complementation of growth on glucose or xylose when a xylose pathway is expressed [44–46]. It has been identified that transporters Hxt2,

Hxt4, Hxt5, Hxt6, Hxt7, and Gal2 are responsible for xylose uptake [47, 48], and Hxt7 and Hxt5 seem to be the main xylose transporters when xylose is the only carbon source or in co-fermentation when glucose reaches low concentrations [47]. Recently, a new transporter HXT11 has been identified, which is capable of simutaneously transporting glucose and xylose [49].

To improve xylose uptake efficiency, the endogenous xylose transporter genes have been engineered. The fusion chimera Hxt36 derived from Hxt3 and Hxt6 was constructed, and its mutant Hxt36-N367A transports xylose with a higher rate and improved affinity, enabling efficient co-consumption of glucose and xylose [50]. A Gal2 mutant (Gal2-N376F) was constructed based on structural modeling, and it showed high specificity for xylose with a moderate transport velocity and completely lost the ability to transport hexoses [46]. The mutant HXT7(F79S) showed improved xylose uptake that allowed *S. cerevisiae* to exhibit significant growth on xylose [51]. Gene shuffling of *HXT1-7* and *GAL2* created two homologous fusion proteins, both consisting of the major central part of Hxt2 and various smaller parts of other Hxt proteins. Both fusions presented increased xylose affinity (8.1 mM) compared to Hxt2 (23.7 mM) [52].

Discovery and mining of specific xylose transporters from natural pentose-assimilating species has also been conducted. These natural microorganisms contain both low-affinity and high-affinity sugar transport systems for xylose uptake, comprising proton symporters and facilitated diffusion transporters. With increasing genomic data and the development of efficient genome mining tools, numerous putative genes for xylose-specific heterologous transporters have been identified. *SUT1*, *SUT2*, *XUT1*, *XUT3*, *XUT4*, *Xyp29*, *SUT3*, *HXT2.6*, and cellobiose permease *QUP2* from *S. stipitis* have been reported to enable the *HXT*-null *S. cerevisiae* strains to grow on xylose [53]. Besides, xylose transporters from other species have also been discovered and verified in *S. cerevisiae*, including arabinose: H^+ symporters (AraE) from *Bacillus subtilis* and *Corynebacterium glutamicum*, KmAXT1 (arabinose-xylose transporter) from *Kluyveromyces marxianus*, PgAXT1 from *Pichia guilliermondii*, etc. [54–56]. However, in all these cases, the uptake efficacy of these heterologous transporters is limited and cannot dramatically improve xylose fermentation rate. Detailed information on xylose transporters has been summarized in a recent review [57].

Apart from expression of xylose transporters, modulation of sugar signaling pathway to regulate xylose uptake is also feasible. CYC8 is a co-repressor in glucose sensing and signaling pathways. A point mutation in *CYC8* (Y353C) gave rise to increased expression levels of *HXTs*, causing a higher transport flux of xylose into the cytoplasm in the presence of glucose [58]. Engineering of the hexose transportation system to relieve glucose repression is still daunting. An alternative has been proposed, in which cellulose in biomass is decomposed into cellobiose (dimer of glucose) instead of glucose with the optimization of the saccharification process before the fermentation step. Cellobiose transporter (CDT-1) and β-glucosidase (gh1-1) from *Neurospora crassa* have been successfully introduced into *S. cerevisiae* to efficiently metabolize cellobiose and xylose simultaneously, which avoids the competition of hexose transporters between glucose and xylose [59, 60].

7.7 Other Targets for Metabolic Engineering

Apart from the genetic modifications on the targets mentioned above, other targets that are located outside the xylose metabolic pathway have also been discovered (Fig. 7.2). In one study, mutations in regulatory pathways were observed in the strain expressing XI pathway after several rounds of adaption. The mutated enzymes contained a component of MAP kinase (MAPK) signaling (HOG1), a regulator of protein kinase A (PKA) signaling (IRA2), and a scaffolding protein for mitochondrial iron-sulfur (Fe-S) cluster biogenesis (ISU1) [61]. A following study verified that the *ira2*Δ *isu1*Δ double deletion led to strains with the highest specific xylose consumption and ethanol production rates but also the lowest biomass [62]. Also, the double deletion enabled the strain to sense both high and low concentrations of glucose, whereas the parental strain could only signal low concentrations of glucose. In other studies, improved xylose metabolism was observed with decreased expression of *ISU1* or addition of iron [63, 64]. These studies validate the role of *ISU1* in regulating xylose metabolism. Similarly, a mutation in *SSK2* was identified in an evolved strain expressing an XI pathway. SSK2 is a member of MAPKKK (MAPK kinase kinases) signaling pathway and interacts with Hog1p by direct phosphorylation to activate the HOG pathway. Deletion of *SSK2* could improve the ability to metabolize xylose in the parental strain without adaptive evolution, suggesting that *SSK2* is a key player in the regulatory network of xylose fermentation [63].

Many metabolic and transcriptional investigations indicate that the reprogramming of glucose signaling pathways contributes to improved xylose consumption. A transcriptome study reveals that Snf1/Mig1-mediated regulation, a part of glucose sensing and repression network, was altered in the evolved strain and might be related to the improvement of xylose utilization [65]. An HXK2p variant (F159Y) had 64% higher catalytic activity in the presence of xylose compared to the wild-type enzyme and is expected to be a key component for increasing the productivity of recombinant xylose-fermenting strains for bioethanol production [66].

Fermentation of xylose alone drastically increases the level of citrate in the TCA cycle and increases the generation of aromatic amino acids tryptophan and tyrosine, strongly supporting the view that carbon starvation is induced [67]. A ^{13}C metabolic flux analysis of recombinant strains reveals that high cell maintenance energy is one of the key bottlenecks in xylose metabolism [68]. Xylose leads to an inefficient metabolic state where the biosynthetic capabilities and energy balance are severely impaired. Thus, modulation of ATP metabolism could facilitate energy balancing and accelerate xylose consumption.

Additionally, systematic analysis of the whole metabolic network reveals other new targets as regulation points for improved xylose metabolism, including trehalose synthase genes, stress responsive genes mediated by Msn2/4p, and genes involved in spore wall metabolism and ammonium uptake [69]. An RNA-seq-based analysis of xylose metabolism in different host strains demonstrates transcription factors involved in homeostasis (Aft1p and Ste12p), regulation of amino acid metabolism (Gcn4p, Gcr2p, and Met4p), and stress responses (Msn2p, Rpn4p,

Sfp1p, and Yap1p) are possible determinants of xylose metabolism in different host strains [70]. These results indicate the complexity of the xylose pathway and the great potential of reprogramming regulatory targets beyond the initial xylose catabolism steps to further improve xylose metabolism.

7.8 New Xylose Metabolic Pathway

The commonly used xylose assimilation pathway, i.e., the XR-XDH pathway or the XI pathway, suffers from its heavy dependence on the non-oxidative PPP and glycolysis and strong interference with them. A promising alternative pathway for xylose metabolism, called the Weimberg pathway, has been found in some archaea or extremophiles [71] (Fig. 7.1). Through this pathway, xylose is initially oxidized by xylose dehydrogenase (XylB, encoded by *xylB*) to xylono-γ-lactone, which is further converted to xylonate by xylono-γ-lactone lactonase (XylC, encoded by *xylC*). Xylonate then undergoes two successive dehydration reactions by xylonate dehydratase (XylD, encoded by *xylD*) and 2-keto-3-deoxy-xylonate dehydratase (XylX, encoded by *xylX*), forming α-ketoglutarate semialdehyde. In the last reaction, α-ketoglutarate semialdehyde is oxidized by α-ketoglutarate semialdehyde dehydrogenase (XylA, encoded by *xylA*) into α-ketoglutarate, which enters the TCA cycle.

The Weimberg pathway is nonphosphorylative and can convert xylose into α-ketoglutarate in five enzymatic steps with a 100 mol% theoretical yield from xylose to α-ketoglutarate [72]. Moreover, the pathway can bypass the interactions with PPP and glycolysis, and channel xylose to α-ketoglutarate in the TCA cycle, which can be used for the production of many value-added chemicals, expanding the end products of xylose metabolism. The Weimberg pathway has been successfully constructed and optimized in *E. coli* for the production of 1,4-butanediol and mesaconate from xylose [72, 73]. Recently, the Weimberg pathway from *Caulobacter crescentus* was cloned and expressed in *S. cerevisiae*, but no obvious growth could be detected on xylose [74]. A functional Weimberg pathway was constructed in *S. cerevisiae* after a series of optimization. Up to 57% of the carbon was metabolized into biomass and carbon dioxide through the combinatorial modifications including the replacement of *xylA* by its orthologue *KsaD* from *Corynebacterium glutamicum*, enhanced expression of the lower Weimberg pathway (*XylD*, *XylX*, and *KsaD*), and deletion of the iron regulation repressor *FRA2* [75]. Although the Weimberg pathway in *S. cerevisiae* is not as efficient as that in *E. coli* or other eukaryotic species, its functional expression opens a new door to utilization of xylose in *S. cerevisiae*.

7.9 Conclusions and Future Perspectives

Efficient utilization of xylose is important for economic transformation of biomass materials to biofuels and chemicals as xylose is the secondly abundant sugar in lignocellulosic hydrolysates (up to 30%). *S. cerevisiae* is commonly used for industrial fermentation and has been one of the most frequently used microorganisms in metabolic engineering. An efficient xylose-utilizing *S. cerevisiae* would be a good platform strain when lignocellulosic hydrolysates are employed as the fermentation feedstocks. In the past few decades, many research groups have taken extensive efforts to engineer *S. cerevisiae* for efficient bioconversion of xylose to ethanol due to the intrinsic ability of this microbe to produce a high amount of ethanol, and global concern on energy security and climate change [76]. These efforts have created many recombinant *S. cerevisiae* strains with very strong xylose fermentation capability.

The optimization of the xylose pathway focuses on genes in the initial pathway (including *Xyl1*, *Xyl2*, *XKS1*, and *XylA*), cofactor imbalance, and genes in the non-oxidative PPP. Combinatorial optimization has been carried out to achieve sufficient and balanced expression of *Xyl1/Xyl2/XKS1* or *XylA/XKS1* to form a strong driving force for xylose metabolism. Multiple methods such as promoter shuffling, DNA assembler, and golden-gate assembly are convenient and feasible methods to obtain an efficient xylose assimilation pathway. In the XR-XDH pathway, the cofactor balancing between XR and XDH is also effective to improve xylose metabolism and decrease xylitol production. Multiple strategies have been used to successfully balance cofactors, including protein engineering of XR or XDH, direct transformation of extra cofactor NADH, and regulation of intracellular cofactor recycling. Combinatorial optimization of the genes in the non-oxidative PPP facilitates improvement of xylose metabolism, as its natural metabolic flux is limited to accommodate the flux from the initial xylose assimilation pathway. Elevated expression of a single gene (e.g., *TAL1*) or multiple genes (e.g., *TAL1, TKL, RPE* and *RKI*) often leads to enhanced xylose metabolism.

In recent years, identification of novel xylose transporters and engineering of xylose transportation are an attractive direction. Multiple hexose transporters with xylose transportation capability have been discovered. Also, engineered hexose transporters with high xylose specificity have been reported although their expression in recombinant *S. cerevisiae* shows limited elevation of xylose uptake rate. It should be noted that direct introduction of a single or multiple transporter may not be sufficient to achieve xylose uptake as fast as glucose, given the delicate and hierarchical structure of glucose sensing and transportation system in *S. cerevisiae*. Hence, it is an option in future engineering to rebuild a similar sensing and delivery system of xylose for its rapid uptake and utilization.

Xylose transportation is also influenced by native signaling pathways in *S. cerevisiae*, such as the SNF1/Mig1p pathway, the cAMP/PKA pathway, and the Snf3p/Rgt2p pathway. These signaling and sensing systems change their behavior when xylose is used as the carbon source. Generally, xylose is not recognized as a fermentable sugar and can have negative impact on cell metabolism, such as inactivation of key pathway enzymes in glycolysis. At present, the whole signaling network has not been figured out although some regulators have been identified by comparative systematic analysis. It is expected that the ongoing studies on signaling mapping will offer new targets and insights to accelerate xylose utilization. All these studies suggest that xylose transportation may be the major obstacle in xylose metabolism in *S. cerevisiae*. Apart from direct engineering attempts to partially overcome this difficulty, fermentation technology of cellobiose and xylose, coupled with optimization of pretreatment and saccharification processes, is also helpful to improve xylose transportation. Finally, construction of an efficient xylose-utilizing platform strain is not an end to the utilization of lignocellulosic hydrolysates. In real practice, the hydrolysates contain many inhibitors, such as furfural and phenolics, which can severely suppress *S. cerevisiae* viability. It is critically essential for xylose-utilizing yeast strains to tolerate those inhibitors and maintain robust utilization capability of sugars in biomass hydrolysates. Such strains have been developed by continuous adaptation in lignocellulosic hydrolysates [77, 78]. Meanwhile, there are abundant lignin degradation products in these hydrolysates, mainly aromatic compounds which can be harmful to cells. Bioconversion of these "toxic compounds" into useful chemicals, albeit still a challenging task, can considerably increase the process cost-effectiveness and lead to comprehensive resource utilization.

References

1. Hahn-Hagerdal, B., Karhumaa, K., Fonseca, C., Spencer-Martins, I., & Gorwa-Grauslund, M. F. (2007). Towards industrial pentose-fermenting yeast strains. *Applied Microbiology and Biotechnology, 74*(5), 937–953.
2. Kim, S. R., Park, Y.-C., Jin, Y.-S., & Seo, J.-H. (2013). Strain engineering of *Saccharomyces cerevisiae* for enhanced xylose metabolism. *Biotechnology Advances, 31*(6), 851–861.
3. Aristidou, A., & Penttila, M. (2000). Metabolic engineering applications to renewable resource utilization. *Current Opinion in Biotechnology, 11*(2), 187–198.
4. Moysés, D. N., Reis, V. C. B., de Almeida, J. R. M., de Moraes, L. M. P., & Torres, F. A. G. (2016). Xylose fermentation by *Saccharomyces cerevisiae*: Challenges and prospects. *International Journal of Molecular Sciences, 17*(3), 207.
5. Kwak, S., & Jin, Y.-S. (2017). Production of fuels and chemicals from xylose by engineered *Saccharomyces cerevisiae*: A review and perspective. *Microbial Cell Factories, 16*(1), 82.
6. Jin, Y.-S., Laplaza, J. M., & Jeffries, T. W. (2004). *Saccharomyces cerevisiae* engineered for xylose metabolism exhibits a respiratory response. *Applied and Environmental Microbiology, 70*(11), 6816–6825.
7. Zha, J., Hu, M.-L., Shen, M.-H., Li, B.-Z., Wang, J.-Y., & Yuan, Y.-J. (2012). Balance of XYL1 and XYL2 expression in different yeast chassis for improved xylose fermentation. *Frontiers in Microbiology, 3*, 355.
8. Chu, B. C., & Lee, H. (2007). Genetic improvement of *Saccharomyces cerevisiae* for xylose fermentation. *Biotechnology Advances, 25*(5), 425–441.
9. Lee, S.-H., Kodaki, T., Park, Y.-C., & Seo, J.-H. (2012). Effects of NADH-preferring xylose reductase expression on ethanol production from xylose in xylose-metabolizing recombinant *Saccharomyces cerevisiae*. *Journal of Biotechnology, 158*(4), 184–191.

10. Matsushika, A., Watanabe, S., Kodaki, T., Makino, K., Inoue, H., Murakami, K., Takimura, O., & Sawayama, S. (2008). Expression of protein engineered $NADP^+$-dependent xylitol dehydrogenase increases ethanol production from xylose in recombinant *Saccharomyces cerevisiae*. *Applied Microbiology and Biotechnology, 81*(2), 243–255.
11. Roca, C., Nielsen, J., & Olsson, L. (2003). Metabolic engineering of ammonium assimilation in xylose-fermenting *Saccharomyces cerevisiae* improves ethanol production. *Applied and Environmental Microbiology, 69*(8), 4732–4736.
12. Suga, H., Matsuda, F., Hasunuma, T., Ishii, J., & Kondo, A. (2013). Implementation of a transhydrogenase-like shunt to counter redox imbalance during xylose fermentation in *Saccharomyces cerevisiae*. *Applied Microbiology and Biotechnology, 97*(4), 1669–1678.
13. Zhang, G.-C., Turner, T. L., & Jin, Y.-S. (2017). Enhanced xylose fermentation by engineered yeast expressing NADH oxidase through high cell density inoculums. *Journal of Industrial Microbiology & Biotechnology, 44*(3), 387–395.
14. Jo, J.-H., Park, Y.-C., Jin, Y.-S., & Seo, J.-H. (2017). Construction of efficient xylose-fermenting *Saccharomyces cerevisiae* through a synthetic isozyme system of xylose reductase from *Scheffersomyces stipitis*. *Bioresource Technology, 241*, 88–94.
15. Wei, N., Xu, H., Kim, S. R., & Jin, Y.-S. (2013). Deletion of FPS1, encoding aquaglyceroporin Fps1p, improves xylose fermentation by engineered *Saccharomyces cerevisiae*. *Applied and Environmental Microbiology, 79*(10), 3193–3201.
16. Karhumaa, K., Fromanger, R., Hahn-Hägerdal, B., & Gorwa-Grauslund, M.-F. (2007). High activity of xylose reductase and xylitol dehydrogenase improves xylose fermentation by recombinant *Saccharomyces cerevisiae*. *Applied Microbiology and Biotechnology, 73*(5), 1039–1046.
17. Kim, S. R., Ha, S.-J., Kong, I. I., & Jin, Y.-S. (2012). High expression of XYL2 coding for xylitol dehydrogenase is necessary for efficient xylose fermentation by engineered *Saccharomyces cerevisiae*. *Metabolic Engineering, 14*(4), 336–343.
18. Jeppsson, M., Träff, K., Johansson, B., Hahn-Hägerdal, B., & Gorwa-Grauslund, M. F. (2003). Effect of enhanced xylose reductase activity on xylose consumption and product distribution in xylose-fermenting recombinant *Saccharomyces cerevisiae*. *FEMS Yeast Research, 3*(2), 167–175.
19. Zha, J., Shen, M., Hu, M., Song, H., & Yuan, Y. (2014). Enhanced expression of genes involved in initial xylose metabolism and the oxidative pentose phosphate pathway in the improved xylose-utilizing *Saccharomyces cerevisiae* through evolutionary engineering. *Journal of Industrial Microbiology & Biotechnology, 41*(1), 27–39.
20. Toivari, M. H., Aristidou, A., Ruohonen, L., & Penttilä, M. (2001). Conversion of xylose to ethanol by recombinant *Saccharomyces cerevisiae*: Importance of xylulokinase (XKS1) and oxygen availability. *Metabolic Engineering, 3*(3), 236–249.
21. Qi, X., Zha, J., Liu, G.-G., Zhang, W., Li, B.-Z., & Yuan, Y.-J. (2015). Heterologous xylose isomerase pathway and evolutionary engineering improve xylose utilization in *Saccharomyces cerevisiae*. *Frontiers in Microbiology, 6*, 1165.
22. Kato, H., Matsuda, F., Yamada, R., Nagata, K., Shirai, T., Hasunuma, T., & Kondo, A. (2013). Cocktail δ-integration of xylose assimilation genes for efficient ethanol production from xylose in *Saccharomyces cerevisiae*. *Journal of Bioscience and Bioengineering, 116*(3), 333–336.
23. Du, J., Yuan, Y., Si, T., Lian, J., & Zhao, H. (2012). Customized optimization of metabolic pathways by combinatorial transcriptional engineering. *Nucleic Acids Research, 40*(18), e142.
24. Latimer, L. N., & Dueber, J. E. (2017). Iterative optimization of xylose catabolism in *Saccharomyces cerevisiae* using combinatorial expression tuning. *Biotechnology and Bioengineering, 114*(6), 1301–1309.
25. Kuyper, M., Harhangi, H. R., Stave, A. K., Winkler, A. A., Jetten, M. S. M., de Laat, W. T. A. M., den Ridder, J. J. J., den Camp, H. J. M. O., van Dijken, J. P., & Pronk, J. T. (2003). High-level functional expression of a fungal xylose isomerase: The key to efficient ethanolic fermentation of xylose by *Saccharomyces cerevisiae*? *FEMS Yeast Research, 4*(1), 69–78.

26. Brat, D., Boles, E., & Wiedemann, B. (2009). Functional expression of a bacterial xylose isomerase in *Saccharomyces cerevisiae*. *Applied and Environmental Microbiology, 75*(8), 2304–2311.
27. Ha, S.-J., Kim, S. R., Choi, J.-H., Park, M. S., & Jin, Y.-S. (2011). Xylitol does not inhibit xylose fermentation by engineered *Saccharomyces cerevisiae* expressing *xylA* as severely as it inhibits xylose isomerase reaction *in vitro*. *Applied Microbiology and Biotechnology, 92*(1), 77–84.
28. Vilela, L. F., de Araujo, V. P. G., Paredes, R. S., Bon, E. P. S., Torres, F. A. G., Neves, B. C., & Eleutherio, E. C. A. (2015). Enhanced xylose fermentation and ethanol production by engineered *Saccharomyces cerevisiae* strain. *AMB Express, 5*, 16.
29. Madhavan, A., Tamalampudi, S., Ushida, K., Kanai, D., Katahira, S., Srivastava, A., Fukuda, H., Bisaria, V. S., & Kondo, A. (2009). Xylose isomerase from polycentric fungus *Orpinomyces*: Gene sequencing, cloning, and expression in *Saccharomyces cerevisiae* for bioconversion of xylose to ethanol. *Applied Microbiology and Biotechnology, 82*(6), 1067–1078.
30. Hector, R. E., Dien, B. S., Cotta, M. A., & Mertens, J. A. (2013). Growth and fermentation of D-xylose by *Saccharomyces cerevisiae* expressing a novel D-xylose isomerase originating from the bacterium *Prevotella ruminicola* TC2-24. *Biotechnology for Biofuels, 6*(1), 84.
31. Katahira, S., Muramoto, N., Moriya, S., Nagura, R., Tada, N., Yasutani, N., Ohkuma, M., Onishi, T., & Tokuhiro, K. (2017). Screening and evolution of a novel protist xylose isomerase from the termite *Reticulitermes speratus* for efficient xylose fermentation in *Saccharomyces cerevisiae*. *Biotechnology for Biofuels, 10*(1), 203.
32. Seike, T., Kobayashi, Y., Sahara, T., Ohgiya, S., Kamagata, Y., & Fujimori, K. E. (2019). Molecular evolutionary engineering of xylose isomerase to improve its catalytic activity and performance of micro-aerobic glucose/xylose co-fermentation in *Saccharomyces cerevisiae*. *Biotechnology for Biofuels, 12*(1), 139.
33. Aeling, K. A., Salmon, K. A., Laplaza, J. M., Li, L., Headman, J. R., Hutagalung, A. H., & Picataggio, S. (2012). Co-fermentation of xylose and cellobiose by an engineered *Saccharomyces cerevisiae*. *Journal of Industrial Microbiology & Biotechnology, 39*(11), 1597–1604.
34. Temer, B., dos Santos, L. V., Negri, V. A., Galhardo, J. P., Magalhães, P. H. M., José, J., Marschalk, C., Corrêa, T. L. R., Carazzolle, M. F., & Pereira, G. A. G. (2017). Conversion of an inactive xylose isomerase into a functional enzyme by co-expression of GroEL-GroES chaperonins in *Saccharomyces cerevisiae*. *BMC Biotechnology, 17*(1), 71.
35. Demeke, M. M., Foulquié-Moreno, M. R., Dumortier, F., & Thevelein, J. M. (2015). Rapid evolution of recombinant *Saccharomyces cerevisiae* for xylose fermentation through formation of extra-chromosomal circular DNA. *PLoS Genetics, 11*(3), e1005010.
36. Zhou, H., Cheng, J.-s., Wang, B. L., Fink, G. R., & Stephanopoulos, G. (2012). Xylose isomerase overexpression along with engineering of the pentose phosphate pathway and evolutionary engineering enable rapid xylose utilization and ethanol production by *Saccharomyces cerevisiae*. *Metabolic Engineering, 14*(6), 611–622.
37. Hou, J., Jiao, C., Peng, B., Shen, Y., & Bao, X. (2016). Mutation of a regulator Ask10p improves xylose isomerase activity through up-regulation of molecular chaperones in *Saccharomyces cerevisiae*. *Metabolic Engineering, 38*, 241–250.
38. Verhoeven, M. D., Lee, M., Kamoen, L., van den Broek, M., Janssen, D. B., Daran, J.-M. G., van Maris, A. J. A., & Pronk, J. T. (2017). Mutations in PMR1 stimulate xylose isomerase activity and anaerobic growth on xylose of engineered *Saccharomyces cerevisiae* by influencing manganese homeostasis. *Scientific Reports, 7*(1), 46155.
39. Son, H., Lee, S. M., & Kim, K. J. (2018). Crystal structure and biochemical characterization of xylose isomerase from *Piromyces* sp. E2. *Journal of Microbiology and Biotechnology, 28*(4), 571–578.
40. Lee, M., Rozeboom, H. J., de Waal, P. P., de Jong, R. M., Dudek, H. M., & Janssen, D. B. (2017). Metal dependence of the xylose isomerase from *Piromyces* sp. E2 explored by activity profiling and protein crystallography. *Biochemistry, 56*(45), 5991–6005.

41. Kobayashi, Y., Sahara, T., Ohgiya, S., Kamagata, Y., & Fujimori, K. E. (2018). Systematic optimization of gene expression of pentose phosphate pathway enhances ethanol production from a glucose/xylose mixed medium in a recombinant *Saccharomyces cerevisiae*. *AMB Express, 8*(1), 139.
42. Xu, H., Kim, S., Sorek, H., Lee, Y., Jeong, D., Kim, J., Oh, E. J., Yun, E. J., Wemmer, D. E., Kim, K. H., Kim, S. R., & Jin, Y.-S. (2016). PHO13 deletion-induced transcriptional activation prevents sedoheptulose accumulation during xylose metabolism in engineered *Saccharomyces cerevisiae*. *Metabolic Engineering, 34*, 88–96.
43. Jojima, T., Omumasaba, C. A., Inui, M., & Yukawa, H. (2010). Sugar transporters in efficient utilization of mixed sugar substrates: Current knowledge and outlook. *Applied Microbiology and Biotechnology, 85*(3), 471–480.
44. Wieczorke, R., Krampe, S., Weierstall, T., Freidel, K., Hollenberg, C. P., & Boles, E. (1999). Concurrent knock-out of at least 20 transporter genes is required to block uptake of hexoses in *Saccharomyces cerevisiae*. *FEBS Letters, 464*(3), 123–128.
45. Young, E. M., Tong, A., Bui, H., Spofford, C., & Alper, H. S. (2014). Rewiring yeast sugar transporter preference through modifying a conserved protein motif. *Proceedings of the National Academy of Sciences, 111*(1), 131–136.
46. Farwick, A., Bruder, S., Schadeweg, V., Oreb, M., & Boles, E. (2014). Engineering of yeast hexose transporters to transport D-xylose without inhibition by D-glucose. *Proceedings of the National Academy of Sciences, 111*(14), 5159–5164.
47. Hamacher, T., Becker, J., Gárdonyi, M., Hahn-Hägerdal, B., & Boles, E. (2002). Characterization of the xylose-transporting properties of yeast hexose transporters and their influence on xylose utilization. *Microbiology, 148*(9), 2783–2788.
48. Young, E., Poucher, A., Comer, A., Bailey, A., & Alper, H. (2011). Functional survey for heterologous sugar transport proteins, using *Saccharomyces cerevisiae* as a host. *Applied and Environmental Microbiology, 77*(10), 3311–3319.
49. Shin, H. Y., Nijland, J. G., de Waal, P. P., de Jong, R. M., Klaassen, P., & Driessen, A. J. M. (2015). An engineered cryptic Hxt11 sugar transporter facilitates glucose-xylose co-consumption in *Saccharomyces cerevisiae*. *Biotechnology for Biofuels, 8*, 176.
50. Nijland, J. G., Shin, H. Y., de Jong, R. M., de Waal, P. P., Klaassen, P., & Driessen, A. J. (2014). Engineering of an endogenous hexose transporter into a specific D-xylose transporter facilitates glucose-xylose co-consumption in *Saccharomyces cerevisiae*. *Biotechnology for Biofuels, 7*(1), 168.
51. Reider Apel, A., Ouellet, M., Szmidt-Middleton, H., Keasling, J. D., & Mukhopadhyay, A. (2016). Evolved hexose transporter enhances xylose uptake and glucose/xylose co-utilization in *Saccharomyces cerevisiae*. *Scientific Reports, 6*, 19512.
52. Nijland, J. G., Shin, H. Y., de Waal, P. P., Klaassen, P., & Driessen, A. J. M. (2018). Increased xylose affinity of Hxt2 through gene shuffling of hexose transporters in *Saccharomyces cerevisiae*. *Journal of Applied Microbiology, 124*(2), 503–510.
53. de Sales, B. B., Scheid, B., Gonçalves, D. L., Knychala, M. M., Matsushika, A., Bon, E. P. S., & Stambuk, B. U. (2015). Cloning novel sugar transporters from *Scheffersomyces* (*Pichia*) *stipitis* allowing D-xylose fermentation by recombinant *Saccharomyces cerevisiae*. *Biotechnology Letters, 37*(10), 1973–1982.
54. Wang, C., Shen, Y., Hou, J., Suo, F., & Bao, X. (2013). An assay for functional xylose transporters in *Saccharomyces cerevisiae*. *Analytical Biochemistry, 442*(2), 241–248.
55. Knoshaug, E. P., Vidgren, V., Magalhaes, F., Jarvis, E. E., Franden, M. A., Zhang, M., & Singh, A. (2015). Novel transporters from *Kluyveromyces marxianus* and *Pichia guilliermondii* expressed in *Saccharomyces cerevisiae* enable growth on L-arabinose and D-xylose. *Yeast, 32*(10), 615–628.
56. Kim, H., Lee, H.-S., Park, H., Lee, D.-H., Boles, E., Chung, D., & Park, Y.-C. (2017). Enhanced production of xylitol from xylose by expression of *Bacillus subtilis* arabinose: H^+ symporter and *Scheffersomyces stipitis* xylose reductase in recombinant *Saccharomyces cerevisiae*. *Enzyme and Microbial Technology, 107*, 7–14.

57. Sharma, N. K., Behera, S., Arora, R., Kumar, S., & Sani, R. K. (2018). Xylose transport in yeast for lignocellulosic ethanol production: Current status. *Journal of Bioscience and Bioengineering, 125*(3), 259–267.
58. Nijland, J. G., Shin, H. Y., Boender, L. G. M., de Waal, P. P., Klaassen, P., & Driessen, A. J. M. (2017). Improved xylose metabolism by a *CYC8* mutant of *Saccharomyces cerevisiae*. *Applied and Environmental Microbiology, 83*(11), e00095–e00017.
59. Zha, J., Li, B.-Z., Shen, M.-H., Hu, M.-L., Song, H., & Yuan, Y.-J. (2013). Optimization of CDT-1 and XYL1 expression for balanced co-production of ethanol and xylitol from cellobiose and xylose by engineered *Saccharomyces cerevisiae*. *PLoS One, 8*(7), e68317.
60. Ha, S.-J., Galazka, J. M., Rin Kim, S., Choi, J.-H., Yang, X., Seo, J.-H., Louise Glass, N., Cate, J. H. D., & Jin, Y.-S. (2011). Engineered *Saccharomyces cerevisiae* capable of simultaneous cellobiose and xylose fermentation. *Proceedings of the National Academy of Sciences, 108*(2), 504–509.
61. Sato, T. K., Tremaine, M., Parreiras, L. S., Hebert, A. S., Myers, K. S., Higbee, A. J., Sardi, M., McIlwain, S. J., Ong, I. M., Breuer, R. J., Avanasi Narasimhan, R., McGee, M. A., Dickinson, Q., La Reau, A., Xie, D., Tian, M., Reed, J. L., Zhang, Y., Coon, J. J., Hittinger, C. T., Gasch, A. P., & Landick, R. (2016). Directed evolution reveals unexpected epistatic interactions that alter metabolic regulation and enable anaerobic xylose use by *Saccharomyces cerevisiae*. *PLoS Genetics, 12*(10), e1006372.
62. Osiro, K. O., Borgström, C., Brink, D. P., Fjölnisdóttir, B. L., & Gorwa-Grauslund, M. F. (2019). Exploring the xylose paradox in *Saccharomyces cerevisiae* through *in vivo* sugar signalomics of targeted deletants. *Microbial Cell Factories, 18*(1), 88.
63. dos Santos, L. V., Carazzolle, M. F., Nagamatsu, S. T., Sampaio, N. M. V., Almeida, L. D., Pirolla, R. A. S., Borelli, G., Corrêa, T. L. R., Argueso, J. L., & Pereira, G. A. G. (2016). Unraveling the genetic basis of xylose consumption in engineered *Saccharomyces cerevisiae* strains. *Scientific Reports, 6*(1), 38676.
64. Tran Nguyen Hoang, P., Ko, J. K., Gong, G., Um, Y., & Lee, S.-M. (2018). Genomic and phenotypic characterization of a refactored xylose-utilizing *Saccharomyces cerevisiae* strain for lignocellulosic biofuel production. *Biotechnology for Biofuels, 11*(1), 268.
65. Shen, Y., Hou, J., & Bao, X. (2013). Enhanced xylose fermentation capacity related to an altered glucose sensing and repression network in a recombinant *Saccharomyces cerevisiae*. *Bioengineered, 4*(6), 435–437.
66. Bergdahl, B., Sandström, A. G., Borgström, C., Boonyawan, T., van Niel, E. W. J., & Gorwa-Grauslund, M. F. (2013). Engineering yeast hexokinase 2 for improved tolerance toward xylose-induced inactivation. *PLoS One, 8*(9), e75055.
67. Matsushika, A., Nagashima, A., Goshima, T., & Hoshino, T. (2013). Fermentation of xylose causes inefficient metabolic state due to carbon/energy starvation and reduced glycolytic flux in recombinant industrial *Saccharomyces cerevisiae*. *PLoS One, 8*(7), e69005.
68. Feng, X., & Zhao, H. (2013). Investigating xylose metabolism in recombinant *Saccharomyces cerevisiae* via ^{13}C metabolic flux analysis. *Microbial Cell Factories, 12*, 114.
69. Matsushika, A., Goshima, T., & Hoshino, T. (2014). Transcription analysis of recombinant industrial and laboratory *Saccharomyces cerevisiae* strains reveals the molecular basis for fermentation of glucose and xylose. *Microbial Cell Factories, 13*, 16.
70. Feng, X., & Zhao, H. (2013). Investigating host dependence of xylose utilization in recombinant *Saccharomyces cerevisiae* strains using RNA-seq analysis. *Biotechnology for Biofuels, 6*(1), 96.
71. Weimberg, R. (1961). Pentose oxidation by *Pseudomonas fragi*. *The Journal of Biological Chemistry, 236*(3), 629–635.
72. Tai, Y. S., Xiong, M., Jambunathan, P., Wang, J., Wang, J., Stapleton, C., & Zhang, K. (2016). Engineering nonphosphorylative metabolism to generate lignocellulose-derived products. *Nature Chemical Biology, 12*(4), 247–253.

73. Bai, W., Tai, Y.-S., Wang, J., Wang, J., Jambunathan, P., Fox, K. J., & Zhang, K. (2016). Engineering nonphosphorylative metabolism to synthesize mesaconate from lignocellulosic sugars in *Escherichia coli. Metabolic Engineering, 38*, 285–292.
74. Wasserstrom, L., Portugal-Nunes, D., Almqvist, H., Sandström, A. G., Lidén, G., & Gorwa-Grauslund, M. F. (2018). Exploring D-xylose oxidation in *Saccharomyces cerevisiae* through the Weimberg pathway. *AMB Express, 8*(1), 33.
75. Borgström, C., Wasserstrom, L., Almqvist, H., Broberg, K., Klein, B., Noack, S., Lidén, G., & Gorwa-Grauslund, M. F. (2019). Identification of modifications procuring growth on xylose in recombinant *Saccharomyces cerevisiae* strains carrying the Weimberg pathway. *Metabolic Engineering, 55*, 1–11.
76. Kwak, S., Jo, J. H., Yun, E. J., Jin, Y.-S., & Seo, J.-H. (2019). Production of biofuels and chemicals from xylose using native and engineered yeast strains. *Biotechnology Advances, 37*(2), 271–283.
77. Parreiras, L. S., Breuer, R. J., Avanasi Narasimhan, R., Higbee, A. J., La Reau, A., Tremaine, M., Qin, L., Willis, L. B., Bice, B. D., Bonfert, B. L., Pinhancos, R. C., Balloon, A. J., Uppugundla, N., Liu, T., Li, C., Tanjore, D., Ong, I. M., Li, H., Pohlmann, E. L., Serate, J., Withers, S. T., Simmons, B. A., Hodge, D. B., Westphall, M. S., Coon, J. J., Dale, B. E., Balan, V., Keating, D. H., Zhang, Y., Landick, R., Gasch, A. P., & Sato, T. K. (2014). Engineering and two-stage evolution of a lignocellulosic hydrolysate-tolerant *Saccharomyces cerevisiae* strain for anaerobic fermentation of xylose from AFEX pretreated corn stover. *PLoS One, 9*(9), e107499.
78. Demeke, M. M., Dietz, H., Li, Y., Foulquié-Moreno, M. R., Mutturi, S., Deprez, S., Den Abt, T., Bonini, B. M., Liden, G., Dumortier, F., Verplaetse, A., Boles, E., & Thevelein, J. M. (2013). Development of a D-xylose fermenting and inhibitor tolerant industrial *Saccharomyces cerevisiae* strain with high performance in lignocellulose hydrolysates using metabolic and evolutionary engineering. *Biotechnology for Biofuels, 6*(1), 89.

Chapter 8
Microbial Lipid Production from Lignocellulosic Biomass Pretreated by Effective Pretreatment

Cui-Luan Ma and Yu-Cai He

8.1 Introduction

A worldwide concern has recently aroused about the soaring depletion of natural resources and degradation of environmental conditions, which is leading to an increased interest in alternative and renewable energy sources [1, 2]. Biomass, consisting of inedible plant material that does not compete with our food production, is regarded as a suitable renewable feedstock [3–6]. From the last decade, there has been an increasing interest in the value-added utilization of lignocellulosic biomass, which can be used as the most abundant, inexpensive, and renewable source for production of platform organic molecules, functional materials, liquid fuels, and value-added chemicals [7–21].

Bioresources comprising over 2.0×10^{12} tons of annual production are potentially the world's largest sustainable and safe source of energy. Very recently, much research has been focused on developing new chemical strategies for the valorization of biomass into liquid biofuels and chemicals [4, 22–31]. Lipids are one kind of value-added energy-rich compounds, which can produce by oleaginous microorganisms using biomass and/or biomass-hydrolysates [16, 32–35]. Microbial lipids are composed of saturated and unsaturated fatty acids with potential use as nutraceuticals, food additives, and biofuels [4, 36–38]. Microbial lipids include

C.-L. Ma · Y.-C. He (✉)
State Key Laboratory of Biocatalysis and Enzyme Engineering, Hubei Collaborative Innovation Center for Green Transformation of Bio-resources, Hubei Key Laboratory of Industrial Biotechnology, School of Life Sciences, Hubei University, Wuhan, People's Republic of China

Advanced Catalysis and Green Manufacturing Collaborative Innovation Center, National–Local Joint Engineering Research Center of Biomass Refining and High–Quality Utilization, School of Pharmacy, Changzhou University, Changzhou, People's Republic of China
e-mail: heyucai2001@163.com; yucaihe_2007@aliyun.com

Z.-H. Liu, A. Ragauskas (eds.), *Emerging Technologies for Biorefineries, Biofuels, and Value-Added Commodities*,
https://doi.org/10.1007/978-3-030-65584-6_8

triacylglycerols (TAGs), glycolipids, phospholipids, and steryl ester, which have many similarities with plant oils [39, 40]. Fatty acid of microbial lipid is composed of Palmitic acid (C16:0), Palmitoleic acid (C16:1), Stearic acid (C18:0), Oleic acid (C18:1), Linoleic acid (C18:2), Linolenic acid (C18:3), etc. The percentage of these saturated and unsaturated fatty acids very much depends on the type of oleaginous microorganisms and growth conditions [40–43].

Biomass sources like energy crops, agriculture and forest residues, sewage sludge, animal and food waste, municipal solid waste, etc. are generally used for energy production [9, 17, 44–46]. Typically, biomass valorization processes to produce both biofuels and/or bio-based chemicals are mainly consisted of three steps: biomass pretreatment, enzymatic saccharification, and fermentation [43, 47]. Pretreatment is considered as a crucial step in lignocellulosic biomass valorization (Fig. 8.1), which can be used for disrupting recalcitrant lignocellulosic structures and removing lignin and hemicelluloses to make cellulose more accessible to the enzymes for efficient conversion into fermentable sugars [48–53]. Although different pretreatments including physical, chemical, physicochemical, biological, or their combination are available [54–58], the development of a suitable pretreatment to avoid or reduce the formation of inhibitors (furfural and/or hydroxymethyl furfural) deserves the great challenge in biofuel production [59, 60].

In this chapter, various biomass pretreatments for effectively improving the enzymatic saccharification of lignocellulosic biomass are introduced. Furthermore, microbial lipid production from lignocellulosic biomass pretreated by effective pretreatment is discussed.

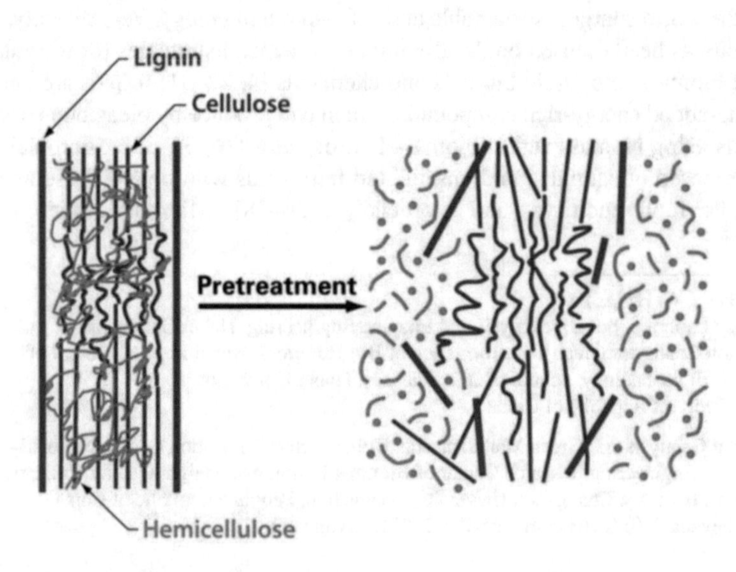

Fig. 8.1 Disruption of lignocellulose structure via pretreatment. (Adapted from Ref. [61, 62])

8.2 Main Components of Lignocellulosic Biomass

Hemicellulose $(C_5H_{10}O_5)_m$, cellulose $(C_6H_{10}O_5)_n$, and lignin $(C_{10}H_{11}O_{3.5})$ are the three major components of lignocellulosic biomass along with small number of other organic compounds such as pectin and protein. Chemical compositions of some lignocellulosic materials are provided in Table 8.1. Most of the biomass is composed of lignocellulose, which is a complex carbohydrate polymer of cellulose, hemicellulose, and lignin [63–65]. The percent composition of cellulose, hemicellulose, and lignin in biomass are in the ranges 30–50%, 20–35%, and 10–20%, respectively.

Cellulose is a linear polysaccharide consisting of β-1,4-linked-*D*-glucose residues, which is made up of crystalline and amorphous region [75–77]. Hemicelluloses (also known as polyose), the second most abundant constituent of lignocellulosic biomass, are polysaccharides in plant cell walls that have β-(1→4)-linked backbones with an equatorial configuration [9, 15, 18, 29]. Hemicelluloses, which have linear and branched structures, include glucomannans, xylans, xyloglucans, mannans, and β-(1→3,1→4)-glucans. It is more energy dense than carbohydrates (hemicelluloses and cellulose) because of its higher carbon-to-oxygen ratio [35, 78]. Lignin is a poly-aromatic non-sugar component typically found in biomass, which confers high mechanical strength and hydrophobicity to plant walls [79]. It is a

Table 8.1 Main components (cellulose, hemicellulose, and lignin) in common lignocellulosic biomass[a]

Lignocellulosic biomass	Cellulose, wt%	Hemicellulose, wt%	Lignin, wt%
Bamboo shoot shell	38.5	23.1	11.4
Cotton seed hairs	80–95	5–20	0
Corncob	45	35	15
Corn straw	42.6	21.3	10–20
Grass	25–40	35–50	10–30
Hardwood stem	40–50	24–40	18–25
Leaves	15–20	80–85	0
Maize stover	37.5	30	10.3
Nutshell	25–30	25–30	30–40
Newspaper	40–55	25–40	18–30
Oat straw	39.4	27.1	20.7
Paper	85–99	0	0–15
Rice straw	32–47	18–28	5.5–24
Rice husk	34.4	29.3	19.2
Softwood stem	45–50	25–35	25–35
Solid cattle manure	1.6–4.7	1.4–3.3	2.7–5.7
Sugarcane bagasse	32–48	19–24	23–32
Switchgrass	45	31.4	12
Wheat straw	33–45	20–32	8–20

[a]Adapted from Refs. [66–74]

highly cross-linked complex aromatic biopolymer formed by polymerization of 4-hydroxyphenylpropanoid monomer units such as syringyl (S), guaiacyl (G) and *p*-hydroxyphenyl (H) units and linked by ether or C–C bonds [61, 79, 80, 81]. The lignin biopolymers are attached to hemicelluloses by covalent bonds creating protection against chemical and biological degradation, inhibiting usability of raw biomass for producing biofuels and biobased chemicals.

8.3 Pretreatments of Lignocellulosic Materials for Enhancing the Production of Microbial Lipids

Various biomass pretreatments including physical (chipping, irradiation, grinding, milling, and pyrolysis) [76, 82–90], chemical (concentrated acid, concentrated alkali, deep eutectic solvent, dilute acid, dilute alkali, ionic liquid, *N*-methylmorpholine-*N*-oxide, ozonolysis, organic solvent, and oxidizing agent) [43, 67, 76, 91–97], physico-chemical (Ammonia fiber explosion, CO_2 explosion, liquid hot water, oxidative pretreatment, sulfite pretreatment, and steam pretreatment) [62, 95, 97, 98–103], biological [67, 75, 104–110], or their combination [99, 110–113] (Fig. 8.2) have been developed for enhancing enzymatic saccharification of biomass and sequential biofuel production (e.g., microbial lipids). The choice of pretreatment technologies that increase the digestibility of cellulose and hemicelluloses to help in cost-effective and eco-friendly conversion of lignocellulosic materials to microbial lipids depends on the compositions of biomass and the generated by-products after pretreatments.

8.3.1 Physical Pretreatment

8.3.1.1 Mechanical Pretreatment

Reduction of biomass particle sizes is a necessary procedure for converting biomass to biofuels. Milling, grinding, and chipping are known as the common mechanical pretreatment techniques for reducing particle sizes [76, 84–88, 90, 114, 115]. Grinding and milling can reduce biomass to 0.20 mm, while chipping can reduce biomass to 10–30 mm [102]. Various milling methods (e.g., knife, ball milling, hammer milling, and attritor milling) can be used for significantly reducing the polymerization degrees of cellulose and lignin, which aid in enzymatic sugar release and subsequent sugar fermentation into biofuels [116–122], wet disk milling (WDM) is used as a popular mechanical pretreatment for treating lignocellulosic biomass because of its low energy consumption [123, 124]. The energy requirements of milling increase with the reduction of biomass particle sizes [82, 125, 126]. This pretreatment is environmentally friendly because no chemicals or reagents were used during the pretreatment; however, it is generally needed to

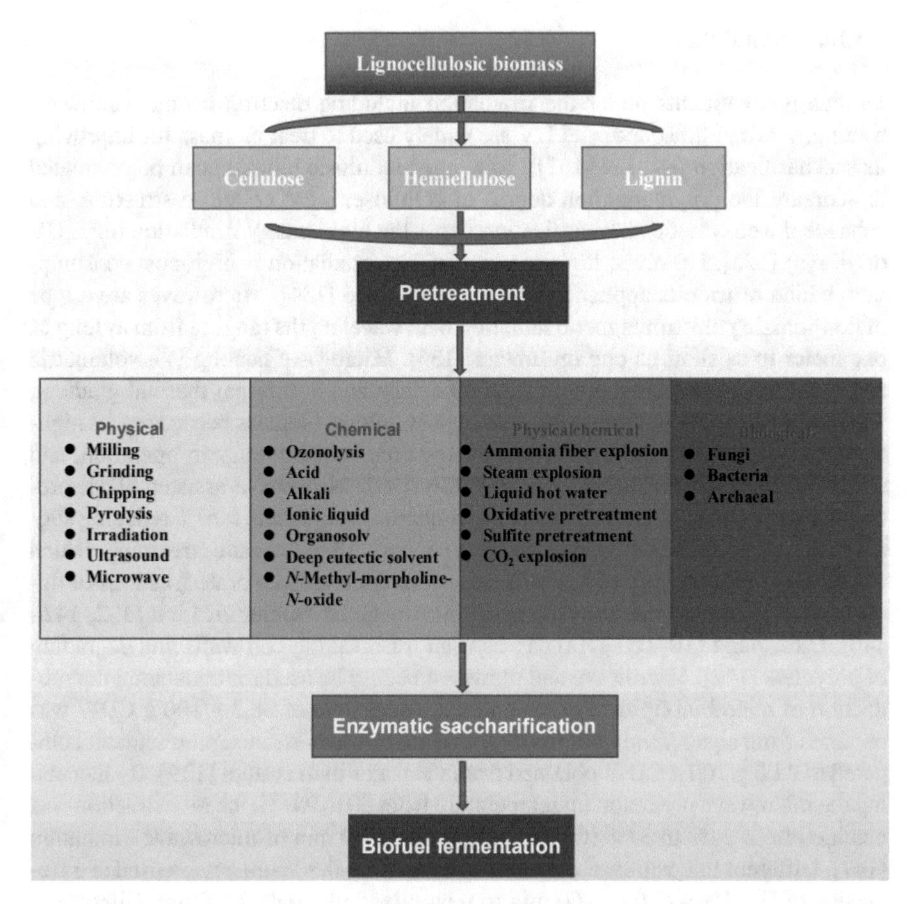

Fig. 8.2 Pretreatment technologies including physical, chemical, physico-chemical, biological, or their combination for enhancing enzymatic saccharification and microbial lipid fermentation

consume a lot of energy. In addition, although the milling can enhance the enzymatic hydrolysis of lignocellulose to a certain extent, it is difficult to thoroughly improve the enzymatic saccharification of biomass because the milling cannot remove lignin effectively, which limited the access of cellulase to cellulose. In most of the cases, a combination of milling and other pretreatments was used to effectively pretreat biomass for enhancing enzymatic saccharification and biofuel production. The milled sorghum stalks were pretreated with 1.25% (w/v) NaOH at 121 °C for 30 min. Commercial cellulases (Cellic C-Tec2 plus Cellic H-Tec2) were complexed for the hydrolysis of pretreated biomass for 48 h. The total lipid of 13.2 g/L and lipid yield of 0.29 g/g were obtained by *Trichosporon oleaginosus* using sorghum stalk hydrolysates as carbon sources [127].

8.3.1.2 Irradiation

Various pretreatments under the irradiation including electron beam, microwave heating, γ-rays, ultrasound, and UV are widely used to treat biomass for improving its saccharification [89, 128–132]). Raw lignocellulosic biomass can be pretreated to decrease the polymerization degree of cellulose, loose cellulose structure, and enhance the enzymatic saccharification under the high-energy irradiation (e.g., UV or γ-rays) [133]. However, high-energy electron irradiation is high-cost consumption, which restrict its application on the large-scale [134]. Microwaves are a type of non-ionizing electromagnetic radiation with wavelengths ranging from as long as one meter to as short as one millimeter [135]. Microwave heating is a volumetric and rapid heating technique with high efficiency and a minimal thermal gradient, which has been widely used to pretreat lignocellulosic biomass because of its high-heating capacity in a short time, low-energy consumption, easy to operation, and minimum formation of inhibitors [102, 136–139]. Microwave-assisted alkali pretreatment of coastal bermudagrass and switchgrass could yield >70% reducing sugars [140, 141]. Ultrasound waves produce cavitation and acoustic streaming, which can alter the morphology of biomass and rupture the carbohydrate fractions in lignocellulosic biomass thereby enhancing enzymatic its saccharification [102, 142–146]. Ultrasound (10–100 kHz) can be used for breaking cell walls and degrading of polymers [142]. Microwave and ultrasound could be used for enhancing the production of microbial lipids. For example, the total lipid of 38.3 g/100 g CDW was obtained from algae *Nannochloropsis* sp. via microwave-assisted pretreatment compared to 23.0 g/100 g CDW obtained from the water-bath system [139]. By increasing the microwave pretreatment temperature from 80 to 95 °C, the oil extraction was obtained from 24% to 33% (dry weight basis) for 30 min of microwave irradiation [147]. Different intensities of ultrasound power (0.1–0.5 W/mL) were used at a frequency of 30 kHz and for 5–60 min to treat mixed microalgal culture. Ultrasound could affect the cell disruption. Moreover, the lipid extraction efficiency could be enhanced under the ultrasound irradiation [148]. At 0.4 kW h/L, up to twofold increase in lipid extraction yields were obtained.

8.3.1.3 Pyrolysis

Pyrolysis is a complex thermochemical conversion process by which a solid or liquid undergoes the thermal degradation into smaller volatile molecules, without interacting with O_2 or any other oxidants [83, 149, 150]. Pyrolysis has also been used for pretreating lignocellulosic biomass in biorefinery processes. The biomass pyrolysis processes typically result in the generation of various products including solids, liquids, and gases [102]. Pyrolysis is found to be more efficient when conducted in the presence of O_2 at lower temperatures [149]. H_2SO_4 (1.0 M) was employed for the pyrolysis of biomass within 2.5 h at 97 °C, and the saccharification was obtained at 85% yield [151]. Coupling lipid fermentation with pyrolysis has been used for converting biomass into lipid. Carboxylic acids, which generated from pyrolysis, were fermented into lipids by *Cryptococcus curvatus* [152]. In pyrolytic liquor, 20 g/L acetate was fermented with *C. curvatus* after neutralization and detoxification to produce ~7.0 g/L dry biomass and 2.0 g/L lipid.

8.3.2 Chemical Pretreatment

8.3.2.1 Alkalic Pretreatment

Alkali-based pretreatment is essentially used for reducing the crystallinity degree, swelling fibers, and removing lignin in corn stover, switchgrass, rice straw, wheat straw, and softwood [95, 153–155]. Alkali can saponify uronic ester linkages between 4-*O*-methyl-*D*-glucuronic acid units in xylan and cleave hydrolysable linkages in lignin and glycosidic bonds of polysaccharides, which causes a reduction in the degree of polymerization and crystallinity, swelling of the fibers, as well as disruption of the lignin structure. The alkali-pretreated lignocellulosic materials are loose, swollen, and porous, which facilitates the cellulose accessibility to enzymes for enhancing their enzymatic saccharification [91, 95]. Various alkaline reagents including oxidative alkali, sodium hydroxide (NaOH), potassium hydroxide (KOH), calcium hydroxide ($Ca(OH)_2$), and ammonia ($NH_3{\cdot}H_2O$) have been widely used to pretreat lignocellulosic materials for enhancing enzymatic saccharification at ambient temperature and pressure [86, 91, 153]. NaOH–CH_3OH solution (NaOH 4.0 g, CH_3OH 80 mL) was used to pretreat 40 g raw corn stover at high solids loading for effectively enhancing its enzymatic saccharification [156]. The hydrolysates were used for lipid production by *Cutaneotrichosporon oleaginosum.* Biomass, lipid content, and lipid yield were 50.7 g/L, 61.7%, and 0.18 g/g, respectively. Dilute alkali (NaOH, 2 wt%) pretreatment of corn stover (10 wt%) was conducted a high-pressure reactor at 121 °C for 20 min, and lipid of *Trichosporon dermatis* 32,903 could reach 20.36 g/L with sugar to lipid yields improved to 0.186 g/g [43]. Compared to mineral acid pretreatment, alkali-based pretreatment is required for a relatively long pretreatment time. Less inhibitors (e.g., acetic acid, hydroxyl acids, and minor amounts of furan aldehydes) form when compared to acid pretreatments [91]. High cost of alkali seriously restricts its application. Aqueous ammonia (AA) can be used for pretreatment of biomass at ambient condition, and the leftover ammonia is an important nitrogen source for the growth of energy microorganisms [157–161]. AA can selectively remove lignin from lignocellulosic materials while most of carbohydrates (hemicelluloses and cellulose) remain in lignocellulose [158, 159, 162, 163]. High enzymatic saccharification was obtained at 73.6% when Pecan Nutshell (PS) pretreated with ethylene glycol–H_2SO_4–water (78:2:20, wt:wt:wt) was further treated by AA (25 wt%) for 1 d at 50 °C. Using hydrolysates (20 g/L) as carbon source, the lipid content 0.44 g lipid/g DCW was achieved in *Rhodococcus opacus* ACCC41043 cells [4].

Alkalic salts (e.g., Na_2CO_3, Na_3PO_4, Na_2S, Na_2SO_3) with its low corrosivity have been employed to pretreat biomass for improving its enzymatic saccharification

[159, 164–169]. Alkalic salts containing sulfite (SO_3^{2-}) can cleave β-benzyl ether, α-benzyl ether, and α-alkyl ether linkages on phenolic lignin units [170]. H_2O_2/ Na_2CO_3 (15% H_2O_2, 40% Na_2CO_3) and Na_2S/ Na_3PO_4(4% Na_3PO_4, 10% sulfidity) were employed to pretreat corn stover at 120 °C for 40–60 min [171, 172], most of carbohydrates (hemicelluloses and cellulose) in pretreated biomass could be saccharified to fermentable sugars. However, it has several disadvantages include the degradation of biomass-derived sugar, large amount of water for post-pretreatment washing of biomass, and high cost for recovering pretreatment chemicals [102]. Using dilute alkali salts (0.4 wt% Na_3PO_4, 0.03 wt% Na_2SO_3) as pretreatment chemicals for treating sugarcane bagasse in an autoclave within 40 min at 110 °C, enzymatic in situ saccharifications of biomass were carried out at 50 °C [173], which avoided the steps for post-pretreatment washing of biomass and recovery of pretreatment chemicals. High saccharification was obtained at 67.6%. Combination pretreatment (BP-AP) by sequential biological treatment with *Galactomyces* sp. and dilute Na_2CO_3 (0.82 wt%) was employed to treat corn stover for improving its enzymatic saccharification. The fermentable sugars containing 25.6 g/L glucose without removal of Na_2CO_3 could be effectively fermented into microbial lipid by *Rhodococcus pyridinivorans* CCZU-B16. Fatty acids rich in C16 and C18 including oleic acid, stearic acid, palmitoleic acid, and palmitic acid were detected in whole-cells [40].

8.3.2.2 Acid Pretreatment

Industrially, various mineral and organic acids are widely used for pretreating biomass [34]. Acid pretreatment can hydrolyze hemicelluloses into monomeric sugars by destroying the polymeric bonds, increasing the availability of cellulose, and thereby enhancing the saccharification. Inorganic acids (e.g., H_2SO_4, HCl, HF, HNO_3, H_3PO_4) are common acid catalysts for acid-catalyzed lignocellulose biomass pretreatment. Concentrated and diluted acids can be employed to destroy the rigid structure of the lignocellulosic material. Out of these acids H_2SO_4 is mostly used because of its low cost and high efficiency in lignin removal [95]. It is performed at 120–210 °C with H_2SO_4 (<4 wt%) at the different pretreatment time from minutes to hours. Although acid pretreatment is cost-effective, it has some of the drawbacks of high reactor cost for their usage, gypsum formation during neutralization after pretreatment, and formation of inhibitory by-products (HMF, furfural, aliphatic carboxylic acids, etc.). Recently, organic carboxylic acids (e.g., acetic acid, fumaric acid, maleic acid, oxalic acid, succinic acid) are considered as alternatives to inorganic acids. Organic acid pretreatment has the advantages including low energy consumption for acid recovery and low equipment corrosions. High recovery of cellulose components in biomass can be obtained by organic acid pretreatment; however, hemicelluloses are recovered at low yields [95]. Dilute acid (H_2SO_4 1 wt%) pretreatment of corn stover (10 wt%) was conducted a high-pressure reactor at 160 °C for 10 min, and lipid of *T. dermatis* 32,903 could reach 11.4 g/L with sugar to lipid yields improved to 0.16 g/g [43]. Corn fiber and sweet sorghum bagasse were pretreated with dilute H_2SO_4 at a severity factor of 1.06 and 1.02. The sweet sorghum bagasse hydrolysates, which were derived from pretreatment at the severity factor of 1.02, were used for microbial lipid of *C. curvatus* at 10.8 g/L with a lipid content of 40% (w/w) [174]. After the pretreatment with 0.25 wt% H_2SO_4 and 2 h, 11.5 g/kg of microbial lipid was obtained with glucan and xylan recovery rate of 82% and 62% [175], respectively.

8.3.2.3 Ionic Liquid Pretreatment

Ionic liquids (ILs), one kind of salts with low melting points (< 100 °C) and high vapor pressure [176, 177], are regarded as being green solvents because of their high thermostability, low toxicity, excellent solvency, nonvolatility, and recyclability [178–187]. Swatloski et al. [188] reported for the first time that imidazolium-based ILs (e.g., [Bmim][C1]) can dissolve cellulose very well. However, many chlorine-free ILs have been developed because chloride-based ILs can be toxic and corrosive [189]. Acetate-based ILs have higher capability to solubilize cellulose [190] and are less toxic and corrosive [112, 177, 191–196]. ILs have tunable capability to dissolve lignocellulosic materials, resulting in reduction of lignin content, increase of surface area, and enhancement of enzymatic saccharification [92, 95, 197, 198]. However, IL pretreatment is costly because of its high prices compared to commercial solvents, which restricts its application on large-scale in the pretreatment of lignocellulose [95, 179, 187, 199, 200]. Hydrolysates from enzymatic saccharification of IL-treated biomass could be effectively employed as carbon source to produce microbial lipids. Simultaneous saccharification and enhanced lipid production (SSELP) were used for converting IL 1-ethyl-3-methylimidazolium acetate (EmimOAc)-pretreated corn stover into lipids. At 5% (w/v) of solid loading, lipid titer could reach 6.2 g/L after 2 d of fermentation by *C. curvatus* cells, and lipid coefficient was 112 mg/g regenerated biomass, or 81 mg/g raw biomass [201]. IL *N*-methylpyrrolidone (NMP)-1-ethyl-3-methyl imidazolium acetate (EmimAc) could dissolve ≥10 wt% corn stover at 140 °C in 1 h. Enzymatic hydrolysis of pretreated corn stover afforded an 83% total reducing sugars yield and 61% glucose yield within 1 d. The hydrolysates without detoxification could be used as carbon sources for producing microbial lipid by *Rhodosporidium toruloides* Y4 [202].

Very recently, ecologically friendly deep eutectic solvents (DESs) are effectively used for pretreating lignocellulosic materials [203, 204], which, can be synthesized by mixing hydrogen bond acceptors (HBAs) and hydrogen bond donors (HBDs) at the appropriate molar ratio and heating this mixture at the moderate temperature (60–100 °C) under stirring for a few hours until a homogeneous clear DES liquid form (100% atom economy). Dissolution of lignocellulosic materials with DESs can be conducted by using glycerol, ethylene glycol, lactic acid, malic acid, malonic acid, formic acid, nicotinic acid, and oxalic acid as hydrogen bond donors combined in a variety of molar ratios with hydrogen bond acceptors alanine, betaine, choline chloride, proline, histidine, and glycine [67, 203, 204]. Lignin and hemicelluloses in biomass can be effectively removed with various DESs [204, 205]. Glucose yields over 80% were obtained after the enzymatic saccharification of biomass pretreatment with cholinium lysinate ([Ch][Lys]). The hydrolysates were fermented directly by *R. toruloides*—with glucose, xylose, acetate, and lactate fully consumed during fermentation [206].

8.3.2.4 Organosolv Pretreatment

Organosolv pretreatment can be effectively used for the extraction of lignin in biomass, which works by breaking the noncovalent bonds between the lignocellulosic components and disrupting the recalcitrant structures [207–209] Recalcitrance. It has the ability to fractionate lignocellulosic biomass into hemicellulose, cellulose, and lignin with high purity, easy solvent recovery, and solvent reuse. Various organic solvents (e.g., acetone, alcohol, amines, dioxane, esters, formaldehyde, phenol, propionic acid) or aqueous-organic solvent system with and without catalyst have been used for pretreating biomass at temperatures ranging from 100 to 250 °C [210]. Low-molecular-weight alcohols (e.g., methanol and ethanol) are often used as solvents for organosolv pretreatment due to their low boiling points and ease of removal. However, the low-boiling-point alcohols can seriously hinder the development of biomass pretreatment process due to their high volatility and flammability under the high-pressure operation [211, 212]. To avoid these drawbacks of low-boiling-point alcohol pretreatments, high-boiling solvents are of great interest. Glycerol and ethylene glycol (EG) are the most widely used high-boiling solvents for treating lignocellulosic biomass [195, 196, 210, 213–215]. Under the microwave irradiation (200 W) at 100 °C for 5 min, the lignin in corn stover could be effectively removed with EG-$HClO_4$-water (88.8:1.2:10, w/w/w) media [211]. Combination pretreatment with EG-H_2SO_4-water (78:2:20, w/w/w) at 130 °C for 0.5 h and AA (25 wt%) at 50 °C for 1 d was employed to improve the enzymatic hydrolysis of Pecan nutshell. The hydrolysates could be effectively fermented into microbial lipids at 17.1 g lipid/g reducing sugars by *R. opacus* ACCC41043 [4].

To help meet the challenge of biomass conversion, *N*-methyl-morpholine-*N*-oxide (NMMO) has attracted substantial research interest for pretreating biomass. NMMO is a hygroscopic compound crystalline at room temperature [216, 217]. It has melting point at 170 °C, implying nonvolatility and nonflammability. NMMO molecules are capable of softening and dissolving cellulose in biomass because of their high polarity of N–O bonds, which disrupt the hydrogen bonds of the cellulose and further form new hydrogen bonds with the solutes [218]. The operation conditions for these pretreatments are much milder (< 100 °C and atmosphere pressure) as compared to the conventional pretreatment methods. NMMO retains all the advantages of the ionic liquids-ability to dissolve a variety of lignocellulosic substrates (up to 20% by weight) without the need to chemically modify them and more than 99% of the solvent can be recovered due to its low vapor pressure [93]. It is also non-toxic and biodegradable as proven by the work of Lenzig researchers (Ramakrishnan et al. 2010). Cellulose withdrawn from NMMO solutions has also generated increased rates of hydrolysis by cellulases thus implying its potential use in pretreating lignocellulosic biomass for biofuels [218].

8.3.3 Physical-Chemical Pretreatment

8.3.3.1 Ammonia Fiber Explosion (AFEX)

AFEX is one kind of alkaline thermal pretreatment with aqueous ammonia ([88, [219–221]). The merit of this pretreatment process is that it does not require small particle size for efficiency, and further, inhibitors are not formed during the pretreatment [149]. AFEX can be carried out at ambient temperature. High saccharification rates (> 90%) based on the carbohydrate content in AFEX-treated biomass can be obtained. During the AFEX pretreatment, ammonia molecules can cause swelling, removal of lignin, and phase change of cellulose crystallinity in lignocellulosic biomass, which promotes in the reactivity of hemicelluloses and cellulose in pretreated biomass [102]. During AFEX pretreatment, no highly inhibitory products (e.g., phenols, furfural, and HMF) generate, which avoids the detoxification step. In addition, aqueous ammonia is easily recycled for reducing overall performance cost. Thus, AFEX is a cost-effective pretreatment technique for treating lignocellulosic biomass if aqueous ammonia can be recovered and recycled for repeated pretreatment. Microbial lipid production from AFEX pretreated and hydrolyzed corn stover using an oleaginous yeast *Lipomyces tetrasporus.* 36.7 g lipids were produced from 1 kg AFEX-pretreated corn stover via SHF at a titer of 8.4 g/L [222].

8.3.3.2 CO_2 Explosion

CO_2 explosion is similar to AFEX. It is a supercritical CO_2-based pretreatment of lignocellulosic biomass [223]. Supercritical CO_2 is considered as one kind of green solvent due to its abundance, low-cost, non-toxicity, non-flammable, and ease to recover [99]. Its critical pressure and critical temperature are 1071 psi and 31 °C, respectively [224]. Various parameters including extraction bed size, performance pressure, performance temperature, and solvent flow can be set to obtain the high yields of specific compounds [225]. The supercritical CO_2 molecules enter lignocellulosic materials at the required temperature and time under the high-performance pressure [99]. Subsequently, the formed H_2CO_3 can hydrolyze hemicelluloses in biomass. The CO_2 release after CO_2 explosion can break the structures of cellulose and hemicelluloses, thereby increasing the surface area of biomass for enhancing its hydrolysis [223, 226]. For CO_2 explosion, no toxin formation makes it an attractive for the pretreatment of biomass [110, 227]. Pretreatment can be used for improving lipid recovery from biomass by disrupting wet cell walls prior to extraction. The extraction of lipid was carried out under the pressurized CO_2 (3500 kPa) [228]. The solubility of CO_2 in *Rhodotorula glutinis* was higher than that of sugar broth media and spent media due to the presence of lipid in *R. glutinis*.

8.3.3.3 Liquid Hot Water (LHW) Pretreatment

LHW pretreatment, commonly known as autohydrolysis, uses pressure to keep water in the liquid state at elevated temperature [95]. It is regarded as an effective pretreatment strategy for selectively recovering hemicelluloses in the liquid stream [54]. In the solid phase, lignin components can be easily recovered with minor losses. No additional chemicals are needed [75]. LHW pretreatment can remove up to 80% of the hemicellulose and to enhance the enzymatic saccharification of pretreated biomass [229]. LHW can be carried out at relatively low pretreatment temperature in the low cost of pretreatment solvent; however, a large amount of water is needed to be recovered in downstream processing [75]. Hot water was used to enhance the microbial lipid production by oleaginous *R. opacus* PD630 and DSM 1069. *R. opacus* PD630 could accumulate lipid from detoxified sweet gum autohydrolysate with the lipid yield of 0.25 g/L of its cell dry weight in lipids while growing on that translates to 0.25 g/L lipid yield, while *R. opacus* DSM 1069 could accumulate lipid from detoxified pine autohydrolysate with the lipid yield of 0.3 g/L [230].

8.3.3.4 Oxidative Pretreatment

Delignification of lignocellulosic biomass can be conducted by using oxidizing agents such as O_2, O_3, H_2O_2, air, or per acetic acid [76, 231–238]. Various reactions including side chain displacements, electrophilic substitution, and oxidative cleavage of aromatic nuclei or cleavage of alkyl aryl ether linkages may happen during the oxidative pretreatment of biomass. The delignification effectiveness is attributed to the high reactivity of oxidizing agents with aromatic rings of lignin in lignocellulosic biomass. The lignin polymers can be oxidized into carboxylic acids, which act as potential inhibitors in the biotransformation or fermentation steps. These inhibitors must be neutralized or removed by additional processes. Oxidative pretreatment can also influence hemicellulose fractions in lignocellulosic biomass. Lignin peroxidase (LiP) can execute the H_2O_2-dependent $C\alpha$-$C\beta$ cleavage of lignin in biomass [239]. H_2O_2 pretreatment undergoes the oxidative delignification by detaching and solubilizing lignin in biomass [240, 241]. Ozonolysis pretreatment can be used for delignification by breaking aromatic rings structures of lignin in bagasse, cotton straw, and wheat straw [88, 242].

Wet oxidation is a simple pretreatment process using air/oxygen along with H_2O or H_2O_2 to fractionate lignocellulosic materials at ≥120 °C [243–246]. The crystalline structure of cellulose in biomass can be opened by wet oxidation [247]. High delignification (65%) is achieved with wheat straw [248]. Alkaline peroxide-assisted wet air oxidation can solubilize 67% of hemicellulose and 88% of lignin in rice husk [249]. Wet oxidation combined with alkali generates the limited formation of fermentation inhibitors (e.g. furan aldehydes and phenolaldehydes), and the main degradation products found from hemicellulose and lignin are carboxylic acids, CO_2, and H_2O. In the wet oxidation, addition of Na_2CO_3 can decrease performance temperature and enhance the removal of hemicellulose, which can avoid the formation of potential inhibitors (e.g., furfurals and HMF) [249]. Hydrolysates could be used to produce microbial lipids from herbaceous lignocellulosic biomass utilizing alkaline hydrogen peroxide pretreatment with NaOH and H_2O_2, which were composed of xylose and glucose as well as acetate and phenolic monomers that could be used as renewable carbon to produce microbial lipids [250].

8.3.3.5 Steam Explosion

Steam explosion is a physico-chemical pretreatment process for the breakdown the lignocellulosic structural components by hot steam (160–260 °C) under the pressure (0.7–4.8 MPa) on the large-scale [75, 98, 101, 102]. Subsequently, the rapid reduction of performance pressure can generate strong shear force in an explosive decompression event, which facilitates the hydrolysis of beta-glycosidic bonds and hydrogen-bonds between the glucose chains, resulting in the degradation of hemicellulose and depolymerization of lignin thereby increasing the cellulose accessibility to cellulases [95, 102]. The advantages of steam explosion include limited use of hazardous chemicals, low energy consumption, and low environmental impact. However, the generation of degradation products from lignin and biomass-derived sugars is unavoidable [75, 251]. Steam explosion pretreatment of microalgae gave the highest lipid extraction yields. The experimental results demonstrate the efficacy and feasibility of the acid catalyzed steam explosion pretreatment, followed by n-hexane lipid extraction. High sugar yields (up to 96%) were obtained with 1.7% H_2SO_4 at 150 °C during steam explosion, and high lipid extraction of exploded microalgae was achieved using n-hexane [252].

8.3.4 Biological Pretreatment (BP)

Lignocellulosic materials are composed of carbohydrate polymers (hemicelluloses and cellulose), coexisting in complex matrices with high carbon content and high aromaticity of highly aromatic biopolymer lignin [38]. Lignin, which is known as the second most abundant terrestrial biopolymers on earth, constitutes ~15–30% of lignocellulosic biomass. Lignin can be underutilized as renewable feedstock for value-added chemicals. The lignin polymer is highly recalcitrant toward chemical and biological degradation due to its molecular architecture. Biomass and its components (e.g., hemicelluloses, lignin, and lignin-derived polyphenols) can be uti-

Table 8.2 Summary of some pretreatment technologies

Pretreatment method	Advantages	Disadvantages
Acid pretreatment	Effective removal of hemicellulose and lignin	High cost, harmful by-products, equipment corrosion requirement of intensive energy
Alkalic pretreatment	Effective removal of lignin and increase of accessible surface areas	Long residence times required, high cost, harmful by-products, requirement of intensive energy
Ammonia fiber explosion (AFEX)	Removal of lignin and hemicellulose to an extent; no generation of inhibitors	Low efficiency for biomass with high lignin content
Biological pretreatment (BP)	Degradation of hemicellulose and lignin to an extent; low energy consumption	Long pretreatment time
CO_2 explosion	Cost-effective; no generation of inhibitors	Low efficiency for removing hemicellulose and lignin; requirement of high cost of high temperature-high pressure equipment and system
Deep eutectic solvent (DES)	Environmentally friendly performance with DES	Lack of economical solutions to DES recycling
Ionic liquid (IL)	Pretreatment by IL with high thermostability, inflammability, low volatility and recyclability; high delignification	Lack of economical solutions to IL recycling
Liquid hot water (LHW)	Requirement of low cost of solvent; generation of minimum inhibitors at low performance temperature	Requirement of a large amount of water; high energy consumption in downstream process
Milling	Environmentally friendly performance without addition of chemicals	High energy consumption, low delignification
Organic solvent	Effective removal of hemicellulose and lignin	Low biomass recovery; solvents need to be drained from the reactor, evaporated, condensed, and recycled; high operation cost
N-methyl-morpholine-*N*-oxide pretreatment (NMMO)	Environmentally friendly performance under below 100 °C	Lack of economical solutions to NMMO recycling
Microwave irradiation	High-heating capacity in a short time, low-energy consumption, easy to operation, and minimum formation of inhibitors	High cost of pretreatment; lack of large-scale equipment
Ozonolysis	Effective removal of hemicellulose and lignin at ambient pressure and temperature	A large amount of ozone required
Pyrolysis	High efficient in the presence of O_2 at lower temperature	High energy consumption

(continued)

Table 8.2 (continued)

Pretreatment method	Advantages	Disadvantages
Steam explosion	Removal of hemicellulose and lignin to an extent, limited use of hazardous chemicals, low energy consumption, and low environmental impact	Generation of degradation products from lignin and biomass-derived sugars at high temperature
Sulfite pretreatment (SPORL)	Removal of hemicellulose and lignin with high efficiency	Degradation of biomass–derived sugar; requirement of large amount of water for post-pretreatment washing of biomass; high cost for recovering pretreatment chemicals
Ultrasonic irradiation	Effective change of the biomass morphology and rupture of carbohydrate fractions in biomass	Lack of large-scale equipment
Wet oxidation	High delignification	High pretreatment temperature

lized via the microbial action by highly efficient bacteria or/and fungi with low-energy consumption, high substrate-specificity, and no generation of undesirable toxic compounds [106, 107, 253–258]. Bacteria and fungi can consume lignin breakdown products and utilize them as carbon sources [16, 33, 35, 259–264], potentiating fuel and chemical production via lignin-consolidated bioprocessing.

Biological pretreatments (BPs) by fungi (e.g., white-rot and brown-rot fungi) have been widely employed as environmentally-friendly approached for pretreating biomass for production of biobased chemicals and biofuels [104, 106, 107, 108, 109, 110, 265–268]. White-rot fungi (e.g., *Trametes versicolor*, *Phlebia radiata*, *Phanerochaete chrysosporium*, *Dichomitus squalen*) are the most effective for delignification in nature [109]. Laccase, manganese peroxidase (MnP), and lignin peroxidase (LiP) are the main extracellular lignin-degrading enzymes of white-rot fungi [255, 269, 270, 271]. LiP catalyzes the homolytic C_a-$C_ß$ cleavage of lignin and depolymerizes methylated lignin in vitro. MnP has the ability to catalyze the oxidation of Mn^{2+} to Mn^{3+} with H_2O_2 on phenolic (or non-phenolic) lignin units. Laccase is a copper-containing enzyme, part of the group of so-called blue oxidase, that catalyzes the one-electron oxidation of aromatic compounds (e.g., phenolics and amines) by oxygen. Brown-rot fungi, on the other hand, slightly oxidize lignin in wood, which preferentially degrades the polysaccharide components [258, 272]. Although pretreatment with fungi can be carried out with simple protocols, low downstream processing costs, low energy-consumption, and no generation of inhibitors to biofuel fermentation, it has several disadvantages, including slow delignification rates, substantial holocellulose loss, and long pretreatment time.

Compared to fungi pretreatment, BP with bacteria harboring oxidases involving lignin depolymerization are more promising candidates for delignification because of their environmental adaptability and immense biochemical versatility [32, 259, 261, 264, 269, 273, 274]. For bacteria pretreatments, a series of enzymes (demethylase, MnP, LiP, catalase, peroxidase, phenol oxidase, etc.) have been isolated and identified [38, 39, 45, 95, 261, 264]. These enzymes in bacteria can catalyze the demethylation, alkyl-aryl cleavage, cross-linking, and Cα–Cβ bonds cleavage, and value-added lignin derivatives, such as protocatechuic acid, vanillic acid, guaiacol, vanillin, and 4-ethoxy-3-methoxybenzaldehyde, can form under the ambient condition [95, 275]. *R. opacus* PD630 metabolized aromatics, gluconate, alkanes, and acetate, to produce microbial lipids [32, 35, 260, 276], *Rhodococcus jostii* RHA1 degraded lignin to a series of phenolics [38, 263]. Degradation of lignin (39.6%, dry weight) was achieved by performing cofermentation with wild *R. opacus* PD630 and *R. jostii* RHA1 $VanA^-$. Fatty acids (C13–C24), especially palmitic acid (C16:0; 35.8%) and oleic acid (C18:1; 47.9%), were accumulated in cells [41]. Co-culture of *R. jostii* RHA1, *R. jostii* RHA1 $VanA^-$, and *R. opacus* PD630 to produce extracellular peroxidases and oxidases for degrading 33.6% of low-molecular weight lignin derived from dilute acid-pretreated poplar wood, and the lipid content in cells was 0.017 g lipid/g DCW [38].

One-step BP is known as a slow pretreatment process that requires careful control (e.g., growth and equipment conditions) [67]. Combination pretreatments including *Sphingobacterium* sp. LD-1 with NaOH/urea [261], *Pleurotus ostreatus* with 2 wt% H_2O_2 [238], and *Echinodontium taxodii* with 0.25 wt% H_2SO_4 [277] were found to have higher pretreatment efficiency and saccharification than single BP, indicating that combination of BP and other pretreatments has high application for improving enzymatic hydrolysis of biomass and biofuels production [67, 261].

8.4 Conclusion and Future Recommendations

In response to growing concerns about environmental sustainability, energy security, and societal sustainability, various renewable lignocellulosic biomasses have been used as inexpensive feedstock for producing biofuels (e.g., microbial lipids) in recent years. The most applicable pretreatment technologies on lignocellulosic materials such as physical, chemical, physico-chemical, biological, or their combinations for biofuel production have been developed [79, 88, 99, 110, 120, 278–285]. The disadvantages and advantages of these common pretreatment technologies for treating lignocellulosic materials are illustrated in Table 8.2. The trend of future research should be directed to address some issues including the increase in the commercialization on large-scale by analysis of economic aspects and application of a suitable method based on the different biomass structures. Thus, the design of suitable research in order to find an efficient combination of the existing methods is recommended [286–288]. In future, the development of cost-effective pretreatment technologies for improving the enzymatic digestion and microbial lipid production deserve in-depth exploration.

Acknowledgments The funding supports of this research by the National Natural Science Foundation of China (No. 21978072), New Engineering Research and Practice project of China (No. E-SWYY20202513), and the Open Project of State Key Laboratory of Biocatalysis and Enzyme Engineering (China) are gratefully acknowledged.

References

1. Jin, M. J., Sousa, L. D., Schwartz, C., He, Y. X., Sarks, C., Gunawan, C., Balan, V., & Dale, B. E. (2016). Toward lower cost cellulosic biofuel production using ammonia based pretreatment technologies. *Green Chemistry, 18*, 957–966.
2. Yang, B., & Wyman, C. E. (2004). Effect of xylan and lignin removal by batch and flowthrough pretreatment on the enzymatic digestibility of corn stover cellulose. *Biotechnology and Bioengineering, 86*, 88–98.
3. Gu, X. H., Dong, W., & He, Y. C. (2011). Detoxification of rapeseed meals by steam explosion. *Journal of the American Oil Chemists, 88*, 1831–1838.
4. Qin, L., Qian, H., & He, Y. (2017). Microbial lipid production from enzymatic hydrolysate of Pecan nutshell pretreated by combined pretreatment. *Applied Biochemistry and Biotechnology, 183*, 1336–1350.
5. Zhang, P., Liao, X., Ma, C., Li, Q., Li, A., & He, Y. (2019). Chemoenzymatic conversion of corncob to furfurylamine via tandem catalysis with tin-based solid acid and transaminase biocatalyst. *ACS Sustainable Chemistry &. Engineering, 7*, 17636–17642.
6. Zheng, Y., Shi, J., Tu, M., & Cheng, Y. S. (2017). Principles and development of Lignocellulosic biomass pretreatment for biofuels. *Advance in Biotechnology, 2*, 1–68.
7. Balat, M., Balat, H., & Öz, C. (2008). Progress in bioethanol processing. *Progress in Energy and Combustion Science, 34*, 551–573.
8. Bozell, J. J., & Petersen, G. R. (2010). Technology development for the production of biobased products from biorefinery carbohydrates—The US department of energy's "top 10" revisited. *Green Chemistry, 12*, 539–554.
9. Di, J., Ma, C., Qian, J., Liao, X., Peng, B., & He, Y. (2018). Chemo-enzymatic synthesis of furfuralcohol from chestnut shell hydrolysate by a sequential acid-catalyzed dehydration under microwave and *Escherichia coli* CCZU-Y10 whole-cells conversion. *Bioresource Technology, 262*, 52–58.
10. Gu, T., Wang, B., Zhang, Z., Wang, Z., Chong, G., Ma, C., Tang, Y. J., & He, Y. (2019). Sequential pretreatment of bamboo shoot shell and biosynthesis of ethyl (*R*)-4-chloro-3-hydroxybutanoate in aqueous-butyl acetate media. *Process Biochemistry, 80*, 112–118.
11. Hallac, B. B., & Ragauskas, A. J. (2011). Analyzing cellulose degree of polymerization and its relevancy to cellulosic ethanol. *Biofuels, Bioproducts and Biorefining, 5*, 215–225.
12. Hamelinck, C. N., Hooijdonk, G. V., & Faaij, A. P. C. (2005). Ethanol from lignocellulosic biomass: Techno-economic performance in short-, middle- and long-term. *Biomass and Bioenergy, 28*, 384–410.
13. He, Y., Jiang, C., Chong, G., Di, J., Wu, Y., Wang, B., Xue, X., & Ma, C. (2017). Chemical-enzymatic conversion of corncob-derived xylose to furfuralcohol by the tandem catalysis with SO_4^{2-}/SnO_2-Kaoline and *E. coli* CCZU-T15 cells in toluene–water media. *Bioresource Technology, 245*, 841–849.
14. Huang, Y., Liao, X., Deng, Y., & He, Y. (2019). Co-catalysis of corncob with dilute formic acid plus solid acid $SO_4^{2}-/SnO_2$-montmorillonite under the microwave for enhancing the biosynthesis of furfuralcohol. *Catalysis Communication, 120*, 38–41.
15. Jiang, C. X., Di, J. H., Su, C., Yang, S. Y., & He, Y. C. (2018). One-pot co-catalysis of corncob with dilute hydrochloric acid and tin-based solid acid for the enhancement of furfural production. *Bioresource Technology, 268*, 315–322.

16. Kurosawa, K., Wewetzer, S. J., & Sinskey, A. J. (2013). Engineering xylose metabolism in triacylglycerol-producing Rhodococcus opacus for lignocellulosic fuel production. *Biotechnology for Biofuels, 6*(1), 1.
17. Ma, Z., Liao, Z., Ma, C., He, Y. C., Gong, C., & Yu, X. (2020). Chemoenzymatic conversion of Sorghum durra stalk into furoic acid by a sequential microwave-assisted solid acid conversion and immobilized whole-cells biocatalysis. *Bioresource Technology*. https://doi.org/10.1016/j.biortech.2020.123474.
18. Xue, X. X., Di, J. H., He, Y. C., Wang, B. Q., & Ma, C. L. (2018). Effective utilization of carbohydrate in corncob to synthesize furfuralcohol by chemical–enzymatic catalysis in toluene–water media. *Applied Biochemistry and Biotechnology, 185*, 42–54.
19. Zhang, R. Q., Ma, C. L., Shen, Y. F., Sun, J. F., Jiang, K., Jiang, Z. B., Dai, Y. J., & He, Y. C. (2020). Enhanced biosynthesis of furoic acid via the effective pretreatment of corncob into furfural in the biphasic media. *Catalysis Letters*. https://doi.org/10.1007/s10562-020-03152-9.
20. Zhu, H., Kong, Q., Cao, X., He, H., Wang, J., & He, Y. (2016). Adsorption of Cr(VI) from aqueous solution by chemically modified natural cellulose. *Desalination and Water Treatment, 57*, 20368–20376.
21. Zhu, H. X., Cao, X. J., He, Y. C., Kong, Q. P., He, H., & Wang, J. (2015). Removal of Cu^{2+} from aqueous solutions by the novel modified bagasse pulp cellulose: Kinetics, isotherm and mechanism. *Carbohydrate Polymers, 129*, 115–126.
22. Baral, N. R., & Shah, A. (2017). Comparative techno-economic analysis of steam explosion, dilute sulfuric acid, ammonia fiber explosion and biological pretreatments of corn Stover. *Bioresource Technology, 232*, 331–343.
23. Bryant, D. N., Firth, E., Kaderbhai, N., Taylor, S., Morris, S. M., Logan, D., Garcia, N., Ellis, A., Martin, S. M., & Gallagher, J. A. (2013). Monitoring real-time enzymatic hydrolysis of Distillers Dried Grains with Solubles (DDGS) by dielectric spectroscopy following hydrothermal pre-treatment by steam explosion. *Bioresource Technology, 12*, 765–768.
24. Hallac, B. B., Sannigrahi, P., Pu, Y., Ray, M., Murphy, R. J., & Ragauskas, A. J. (2010). Effect of ethanol Organosolv pretreatment on enzymatic hydrolysis of *Buddleja davidii* stem biomass. *Industrial & Engineering Chemistry Research, 49*, 1467–1472.
25. He, Y., Jiang, C., Chong, G., Di, J., & Ma, C. (2018). Biological synthesis of 2,5-bis(hydroxymethyl)furan from biomass-derived 5-hydroxymethylfurfural by E. coli CCZU-K14 whole cells. *Bioresource Technology, 247*, 1215–1220.
26. He, Y., Li, X., Xue, X., Swita, M. S., Schmidt, A. J., & Yang, B. (2017). Biological conversion of the aqueous wastes from hydrothermal liquefaction of algae and pine wood by *Rhodococci*. *Bioresource Technology, 224*, 457–464.
27. Peng, B., Ma, C. L., Zhang, P. Q., Wu, C. Q., Wang, Z. W., Li, A. T., He, Y. C., & Yang, B. (2019). An effective hybrid strategy for converting rice straw to furoic acid by tandem catalysis via Sn-sepiolite combined with recombinant *E. coli* whole cells harboring horse liver alcohol dehydrogenase. *Green Chemistry, 21*, 5914–5923.
28. Wang, Z. W., Gong, C. J., & He, C. J. (2020). Improved biosynthesis of 5-hydroxymethyl-2-furancarboxylic acid and furoic acid from biomass-derived furans with high substrate tolerance of recombinant *Escherichia coli* HMFOMUT whole-cells. *Bioresource Technology, 303*, 122930.
29. Xue, X. X., Ma, C. L., Di, J. H., Huo, X. Y., & He, Y. C. (2018). One-pot chemo-enzymatic conversion of D-xylose to furfuralcohol by sequential dehydration with oxalic acid plus tin-based solid acid and bioreduction with whole-cells. *Bioresource Technology, 268*, 292–299.
30. Zhang, J., Zhuang, J., Lin, L., Liu, S., & Zhang, Z. (2012). Conversion of *D*-xylose into furfural with mesoporous molecular sieve MCM-41 as catalyst and butanol as the extraction phase. *Biomass and Bioenergy, 39*, 73–77.
31. Zhou, Z., Ju, X., Zhou, M., Xu, X., & Li, L. (2019). An enhanced ionic liquid-tolerant immobilized cellulase system via hydrogel microsphere for improving in situ saccharification of biomass. *Bioresource Technology, 294*, 122146.

32. Kosa, M., & Ragauskas, A. J. (2012). Bioconversion of lignin model compounds with oleaginous Rhodococci. *Applied Microbiology and Biotechnology, 93*(2), 891–900.
33. Wang, B., Rezenom, Y. H., Cho, K. C., Tran, J. L., Lee, D. G., Russell, D. H., Gill, J. J., Young, R., & Chu, K. H. (2014). Cultivation of lipid-producing bacteria with lignocellulosic biomass: Effects of inhibitory compounds of lignocellulosic hydrolysates. *Bioresource Technology, 161*, 162–170.
34. Ye, S., & Cheng, J. (2002). Hydrolysis of Lignocellulosic materials for ethanol production: A review. *ChemInform, 83*, 1–11.
35. Zhao, C., Xie, S., Pu, Y., Zhang, R., Huang, F., Ragauskas, A. J., & Yuan, J. S. (2016). Synergistic enzymatic and microbial lignin conversion. *Green Chemistry, 18*(5), 1306–1312.
36. Huang, C., Chen, X.F., Lian, X., Chen, X.D., & Ma, L.L. (2012). Oil production by the yeast *Trichosporon dermatis* cultured in enzymatic hydrolysates of corncobs. *Bioresource Technology, 110*, 711–714.
37. Karmakar, A., Karmakar, S., & Mukherjee, S. (2010). Properties of various plants and animals feedstocks for biodiesel production. *Bioresource Technology, 101*, 7201–7210.
38. Li, X., He, Y., Zhang, L., Xu, Z., Ben, H., Gaffrey, M. J., Yang, Y., Yang, S., Yuan, J. S., Qian, W. J., & Yang. (2019). Discovery of potential pathways for biological conversion of poplar wood into lipids by co-fermentation of *Rhodococci* strains. *Biotechnology for Biofuels, 12*, 1856.
39. Chong, G. G., Huang, X. J., Di, J. H., Xu, D. Z., He, Y. C., Pei, Y. N., Tang, Y. J., & Ma, C. L. (2018). Biodegradation of alkali lignin by a newly isolated *Rhodococcus pyridinivorans* CCZU-B16. *Bioprocess and Biosystems Engineering, 41*, 501–510.
40. Huang, X., Ding, Y., Liao, X., Peng, B., He, Y., & Ma, C. (2018). Microbial lipid production from enzymatic hydrolysate of corn stover pretreated by combining with biological pretreatment and alkalic salt soaking. *Industrial Crops and Products, 2018*(124), 487–494.
41. He, Y. C., Ding, Y., Ma, C. L., Di, J. H., Jiang, C. L., & Li, A. T. (2017). One-pot conversion of biomass-derived xylose to furfuralcohol by a chemo-enzymatic sequential acid-catalyzed dehydration and bioreduction. *Green Chemistry, 19*, 3844–3850.
42. He, Y., Li, X., Ben, H., Xue, X., & Yang, B. (2017). Lipid production from dilute alkali corn stover lignin by *Rhodococcus* strains. *ACS Sustainable Chemistry & Engineering, 5*, 2302–2311.
43. Yu, Y., Xu, Z., Chen, S., & Jin, M. (2020). Microbial lipid production from dilute acid and dilute alkali pretreated corn stover via *Trichosporon dermatis*. *Bioresource Technology, 295*, 122253.
44. Achinas, S., & Euverink, G. J. W. (2016). Consolidated briefing of biochemical ethanol production from lignocellulosic biomass. *Electronic Journal of Biotechnology, 23*, 44–53.
45. Chong, G., Di, J., Ma, C., Wang, D., Zhang, P., Zhu, J., & He, Y. (2018). Enhanced bioreduction synthesis of ethyl (*R*)-4-chloro-3-hydroybutanoate by alkalic salt pretreatment. *Bioresource Technology, 261*, 196–205.
46. Yu, I. K. M., & Tsang, D. C. W. (2017). Conversion of biomass to hydroxymethylfurfural: A review of catalytic systems and underlying mechanisms. *Bioresource Technology, 238*, 716–732.
47. Dias, M. O. S., Ensinas, A. V., Nebra, S. A., Maciel Filho, R., Rossell, C. E. V., & Maciel, M. R. W. (2009). Production of bioethanol and other bio-based materials from sugarcane bagasse: Integration to conventional bioethanol production process. *Chemical Engineering Research and Design, 87*, 1206–1216.
48. Fatih Demirbas, M. (2009). Biorefineries for biofuel upgrading: A critical review. *Applied Energy, 86*, S151–S161.
49. Foston, M., Katahira, R., Gjersing, E., Davis, M. F., & Ragauskas, A. J. (2012). Solid-state selective ^{13}C excitation and spin diffusion NMR to resolve spatial dimensions in plant cell walls. *Journal of Agricultural and Food Chemistry, 60*, 1419–1427.
50. Kim, J. Y., Lee, H. W., Lee, S. M., Jae, J., & Park, Y. K. (2019). Overview of the recent advances in lignocellulose liquefaction for producing biofuels, bio-based materials and chemicals. *Bioresource Technology, 279*, 373–384.

51. Le, R. K., Das, P., Mahan, K. M., Anderson, S. A., Wells, T., Yuan, J. S., & Ragauskas, A. J. (2017). Utilization of simultaneous saccharification and fermentation residues as feedstock for lipid accumulation in *Rhodococcus opacus*. *AMB Express, 7*, 185.
52. Taherzadeh, M. J., & Keikhosro, K. (2008). Pretreatment of lignocellulosic wastes to improve ethanol and biogas production: A review. *International Journal of Molecular Sciences, 9*, 1621.
53. Yan, L., Zhang, L., & Yang, B. (2014). Enhancement of total sugar and lignin yields through dissolution of poplar wood by hot water and dilute acid flowthrough pretreatment. *Biotechnology for Biofuels, 7*(1), 76.
54. Abdullah, R., Ueda, K., & Saka, S. (2014). Hydrothermal decomposition of various crystalline celluloses as treated by semi-flow hot-compressed water. *Journal of Wood Science, 60*, 278–286.
55. He, Y. C., Zhang, D. P., Di, J. H., Wu, Y. Q., Tao, Z. C., Liu, F., Zhang, Z. J., Chong, G. G., Ding, Y., & Ma, C. L. (2016). Effective pretreatment of sugarcane bagasse with combination pretreatment and its hydrolyzates as reaction media for the biosynthesis of ethyl (*S*)-4-chloro-3-hydroxybutanoate by whole cells of *E. coli* CCZU-K14. *Bioresource Technology, 211*, 720–726.
56. He, Y. C., Ding, Y., Xue, Y. F., Yang, B., Liu, F., Wang, C., Zhu, Z. Z., Qing, Q., Wu, H., Zhu, C., Tao, Z. C., & Zhang, D. P. (2015). Enhancement of enzymatic saccharification of corn stover with sequential Fenton pretreatment and dilute NaOH extraction. *Bioresource Technology, 193*, 324–330.
57. Ramesh, D., Muniraj, I. K., Thangavelu, K., & Karthikeyan, S. (2018). Chapter 2 Pretreatment of lignocellulosic Biomass feedstocks for biofuel production. IGI Global.
58. Rosgaard, L., Pedersen, S., & Meyer, A. S. (2007). Comparison of different pretreatment strategies for enzymatic hydrolysis of wheat and barley straw. *Applied Biochemistry and Biotechnology, 143*, 284–296.
59. Bhatt, S. M., & Shilpa. (2014). Lignocellulosic feedstock conversion, inhibitor detoxification and cellulosic hydrolysis – A review. *Biofuels, 5*, 633–649.
60. He, Y., Jiang, C., Jiang, J., Di, J., Liu, F., Ding, Y., Qing, Q., & Ma, C. (2017). One-pot chemo-enzymatic synthesis of furfuralcohol from xylose. *Bioresource Technology, 238*, 698–705.
61. Bhatia, L., Johri, S., & Ahmad R. (2012). An economic and ecological perspective of ethanol production from renewable agro waste: a review. *AMB Express 2,* 65.
62. Mosier, N., Wyman, C., Dale, B., Elander, R., Lee, Y. Y., Holtzapple, M., & Ladisch, M. (2005). Features of promising technologies for pretreatment of lignocellulosic biomass. *Bioresource Technology, 96*, 673–686.
63. Anwar, Z., Gulfraz, M., & Irshad, M. (2014). Agro-industrial lignocellulosic biomass a key to unlock the future bio-energy: A brief review. *Journal of Radiation Research and Applied Science, 7*(2), 163–173.
64. Pu, Y., Kosa, M., Kalluri, U. C., Tuskan, G. A., & Ragauskas, A. J. (2011). Challenges of the utilization of wood polymers: How can they be overcome? *Applied Microbiology and Biotechnology, 91*, 1525–1536.
65. Ralph, J., Lundquist, K., Brunow, G., Lu, F., Kim, H., Schatz, P. F., Marita, J. M., Hatfield, R. D., Ralph, S. A., Christensen, J. H., & Boerjan, W. (2004). Lignins: Natural polymers from oxidative coupling of 4-hydroxyphenyl- propanoids. *Phytochemistry Reviews, 3*, 29–60.
66. Chandra, R., Takeuchi, H., & Hasegawa, T. (2012). Methane production from lignocellulosic agricultural crop wastes: A review in context to second generation of biofuel production. *Renewable and Sustainable Energy Reviews, 16*, 1462–1476.
67. Dai, Y., Zhang, H. S., Huan, B., & He, Y. C. (2017). Enhancing the enzymatic saccharification of bamboo shoot shell by sequential biological pretreatment with *Galactomyces* sp. CCZU11-1 and deep eutectic solvent extraction. *Bioprocess and Biosystems Engineering, 40*, 1427–1436.

68. Jørgensen, H., Kristensen, J. B., & Felby, C. (2007). Enzymatic conversion of lignocellulose into fermentable sugars: Challenges and opportunities. *Biofuels, Bioproducts and Biorefining, 1*, 119–134.
69. Kaparaju, P., Serrano, M., Thomsen, A. B., Kongjan, P., & Angelidaki, I. (2009). Bioethanol, biohydrogen and biogas production from wheat straw in a bio refifinery concept. *Journal of Bioresource Technology, 100*, 2562–2568.
70. Mahvi, A. H., Maleki, A., & Eslami, A. (2004). Potential of rice husk and rice husk ash for phenol removal in aqueous systems. *American Journal of Applied Sciences, 1*(4), 321–326.
71. Nigam, P. S., Gupta, N., & Anthwal, A. (2009). Pre-treatment of agro-industrial residues. In P. S. Nigam & A. Pandey (Eds.), *Biotechnology for agro-industrial residues utilization* (1st ed., pp. 13–33). Dordrecht: Springer.
72. Saha, B. C., & Cotta, M. A. (2006). Ethanol production from alkaline peroxide pretreated enzymatically saccharifified wheat straw. *Biotechnology Progress, 22*, 449–453.
73. Saini, J. K., Saini, R., & Tewari, L. (2015). Lignocellulosic agriculture wastes as biomass feedstocks for second-generation bioethanol production: Concepts and recent developments. *3 Biotech, 5*, 337–353.
74. Tye, Y. Y., Lee, K. T., Abdullah, W. N. W., & Leh, C. P. (2016). The world availability of nonwood lignocellulosic biomass for the production of cellulosic ethanol and potential pretreatments for the enhancement of enzymatic saccharification. *Renewable Sustainable Energy Reviews, 60*, 155–172.
75. Agbor, V. B., Cicek, N., Sparling, R., Berlin, A., & Levin, D. B. (2011). Biomass pretreatment: Fundamentals toward application. *Biotechnology Advances, 29*, 675–685.
76. Alvira, P., Tomáspejó, E., Ballesteros, M., Negro, M. J., & Pandey, A. (2010). Pretreatment technologies for an efficient bioethanol production process based on enzymatic hydrolysis: A review. *Bioresource technology, 101*, 4851–4861.
77. Lawoko, M., Henriksson, G., & Gellerstedt, G. (2005). Structural differences between the lignin−carbohydrate complexes present in wood and in chemical pulps. *Biomacromolecules, 6*, 3467–3473.
78. Ayyachamy, M., Cliffe, F. E., Coyne, J. M., Collier, J., & Tuohy, M. G. (2013). Lignin: Untapped biopolymers in biomass conversion technologies. *Biomass Conversion and Biorefinery, 3*(3), 255–269.
79. Yang, B., Tao, L., & Wyman, C. E. (2018). Strengths, challenges, and opportunities for hydrothermal pretreatment in lignocellulosic biorefineries. *Biofuels, Bioproducts and Biorefining, 12*(1), 125–138.
80. Kosa, M., & Ragauskas, A. J. (2013). Lignin to lipid bioconversion by oleaginous Rhodococci. *Green Chemistry, 15*(8), 2070–2074.
81. Laskar, D. D., Yang, B., Wang, H., & Lee, J. (2013). Pathways for biomass-derived lignin to hydrocarbon fuels. *Biofuels, Bioproducts and Biorefining, 7*(5), 602–626.
82. Bitra, V. S. P., Womac, A. R., Igathinathane, C., Miu, P. I., Yang, Y. T., Smith, D. R., Chevanan, N., & Sokhansanj, S. (2009). Direct measures of mechanical energy for knife mill size reduction of switchgrass, wheat straw, and corn stover. *Bioresource Technology, 100*, 6578–6585.
83. Case, P. A., Truong, C., Wheeler, M. C., & DeSisto, W. J. (2015). Calcium-catalyzed pyrolysis of lignocellulosic biomass components. *Bioresource Technology, 192*, 247–252.
84. Chang, V. S., Burr, B., & Holtzapple, M. T. (1997). *Lime pretreatment of Switchgrass*. Clifton: Humana Press.
85. Lin, Z., Huang, H., Zhang, H., Zhang, L., Yan, L., & Chen, J. (2010). Ball milling pretreatment of corn stover for enhancing the efficiency of enzymatic hydrolysis. *Applied Biochemistry and Biotechnology, 162*, 1872–1880.
86. Millett, M. A., Baker, A. J., & Satter, L. D. (1976). Physical and chemical pretreatments for enhancing cellulose saccharification. *Biotechnology & Bioengineering Symposium, 6*, 125.
87. Paudel, S. R., Banjara, S. P., Choi, O. K., Park, K. Y., Kim, Y. M., & Lee, J. W. (2017). Pretreatment of agricultural biomass for anaerobic digestion: Current state and challenges. *Bioresource Technology, 245*, 1194–1205.

88. Sun, Y., & Cheng, J. (2002). Hydrolysis of lignocellulosic materials for ethanol production: A review. *Bioresource Technology, 83*, 1–11.
89. Yang, C. P., Shen, Z. Q., Yu, G. C., et al. (2008). Effect and aftereffect of gamma radiation pretreatment on enzymatic hydrolysis of wheat straw. *Bioresourc Technology, 99*, 6240–6245.
90. Zakaria, M. R., Fujimoto, S., Hirata, S., & Hassan, M. A. (2014). Ball milling pretreatment of oil palm biomass for enhancing enzymatic hydrolysis. *Applied Biochemistry and Biotechnology, 173*, 1778–1789.
91. Bali, G., Meng, X., Deneff, J. I., Sun, Q., & Ragauskas, A. J. (2015). The effect of alkaline pretreatment methods on cellulose structure and accessibility. *ChemSusChem, 8*, 275–279.
92. Elgharbawy, A. A., Alam, M. Z., Moniruzzaman, M., & Goto, M. (2016). Ionic liquid pretreatment as emerging approaches for enhanced enzymatic hydrolysis of lignocellulosic biomass. *Biochemical Engineering Journal, 109*, 252–267.
93. Kuo, C. H., & Lee, C. K. (2009). Enhanced enzymatic hydrolysis of sugarcane bagasse by *N*-methylmorpholine-*N*-oxide pretreatment. *Bioresource Technology, 100*, 866–871.
94. Quesada, J., Rubio, M., & Gómez, D. (1999). Ozonation of lignin rich solid fractions from corn stalks. *Journal of Wood Chemistry & Technology, 19*, 115–137.
95. Rabemanolontsoa, H., & Saka, S. (2016). Various pretreatments of lignocellulosics. *Bioresource Technology, 199*, 83–91.
96. Silverstein, R. A., Chen, Y., Sharma-Shivappa, R. R., Boyette, M. D., & Osborne, J. (2007). A comparison of chemical pretreatment methods for improving saccharification of cotton stalks. *Bioresource Technology, 98*, 3000–3011.
97. Veluchamy, C., & Kalamdhad, A. S. (2017). Influence of pretreatment techniques on anaerobic digestion of pulp and paper mill sludge: A review. *Bioresource Technology, 245*, 1206–1219.
98. Grous, W. R., Converse, A. O., & Grethlein, H. E. (1986). Effect of steam explosion pretreatment on pore size and enzymatic hydrolysis of poplar. *Enzyme & Microbial Technology, 8*, 274–280.
99. Hendriks, A. T. W. M., & Zeeman, G. (2009). Pretreatments to enhance the digestibility of lignocellulosic biomass. *Bioresource Technology, 100*, 10–18.
100. Idrees, M., Adnan, A., & Qureshi, F. A. (2013). Optimization of sulfide/sulfite pretreatment of lignocellulosic biomass for lactic acid production. *BioMed Research International, 2013*, 934171.
101. Jacquet, N., Vanderghem, C., Danthine, S., Quiévy, N., Blecker, C., Devaux, J., & Paquot, M. (2012). Influence of steam explosion on physicochemical properties and hydrolysis rate of pure cellulose fibers. *Bioresource Technology, 121*, 221–227.
102. Kumar, A. K., & Sharma, S. (2017). Recent updates on different methods of pretreatment of lignocellulosic feedstocks: A review. *Bioresources and Bioprocessing, 4*, 7.
103. Stanmore, B. R. (2010). Generation of energy from sugarcane bagasse by thermal treatment. *J Waste Biomass Valoriz, 1*, 77–89.
104. Aguiar, A., & Ferraz, A. (2008). Relevance of extractives and wood transformation products on the biodegradation of Pinus taeda by *Ceriporiopsis subvermispora*. *International Biodeterioration & Biodegradation, 61*, 182–188.
105. Ge, X., Matsumoto, T., Keith, L., & Li, Y. (2015). Fungal pretreatment of Albizia chips for enhanced Biogas production by solid-state anaerobic digestion. *Energy & Fuels, 29*, 200–204.
106. Kandhola, G., Djioleu, A., Carrier, D. J., & Kim, J. W. (2017). Pretreatments for enhanced enzymatic hydrolysis of Pinewood: A review. *BioEnergy Research, 10*, 1138–1154.
107. Kandhola, G., Djioleu, A., Carrier, D. J., & Kim, J.-W. (2017). Pretreatments for enhanced enzymatic hydrolysis of pinewood: A review. *BioEnergy Research, 5*, 1–17.
108. Mäkelä, M. R., Donofrio, N., & de Vries, R. P. (2014). Plant biomass degradation by fungi. *Fungal Genetics and Biology, 72*, 2–9.

109. Ryu, S.-H., Cho, M.-K., Kim, M., Jung, S.-M., & Seo, J.-H. (2013). Enhanced lignin biodegradation by a laccase-overexpressed white-rot fungus *Polyporus brumalis* in the pretreatment of Wood chips. *Applied Biochemistry and Biotechnology, 171*, 1525–1534.
110. Sindu, R., Binod, P., & Pandey, A. (2016). Biological pretreatment of lignocellulosic biomass – An overview. *Bioresource Technology, 199*, 76–82.
111. Brodeur, G., Yau, E., Badal, K., Collier, J., Ramachandran, K. B., & Ramakrishnan, S. (2011). Chemical and physicochemical pretreatment of lignocellulosic biomass: A review. *Enzyme Research, 2011*, 787–532.
112. He, Y. C., Tao, Z. C., Di, J. H., Chen, L., Zhang, L. B., Zhang, D. P., Chong, G. G., Liu, F., Ding, Y., Jiang, C. X., & Ma, C. L. (2016). Effective asymmetric bioreduction of ethyl 4-chloro-3-oxobutanoate to ethyl (*R*)-4-chloro-3-hydroxybutanoate by recombinant *E. coli* CCZU-A13 in [Bmim]PF_6–hydrolyzate media. *Bioresource Technology, 214*, 414–418.
113. Sun, F., Wang, L., Hong, J., Ren, J., Du, F., Hu, J., Zhang, Z., & Zhou, B. (2015). The impact of glycerol organosolv pretreatment on the chemistry and enzymatic hydrolyzability of wheat straw. *Bioresource Technology, 187*, 354–361.
114. Cadoche, L., & López, G. D. (1989). Assessment of size reduction as a preliminary step in the production of ethanol from lignocellulosic wastes. *Biological Wastes, 30*, 153–157.
115. Kim, H. J., Chang, J. H., Jeong, B. Y., & Jin, H. L. (2013). Comparison of milling modes as a pretreatment method for cellulosic biofuel production. *Journal of Clean Energy Technologies, 1*, 45–48.
116. Himmel, M., Tucker, M., Baker, J., Rivard, C., Oh, K., & Grohmann, K. (1985). Comminution of biomass: Hammer and knife mills. In *Biotechnology and bioengineering* (symposium no 15) (pp. 39–57). New York: Wiley.
117. Kim, H., & Ralph, J. (2010). Solution-state 2D NMR of ball-milled plant cell wall gels in DMSO-d6/pyridine-d5. *Organic and Biomolecular Chemistry, 8*, 576–591.
118. Maier, G., Zipper, P., Stubicar, M., & Schurz, J. (2005). Amorphization of different cellulose samples by ball milling. *Cellulose Chemistry and Technology, 39*, 167–177.
119. Millett, M. A., Baker, A. J., Feist, W. C., Mellenberger, R. W., & Satter, L. D. (1970). Modifying wood to increase its in vitro digestibility. *Journal of Animal Science, 31*, 781–788.
120. Tassinari, T., Macy, C., Spano, L., & Ryu, D. D. Y. (1980). Energy requirements and process design considerations in compression-milling pretreatment of cellulosic wastes for enzymic hydrolysis. *Biotechnology and Bioengineering, 22*, 1689–1705.
121. Wu, Z. H., Sumimoto, M., & Tanaka, H. (1995). Mechanochemistry of lignin. XIII. Generation of oxygen-containing radicals in the aqueous media of mechanical pulping. *Journal of Wood Chemistry and Technology, 15*, 27–42.
122. Yu, M., Womac, A. R., Igathinathane, C., Ayers, P. D., & Buschermohle, M. (2006). Switchgrass ultimate stresses at typical biomass conditions available for processing. *Biomass & Bioenergy, 30*, 214–219.
123. Hideno, A., Inoue, H., Tsukahara, K., Fujimoto, S., Minowa, T., Inoue, S., Endo, T., & Sawayama, S. (2009). Wet disk milling pretreatment without sulfuric acid for enzymatic hydrolysis of rice straw. *Bioresource Technology, 100*, 2706–2711.
124. Silva, A. S. D., Inoue, H., Endo, T., Yano, S., & Bon, E. P. S. (2010). Milling pretreatment of sugarcane bagasse and straw for enzymatic hydrolysis and ethanol fermentation. *Bioresource Technology, 101*, 7402–7409.
125. Himmel, M. E., Ding, S. Y., Johnson, D. K., Adney, W. S., Nimlos, M. R., Brady, J. W., & Foust, T. D. (2007). Biomass recalcitrance: Engineering plants and enzymes for biofuels production. *Science, 315*, 804–807.
126. Mani, S., Tabil, L. G., & Sokhansanj, S. (2004). Grinding performance and physical properties of wheat and barley straws, corn stover and switchgrass. *Biomass & Bioenergy, 27*, 339–352.
127. Lee, J. E., Vadlani, P. V., & Min, D. (2017). Sustainable production of microbial lipids from lignocellulosicbiomass using Oleaginous yeast cultures. *Journal of Sustainable Bioenergy Systems, 7*, 74871.

128. Bak, J. S., Ko, J. K., Han, Y. H., Lee, B. C., Choi, I. G., & Kim, K. H. (2009). Improved enzymatic hydrolysis yield of rice straw using electron beam irradiation pretreatment. *Bioresource Technology, 100*, 1285–1290.
129. Dunlap, C. E., & Chiang, L. C. (1980). Cellulose degradation-a common link. In M. L. Shuler (Ed.), *Utilization and recycle of agricultural wastes and residues* (pp. 19–65). Boca Raton: CRC Press.
130. Kapoor, K., Garg, N., Garg, R. K., Varshney, L., & Tyagi, A. K. (2017). Study the effect of gamma radiation pretreatment of sugarcane bagasse on its physcio-chemical morphological and structural properties. *Radiation Physics and Chemistry, 141*, 190–195.
131. Ma, H., Liu, W. W., Chen, X., Wu, Y., & Yu, Z. (2009). Enhanced enzymatic saccharification of rice straw by microwave pretreatment. *Bioresource Technology, 100*, 1279–1284.
132. Velmurugan, R., & Muthukumar, K. (2011). Utilization of sugarcane bagasse for bioethanol production: Sono-assisted acid hydrolysis approach. *Bioresource Technology, 102*, 7119–7123.
133. Takács, E., Wojnárovits, L., Földváry, C., Hargittai, P., Borsa, J., & Sajó, I. (2000). Effect of combined gamma-irradiation and alkali treatment on cotton–cellulose. *Radiation Physics & Chemistry, 57*, 399–403.
134. Galbe, M., & Zacchi, G. (2007). *Pretreatment of Lignocellulosic materials for efficient bioethanol production*. Berlin/Heidelberg: Springer.
135. Singh, R., Krishna, B. B., Kumar, J., & Bhaskar, T. (2016). Opportunities for utilization of non-conventional energy sources for biomass pretreatment. *Bioresource Technology, 199*, 398–407.
136. Chen, X., Yu, J., Zhang, Z., & Lu, C. (2011). Study on structure and thermal stability properties of cellulose fibres from rice straw. *Journal of Carbohydrate Polymers, 85*, 245–250.
137. Chen, W., Yu, H., Liu, Y., Chen, P., Zhang, M., & Hai, Y. (2011). Individualization of cellulose nanofibers from wood using high-intensity ultrasonication combined with chemical pretreatments. *Carbohydrate Polymers, 83*, 1804–1811.
138. Lu, X., Bo, X., Zhang, Y., & Angelidaki, I. (2011). Microwave pretreatment of rape straw for bioethanol production: Focus on energy efficiency. *Bioresource Technology, 102*, 7937.
139. Wahidin, A. I., & Shaleh, S. R. M. (2014). Rapid biodiesel production using wet microalgae via microwave irradiation. *Energy Conversion and Management, 84*, 227–233.
140. Hu, Z., & Wen, Z. (2008). Enhancing enzymatic digestibility of switchgrass by microwave-assisted alkali pretreatment. *Biochemical Engineering Journal, 38*, 369–378.
141. Keshwani, D. R., & Cheng, J. J. (2010). Microwave-based alkali pretreatment of switchgrass and coastal bermudagrass for bioethanol production. *Biotechnology Progress, 26*, 644–652.
142. Gogate, P. R., Sutkar, V. S., & Pandit, A. B. (2011). Sonochemical reactors: Important design and scale up considerations with a special emphasis on heterogeneous systems. *Chemical Engineering Journal, 166*, 1066–1082.
143. Montalbo-Lomboy, M., Johnson, L., Khanal, S. K., Leeuwen, J. V., & Grewell, D. (2010). Sonication of sugary-2 corn: A potential pretreatment to enhance sugar release. *Bioresource Technology, 101*, 351–358.
144. Rehman, M. S. U., Kim, I., Chisti, Y., & Han, J. I. (2013). Use of ultrasound in the production of bioethanol from lignocellulosic biomass. *Energy Education Science & Technology, 30*, 1391–1410.
145. Tang, A., Zhang, H., Gang, C., Xie, G., & Liang, W. (2005). Influence of ultrasound treatment on accessibility and regioselective oxidation reactivity of cellulose. *Ultrasonics Sonochemistry, 12*, 467.
146. Yachmenev, V., Condon, B., Klasson, T., & Lambert, A. (2009). Acceleration of the enzymatic hydrolysis of corn stover and sugar cane bagasse celluloses by low intensity uniform ultrasound. *Journal of Biobased Materials & Bioenergy, 3*, 25–31.
147. Balasubramanian, J. D. A., Kanitkar, A., & Boldor, D. (2011). Oil extraction from Scenedesmus obliquus using a continuous microwave system – Design, optimization, and quality characterization. *Bioresource Technology, 102*, 3396–3403.

148. Keris-Sen, U. D., Sen, U., Soydemir, G., & Gurol, M. D. (2014). An investigation of ultrasound effect on microalgal cell integrity and lipid extraction efficiency. *Bioresource Technology, 152*, 407–413.
149. Kumar, P., Barrett, D. M., Delwiche, M. J., & Stroeve, P. (2009). Methods for pretreatment of lignocellulosic biomass for efficient hydrolysis and biofuel production. *Industrial & Engineering Chemistry Research, 48*, 3713–3729.
150. Roy, P., & Dias, G. (2017). Prospects for pyrolysis technologies in the bioenergy sector: A review. *Renewable and Sustainable Energy Reviews, 77*, 59–69.
151. Fan, L. T., Gharpuray, M. M., & Lee, Y. H. (1987). *Cellulose hydrolysis* (Biotechnology monographs. Volume 3). New York: Springer.
152. Lian, J., Garcia-Perez, M., Coates, R., Wu, H., & Chen, S. (2012). Yeast fermentation of carboxylic acids obtained from pyrolytic aqueous phases for lipid production. *Bioresource Technology, 118*, 177–186.
153. Janu, K. U., Sindhu, R., Binod, P., Kuttiraja, M., Sukumaran, R. K., & Pandey, A. (2011). Studies on physicochemical changes during alkali pretreatment and optimization of hydrolysis conditions to improve sugar yield from bagasse. *Journal Ofentific & Industrial Research, 70*, 952–958.
154. McMillan, J. D. (1994). Pretreatment of lignocellulosic biomass. In *Enzymatic conversion of biomass for fuels production* (Vol. 566, pp. 292–324). Washington, DC: American Chemical Society.
155. Teramoto, Y., Lee, S. H., & Endo, T. (2008). Pretreatment of woody and herbaceous biomass for enzymatic saccharification using sulfuric acid-free ethanol cooking. *Bioresource Technology, 99*, 8856–8863.
156. Gong, Z., Wang, X., Yuan, W., Wang, Y., Zhou, W., Wang, G., & Liu, Y. (2020). Fed-batch enzymatic hydrolysis of alkaline organosolv-pretreated corn stover facilitating high concentrations and yields of fermentable sugars for microbial lipid production. *Biotechnology for Biofuels, 13*, 613.
157. Aita, G. A., Salvi, D. A., & Walker, M. S. (2011). Enzyme hydrolysis and ethanol fermentation of dilute ammonia pretreated energy cane. *Bioresource Technology, 102*, 4444–4448.
158. Chong, G. G., He, Y. C., Liu, Q. X., Kou, X. Q., & Qing, Q. (2017). Sequential aqueous ammonia extraction and LiCl/N,N-Dimethyl formamide pretreatment for enhancing enzymatic saccharification of winterbamboo shoot shell. *Applied Biochemistry and Biotechnology, 182*, 1341–1357.
159. Chong, G. G., He, Y. C., Liu, Q. X., Kou, X. Q., Huang, X. J., Di, J. H., & Ma, C. L. (2017). Effective enzymatic in situ saccharification of bamboo shoot shell pretreated by dilute alkalic salts sodium hypochlorite/sodium sulfide pretreatment under the autoclave system. *Bioresource Technology, 241*, 726–734.
160. Gupta, R., & Lee, Y. Y. (2010). Investigation of biomass degradation mechanism in pretreatment of switchgrass by aqueous ammonia and sodium hydroxide. *Bioresource Technology, 101*, 8185.
161. Yoo, C. G., Nghiem, N. P., Hicks, K. B., & Kim, T. H. (2011). Pretreatment of corn stover using low-moisture anhydrous ammonia (LMAA) process. *Bioresource Technology, 102*, 10028–10034.
162. Liu, Z., Padmanabhan, S., Cheng, K., Schwyter, P., Pauly, M., Bell, A. T., & Prausnitz, J. M. (2013). Aqueous-ammonia delignification of miscanthus followed by enzymatic hydrolysis to sugars. *Bioresource Technology, 135*, 23–29.
163. Pryor, S. W., Karki, B., & Nahar, N. (2012). Effect of hemicellulase addition during enzymatic hydrolysis of switchgrass pretreated by soaking in aqueous ammonia. *Bioresource Technology, 123*, 620–626.
164. Kumar, L., Chandra, R., & Saddler, J. (2011). Influence of steam pretreatment severity on post-treatments used to enhance the enzymatic hydrolysis of pretreated softwoods at low enzyme loadings. *Biotechnology and Bioengineering, 108*, 2300–2311.

165. Liu, H., Pang, B., Zhou, J., Han, Y., Lu, J., Li, H., & Wang, H. (2016). Comparative study of pretreated corn stover for sugar production using cotton pulping black liquor (CPBL) instead of sodium hydroxide. *Industrial Crops and Products, 84*, 97–103.
166. Mendes, F. M., Siqueira, G., Carvalho, W., Ferraz, A., & Milagres, A. M. (2011). Enzymatic hydrolysis of Chemithermomechanically pretreated sugarcane bagasse and samples with reduced initial lignin content. *Biotechnology Progress, 27*, 395–401.
167. Mendes, F. M., Heikkilä, E., Fonseca, M. B., Milagres, A. M. F., Ferraz, A., & Fardim, P. (2015). Topochemical characterization of sugar cane pretreated with alkaline sulfite. *Industrial Crops and Products, 69*, 60–67.
168. Xu, H., Li, B., & Mu, X. (2016). Review of alkali-based pretreatment to enhance enzymatic Saccharification for Lignocellulosic biomass conversion. *Industrial & Engineering Chemistry Research, 55*, 8691–8705.
169. Zhang, D. S., Yang, Q., Zhu, J. Y., & Pan, X. J. (2013). Sulfite (SPORL) pretreatment of switchgrass for enzymatic saccharification. *Bioresource Technology, 129*, 127–134.
170. Yang, L., Cao, J., Mao, J., & Jin, Y. (2013). Sodium carbonate–sodium sulfite pretreatment for improving the enzymatic hydrolysis of rice straw. *Industrial Crops and Products, 43*, 711–717.
171. Gong, W., Liu, C., Mu, X., Du, H., Lv, D., Li, B., & Han, S. (2015). Hydrogen peroxide-assisted sodium carbonate pretreatment for the enhancement of enzymatic saccharification of corn stover. *ACS Sustainable Chemistry & Engineering, 3*, 3477–3485.
172. Qing, Q., Zhou, L. L., Guo, Q., Huang, M. Z., He, Y. C., Wang, L. Q., & Zhang, Y. (2016). A combined sodium phosphate and sodium sulfide pretreatment for enhanced enzymatic digestibility and delignification of corn stover. *Bioresource Technology, 218*, 209–216.
173. Jiang, C. X., He, Y. C., Chong, G. G., Di, J. H., Tang, Y. J., & Ma, C. L. (2017). Enzymatic in situ saccharification of sugarcane bagasse pretreated with low loading of alkalic salts Na_2SO_3/Na_3PO_4 by autoclaving. *Journal of Biotechnology, 259*, 73–82.
174. Liang, Y., Jarosz, K., Wardlow, A. T., Zhang, J., & Cui, Y. (2014). Lipid production by *Cryptococcus curvatus* on hydrolysates derived from corn fiber and sweet sorghum bagasse following dilute acid pretreatment. *Applied Biochemistry and Biotechnology, 173*, 2086–2098.
175. Cai, D., Dong, Z., Wang, Y., Chen, C., Li, P., Qin, P., Wang, Z., & Tan, T. W. (2016). Biorefinery of corn cob for microbial lipid and bio-ethanol production: An environmental friendly process. *Bioresource Technology, 211*, 677–684.
176. Andanson, J. M., & Costa Gomes, M. F. (2015). Thermodynamics of cellulose dissolution in an imidazolium acetate ionic liquid. *Chemical Communications, 51*, 4485–4487.
177. Li, Q., He, Y. C. X., Jun, G., Xu, X., Yang, J. M., & Li, L. Z. (2009). Improving enzymatic hydrolysis of wheat straw using ionic liquid 1-ethyl-3-methyl imidazolium diethyl phosphate pretreatment. *Bioresource Technology, 100*, 3570–3575.
178. Brandt, A., Ray, M. J., To, T. Q., Leak, D. J., Murphy, R. J., & Welton, T. (2011). Ionic liquid pretreatment of lignocellulosic biomass with ionic liquid–water mixtures. *Green Chemistry, 13*, 2489–2499.
179. He, Y. C., Liu, F., Gong, L., Di, J. H., Ding, Y., Ma, C. L., Zhang, D. P., Tao, Z. C., Wang, C., & Yang, B. (2016). Enzymatic in situ saccharification of chestnut shell with high ionic liquid-tolerant cellulases from *Galactomyces* sp. CCZU11-1 in a biocompatible ionic liquid-cellulase media. *Bioresource Technology, 201*, 133–139.
180. Li, C., Knierim, B., Manisseri, C., Arora, R., Scheller, H. V., Auer, M., et al. (2010). Comparison of dilute acid and ionic liquid pretreatment of switchgrass: Biomass recalcitrance, delignification and enzymatic saccharification. *Bioresource Technology, 101*, 4900–4906.
181. Li, X., Kim, T. H., & Nghiem, N. P. (2010). Bioethanol production from corn stover using aqueous ammonia pretreatment and two-phase simultaneous saccharifification and fermentation (TPSSF). *Bioresource Technology, 101*, 5910–5916.
182. Pinkert, A., Marsh, K. N., Pang, S. S., & Staiger, M. P. (2009). Ionic liquids and their interaction with cellulose. *Chemical Reviews, 109*, 6712–6728.

183. Shill, K., Padmanabhan, S., Xin, Q., Prausnitz, J. M., Clark, D. S., & Blanch, H. W. (2011). Ionic liquid pretreatment of cellulosic biomass: Enzymatic hydrolysis and ionic liquid recycle. *Biotechnology and Bioengineering, 108*, 511–520.
184. Singh, S., Simmons, B. A., & Vogel, K. P. (2009). Visualization of biomass solubilization and cellulose regeneration during ionic liquid pretreatment of switchgrass. *Biotechnology and Bioengineering, 104*, 68–75.
185. Silva, S. S., Mano, J. F., & Reis, R. L. (2017). Ionic liquids in the processing and chemical modification of chitin and chitosan for biomedical applications. *Green Chemistry, 19*, 1208–1220.
186. Xu, J. X., Xiong, P., & He, B. F. (2016). Advances in improving the performance of cellulase in ionic liquids for lignocellulose biorefinery. *Bioresource Technology, 200*, 961–970.
187. Zhang, Q., Hu, J., & Lee, D. J. (2017). Pretreatment of biomass using ionic liquids: Research updates. *Renewable Energy, 11*, 77–84.
188. Swatloski, R. P., Spear, S. K., Holbrey, J. D., & Rogers, R. D. (2002). Dissolution of cellose with ionic liquids. *Journal of the American Chemical Society, 124*, 4974–4975.
189. Gurau, G., Wang, H., Qiao, Y., Lu, X., Zhang, S., & Rogers Robin, D. (2012). Chlorine-free alternatives to the synthesis of ionic liquids for biomass processing. *Pure and Applied Chemistry, 84*(3), 745.
190. Kosan, B., Michels, C., & Meister, F. (2008). Dissolution and forming of cellulose with ionic liquids. *Cellulose, 15*(1), 59–66.
191. Aid, T., Hyvarinen, S., Vaher, M., Koel, M., & Mikkola, J. P. (2016). Saccharification of lignocellulosic biomasses via ionic liquid pretreatment. *Industrial Crops and Products, 92*, 336–341.
192. Chang, K. L., Chen, X. M., Wang, X. Q., Han, Y. J., Potprommanee, L., Liu, J. Y., Liao, Y. L., Ning, X. A., Sun, S. Y., & Huang, Q. (2017). Impact of surfactant type for ionic liquid pretreatment on enhancing delignification of rice straw. *Bioresource Technology, 227*, 388–392.
193. Clough, M. T., Geyer, K., Hunt, P. A., Son, S., Vagt, U., & Welton, T. (2015). Ionic liquids: Not always innocent solvents for cellulose. *Green Chemistry, 17*, 231–243.
194. Sun, N., Rahman, M., Qin, Y., Maxim, M. L., Rodriguez, H., & Rogers, R. D. (2009). Complete dissolution and partial delignification of wood in the ionic liquid 1-ethyl-3-methylimidazolium acetate. *Green Chemistry, 11*, 646–655.
195. Zhao, H., Jones, C. L., Baker, G. A., Xia, S., Olubajo, O., & Person, V. N. (2009). Regenerating cellulose from ionic liquids for an accelerated enzymatic hydrolysis. *Journal of Biotechnology, 139*, 47–54.
196. Zhao, X., Cheng, K., & Liu, D. (2009). Organosolv pretreatment of lignocellulosic biomass for enzymatic hydrolysis. *Applied Microbiology & Biotechnology, 82*, 815.
197. He, Y. C., Liu, F., Gong, L., Zhu, Z. Z., Ding, Y., Wang, C., Xue, Y. F., Rui, H., Tao, Z. C., Zhang, D. P., & Ma, C. L. (2015). Significantly improving enzymatic saccharification of high crystallinity index's corn stover by combining ionic liquid [Bmim]Cl–HCl–water media with dilute NaOH pretreatment. *Bioresource Technology, 189*, 421–425.
198. Zhang, J., Feng, L., Wang, D., Zhang, R., Liu, G., & Cheng, G. (2014). Thermogravimetric analysis of lignocellulosic biomass with ionic liquid pretreatment. *Bioresource Technology, 153*, 379–382.
199. de Oliveira, H. F., & Rinaldi, R. (2015). Understanding cellulose dissolution: Energetics of interactions of ionic liquids and cellobiose revealed by solution microcalorimetry. *ChemSusChem, 8*, 1577.
200. Kanbayashi, T., & Miyafuji, H. (2015). Topochemical and morphological characterization of wood cell wall treated with the ionic liquid, 1-ethylpyridinium bromide. *Planta, 242*, 509–518.
201. Gong, Z., Shen, H., Wang, Q., Yang, X., Xie, H., & Zhao, Z. K. (2013). Efficient conversion of biomass into lipids by using the simultaneous saccharification and enhanced lipid production process. *Biotechnology for Biofuels, 6*, 36.

202. Xie, H., Shen, H., Gong, Z., Wang, Q., & Zhao, Z. K. (2012). Enzymatic hydrolysates of corn stover pretreated by a *N*-methylpyrrolidone–ionic liquid solution for microbial lipid production. *Green Chemistry, 14*, 1202–1210.
203. Procentese, A., Johnson, E., Orr, V., Campanile, A. G., Wood, J. A., Marzocchella, A., & Rehmann, F. (2015). Deep eutectic solvent pretreatment and subsequent saccharification of corncob. *Bioresource Technology, 92*, 31–36.
204. Xu, G. C., Ding, J. C., Han, R. Z., Dong, J. J., & Ni, Y. (2016). Enhancing cellulose accessibility of corn stover by deep eutectic solvent pretreatment for butanol fermentation. *Bioresource Technology, 203*, 364–369.
205. Zhang, W. C., Xia, S. Q., & Ma, P. S. (2016). Facile pretreatment of lignocellulosic biomass using deep eutectic solvents. *Bioresource Technology, 219*, 1–5.
206. Sundstrom, E., Yaegashi, J., Yan, J., Masson, F., Papa, G., Rodriguez, A., Mirsiaghi, M., Liang, L., He, Q., Tanjore, D., Pray, T. R., Singh, S., Simmons, B., Sun, N., Magnuson, J., & Gladden, J. (2018). Demonstrating a separation-free process coupling ionic liquid pretreatment, saccharification, and fermentation with *Rhodosporidium toruloides* to produce advanced biofuels. *Green Chemistry, 20*, 2870–2879.
207. Koo, B. W., Min, B. C., Gwak, K. S., Lee, S. M., Choi, J. W., Yeo, H., & Choi, I. G. (2012). Structural changes in lignin during organosolv pretreatment of Liriodendron tulipifera and the effect on enzymatic hydrolysis. *Biomass & Bioenergy, 42*, 24–32.
208. Mesa, L., González, E., Cara, C., González, M., Castro, E., & Mussatto, S. I. (2011). The effect of organosolv pretreatment variables on enzymatic hydrolysis of sugarcane bagasse. *Chemical Engineering Journal, 168*, 1157–1162.
209. Qing, Q., Zhou, L. L., Guo, Q., Gao, X. H., Zhang, Y., He, Y. C., & Zhang, Y. (2017). Mild alkaline presoaking and organosolv pretreatment of corn stover and their impacts on corn stover composition, structure, and digestibility. *Bioresource Technology, 233*, 284–290.
210. Ostovareh, S., Karimi, K., & Zamani, A. (2015). Efficient conversion of sweet sorghum stalks to biogas and ethanol using organosolv pretreatment. *Industrial Crops and Products, 66*, 170–177.
211. He, Y. C., Liu, F., Gong, L., Lu, T., Ding, Y., Zhang, D. P., Qing, Q., & Zhang, Y. (2015). Improving enzymatic hydrolysis of corn stover pretreated by ethylene glycol-perchloric acid-water mixture. *Applied Biochemistry and Biotechnology, 175*, 1306–1317.
212. Liu, J., Takada, R., Karita, S., Watanabe, T., Honda, Y., & Watanabe, T. (2010). Microwave-assisted pretreatment of recalcitrant softwood in aqueous glycerol. *Bioresource Technology, 101*, 9355–9360.
213. He, Y. C., Liu, F., Di, J. H., Ding, Y., Tao, Z. C., Zhu, Z. Z., Wu, Y. Q., Chen, L., Wang, C., Xue, Y. F., Chong, G. G., & Ma, C. L. (2016). Effective enzymatic saccharification of dilute NaOH extraction of chestnut shell pretreated by acidified aqueous ethylene glycol media. *Industrial Crops and Products, 81*, 129–138.
214. Novo, L. P., Gurgel, L. V. A., Marabezi, K., & da Silva Curvelo, A. A. (2011). Delignification of sugarcane bagasse using glycerol-water mixtures to produce pulps for saccharification. *Bioresource Technology, 102*, 10040–10046.
215. Zhang, T., Zhou, Y. J., Liu, D. L., & Petrus, L. (2007). Qualitative analysis of products formed during the acid catalyzed liquefaction of bagasse in ethylene glycol. *Bioresource Technology, 98*, 1454–1459.
216. Biganska, O., & Navard, P. (2009). Morphology of cellulose objects regenerated from cellulose-*N*-methy morpholine *N*-oxide-water solutions. *Cellulose, 16*, 179–188.
217. Li, Q., Ji, G. S., Tang, Y. B., Gu, X. D., Fei, J. J., & Jiang, H. Q. (2012). Ultrasound-assisted compatible in situ hydrolysis of sugarcane bagasse in cellulase-aqueous–*N*-methylmorpholine-N-oxide system for improved saccharification. *Bioresource Technology, 107*, 251–257.
218. He, Y. C., Xia, D. Q., Ma, C. L., Gong, L., Gong, T., Wu, M. X., Zhang, Y., Tang, Y. J., Xu, J. H., & Liu, Y. Y. (2013). Enzymatic saccharification of sugarcane baggage by *N*-methylmorpholine-*N*-oxide-tolerant cellulase from a newly isolated *Galactomyces* sp. CCZU11-1. *Bioresource Technology, 135*, 18–22.

219. Alizadeh, H., Teymouri, F., Gilbert, T. I., & Dale, B. E. (2005). Pretreatment of switchgrass by ammonia fiber explosion (AFEX). *Applied Biochemistry and Biotechnology, 121–124*, 1133.
220. Bals, B., Wedding, C., Balan, V., Sendich, E., & Dale, B. (2011). Evaluating the impact of ammonia fibre expansion (AFEX) pretreatment conditions on the cost of ethanol production. *Bioresource Technology, 102*, 1277–1283.
221. Kim, T. H., & Lee, Y. Y. (2005). Pretreatment of corn stover by soaking in aqueous ammonia. *Applied Biochemistry and Biotechnology, 124*, 1119–1131.
222. Xue, Y. P., Jin, M., Orjuela, A., Slininger, P. J., Dien, B. S., Dale, B. E., & Balan, V. (2015). Microbial lipid production from AFEX™ pretreated corn stover. *RSC Advances, 5*, 28725–28734.
223. Kyoungheon, K., & Hong, J. (2001). Supercritical CO2 pretreatment of lignocellulose enhances enzymatic cellulose hydrolysis. *Bioresource Technology, 77*, 139–144.
224. Gu, T., Held, M. A., & Faik, A. (2013). Supercritical CO_2 and ionic liquids for the pretreatment of lignocellulosic biomass in bioethanol production. *Environ Technol (United Kingdom), 34*, 1735–1749.
225. Duarte, S. H., dos Santos, P., Michelon, M., de Pinho Oliveira, S. M., Martínez, J., & Maugeri, F. (2017). Recovery of yeast lipids using different cell disruption techniques and supercritical CO_2 extraction. *Biochemical Engineering Journal, 125*, 230–237.
226. Zheng, Y., Lin, H. M., Wen, J., Cao, N., Yu, X., & Tsao, G. T. (1995). Supercritical carbon dioxide explosion as a pretreatment for cellulose hydrolysis. *Biotechnology Letters, 17*, 845–850.
227. Srinivasan, N., & Ju, L. K. (2010). Pretreatment of guayule biomass using supercritical carbon dioxide-based method. *Bioresource Technology, 101*, 9785–9791.
228. Howlader, M. S., French, W. T., Shields-Menard, S. A., Amirsadeghi, M., Green, M., & Rai, N. (2017). Microbial cell disruption for improving lipid recovery using pressurized CO_2: Role of CO_2 solubility in cell suspension, sugar broth, and spent media. *Biotechnology Progress, 33*, 737–748.
229. Laser, M., Schulman, D., Allen, S. G., Lichwa, J., Antal, M. J., & Lynd, L. R. (2002). A comparison of liquid hot water and steam pretreatments of sugar cane bagasse for conversion to ethanol. *Bioresource Technology, 81*, 33–44.
230. Wei, Z., Zeng, G., Huang, F., Kosa, M., Sun, Q., Meng, X., Huang, D., & Ragauskas, A. J. (2015). Microbial lipid production by oleaginous *Rhodococci* cultured in lignocellulosic autohydrolysates. *Applied Microbiology and Biotechnology, 99*, 7369–7377.
231. Banerjee, S., Sen, R., Pandey, R. A., Chakrabarti, T., Satpute, D., Giri, B. S., & Mudliar, S. (2009). Evaluation of wet air oxidation as a pretreatment strategy for bioethanol production from rice husk and process optimization. *Biomass Bioenerg, 33*, 1680–1686.
232. Bjerre, A. B., Olesen, A. B., Fernqvist, T., Plöger, A., & Schmidt, A. S. (1996). Pretreatment of wheat straw using combined wet oxidation and alkaline hydrolysis resulting in convertible cellulose and hemicellulose. *Biotechnology and Bioengineering, 49*, 568–577.
233. Hammel, K. E., Kapich, A. N., Jensen, K. A., Jr., & Ryan, Z. C. (2002). Reactive oxygen species as agents of wood decay by fungi. *Enzyme and Microbial Technology, 30*, 445–453.
234. Lucas, M., Hanson, S. K., Wagner, G. L., Kimball, D. B., & Rector, K. D. (2012). Evidence for room temperature delignification of wood using hydrogen peroxide and manganese acetate as a catalyst. *Bioresource Technology, 119*, 174–180.
235. Martín, C., Thomsen, M. H., Hauggaard-Nielsen, H., & Thomsen, A. B. (2008). Wet oxidation pretreatment, enzymatic hydrolysis and simultaneous saccharification and fermentation of clover–ryegrass mixtures. *Bioresource Technology, 99*, 8777–8782.
236. Nakamura, Y., Daidai, M., & Kobayashi, F. (2004). Ozonolysis mechanism of lignin model compounds and microbial treatment of organic acids produced. *Water Science & Technology A Journal of the International Association on Water Pollution Research, 50*, 167.
237. Saha, B. C., & Cotta, M. A. (2007). Enzymatic saccharification and fermentation of alkaline peroxide pretreated rice hulls to ethanol. *Enzyme & Microbial Technology, 41*, 528–532.

238. Yu, J., Zhang, J. B., He, J., Liu, Z. D., & Yu, Z. N. (2009). Combinations of mild physical or chemical pretreatment with biological pretreatment for enzymatic hydrolysis of rice hull. *Bioresource Technology, 100*, 903–908.
239. Cao, W. X., Sun, C., Liu, R. H., Yin, R. Z., & Wu, X. W. (2012). Comparison of the effects of five pretreatment methods on enhancing the enzymatic digestibility and ethanol production from sweet sorghum bagasse. *Bioresource Technology, 111*, 215–221.
240. Azzam, A. M. (1989). Pretreatment of cane bagasse with alkaline hydrogen peroxide for enzymatic hydrolysis of cellulose and ethanol fermentation. *Journal of Environmental Science and Health, Part B: Pesticides, Food Contaminants, and Agricultural Wastes; (USA), 24*, 421–433.
241. Sheikh, M. M. I., Kim, C. H., Park, H. H., Nam, H. G., Lee, G. S., Jo, H. S., Lee, J. Y., & Kim, J. W. (2015). A synergistic effect of pretreatment on cell wall structural changes in barley straw (Hordeum vulgare L) for efficient bioethanol production. *Journal of Science of Food and Agriculture, 95*, 843–850.
242. Shi, Y., Huang, C., Rocha, K. C., El-Din, M. G., & Liu, Y. (2015). Treatment of oil sands process-affected water using moving bed biofilm reactors: With and without ozone pretreatment. *Bioresource Technology, 192*, 219–227.
243. Arvaniti, E., Bjerre, A. B. A., & Schmidt, J. E. (2012). Wet oxidation pretreatment of rape straw for ethanol production. *Biomass and Bioeenergy, 39*, 94–105.
244. Chaturvedi, V., & Verma, P. (2013). An overview of key pretreatment processes employed for bioconversion of lignocellulosic biomass into biofuels and value added products. *Biotech, 3*, 415–431.
245. Szijártó, N., Kádár, Z., Varga, E., Thomsen, A. B., Costaferreira, M., & Réczey, K. (2009). Pretreatment of reed by wet oxidation and subsequent utilization of the pretreated fibers for ethanol production. *Applied Biochemistry and Biotechnology, 155*, 83–93.
246. Varga, E., Schmidt, A. S., Réczey, K., & Thomsen, A. B. (2003). Pretreatment of corn Stover using wet oxidation to enhance enzymatic digestibility. *Applied Biochemistry & Biotechnology, 104*, 37–50.
247. Panagiotou, G., & Olsson, L. (2007). Effect of compounds released during pretreatment of wheat straw on microbial growth and enzymatic hydrolysis rates. *Biotechnology and Bioengineering, 96*, 250–258.
248. Klinke, H. B., Ahring, B. K., Schmidt, S. S., & Thomsen, A. B. (2002). Characterization of degradation products from alkaline wet oxidation of wheat straw. *Bioresource Technology, 82*, 15–26.
249. Banerjee, S., Sen, R., Mudliar, S., Pandey, R. A., Chakrabarti, T., & Satpute, D. (2011). Alkaline peroxide assisted wet air oxidation pretreatment approach to enhance enzymatic convertibility of rice husk. *Biotechnology Progress, 27*, 691–697.
250. Crowe, J. D., Li, M., Williams, D. L., Smith, A. D., Liu, T., & Hodge, D. B. (2019). Alkaline and alkaline-oxidative pretreatment and hydrolysis of herbaceous biomass for growth of Oleaginous microbes. *Microbial Lipid Production, 1995*, 173–182.
251. Tengborg, C., Galbe, M., & Zacchi, G. (2001). Reduced inhibition of enzymatic hydrolysis of steam-pretreated softwood. *Enzyme and Microbial Technology, 28*, 835–844.
252. Lorente, E., Farriol, X., & Salvadó, J. (2015). Steam explosion as a fractionation step in biofuel production from microalgae. *Fuel Processing Technology, 131*, 93–98.
253. Aguiar, A., Gavioli, D., & Ferraz, A. (2013). Extracellular activities and wood component losses during Pinus taeda biodegradation by the brown-rot fungus *Gloeophyllum trabeum*. *International Biodeterioration & Biodegradation, 82*, 187–191.
254. Cianchetta, S., Maggio, B. D., Burzi, P. L., & Galletti, S. (2014). Evaluation of selected white-rot fungal isolates for improving the sugar yield from wheat straw. *Applied Biochemistry and Biotechnology, 173*, 609–623.
255. Guerra, A., Mendonça, R., & Ferraz, A. (2003). Molecular weight distribution of wood components extracted from Pinus taeda biotreated by *Ceriporiopsis subvermispora*. *Enzyme and Microbial Technology, 33*, 12–18.

256. Koray Gulsoy, S., & Eroglu, H. (2011). Biokraft pulping of European black pine with *Ceriporiopsis subvermispora*. *International Biodeterioration & Biodegradation, 65*, 644–648.
257. Larran, A., Jozami, E., Vicario, L., Feldman, S. R., Podestá, F. E., & Permingeat, H. R. (2015). Evaluation of biological pretreatments to increase the efficiency of the saccharification process using *Spartina argentinensis* as a biomass resource. *Bioresource Technology, 194*, 320–325.
258. Monrroy, M., Ortega, I., Ramírez, M., Baeza, J., & Freer, J. (2011). Structural change in wood by brown rot fungi and effect on enzymatic hydrolysis. *Enzyme and Microbial Technology, 49*, 472–477.
259. Ahmad, M., Taylor, C. R., Pink, D., Burton, K., Eastwood, D., Bending, G. D., & Bugg, T. D. (2010). Development of novel assays for lignin degradation: Comparative analysis of bacterial and fungal lignin degraders. *Molecular BioSystems, 6*(5), 815–821.
260. Chen, Y., Ding, Y., Yang, L., Yu, J., Liu, G., Wang, X., Zhang, S., Yu, D., Song, L., & Zhang, H. (2013). Integrated omics study delineates the dynamics of lipid droplets in Rhodococcus opacus PD630. *Nucleic Acids Rresearch, 42*(2), 1052–1064.
261. Dai, Y. Z., Si, M. Y., Chen, Y. H., Zhang, N. L., Zhou, M., Liao, Q., Shi, D. Q., & Liu, Y. N. (2015). Combination of biological pretreatment with NaOH/Urea pretreatment at cold temperature to enhance enzymatic hydrolysis of rice straw. *Bioresource Technology, 198*, 725–731.
262. Guillén, F., Martínez, M. J., Gutiérrez, A., & Del Rio, J. (2005). Biodegradation of lignocellulosics: Microbial, chemical, and enzymatic aspects of the fungal attack of lignin. *International Microbiology, 8*, 195–204.
263. Sainsbury, P. D., Hardiman, E. M., Ahmad, M., Otani, H., Seghezzi, N., Eltis, L. D., & Bugg, T. D. H. (2013). Breaking down lignin to high-value chemicals: The conversion of lignocellulose to vanillin in a gene deletion mutant of Rhodococcus jostii RHA1. *ACS Chemical Biology, 8*(10), 151–156.
264. Salvachúa, D., Karp, E. M., Nimlos, C. T., Vardon, D. R., & Beckham, G. T. (2015). Towards lignin consolidated bioprocessing: Simultaneous lignin depolymerization and product generation by bacteria. *Green Chemistry, 17*, 4951–4967.
265. Leonowicz, A., Matuszewska, A., Luterek, J., Ziegenhagen, D., Wojtaś-Wasilewska, M., Cho, N. S., Hofrichter, M., & Rogalski, J. (1999). Biodegradation of lignin by white rot Fungi. *Fungal Genetics and Biology, 27*, 175–185.
266. Pérez, J., Muñozdorado, J., de la Rubia, T., & Martínez, J. (2002). Biodegradation and biological treatments of cellulose, hemicellulose and lignin: An overview. *International Microbiology the Official Journal of the Spanish Society for Microbiology, 5*, 53–63.
267. Tien, M., & Kirk, T. (1983). Lignin-degrading enzyme from the hymenomycete Phanerochaete chrysosporium Burds. *Science, 221*, 661.
268. Tišma, M., Planinić, M., Bucić-Kojić, A., Panjičko, M., Zupančič, G. D., & Zelić, B. (2018). Corn silage fungal-based solid-state pretreatment for enhanced biogas production in anaerobic co-digestion with cow manure. *Bioresource Technology, 253*, 220–226.
269. Aguiar, A., Souza-Cruz, P. B. D., & Ferraz, A. (2006). Oxalic acid, Fe^{3+}-reduction activity and oxidative enzymes detected in culture extracts recovered from Pinus taeda wood chips biotreated by *Ceriporiopsis subvermispora*. *Enzyme and Microbial Technology, 38*, 873–878.
270. Sanchez, C. (2009). Lignocellulosic residues: Biodegradation and bioconversion by fungi. *Biotechnology Advances, 27*, 185–194.
271. Saratale, G. D., Chien, L. J., & Chang, J. S. (2010). Enzymatic treatment of lignocellulosic wastes for anaerobic digestion and bioenergy production. In *Environmental anaerobic technology applications and new developments* (pp. 279–308). London: World Scientific Pub. Co. Inc.
272. Schilling, J. S., Tewalt, J. P., & Duncan, S. M. (2009). Synergy between pretreatment lignocellulose modifications and saccharification efficiency in two brown rot fungal systems. *Applied Microbiology and Biotechnology, 84*, 465.

273. de Gonzalo, G. D. I., Habib, M. H., & Fraaije, M. W. (2016). Bacterial enzymes involved in lignin degradation. *Journal of Biotechnology, 236*, 110–119.
274. Ma, K., & Ruan, Z. (2015). Production of a lignocellulolytic enzyme system for simultaneous bio-delignification and saccharification of corn stover employing co-culture of fungi. *Bioresource Technology, 175*, 586–593.
275. Godden, B., Ball, A. S., Helvenstein, P., Mccarthy, A. J., & Penninckx, M. J. (1992). Towards elucidation of the lignin degradation pathway in actinomycetes. *Journal of General Microbiology, 138*, 2441–2448.
276. Wältermann, M., Luftmann, H., Baumeister, D., Kalscheuer, R., & Steinbüchel, A. (2000). *Rhodococcus opacus* strain PD630 as a new source of high-value single-cell oil? Isolation and characterization of triacylglycerols and other storage lipids. *Microbiology, 146*(5), 1143–1149.
277. Ma, F., Yang, N., Xu, C., Yu, H., Wu, J., & Zhang, X. (2010). Combination of biological pretreatment with mild acid pretreatment for enzymatic hydrolysis and ethanol production from water hyacinth. *Bioresource Technology, 101*, 9600–9604.
278. Asadi, N., & Zilouei, H. (2017). Optimization of organosolv pretreatment of rice straw for enhanced biohydrogen production using Enterobacter aerogenes. *Bioresource Technology, 227*, 335–344.
279. Feng, R., Zaidi, A. A., Zhang, K., & Shi, Y. (2018). Optimization of microwave pretreatment for biogas enhancement through anaerobic digestion of microalgal biomass. *Periodica Polytechnica, Chemical Engineering, 63*, 65–72.
280. Koupaie, E. H., Dahadha, S., BazyarLakeh, A. A., Azizi, A., & Elbeshbishy, E. (2019). Enzymatic pretreatment of lignocellulosic biomass for enhanced biomethane production – A review. *Journal of Environmental Management, 233*, 774–784.
281. Hashemi, S. S., Karimi, K., & Mirmohamadsadeghi, S. (2019). Hydrothermal pretreatment of safflower straw to enhance biogas production. *Energy, 172*, 545–554.
282. Houtman, C. J., Maligaspe, E., Hunt, C. G., Fernández-Fueyo, E., Martínez, A. T., & Hammel, K. E. (2018). Fungal lignin peroxidase does not produce the veratryl alcohol cation radical as a diffusible ligninolytic oxidant. *The Journal of Biological Chemistry, 293*, 4702–4712.
283. Parveen, K., Diane, M., Barrett, M., Delwiche, J., & Pieter, S. (2009). Methods for pretreatment of lignocellulosic biomass for efficient hydrolysis and biofuel production. *Industrial and Engineering Chemistry Research, 48*, 3713–3729.
284. Taherdanak, M., Zilouei, H., & Karimi, K. (2016). The inflfluence of dilute sulfuric acid pretreatment on biogas production form wheat plant. *International Journal of Green Energy, 13*, 1129–1134.
285. Wright, J. D. (1988). Ethanol from biomass by enzymatic hydrolysis. *Chemical Engineering Progress, 84*, 8.
286. Wyman, C. E., Dale, B. E., Elander, R. T., Holtzapple, M., Ladisch, M. R., & Lee, Y. Y. (2005). Coordinated development of leading biomass pretreatment technologies. *Bioresource Technology, 96*, 1959–1966.
287. Wyman, C. E., Dale, B. E., Elander, R. T., Holtzapple, M., Ladisch, M. R., & Lee, Y. Y. (2005). Comparative sugar recovery data from laboratory scale application of leading pretreatment technologies to corn stover. *Bioresource Technology, 96*, 2026–2032.
288. Xu, Z. Y., & Huang, F. (2014). Pretreatment methods for bioethanol production. *Applied Biochemistry and Biotecnology, 174*, 43–62.

Chapter 9
Metabolic Engineering of Yeast for Enhanced Natural and Exotic Fatty Acid Production

Wei Jiang, Huadong Peng, Rodrigo Ledesma Amaro, and Victoria S. Haritos

Acronyms and Abbreviations

'tesA	Truncated *E. coli* thioesterase
ACC1	Acetyl-CoA carboxylase
*ACC1***	Acetyl-CoA carboxylase carrying two mutations ser659ala and ser1157ala
ACL1,2	ATP-citrate synthase subunit 1,2
ACS	Acetyl-coA synthetase
ADH	Alcohol dehydrogenase
ALD6	Native aldehyde dehydrogenase isoform 6
ARE1, 2	Sterol O-acyltransferase
AtCLO1	Caleosin, lipid droplet stabilization protein from *Arabidopsis thaliana*
DGAT	acyl-CoA: Diacylglycerol acyltransferase
FA	Fatty acid
FAA1, 4	Long-chain fatty acyl-CoA synthetase
FAA2	Medium-chain fatty acyl-CoA synthetase
FAME	Fatty acid methyl ester
FAS1	Fatty acid synthase subunit β

Wei Jiang and Huadong Peng contributed equally.

W. Jiang
Department of Chemical Engineering, Monash University, Clayton, VIC, Australia

Imperial College Centre for Synthetic Biology and Department of Bioengineering, Imperial College London, London, UK

H. Peng · R. Ledesma Amaro
Imperial College Centre for Synthetic Biology and Department of Bioengineering, Imperial College London, London, UK

V. S. Haritos (✉)
Department of Chemical Engineering, Monash University, Clayton, VIC, Australia
e-mail: victoria.haritos@monash.edu

Z.-H. Liu, A. Ragauskas (eds.), *Emerging Technologies for Biorefineries, Biofuels, and Value-Added Commodities*,
https://doi.org/10.1007/978-3-030-65584-6_9

FAS2	Fatty acid synthase subunit α
FAT1	Very-long-chain fatty acid transport protein
FFA(s)	Free fatty acid(s)
GPAT	Glycerol 3-phosphate acyltransferase
GUT2	Glycerol-3-phosphate dehydrogenase gene
LPAT	Lysophosphatidate acyltransferase
MAE	Malic acid transport protein
MFE1	Peroxisomal multifunctional enzyme
MmACL	ACL from *Mus musculus*
PAP	Phosphatidate phosphatase
PEX10	Peroxisome biogenesis factor 10
POX	Peroxisomal β-oxidation
PXA1	Subunit of heterodimeric peroxisomal ABC transport complex
RtFAS	Fatty acid synthetase from *R. toruloides*
SCD	Acyl-CoA desaturase
SeACS^{L641p}	Acetyl-CoA synthetase with L641P mutation, derived from *Salmonella enterica*
TAG	Triacylglycerol
TGL3–5	Triacylglycerol lipase 3–5
WT	Wild type

9.1 Introduction

The economic, environmental, and social sustainability problems caused by the dependency on petroleum have motivated a global shift to renewable, sustainable, and green alternative energy sources [1]. Replacement of crude oil-derived fuels and chemicals by the production of biofuels and bioproducts can be an effective strategy to reduce pollution and carbon dioxide emissions [2]. In particular, a useful feedstock for biofuels and bioproducts are lipids, consisting mainly of triacylglycerols, as they have high energy density and are readily converted to mono-alkyl esters for use as a diesel substitute. While lipids have excellent commercial utility, they are relatively expensive and in short supply, as there are important applications for lipids in food processing and oleochemical manufacture. In 2018, the natural fatty acid global market was valued at nearly $13.5 billion and expected to reach $17.5 billion in 2023 with a compound annual growth rate (CAGR) of 5.4% [3] (BCC Research LLC, 2019). Therefore, new sources of cost-effective lipids for the production of fuels and chemicals will be in increasingly high demand.

Lipids produced by microbes have huge potential to satisfy the growing demand for bio-based energy-dense hydrocarbons and related natural products [4], especially where their production is based on non-food carbon sources such as lignocellulosic sugars or by-product streams from biorefineries. Recent advances in microbial metabolic engineering and process technologies have brought us closer to cost-effective yields and diversity of oleaginous products that can support this growing market [5].

Compared with the ubiquitous bacterium *Escherichia coli*, yeasts like *Saccharomyces cerevisiae* and *Yarrowia lipolytica* are more effective hosts for lipid production because they synthesize C16-18 carbon chain fatty acids very efficiently requiring just two fatty acid synthases, whereas *E. coli* requires ten enzymes to reach the same endpoint [1, 6]. Furthermore, yeast can store large quantities of fatty acid internally as triacylglycerol. *S. cerevisiae* is a widely used industrial yeast due to its robustness and good tolerance of harsh industrial conditions [7] and its long history of use in large-scale fermentation to produce ethanol and beverages [8]. The oleaginous yeast *Y. lipolytica* has also had wide use in biotechnology and has several advantages over *S. cerevisiae* in that it naturally stores substantially more lipid within the cell and utilizes a broad range of low-cost feedstocks such as glycerol. The yeast holds generally recognized as safe (GRAS) status for the production of citric acid and has been explored for the production of sugar derivatives and nonnative, lipid products such as β-carotene and lycopene [9–15].

Another active area of research is in the production of exotic fatty acids and derivatives in yeasts as feedstocks in the production of fine chemicals, medicines, detergents and soaps, lubricants, cosmetics, and skin care products [16]. These lipids are not naturally present in yeast but are produced through the introduction of genes sourced from other organisms. Exotic fats include fatty alcohols and esters and unusual fatty acids such as those with modifications to fatty acid chain length, polyunsaturation, or added functional groups.

In recent years, the emerging synthetic biology field has brought new vitality into the development of microbial cell factories providing more powerful tools and methods to modify the microbial metabolic pathways [17]. To date, both natural lipids and lipid derivatives have been successfully produced through the benefits of synthetic biology and metabolic engineering in impressive yields. While most of the basic research in yeast lipid engineering have used purified sugars as carbon feedstocks, it has also been shown that lignocellulosic-derived sugars and other biorefinery by-products will also be effective substrates for these organisms.

In one concept of a biorefinery, cheap, plentiful biomass can be deconstructed to produce lignocellulosic sugars that are used as feedstocks for microbial oil production leaving lignin and hemicellulose sugars which can be further converted into products. For example, Wei et al. tested loblolly pine and sweetgum autohydrolysates after detoxification as feedstocks for lipid production via the oleaginous bacterium, *Rhodococcus opacus,* and achieved 0.25–0.31 g/L lipid titer [18]. Slininger et al. screened and identified three oleaginous yeasts that could utilize raw enzyme hydrolysates of ammonia fiber expansion (AFEX)-pretreated corn stover and acid-pretreated switchgrass as feedstocks for lipid production, and the lipid titer reached 25–30 g/L (39–45% of the theoretical yield) [19]. Here, we review recent progress in the application of synthetic biology and metabolic engineering focusing on yeasts *S. cerevisiae* and *Y. lipolytica,* as cell factories to produce lipids and higher-value fatty acid derivatives as part of a biorefinery.

9.2 Microbial Lipids from Lignocellulose-Derived Substrates

Microorganisms that can use lignocellulosic-derived substrates such as glucose, xylose, glycerol, and acetic acid for the production of microbial lipids are important for the utilization of biorefinery streams. Here, we review recent promising microbial lipid production research featuring the yeasts *S. cerevisiae* and *Y. lipolytica* cultured using lignocellulose-derived substrates.

Apart from glucose, the other two major sugars from lignocellulosic biomass are xylose and arabinose. For *S. cerevisiae* to be a more competitive chassis for the biotechnology industry, it is important to extend its growth substrates beyond glucose. The Pronk group has adopted metabolic engineering strategies, laboratory evolution, and co-culture approaches to enable *S. cerevisiae* to use xylose as a carbon source and improved ethanol fermentation performance using mixed sugars including glucose-xylose-arabinose [20, 21]. Ionic-liquid-pretreated switchgrass and sorghum were used as feedstocks for fatty alcohol production by *S. cerevisiae* engineered with 11 genetic modifications compared with the parent BY4741 strain, and the fatty alcohol titer reached 0.7 g/L in shaker flasks [22].

Also, a series of lignocellulose substrates were assessed for growth of engineered *Y. lipolytica* for microbial lipid production in bioreactors. For example, Li and Alper used xylose as carbon source bringing lipid production to 15 g/L [23]; Rakicha et al. improved lipid titer to 24.2 g/L using molasses/glycerol as feedstocks [24], Ledesma-Amaro et al. further improved lipid titer to 50.5 g/L using xylose and glycerol as substrates [25], and Niehus et al. achieved a very high lipid titer of 16.5 g/L and showed *Y. lipolytica* was tolerant to the toxicity of xylose-rich agave bagasse hydrolysate [26]. Furthermore, in a semicontinuous system, the high-density cell culture of *Y. lipolytica* was assessed using 3% acetic acid as a carbon source. The acetic acid was consumed completely, and yeast achieved a lipid titer of 115 g/L, yield of 0.16 g/g, and productivity of 0.8 $g{\cdot}L^{-1}{\cdot}h^{-1}$, respectively [27]. In further examples, Slininger et al. screened and identified three oleaginous yeasts that could use non-detoxified enzyme hydrolysates of ammonia fiber expansion (AFEX)-pretreated corn stover and acid-pretreated switchgrass as feedstocks for lipid production. The highest lipid yield reached 25–30 g/L, 39–45% of the theoretical yield [19].

As these successful attempts in production of microbial lipids show, biorefinery by-product streams such as non-glucose sugars, glycerol, and acetic acid are feasible cheap substrates for microbial lipids production. An ongoing challenge of the biorefinery concept is to develop robust microbial cell factories with greater productivity and cost-effectiveness. Recent research toward this aim is reviewed in the following section focusing on progress with *S. cerevisiae* and *Y. lipolytica*.

9.3 Metabolic Engineering Strategies and Recent Progress Toward Improved Yeast Lipid Production

In general, natural lipid production with yeast can be enhanced by (1) increasing fatty acid (FA) biosynthesis, such as by "pushing" carbon flux toward precursor acetyl-CoA and malonyl-CoA pools; (2) "blocking" competing pathways that consume lipids or free fatty acids, such as beta-oxidation; (3) balancing cofactor requirements and enzyme activity to deliver a steady NADPH supply to support fatty acid synthase activity; and (4) secreting free fatty acids into culture media or sequestering nascent lipids within lipid droplet to avoid toxicity [28–30]. This section highlights the recent advances in engineering efforts to increase lipid production in *S. cerevisiae* and *Y. lipolytica* yeast. The key genes/enzymes that have been engineered to improve lipid production and their cellular locations are shown in Fig. 9.1.

9.3.1 Lipid Metabolic Engineering of S. cerevisiae

Storage lipids make up no more than 10% dry cell weight (DCW) in wild-type *S. cerevisiae*, while they are accumulated to a much higher degree in some oleaginous yeast like *Y. lipolytica* [31]. Well-targeted single gene or pathway modifications in yeast normally lead to increased lipid content though improvement is limited. For more considerable improvement in lipid production, it is necessary to combine multiple approaches including synthetic biology, metabolic engineering, protein/enzyme engineering, adaptive laboratory evolution, machine learning, etc. Several successful attempts have been undertaken by researchers to enhance extracellular and intracellular lipid production in *S. cerevisiae*, and here, only recent examples with promising lipid yield are summarized.

Successful approaches to enhancing fatty acid (FA) biosynthesis include the overexpression of *ACC1* (or *ACC1***) [32, 33], *ACS1, FAS1,* and *FAS2* [6, 34–36], blocking FA competing pathways by deleting genes in beta-oxidation such as *POX1* and *POX2* [37, 38] and assessing the effects of lipid accumulation and storage genes such as *DGAT1* and *PDAT1* [39–41]. In terms of intracellular free fatty acid (FFA) accumulation, as opposed to esterified fatty acids, Valle-Rodriguez et al. (2014) deleted *DGA1, LRO1, ARE1*, and *ARE2* to block formation of neutral lipid and deleted *POX1* to avoid FA degradation whereby the engineered *S. cerevisiae* reached 1.5% intracellular FFA by DCW, fivefold higher than control [37].

S. cerevisiae BY4742 produced 17% DCW TAG was produced by overexpressing genes coding for FAS, ACC, and DGA [6]. Then, introducing ATP-citrate lyase (ACL) from the metabolism of an oleaginous microorganism to *S. cerevisiae* and disrupting isocitrate dehydrogenase genes *IDH1* and *IDH2*, they could increase the total fatty acids to 21% [42]. Peng et al. (2018) strengthened three steps of lipid production including FA biosynthesis (*Ald6-SEACSL641P, ACC1***), lipid accumulation (*DGAT1*), and lipid sequestration (Δ*TGL3*, *AtCLO1*) and achieved 8.0% DCW (2.6-fold than control) and 0.3 g/L lipid (4.6-fold than control) in a two-stage bioprocess [30]. Notably, the Nielsen group implemented a comprehensive strategy to increase TAG accumulation and reached 254 mg TAG/g DCW in *S. cerevisiae.* The strategy included increasing acetyl-CoA supply (*ACC1***), improving lipid accumulation (*PAH1* and *DGA1*), blocking lipid degradation (Δ*TGL3*, 4, 5, Δ*POX1*, Δ*PXA1*), sterol synthesis (Δ*ARE1*), glycerol-3-phosphate utilization (Δ*GUT2*) [43].

For secreted free fatty acid (FFA) production, Li et al. (2014) disrupted β-oxidation, deleted acyl-CoA synthetase, and overexpressed thioesterases and *ACC1* in *S. cerevisiae* to achieve 140 mg/L [44]. The Da Silva group achieved 2.2 g/L extracellular FFAs through disrupted neutral lipid recycle in *S. cerevisiae*

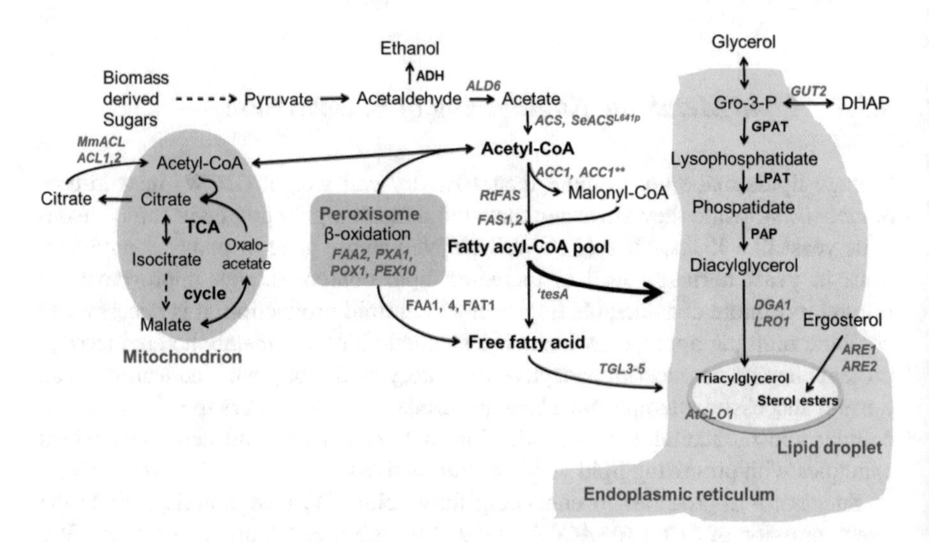

Fig. 9.1 Main metabolic pathways, control points, and organelles from sugars to lipid in yeast *S. cerevisiae*. The key genes/enzymes that have been engineered to impact lipid production have been highlighted in red text (*ADH* alcohol dehydrogenase, *ALD6* cytosolic aldehyde dehydrogenase 6, *ACS* acetyl-coA synthetase, *SeACS^{L641p}* acetyl-CoA synthetase with L641P mutation, derived from *Salmonella enterica*, *ACC1* acetyl-CoA carboxylase, *ACC1*** acetyl-CoA carboxylase carrying two mutations ser659ala and ser1157ala, *RtFAS* fatty acid synthetase from *R. toruloides*, *FAS1, 2* fatty acid synthetase, *ACL1, 2* ATP-citrate synthase subunit 1, 2, *MmACL* ACL from *Mus musculus, FAA2* medium-chain fatty acyl-CoA synthetase, *PXA1* subunit of heterodimeric peroxisomal ABC transport complex, *POX1* fatty-acyl coenzyme A oxidase, *MEF1* peroxisomal multifunctional enzyme, *PEX10* peroxisome biogenesis factor 10, *FAA1, 4* long-chain fatty acyl-CoA synthetase, *FAT1* very-long-chain fatty acyl-CoA synthetase, '*tesA* truncated *E. coli* thioesterase, *TGL3-5* triacylglycerol lipase genes, *GUT2* glycerol-3-phosphate dehydrogenase gene, *GPAT* glycerol 3-phosphate acyltransferase, *LPAT* lysophosphatidate acyltransferase, *PAP* phosphatidate phosphatase, *DGA1*, *LRO1* diacylglycerol acyltransferase, *ARE1, 2* sterol O-acyltransferase genes, *AtCLO1* caleosin, lipid droplet stabilization protein from *Arabidopsis thaliana*)

including disruption of β-oxidation (△*FAA2*, *PXA1*, *POX1*), acyl-CoA synthetase genes (*FAA1, FAA4, FAT1*), and coexpression of *DGA1* and *TGL3* [45]. Zhou et al. (2016) reached 10.4 g/L extracellular FFAs by enhancing acetyl-CoA supply, malonyl-CoA pathway, and fatty acid synthase expression and blocking fatty acid activation and degradation. The Nielsen group further engineering efforts to reprogram yeast metabolism from alcohol fermentation to lipogenesis whereby they constructed an impressive FFA-producing yeast delivering up to 33.4 g/L FFAs. The metabolic engineering included increasing cytosolic acetyl-CoA and NADPH supplies, redistributing carbon flux toward fatty acid biosynthesis, abolishing ethanol production pathway, mutating pyruvate kinase, and directing evolution [46].

9.3.2 Lipid Metabolic Engineering of **Y. lipolytica**

Due to the similarity of lipid metabolism between yeasts *S. cerevisiae* and *Y. lipolytica*, general metabolic strategies to enhance lipid production are transferable. Similar to the effectiveness of lipid pathway engineering in *S. cerevisiae*, there has been much progress in increasing lipid production in *Y. lipolytica*. Here, recent successful examples with promising lipid yields or addressing the key bottleneck metabolic issues have been addressed.

The Stephanopoulos group used lipid pathway engineering in *Y. lipolytica* to markedly improve production; their strategies have ranged the introduction of multiple gene combinations to the analysis of cellular physiological issues. Tai and Stephanopoulos (2013) firstly identified a more efficient promoter (intron-containing *TEF*) to assist heterologous gene expression by 17-fold and then improved *ACC1* and *DGA1* expression to increase lipid to 61.7% DCW, 0.270 g/g lipid yield, and 0.253 g L^{-1} h^{-1} lipid productivity [47]. Qiao et al. (2015) successfully identified the Δ9 stearoyl-CoA desaturase (SCD), which was overexpressed to avoid the repression of acetyl-CoA carboxylase via increasing fatty-acyl-CoA desaturation. Meanwhile, simultaneous overexpression of *SCD, ACC1,* and *DGA1* in *Y. lipolytica* achieved improved cell growth and increased tolerance to sugars plus a high-level lipid titer of 55 g/L and high carbon to lipid conversion yield (84.7% of theoretical maximal yield) [48]. Further, Qiao et al. (2017) successfully demonstrated that redox engineering via the modulation of the NADPH recovery pathway in *Y. lipolytica* increased lipid accumulation to 98.9 g/L measured as fatty acid methyl ester (FAME) [49]. Furthermore, Xu et al. (2017) employed a semicontinuous fermentation mode to bring the lipid titer of 115 g/L with an engineered *Y. lipolytica* (PO1g: *ACC1*, *DGA1*) and acetic acid as substrates [27].

Cellular oxidative stress defense pathways were investigated in *Y. lipolytica* to determine their impact on lipid production. Additional glutathione disulfide reductase to reduce oxidative stress, glucose-6-phosphate dehydrogenase for NADPH recycling and an engineered aldehyde dehydrogenase with broad substrate range were introduced into the yeast which proved to be efficient solutions to combat reactive oxygen and aldehyde stress in *Y. lipolytica*. The lipid titer reached 72.7 g/L and oil content 84.4% [50].

A comprehensive overexpression strategy in *Y. lipolytica* was adopted by the Alper group to improve lipid production. Blazeck et al. (2014) improved lipid production titer to 25 g/L using metabolic engineering strategies that included enhancing TAG biosynthesis (*DGA1, 2*), increasing acetyl-CoA (*ACL1, 2*), increasing NADPH cofactor supply (MAE), inhibiting the TCA cycle, increasing the citric acid level (ΔAMPD), and preventing beta-oxidation and peroxisome biogenesis (knockout of *mfe1, pex10*) [51]. Based on the engineered strains, Liu et al. (2015) identified a mutant Mga2p regulator in *Y. lipolytica*, which increased unsaturated fatty acid biosynthesis, possibly due to reduced feedback inhibition of ACC or reduced degradation of the stearoyl-CoA desaturase. Also, the mutant strain containing Mga2p maintained a high lipid titer (25 g/L) [52]. Furthermore, Liu et al. adopted a laboratory adaptive evolution approach to further screen for a super lipid producer strain with 87.1% DCW and 39.1 g/L lipid production [53].

Further examples of *Y. lipolytica* metabolic engineering with promising lipid yield include Ledesma-Amaro et al. (2016) who tested two synthetic approaches, firstly redirecting carbon flux to neutral lipids and, secondly, by mimicking a bacterial system to produce free FFAs. One optimal strain engineered to overexpress lipases that convert lipids to FFAs, and prevented the formation of CoA esters and β-oxidation of fats, produced up to 20.8 g/L lipids in a 5 L bioreactor [54]. Meanwhile, Ledesma-Amaro et al. (2016b) engineered PO1d strain with the following interventions: *Δpox1–6, ΔTGL4, GDP1, DGA2, ssXR, ssXDH,* and *ylXK*. Using xylose/glycerol as substrates, the lipid titer reached 50.5 g/L [25]. Friedlander et al. (2016) enhanced lipid accumulation and sequestration in *Y. lipolytica* by overexpression of both *DGA1* from *Rhodosporidium toruloides* and *DGA2* from *Claviceps purpurea*, plus deleted a key lipase (TGL3). The final engineered strain NS432 achieved 77% lipid content and 0.21 g lipid per g glucose yield in batch fermentation and 85 g/L lipid in fed-batch glucose fermentation [55]. Besides, ^{13}C-metabolic flux analysis was employed to understand whether the malic enzyme contributes to lipogenic NADPH production in *Y. lipolytica*, and the oxidative pentose phosphate pathway was proved to be the primary source of NADPH for lipid overproduction from glucose [56] (Table 9.1).

9.4 Exotic Fatty Acid/Alcohol Production in Engineered Yeast

9.4.1 Short- and Medium-Chain Fatty Acids

Short-chain fatty acids (SCFAs), where the carbon chain length is less than 10, are important industrial products as they can be used as gasoline and jet fuel precursors and intermediates in the synthesis of alkenes [59]. Producing SCFAs in common

Table 9.1 Metabolic engineering gene knockout or overexpression strategies for increased lipids or FFAs in *S. cerevisiae* and *Y. lipolytica*

Goals and genetic modification	Remarks/achievements	Host	References
↑ FA biosynthesis and accumulation: ↑ *ACC1*, ↑ *FAS1*, ↑ *FAS2*, ↑ *DGA1*	>17% DCW lipids, ↑ 4 × than WT	Sc	[6]
Disrupt β-oxidation: *ΔFAA2, ΔPXA1, ΔPOX1;* Δ acyl-CoA synthetase genes: *ΔFAA1, ΔFAA4, ΔFAT1;* increase triacylglycerol synthesis but increase rate of hydrolysis to FFA: ↑ *DGA1*, ↑ *TGL3*	2.2 g/L extracellular FFA, 4.2-fold higher than previous reported, fed-batch	Sc	[45]
↑ FA biosynthesis: ↑ *ALD6-SEACS*^{L641P}, ↑ *ACC1***; ↑ lipid accumulation, sequestration: ↑ *DGAT1*, *ΔTGL3*, ↑ At*CLO1*	0.3 g/L lipid, two-stage bioprocess in flask, 4.6-fold than control	Sc	[30]
↑ acetyl-CoA pathway: ↑ *RtME*, ↑ *MDH3*, ↑ *CTP1*, ↑ *MmACL;* ↑ fatty acid synthase (FAS): ↑*RtFAS;* ↑ malonyl-CoA: ↑*ACC1;* block FA activation and degradation: Δ*POX1*, Δ*FAA1, 4;* ↑ secrete FFA: ↑ *tesA*	10.4 g/L extracellular FFA, fed-batch	Sc	[57]
↑ acetyl-CoA supply: ↑ *ACC1***; ↑ lipid accumulation: ↑ *PAH1*, ↑ *DGA1;* block lipid degradation: Δ*TGL3–5*, Δ*POX1*, Δ*PXA1*, Δ*FAA2;* block sterol synthesis: Δ*ARE1*, Δ*GUT2*	254 mg TAG/g DCW, 27.4% of the maximal theoretical yield	Sc	[43]
↑ cytosolic acetyl-CoA, ↑ NADPH supply, ↑ FA biosynthesis, Δ ethanol pathway, mutate pyruvate kinase and direct evolution	33.4 g/L extracellular FFA, the highest titer reported to date in Sc	Sc	[46]
↑ acetyl-CoA supply and lipid formation: ↑ YpTEF-*ACC1*, ↑ *DGA1*	61.7% lipid content, 0.270 g/g lipid yield, 0.253 g/L/h lipid productivity	Yl	[47]
↑ FA biosynthesis: ↑ *DGA1, 2;* ↑ acetyl-CoA: ↑ *ACL1, 2;* ↑ NADPH cofactor supply: ↑ *MAE;* ↑ citric acid level: ΔAMPD; ↓ TCA cycle, ↓ β-oxidation, peroxisome: Δ*MFE1*, Δ*PEX10*	90% lipid content, 25 g/L lipid, fed-batch	Yl	[51]
↑ fatty acid synthesis and triacylglycerol synthesis: *SCD*, ↑ *ACC1*, ↑ *DGA1*	55 g/L lipid titer, 84.7% of theoretical maximal yield, fed-batch	Yl	[48]
Adaptive laboratory evolution and metabolic engineering	87.1% DCW, 39.1 g/L lipid, fed-batch	Yl	[53]
Mutant Mga2p, ΔPEX10 β-oxidation knockout, increase lipid formation ↑ *DGA1*	25 g/L lipid, fed-batch	Yl	[52]
Engineering fatty acyl-ACP/ acyl-CoA, thioesterase, rewiring acetyl-CoA pathway	9.67 g/L FFA, 66.4 g/L TAGs, fed-batch	Yl	[58]

(continued)

Table 9.1 (continued)

Goals and genetic modification	Remarks/achievements	Host	References
Increase flux toward triacylglycerol synthesis but increase rate of hydrolysis to FFA which are secreted, block FA oxidation *Δfaa1, Δmfe1, ↑DGA2 ↑TLG4 ↑klTGL3*	20.8 g/L lipids, fed-batch	Yl	[54]
PO1d: Δpox1–6, ΔTGL4, ↑GDP1, ↑DGA2, ↑ssXR, ↑ssXDH, ↑ylXK	50.5 g/L, 42% lipid content, xylose/glycerol as substrates	Yl	[25]
Engineering oxidative stress defense pathways	72.7 g/L, oil content 84.4%, fed-batch	Yl	[50]
Increase lipid by increasing flux towards triacylglycerol formation, block lipase-catalyzed lipid hydrolysis: ↑ *DGA1* from *Rhodosporidium toruloides* ↑ *DGA2* from *Claviceps purpurea, ΔTGL3*	85 g/L lipid, fed-batch	Yl	[55]
Cytosolic redox engineering, ↑ NADPH and acetyl-CoA supply	98.9 g/L FAME, fed-batch	Yl	[49]
↑ fatty acid synthesis and triacylglycerol synthesis: PO1g: ↑ *ACC1*, ↑ *DGA1*	115 g/L, 0.16 g/g, 0.8 g/(L h), semicontinuous, acetic acid as substrates, the highest titer reported to date in Yl	Yl	[27]

Symbols and prefixes: Sc, *S.cerevisiae*, Yl, *Y.lipolytica*, "↑": overexpression or heterologous expression, increase; "↓": downregulation or reduce; "△": deletion or knockout, "×": times by folds

biotechnological organisms is challenging as they do not natively produce short-chain fatty acids but prefer chain length range between C14 and C18 as these are primarily precursors for the formation of cellular membranes to support cell homeostasis [60]. Beyond the challenge of producing substantial SCFA within the cells, the potential cytotoxicity due to SCFAs' capacity to damage cell membranes needs to be addressed [61].

The first challenge is that the acyl carrier protein (ACP) and a phosphopantetheine transferase (PPT) are too large for the natural fatty acid synthase (FAS) of *S. cerevisiae* to passively diffuse into for elongation [62–64]. Also, the size of the short-chain thioesterases (TE) cleaving the elongating fatty acid is more than 9 kDa [65, 66]. To overcome these issues, Leber and Da Silva (2014) [67] expressed the FAS from *Homo sapiens* (hFAS); two heterologous TEs from *Cuphea palustris,* a plant that naturally produces SCFA, and *Rattus norvegicus*; and PTTs from *E. coli* and *Bacillus subtilis* in *S. cerevisiae,* respectively. Compared with native yeast, C8 levels were increased by 17-fold by overexpression of hFAS. Linking hFAS with heterologous TEs further improved the yield of C8 by four- and nine-fold. After introducing heterologous PPTs, total SCFA titers and C8 titers could reached 111 mg/L and 82 mg/L, respectively. In 2015, the freestanding thioesterase (HTEII) in *H. sapiens* was found to have a primary chain length selectivity for octanoic acid. HTEII was fused to hFAS and PTTs from *H. sapiens* was expressed in *S. cerevisiae.* Also, β-oxidation was fully disrupted. Finally, hexanoic and octanoic acid levels were increased by eight- and 79-fold over the parent strain with hFAS only [68].

Zhu et al. achieved the production of >1 g/L extracellular SCFA (C6-C12) in *S. cerevisiae*, a more than 250-fold improvement over the original strain. To achieve this, they engineered both the endogenous FAS and an orthogonal bacterial type I FAS and performed directed evolution on the membrane transporter *Tpo1*. They further developed the strain via adaptive laboratory evolution and metabolic flux control to markedly improve the SCFA production [69].

Meanwhile, Xu et al. (2016) demonstrated the specific structure of fungal type I FAS in *Y. lipolytica*. Then, they swapped malonyl/palmitoyl transacylase domain in FAS1 and fused the truncated FAS1 with smaller TE to improve medium-chain fatty acid production, which resulted in remarkably increasing C12 and C14 portions of fatty acids to 29.2% and 7.5%, respectively [70] .

9.4.2 Fatty Acid Esters and Alcohols

Fatty acid ethyl esters (FAEEs) are an attractive diesel oil alternative with high energy density and low toxicity to the production host (Zhang et al., 2012; Zhou et al., 2014). Acyl-CoAs formed within the cell can be condensed by wax ester synthase/acyl-CoA:diacylglycerol acyltransferase with ethanol to synthesize FAEEs. In order to improve FAEE yield, the pathway for the intermediate acyl-CoAs is enhanced by metabolic engineering. Shi et al. screened five wax ester synthases for FAEE biosynthesis; a candidate obtained from *Marinobacter hydrocarbonoclasticus* gave 6.3 mg/L FAEE titer (Shi et al., 2012). With integration of this wax synthetase into the *S. cerevisiae* genome, FAEE yield improved to 34 mg/L (Shi et al., 2014b).

In addition, reducing competition for acyl-CoAs from non-lipid pathways was shown to improve FAEE productivity. For example, Valle Rodriguez et al. blocked β-oxidation, sterol esters, and TAG biosynthesis in *S. cerevisiae* to yield 17.2 mg/L in the mutant strain, threefold higher than the wild-type strain (Valle-Rodríguez et al., 2014). As NADPH and acetyl-CoA are required to synthesize acyl-CoA, De Jong et al. 2014 upregulated ethanol degradation and constructed a phosphoketolase pathway to increase flux of acetyl-CoA and NADPH, which can improve the pool of acyl-CoA. Alcohol dehydrogenase *Adh2*, the *Salmonella enterica* acetyl-CoA synthetase variant *SeACS* (L641P), and acetaldehyde dehydrogenase *Ald6* were overexpressed to accelerate ethanol degradation, which improved threefold FAEE yield (Starai et al., 2005). The overexpression of ACC1 also contributed to the accumulation of acetyl-CoA, whereby FAEE production reached 8.2 mg/L (Shi et al., 2012). *Y. lipolytica* has also been developed as a host for FAEE production by similar metabolic engineering strategies. An efficient FAEE biosynthetic pathway was constructed by expression of heterologous wax ester synthase gene with codon optimization for *Y. lipolytica* and under strong promoters. In addition, carbon flux was redirected toward the FAEE biosynthesis pathway by modifying the acetyl-CoA node, and β-oxidation was deleted by *PEX10* knockout. Finally, the engineered strains coupled with the exogenous optimized ethanol concentration can produce an extracellular FAEE yield of 1.18 g/L via shake-flask fermentation [71] .

Fatty alcohols have applications in detergents, medicine, cosmetics, and biofuels (Beller et al., 2015). In yeast, fatty alcohol can be obtained by the reduction of a fatty aldehyde intermediate or directly synthesized by fatty acyl-CoAs that undergo reduction via the action of a bifunctional fatty acyl-CoA reductase (Willis et al., 2011). The expression of fatty acyl-CoA reductase from mouse in *S. cerevisiae* resulted in 47.4 mg/L of fatty alcohols (Sangwallek et al., 2013). To further improve fatty alcohol yield, a mouse fatty acid reductase MmFar1p (NADPH-dependent) with high activity was expressed in *S. cerevisiae*. Also, diacylglycerol acyltransferase1 DGA1, fatty aldehyde dehydrogenase HFD1, and medium-chain alcohol dehydrogenase ADH6 were deleted to redirect carbon flux toward fatty alcohols instead of toward TAG, FFA, and ethanol. Further, a mutant acetyl-CoA carboxylase was overexpressed to increase acetyl-CoA flux. The Δ9-desaturase *OLE1* was overexpressed to increase membrane fluidity and access of MmFar1p to the substrate. The final strain containing 11 genetic modifications than parent BY4741 strain produced 1.2 g/L fatty alcohols in shake flasks from glucose (d'Espaux et al., 2017).

9.4.3 Ricinoleic Fatty Acids

Ricinoleic acid (RA) accounts for around 90% of the total fatty acid in castor seeds [72]. Because of its specific structure, RA can be a substrate for double bond and hydroxyl-group reactions and, therefore, an important natural raw material for the chemical industry [73]. RA and its derivatives have broad commercial applications, including food, textile, paper, plastics, perfumes, cosmetics, paints, inks and lubricants, and biofuels [74, 75]. Although RA is the major component of castor seeds, the castor plant has many serious challenges in its production. In addition, the process of extracting RA from the castor seeds is complicated [76].

To date, RA biosynthesis has been most successful in *Y. lipolytica* although a major challenge is that the hydroxylated ricinoleic acid is formed at the *sn*-2 position of phosphatidylcholine (PC) in membranes when the Δ12 hydroxylase (FAH12) from castor is expressed. As *Y. lipolytica* accumulates high amounts of oleic acid, the substrate for FAH12, it provides a direct precursor for RA synthesis. Bressy et al. (2014) [77] expressed the castor FAH12 in *Y. lipolytica* which resulted in 7% RA of the total fatty acid; however, when two copies of the *Claviceps purpurea* hydroxylase CpFAH12 were expressed in a modified strain, RA content increased to 35% of the total lipids. Next, they deleted six *POX* genes to prevent β-oxidation of fatty acids, the native Δ12-desaturase which converts oleic acid to linoleic acid and *DGA1* and *DGA2* which form TAG via the addition of acyl-CoA to the glycerol backbone. In the final version, the native *Y. lipolytica* PDAT acyltransferase (Lro1p) was overexpressed, and RA yield reached 43% of total fatty acid and over 60 mg/g of dry cell weight in small scale-cultures and up to 12 g/L and 60% of total lipids when supplemented with 24 g/L of oleic acid at 10 L bioreactor scale (Fig. 9.2).

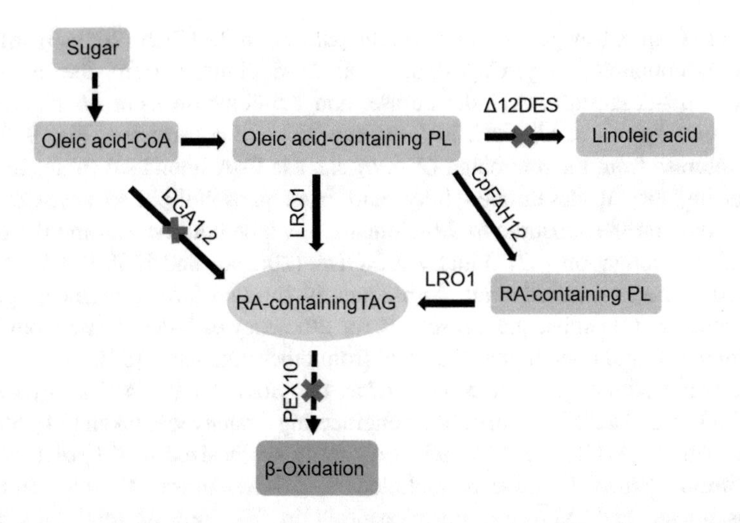

Fig. 9.2 Strategy to maximize RA production in *Y. lipolytica* through metabolic engineering

9.4.4 Long-Chain Polyunsaturated Fatty Acids

The most common carbon chain length of yeast fatty acids is 16–18, whereas a group of valuable long-chain polyunsaturated fatty acids (LC-PUFAs) has carbon chain lengths of 20–24 and includes multiple double bonds in a methylene interrupted pattern. Two main categories of desaturation of the fatty acid carbon chain are known as omega-6 (n-6) and omega-3 (n-3), and the numbering is determined by the position of the first double bond from the methyl end group of the fatty chain [78]. Omega-6 LC-PUFA can be precursors to the eicosanoids, a group of powerful bioactive molecules that include prostaglandins and thromboxane. The omega-3 PUFAs are important human dietary fatty acids that can regulate the immune system, blood clots, neurotransmitters, and cholesterol metabolism and adjust membrane phospholipids of both the brain and the retina [79]. Although LC-PUFA can have a positive effect on health, these LC-PUFAs cannot be synthesized in the human body and so are required to be taken via the diet [80]. Currently, dietary omega-3 LC-PUFAs, especially eicosapentaenoic acid (EPA) and docosahexaenoic acid (DHA), are obtained mainly from fish oil, and due to fish stock depletion and an increasing demand, obtaining alternative sources is becoming necessary [76]. Both *S. cerevisiae* and *Y. lipolytica* have been engineered to produce LC-PUFA.

LC-PUFAs are biosynthesized in cells by a series of alternating fatty acid desaturations and carbon chain elongation. In yeast, the elongation step occurs in the acyl-CoA pool but the special desaturases introduced into the cells undertake desaturation of the phospholipid-linked fatty acids, which causes an acyl exchange bottleneck and reduces yield [81]. Also, in the final step, a double bond is introduced between carbon 5 and 6 in dihomo-γ-linolenic acid (DHGLA, 20:3ω6) and eicosatetraenoic acid (ETA 20:4ω3) (Fig. 9.3a) by a Δ5-desaturase belonging to "front-end" desaturase family. Key genes relating to the pathway of DHGLA (20:3ω6) and ETA (20:4ω3), containing acyl-CoA-dependent Δ6-desaturase from the microalga *O. tauri;* Δ9-desaturase, Δ12-desaturase, and Δ6-elongase from *M. alpina*; and ω3-desaturase from *S. kluyveri*, in *S. cerevisiae* were constructed (Fig. 9.3). The Δ6-desaturase from the microalga *O. tauri* can use CoA-bound substrates to avoid transferring the Δ6-desaturated fatty acid from phospholipid to acyl-CoA and directly pass on the substrate to Δ6-elongation, which could overcome the bottleneck of Δ6-elongation [82]. Finally, DHGLA (20:3ω6) and ETA (20:4ω3) were obtained in the engineered strain. Subsequently, through further engineering, EPA was synthesized. In subsequent research, the efficiency of 5-desaturase from *P. tetraurelia* was found to be higher than that from other organisms [83].

Compared with engineered *S. cerevisiae*, the yield of EPA in *Y. lipolytica* was much higher, and a different metabolic engineering strategy was taken [84]. Starting with linoleic acid (C18:2 n-6) which is naturally synthesized in *Y. lipolytica* wild-type strain, genes introduced included a Δ6-desaturase, C18/20 elongase, Δ5-desaturase, and Δ17-desaturase resulting in 3% EPA of total fatty acids. Subsequently, overexpression of a C16/18 elongase from *M. alpina,* introducing a Δ12 desaturase from *Fusarium moniliforme*, increased gene copy numbers, and promoter optimization resulted in 40% EPA of total fatty acids. However, a large amount of γ-linolenic acid (C18:3ω6) was also accumulated via this strategy, which showed that the conversion of GLA to DHGLA (C20:3ω6) was rate limiting. Therefore, Δ9 pathway was constructed by introducing Δ9-elongase, Δ8-desaturase, Δ5-desaturase, and a Δ17 desaturase sourced from a range of organisms to avoid the buildup of GLA (Fig. 9.3). In the same way, they integrated multiple copies of the genes after codon optimization with strong promoters. In order to reduce the consumption of LC-PUFA by β-oxidation, PEX10 was deleted. The final strains contained 30 copies of nine different genes, and the yield of EPA was 56.6% of the total, which can be used as a commercial product produced by metabolically engineered yeast to take the place of that derived from fish [85].

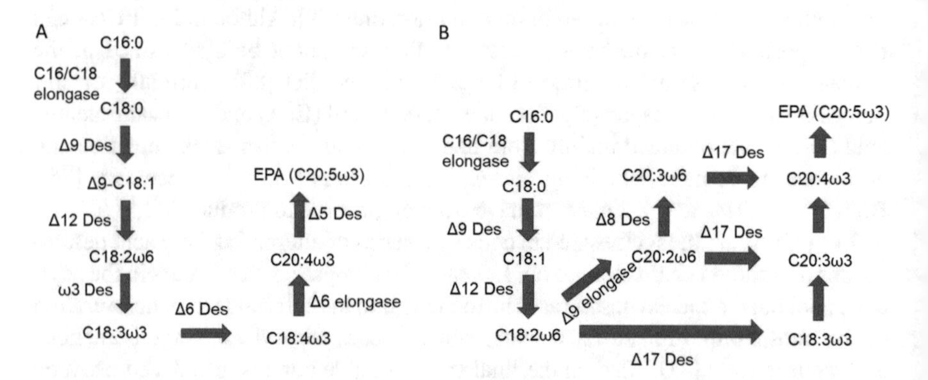

Fig. 9.3 Engineered EPA biosynthetic pathway in (**a**) *S. cerevisiae* and (**b**) *Y. lipolytica*

9.4.5 Cyclopropane Fatty Acids

Cyclopropane fatty acids (CFAs) are naturally occurring saturated fatty acids that possess a strained three-membered ring within the fatty acid chain. They have been found in bacteria [86, 87], some fungi [88], plants [89, 90], and parasites [91]. This fatty acid has potential high value as an equivalent compound to isostearic acid which has industrial application in the lubrication and oleochemical industries [92]. Cyclopropane fatty acids have unique characteristics such as ring opening by hydrogenation to produce methyl branched-chain fatty acid, which combines the chemical and physical properties of unsaturated fatty acid with oxidative stability of saturated fatty acids [93]. There has been recent interest and research into building microbial cell factories for the production of CFAs, including both *S. cerevisiae* and *Y. lipolytica*.

Peng et al. (2018) expressed the *E. coli CFA* gene in *S. cerevisiae* that had been engineered for higher fatty acid (FA) biosynthesis, lipid production, and sequestration. *TGL3*, encoding triglyceride lipase 3, the main enzyme responsible for hydrolyzing CFA from TAG, was knocked out to block CFA loss from the lipid droplet. The highest CFA yield was 12 mg/g dry cell weight (DCW) which was four-fold above the strain expressing *E. coli CFA* gene only and up to 68.3 mg/L in a two-stage bioprocess [30, 94]. *Y. lipolytica* has also been engineered for the production of CFAs. *E. coli* was the preferred candidate from among a range of *CFA* genes screened from bacteria and selected plants for expression as it provided good yield and both C17 and C19 cyclopropane products [95]. Blocking β-oxidation by knocking out *PEX10* and *MFE1,* overexpression of *DGA1*, and increasing the genomic copy number of the *E. coli CFA* gene were successful strategies to produce cyclopropane fatty acids in *Y. lipolytica* [96]. A further strain was constructed by mutating regulatory protein encoded by MGA2 paired with *DGA1* overexpression and *CFA* expression, which produced 200 mg/L of C19:0 CFA in small-scale fermentation. Moreover, more than 3 g/L of C19:0 CFA was achieved in bioreactor fermentation, which accounted for up to 32.7% of total lipids [96].

9.5 Conclusion and Outlook

With increasing interest globally toward sustainable industrial production, microbial lipids are attracting significant attention due to their energy density and versatility plus the prospect of obtaining microbial lipids with a broad range of functionalities. Microbes require carbon as a key feedstock for growth, and lignocellulose is a rich and sustainable resource that has enormous potential as a substrate for microbial lipid production [97]. Within biorefineries, microbial lipid factories can play a central role in converting renewable substrates into versatile lipid products. Regarding microbial lipid factories, yeasts such as *S. cerevisiae* and *Y. lipolytica* have been the most widely investigated. With advances in synthetic biology and metabolic engineering, more tools and approaches have become available to support and enhance the introduction of gene modifications [76]. Although high performing strains with good tolerance to stressful environments and efficient lipid conversion rates can be obtained via synthetic biology and metabolic engineering strategies, the quality and availability of the sugar feedstock remain one of the important limiting factors for microbial factories. Despite the development of successful lignocellulosic sugar production from raw materials in biorefineries, the volume is not currently sufficient to meet the growing demand for bioproduction. Therefore, further research and development for low-cost and efficient production of non-food sugar sources needs to be undertaken to ensure these supplies [98–100].

There are some limitations to what can be achieved in yeast biofactories through metabolic engineering and synthetic biology. Cellular metabolic burden is a long-standing problem in biotechnology which was first noticed by metabolic engineers in the 1970s and 1980s when they attempted to overexpress proteins for desired products, and they found the cell growth reduced and mutation rates increased after the overexpression of protein [101–103]. The metabolic burden can be caused by an imbalance of energy molecules (e.g., NAD(P)H and ATP) or redirection of carbon building blocks away from essential cellular processes, for example. A cell's carbon and energy resource distribution have been optimized to reach equilibrium states by a natural evolution [101] and modification and manipulation of these via metabolic engineering and synthetic biology alter the natural balance. There are a number of effective strategies available and in development to address these specific issues of carbon and energy imbalance in engineered strains [104].

In short, much progress in metabolic engineering of yeast for enhanced lipid production has been made, and while natural fatty acid production levels are reportedly close to commercial realization, there is more research needed for exotic lipid production to improve productivity and purity. Furthermore, the engineering strategies and modifications that have been shown to be highly effective in laboratory strains now need to undergo development and translation to their industrial environments. The fast development of stable strains useful for the industrial environment is still challenging [7]. Therefore, further research is needed to overcome the many challenges to bring microbial lipid factories to commercial reality. With more technologies and strategies becoming integrated in the future, such as systems biology, protein engineering, and "omics" analysis, these can enrich the progress to date and help obtain the goals for producing fatty acid-derived biofuels and bioproducts in an affordable and sustainable manner.

Acknowledgments WJ is financially supported by Monash University for Monash Graduate Scholarship (MGS) and Monash International Tuition Scholarships (MITS), and Graduate Research International Travel Award (GRITA). WJ thanks RLA for providing an opportunity to work within his group at Imperial College Centre for Synthetic Biology. RLA and HP received funding from BBSRC (BB/R01602X/1).

Authors' Contributions VH determined the book chapter design, WJ drafted Sects. 9.1, 9.4, and 9.5 and HP drafted Sects. 9.2 and 9.3. VH, WJ, HP, and RLA revised the manuscript, proofread, and approved the final manuscript.

References

1. Zhang, F., Rodriguez, S., & Keasling, J. D. (2011). Metabolic engineering of microbial pathways for advanced biofuels production. *Current Opinion in Biotechnology, 22*(6), 775–783.
2. Fortman, J., Chhabra, S., Mukhopadhyay, A., Chou, H., Lee, T. S., Steen, E., & Keasling, J. D. (2008). Biofuel alternatives to ethanol: Pumping the microbial well. *Trends in Biotechnology, 26*(7), 375–381.
3. BBC Publishing (2019). *Oleochemical fatty acids: Global markets to 2023, report highlights.*
4. Karmee, S. K., Linardi, D., Lee, J., & Lin, C. S. K. (2015). Conversion of lipid from food waste to biodiesel. *Waste Management, 41*, 169–173.
5. Zhou, Y. J., Kerkhoven, E. J., & Nielsen, J. (2018). Barriers and opportunities in bio-based production of hydrocarbons. *Nature Energy, 3*, 925.
6. Runguphan, W., & Keasling, J. D. (2014). Metabolic engineering of Saccharomyces cerevisiae for production of fatty acid-derived biofuels and chemicals. *Metabolic Engineering, 21*, 103–113.
7. Hong, K.-K., & Nielsen, J. (2012). Metabolic engineering of Saccharomyces cerevisiae: A key cell factory platform for future biorefineries. *Cellular and Molecular Life Sciences, 69*(16), 2671–2690.
8. Mussatto, S. I., Dragone, G., Guimarães, P. M., Silva, J. P. A., Carneiro, L. M., Roberto, I. C., Vicente, A., Domingues, L., & Teixeira, J. A. (2010). Technological trends, global market, and challenges of bio-ethanol production. *Biotechnology Advances, 28*(6), 817–830.
9. Abdel-Mawgoud, A. M., Markham, K. A., Palmer, C. M., Liu, N., Stephanopoulos, G., & Alper, H. S. (2018). Metabolic engineering in the host Yarrowia lipolytica. *Metabolic Engineering, 50*, 192–208.
10. Kamzolova, S. V., & Morgunov, I. G. (2017). Metabolic peculiarities of the citric acid overproduction from glucose in yeasts Yarrowia lipolytica. *Bioresource Technology, 243*, 433–440.
11. Rymowicz, W., Rywińska, A., & Marcinkiewicz, M. (2009). High-yield production of erythritol from raw glycerol in fed-batch cultures of Yarrowia lipolytica. *Biotechnology Letters, 31*(3), 377–380.
12. Janek, T., Dobrowolski, A., Biegalska, A., & Mirończuk, A. M. (2017). Characterization of erythrose reductase from Yarrowia lipolytica and its influence on erythritol synthesis. *Microbial Cell Factories, 16*(1), 118.
13. Carly, F., Vandermies, M., Telek, S., Steels, S., Thomas, S., Nicaud, J.-M., & Fickers, P. (2017). Enhancing erythritol productivity in Yarrowia lipolytica using metabolic engineering. *Metabolic Engineering, 42*, 19–24.
14. Kildegaard, K. R., Adiego-Pérez, B., Belda, D. D., Khangura, J. K., Holkenbrink, C., & Borodina, I. (2017). Engineering of Yarrowia lipolytica for production of astaxanthin. *Synthetic and Systems Biotechnology, 2*(4), 287–294.
15. Ma, T., Shi, B., Ye, Z., Li, X., Liu, M., Chen, Y., Xia, J., Nielsen, J., Deng, Z., & Liu, T. (2019). Lipid engineering combined with systematic metabolic engineering of Saccharomyces cerevisiae for high-yield production of lycopene. *Metabolic Engineering, 52*, 134–142.
16. Liu, R., Zhu, F., Lu, L., Fu, A., Lu, J., Deng, Z., & Liu, T. (2014). Metabolic engineering of fatty acyl-ACP reductase-dependent pathway to improve fatty alcohol production in Escherichia coli. *Metabolic Engineering, 22*, 10–21.
17. Kondo, A., Ishii, J., Hara, K. Y., Hasunuma, T., & Matsuda, F. (2013). Development of microbial cell factories for bio-refinery through synthetic bioengineering. *Journal of Biotechnology, 163*(2), 204–216.
18. Wei, Z., Zeng, G. M., Huang, F., Kosa, M., Sun, Q. N., Meng, X. Z., Huang, D. L., & Ragauskas, A. (2015). Microbial lipid production by oleaginous Rhodococci cultured in lignocellulosic autohydrolysates. *Applied Microbiology and Biotechnology, 99*(17), 7369–7377.
19. Slininger, P. J., Dien, B. S., Kurtzman, C. P., Moser, B. R., Bakota, E. L., Thompson, S. R., O'Bryan, P. J., Cotta, M. A., Balan, V., Jin, M., Sousa, L. d. C., & Dale, B. E. (2016).

Comparative lipid production by oleaginous yeasts in hydrolyzates of lignocellulosic biomass and process strategy for high titers. *Biotechnology and Bioengineering, 113*(8), 1676–1690.
20. Verhoeven, M. D., de Valk, S. C., Daran, J. G., van Maris, A. J. A., & Pronk, J. T. (2018). Fermentation of glucose-xylose-arabinose mixtures by a synthetic consortium of single-sugar-fermenting Saccharomyces cerevisiae strains. *FEMS Yeast Research, 18*(8), 1–12.
21. Papapetridis, I., Verhoeven, M. D., Wiersma, S. J., Goudriaan, M., van Maris, A. J. A., & Pronk, J. T. (2018). Laboratory evolution for forced glucose-xylose co-consumption enables identification of mutations that improve mixed-sugar fermentation by xylose-fermenting Saccharomyces cerevisiae. *FEMS Yeast Research, 18*(6), foy056.
22. d'Espaux, L., Ghosh, A., Runguphan, W., Wehrs, M., Xu, F., Konzock, O., Dev, I., Nhan, M., Gin, J., Apel, A. R., Petzold, C. J., Singh, S., Simmons, B. A., Mukhopadhyay, A., Martin, H. G., & Keasling, J. D. (2017). Engineering high-level production of fatty alcohols by Saccharomyces cerevisiae from lignocellulosic feedstocks. *Metabolic Engineering, 42*, 115–125.
23. Li, H., & Alper, H. S. (2016). Enabling xylose utilization in Yarrowia lipolytica for lipid production. *Biotechnology Journal, 11*(9), 1230–1240.
24. Rakicka, M., Lazar, Z., Dulermo, T., Fickers, P., & Nicaud, J. M. (2015). Lipid production by the oleaginous yeast Yarrowia lipolytica using industrial by-products under different culture conditions. *Biotechnology for Biofuels, 8*, 104.
25. Ledesma-Amaro, R., Lazar, Z., Rakicka, M., Guo, Z., Fouchard, F., Coq, A. C., & Nicaud, J. M. (2016). Metabolic engineering of Yarrowia lipolytica to produce chemicals and fuels from xylose. *Metabolic Engineering, 38*, 115–124.
26. Niehus, X., Crutz-Le Coq, A. M., Sandoval, G., Nicaud, J. M., & Ledesma-Amaro, R. (2018). Engineering Yarrowia lipolytica to enhance lipid production from lignocellulosic materials. *Biotechnology for Biofuels, 11*, 1.
27. Xu, J. Y., Liu, N., Qiao, K. J., Vogg, S., & Stephanopoulos, G. (2017). Application of metabolic controls for the maximization of lipid production in semicontinuous fermentation. *Proccedings of the National Academy of Sciences of the United States of America, 114*(27), E5308–E5316.
28. Pfleger, B. F., Gossing, M., & Nielsen, J. (2015). Metabolic engineering strategies for microbial synthesis of oleochemicals. *Metabolic Engineering, 29*, 1–11.
29. Yan, Q., & Pfleger, B. F. (2019). Revisiting metabolic engineering strategies for microbial synthesis of oleochemicals. *Metabolic Engineering, 58*, 35–46.
30. Peng, H., He, L., & Haritos, V. S. (2018). Metabolic engineering of lipid pathways in Saccharomyces cerevisiae and staged bioprocess for enhanced lipid production and cellular physiology. *Journal of Industrial Microbiology & Biotechnology, 45*(8), 707–717.
31. Li, Y., Zhao, Z. K., & Bai, F. (2007). High-density cultivation of oleaginous yeast Rhodosporidium toruloides Y4 in fed-batch culture. *Enzyme and Microbial Technology, 41*(3), 312–317.
32. Shi, S., Chen, Y., Siewers, V., & Nielsen, J. (2014). Improving production of Malonyl coenzyme A-Derived metabolites by abolishing Snf1-dependent regulation of Acc1. *MBio, 5*(3), e01130-14.
33. Choi, J. W., & Da Silva, N. A. (2014). Improving polyketide and fatty acid synthesis by engineering of the yeast acetyl-CoA carboxylase. *Journal of Biotechnology, 187*, 56–59.
34. Chen, Y., Daviet, L., Schalk, M., Siewers, V., & Nielsen, J. (2013). Establishing a platform cell factory through engineering of yeast acetyl-CoA metabolism. *Metabolic Engineering, 15*, 48–54.
35. Chen, F., Zhou, J., Shi, Z., Liu, L., Du, G., & Chen, J. (2010). Effect of acetyl-CoA synthase gene overexpression on physiological function of Saccharomyces cerevisiae. *Wei Sheng Wu Xue Bao = Acta Microbiologica Sinica, 50*(9), 1172–1179.
36. Shiba, Y., Paradise, E. M., Kirby, J., Ro, D.-K., & Keasling, J. D. (2007). Engineering of the pyruvate dehydrogenase bypass in Saccharomyces cerevisiae for high-level production of isoprenoids. *Metabolic Engineering, 9*(2), 160–168.

37. Valle-Rodriguez, J. O., Shi, S. B., Siewers, V., & Nielsen, J. (2014). Metabolic engineering of Saccharomyces cerevisiae for production of fatty acid ethyl esters, an advanced biofuel, by eliminating non-essential fatty acid utilization pathways. *Applied Energy, 115*, 226–232.
38. Chen, L., Zhang, J., & Chen, W. N. (2014). Engineering the Saccharomyces cerevisiae β-oxidation pathway to increase medium chain fatty acid production as potential biofuel. *PLoS One, 9*(1), e84853: 1-10.
39. Greer, M. S., Truksa, M., Deng, W., Lung, S. C., Chen, G. Q., & Weselake, R. J. (2015). Engineering increased triacylglycerol accumulation in Saccharomyces cerevisiae using a modified type 1 plant diacylglycerol acyltransferase. *Applied Microbiology and Biotechnology, 99*(5), 2243–2253.
40. Dahlqvist, A., Stahl, U., Lenman, M., Banas, A., Lee, M., Sandager, L., Ronne, H., & Stymne, H. (2000). Phospholipid: Diacylglycerol acyltransferase: An enzyme that catalyzes the acyl-CoA-independent formation of triacylglycerol in yeast and plants. *Proccedings of the National Academy of Sciences of the United States of America, 97*(12), 6487–6492.
41. Peng, H., Moghaddam, L., Brinin, A., Williams, B., Mundree, S., & Haritos, V. S. (2018). Functional assessment of plant and microalgal lipid pathway genes in yeast to enhance microbial industrial oil production. *Biotechnology and Applied Biochemistry, 65*(2), 138–144.
42. Tang, X., Feng, H., & Chen, W. N. (2013). Metabolic engineering for enhanced fatty acids synthesis in Saccharomyces cerevisiae. *Metabolic Engineering, 16*, 95–102.
43. Ferreira, R., Teixeira, P. G., Gossing, M., David, F., Siewers, V., & Nielsen, J. (2018). Metabolic engineering of Saccharomyces cerevisiae for overproduction of triacylglycerols. *Metabolic Engineering Communications, 6*, 22–27.
44. Li, X., Guo, D., Cheng, Y., Zhu, F., Deng, Z., & Liu, T. (2014). Overproduction of fatty acids in engineered Saccharomyces cerevisiae. *Biotechnology and Bioengineering, 111*(9), 1841–1852.
45. Leber, C., Polson, B., Fernandez-Moya, R., & Da Silva, N. A. (2015). Overproduction and secretion of free fatty acids through disrupted neutral lipid recycle in Saccharomyces cerevisiae. *Metabolic Engineering, 28*, 54–62.
46. Yu, T., Zhou, Y. J., Huang, M., Liu, Q., Pereira, R., David, F., & Nielsen, J. (2018). Reprogramming yeast metabolism from alcoholic fermentation to lipogenesis. *Cell, 174*, 1549.
47. Tai, M., & Stephanopoulos, G. (2013). Engineering the push and pull of lipid biosynthesis in oleaginous yeast Yarrowia lipolytica for biofuel production. *Metabolic Engineering, 15*, 1–9.
48. Qiao, K., Imam Abidi, S. H., Liu, H., Zhang, H., Chakraborty, S., Watson, N., Kumaran Ajikumar, P., & Stephanopoulos, G. (2015). Engineering lipid overproduction in the oleaginous yeast Yarrowia lipolytica. *Metabolic Engineering, 29*(0), 56–65.
49. Qiao, K., Wasylenko, T. M., Zhou, K., Xu, P., & Stephanopoulos, G. (2017). Lipid production in Yarrowia lipolytica is maximized by engineering cytosolic redox metabolism. *Nature Biotechnology, 35*(2), 173–177.
50. Xu, P., Qiao, K., & Stephanopoulos, G. (2017). Engineering oxidative stress defense pathways to build a robust lipid production platform in Yarrowia lipolytica. *Biotechnology and Bioengineering, 114*(7), 1521–1530.
51. Blazeck, J., Hill, A., Liu, L., Knight, R., Miller, J., Pan, A., Otoupal, P., & Alper, H. S. (2014). Harnessing Yarrowia lipolytica lipogenesis to create a platform for lipid and biofuel production. *Nature Communications, 5*, 3131.
52. Liu, L. Q., Markham, K., Blazeck, J., Zhou, N. J., Leon, D., Otoupal, P., & Alper, H. S. (2015). Surveying the lipogenesis landscape in Yarrowia lipolytica through understanding the function of a Mga2p regulatory protein mutant. *Metabolic Engineering, 31*, 102–111.
53. Liu, L., Pan, A., Spofford, C., Zhou, N., & Alper, H. S. (2015). An evolutionary metabolic engineering approach for enhancing lipogenesis in Yarrowia lipolytica. *Metabolic Engineering, 29*, 36–45.
54. Ledesma-Amaro, R., Dulermo, R., Niehus, X., & Nicaud, J.-M. (2016). Combining metabolic engineering and process optimization to improve production and secretion of fatty acids. *Metabolic Engineering, 38*, 38–46.

55. Friedlander, J., Tsakraklides, V., Kamineni, A., Greenhagen, E. H., Consiglio, A. L., MacEwen, K., Crabtree, D. V., Afshar, J., Nugent, R. L., & Hamilton, M. A. (2016). Engineering of a high lipid producing Yarrowia lipolytica strain. *Biotechnology for Biofuels, 9*(1), 1.
56. Wasylenko, T. M., Ahn, W. S., & Stephanopoulos, G. (2015). The oxidative pentose phosphate pathway is the primary source of NADPH for lipid overproduction from glucose in Yarrowia lipolytica. *Metabolic Engineering, 30*, 27–39.
57. Zhou, Y. J., Buijs, N. A., Zhu, Z., Qin, J., Siewers, V., & Nielsen, J. (2016). Production of fatty acid-derived oleochemicals and biofuels by synthetic yeast cell factories. *Nature Communications, 7*, 11709.
58. Xu, P., Qiao, K. J., Ahn, W. S., & Stephanopoulos, G. (2016). Engineering Yarrowia lipolytica as a platform for synthesis of drop-in transportation fuels and oleochemicals. *Proccedings of the National Academy of Sciences of the United States of America, 113*(39), 10848–10853.
59. Peralta-Yahya, P. P., Zhang, F., Del Cardayre, S. B., & Keasling, J. D. (2012). Microbial engineering for the production of advanced biofuels. *Nature, 488*(7411), 320.
60. Beld, J., Lee, D. J., & Burkart, M. D. (2015). Fatty acid biosynthesis revisited: Structure elucidation and metabolic engineering. *Molecular BioSystems, 11*(1), 38–59.
61. Jarboe, L. R., Royce, L. A., & Liu, P. (2013). Understanding biocatalyst inhibition by carboxylic acids. *Frontiers in Microbiology, 4*, 272.
62. Lomakin, I. B., Xiong, Y., & Steitz, T. A. (2007). The crystal structure of yeast fatty acid synthase, a cellular machine with eight active sites working together. *Cell, 129*(2), 319–332.
63. Mootz, H. D., Finking, R., & Marahiel, M. A. (2001). 4′-Phosphopantetheine transfer in primary and secondary metabolism of Bacillus subtilis. *Journal of Biological Chemistry, 276*(40), 37289–37298.
64. White, S. W., Zheng, J., Zhang, Y.-M., & Rock, C. O. (2005). The structural biology of type II fatty acid biosynthesis. *Annual Review of Biochemistry, 74*, 791–831.
65. Buchbinder, J. L., Witkowski, A., Smith, S., & Fletterick, R. J. (1995). Crystallization and preliminary diffraction studies of thioesterase II from rat mammary gland. *Proteins: Structure, Function, and Bioinformatics, 22*(1), 73–75.
66. Dehesh, K., Edwards, P., Hayes, T., Cranmer, A. M., & Fillatti, J. (1996). Two novel thioesterases are key determinants of the bimodal distribution of acyl chain length of Cuphea palustris seed oil. *Plant Physiology, 110*(1), 203–210.
67. Leber, C., & Da Silva, N. A. (2014). Engineering of Saccharomyces cerevisiae for the synthesis of short chain fatty acids. *Biotechnology and Bioengineering, 111*(2), 347–358.
68. Leber, C., Choi, J. W., Polson, B., & Da Silva, N. A. (2016). Disrupted short chain specific β-oxidation and improved synthase expression increase synthesis of short chain fatty acids in Saccharomyces cerevisiae. *Biotechnology and Bioengineering, 113*(4), 895–900.
69. Zhu, Z., Hu, Y., Teixeira, P. G., Pereira, R., Chen, Y., Siewers, V., & Nielsen, J. (2020). Multidimensional engineering of Saccharomyces cerevisiae for efficient synthesis of medium-chain fatty acids. *Nature Catalysis, 3*(1), 64–74.
70. Xu, P., Qiao, K., Ahn, W. S., & Stephanopoulos, G. (2016). Engineering Yarrowia lipolytica as a platform for synthesis of drop-in transportation fuels and oleochemicals. *Proceedings of the National Academy of Sciences, 113*(39), 10848–10853.
71. Gao, Q., Cao, X., Huang, Y.-Y., Yang, J.-L., Chen, J., Wei, L.-J., & Hua, Q. (2018). Overproduction of fatty acid ethyl esters by the oleaginous yeast Yarrowia lipolytica through metabolic engineering and process optimization. *ACS Synthetic Biology, 7*(5), 1371–1380.
72. Yamamoto, K., Kinoshita, A., & Shibahara, A. (2008). Ricinoleic acid in common vegetable oils and oil seeds. *Lipids, 43*(5), 457–460.
73. Mander, L., & Liu, H.-W. (2010). *Comprehensive natural products II: Chemistry and biology* (Vol. 1). Boston: Elsevier.
74. Kılıç, M., Uzun, B. B., Pütün, E., & Pütün, A. E. (2013). Optimization of biodiesel production from castor oil using factorial design. *Fuel Processing Technology, 111*, 105–110.
75. Ogunniyi, D. S. (2006). Castor oil: A vital industrial raw material. *Bioresource Technology, 97*(9), 1086–1091.

76. Ledesma-Amaro, R., & Nicaud, J.-M. (2016). Yarrowia lipolytica as a biotechnological chassis to produce usual and unusual fatty acids. *Progress in Lipid Research, 61*, 40–50.
77. Béopoulos, A., Verbeke, J., Bordes, F., Guicherd, M., Bressy, M., Marty, A., & Nicaud, J.-M. (2014). Metabolic engineering for ricinoleic acid production in the oleaginous yeast Yarrowia lipolytica. *Applied Microbiology and Biotechnology, 98*(1), 251–262.
78. Venegas-Calerón, M., Sayanova, O., & Napier, J. A. (2010). An alternative to fish oils: Metabolic engineering of oil-seed crops to produce omega-3 long chain polyunsaturated fatty acids. *Progress in Lipid Research, 49*(2), 108–119.
79. Salas Lorenzo, I., Chisaguano Tonato, A. M., de la Garza Puentes, A., Nieto, A., Herrmann, F., Dieguez, E., Castellote, A. I., López-Sabater, M. C., Rodríguez-Palmero, M., & Campoy, C. (2019). The effect of an infant formula supplemented with AA and DHA on fatty acid levels of infants with different FADS genotypes: The COGNIS study. *Nutrients, 11*(3), 602.
80. Blondeau, N., Lipsky, R. H., Bourourou, M., Duncan, M. W., Gorelick, P. B., & Marini, A. M. (2015). Alpha-linolenic acid: An omega-3 fatty acid with neuroprotective properties—Ready for use in the stroke clinic? *BioMed Research International, 2015*, 519830.
81. Domergue, F., Abbadi, A., Ott, C., Zank, T. K., Zähringer, U., & Heinz, E. (2003). Acyl carriers used as substrates by the desaturases and elongases involved in very long-chain polyunsaturated fatty acids biosynthesis reconstituted in yeast. *Journal of Biological Chemistry, 278*(37), 35115–35126.
82. Domergue, F., Abbadi, A., Zähringer, U., Moreau, H., & Heinz, E. (2005). In vivo characterization of the first acyl-CoA Δ6-desaturase from a member of the plant kingdom, the microalga Ostreococcus tauri. *Biochemical Journal, 389*(2), 483–490.
83. Tavares, S., Grotkjær, T., Obsen, T., Haslam, R. P., Napier, J. A., & Gunnarsson, N. (2011). Metabolic engineering of Saccharomyces cerevisiae for production of eicosapentaenoic acid, using a novel Δ5-desaturase from Paramecium tetraurelia. *Applied and Environmental Microbiology, 77*(5), 1854–1861.
84. Xie, D., Jackson, E. N., & Zhu, Q. (2015). Sustainable source of omega-3 eicosapentaenoic acid from metabolically engineered Yarrowia lipolytica: From fundamental research to commercial production. *Applied Microbiology and Biotechnology, 99*(4), 1599–1610.
85. Xue, Z., Sharpe, P. L., Hong, S.-P., Yadav, N. S., Xie, D., Short, D. R., Damude, H. G., Rupert, R. A., Seip, J. E., & Wang, J. (2013). Production of omega-3 eicosapentaenoic acid by metabolic engineering of Yarrowia lipolytica. *Nature Biotechnology, 31*(8), 734.
86. Barry, C., 3rd, Lee, R. E., Mdluli, K., Sampson, A. E., Schroeder, B. G., Slayden, R. A., & Yuan, Y. (1998). Mycolic acids: Structure, biosynthesis and physiological functions. *Progress in Lipid Research, 37*(2–3), 143.
87. Grogan, D. W., & Cronan, J. E. (1997). Cyclopropane ring formation in membrane lipids of bacteria. *Microbiology and Molecular Biology Reviews, 61*(4), 429–441.
88. Law, J. H. (1971). Biosynthesis of cyclopropane rings. *Accounts of Chemical Research, 4*(6), 199–203.
89. Bao, X., Katz, S., Pollard, M., & Ohlrogge, J. (2002). Carbocyclic fatty acids in plants: Biochemical and molecular genetic characterization of cyclopropane fatty acid synthesis of Sterculia foetida. *Proceedings of the National Academy of Sciences, 99*(10), 7172–7177.
90. Bao, X., Thelen, J. J., Bonaventure, G., & Ohlrogge, J. B. (2003). Characterization of cyclopropane fatty-acid synthase from Sterculia foetida. *Journal of Biological Chemistry, 278*(15), 12846–12853.
91. Rahman, M. D., Ziering, D. L., Mannarelli, S. J., Swartz, K. L., Huang, D. S., & Pascal, R. A., Jr. (1988). Effects of sulfur-containing analogs of stearic acid on growth and fatty acid biosynthesis in the protozoan Crithidia fasciculata. *Journal of Medicinal Chemistry, 31*(8), 1656–1659.
92. Schmid, K. M.. (1999). *Cyclopropane fatty acid expression in plants*. Google Patents.
93. Gontier, E., Thomasset, B., Wallington, E., & Wilmer, J. (2008) *Plant cyclopropane fatty acid synthase genes and uses thereof*. Google Patents.

94. Peng, H., He, L., & Haritos, V. S. (2019). Enhanced production of high-value cyclopropane fatty acid in yeast engineered for increased lipid synthesis and accumulation. *Biotechnology Journal, 14*(4), 1800487.
95. Czerwiec, Q., Idrissitaghki, A., Imatoukene, N., Nonus, M., Thomasset, B., Nicaud, J. M., & Rossignol, T. (2019). Optimization of cyclopropane fatty acids production in Yarrowia lipolytica. *Yeast, 36*(3), 143–151.
96. Markham, K. A., & Alper, H. S. (2018). Engineering Yarrowia lipolytica for the production of cyclopropanated fatty acids. *Journal of Industrial Microbiology & Biotechnology, 45*(10), 881–888.
97. Shields-Menard, S. A., Amirsadeghi, M., French, W. T., & Boopathy, R. (2018). A review on microbial lipids as a potential biofuel. *Bioresource Technology, 259*, 451–460.
98. Chandel, A. K., & Singh, O. V. (2011). Weedy lignocellulosic feedstock and microbial metabolic engineering: Advancing the generation of 'biofuel'. *Applied Microbiology and Biotechnology, 89*(5), 1289–1303.
99. Elkins, J. G., Raman, B., & Keller, M. (2010). Engineered microbial systems for enhanced conversion of lignocellulosic biomass. *Current Opinion in Biotechnology, 21*(5), 657–662.
100. Wilson, D. B. (2011). Microbial diversity of cellulose hydrolysis. *Current Opinion in Microbiology, 14*(3), 259–263.
101. Glick, B. R. (1995). Metabolic load and heterologous gene expression. *Biotechnology Advances, 13*(2), 247–261.
102. Colletti, P. F., Goyal, Y., Varman, A. M., Feng, X., Wu, B., & Tang, Y. J. (2011). Evaluating factors that influence microbial synthesis yields by linear regression with numerical and ordinal variables. *Biotechnology and Bioengineering, 108*(4), 893–901.
103. Poust, S., Hagen, A., Katz, L., & Keasling, J. D. (2014). Narrowing the gap between the promise and reality of polyketide synthases as a synthetic biology platform. *Current Opinion in Biotechnology, 30*, 32–39.
104. Wu, G., Yan, Q., Jones, J. A., Tang, Y. J., Fong, S. S., & Koffas, M. A. (2016). Metabolic burden: Cornerstones in synthetic biology and metabolic engineering applications. *Trends in Biotechnology, 34*(8), 652–664.

Chapter 10
Advanced Fermentation Strategies to Enhance Lipid Production from Lignocellulosic Biomass

Qiang Fei, Yunyun Liu, Haritha Meruvu, Ziyue Jiao, and Rongzhan Fu

10.1 Introduction

Lignocellulosic-derived lipids produced by oleaginous microorganisms have been receiving attention lately owing to their abundance, low costs, and renewability. However, the economic feasibility of cellulosic lipids has been hampered due to the lack of dedicated technologies necessary for engineering microbial strains to meet industrial purposes. Tailor-made techniques applying fermentation strategies could address such issues and improve lipid yields of microbial fermentation volumetrically by improving cell growth, carbon conversion efficacies, and tolerance to undesirable inhibitors [1].

The dry cellular weight (more than 20%) of oleaginous microorganisms is comprised of triacylglycerol (TAG) that can be catalyzed into fatty acid methyl esters (FAME) and fatty acid ethyl esters (FAEE) considered as precursors for biodiesel production during transesterification [2]. Lipid production can be associated with nutrient limitation, and preferential utilization of abundantly provided carbon sources over limited nitrogen/phosphate supplies has been applied and investigated [3]. Lipid biosynthesis in oleaginous microorganisms commences once nitrogen reserves are exhausted, causing activation of adenosine monophosphate (AMP)

Q. Fei (✉) · H. Meruvu · Z. Jiao
School of Chemical Engineering and Technology, Xi'an Jiaotong University,
Xi'an, Shaanxi, China
e-mail: feiqiang@xjtu.edu.cn

Y. Liu (✉)
College of Mechanical and Electrical Engineering, Shaanxi University of Science & Technology, Xi'an, China
e-mail: liuyu282009@126.com

R. Fu
School of Chemical Engineering, Northwest University, Xi'an, China

Z.-H. Liu, A. Ragauskas (eds.), *Emerging Technologies for Biorefineries, Biofuels, and Value-Added Commodities*,
https://doi.org/10.1007/978-3-030-65584-6_10

deaminase to subsequently decrease the cellular AMP content. This sets a chain reaction, diminishing mitochondrial isocitrate dehydrogenase activity (dependent on AMP levels) within the tricarboxylic acid cycle, and resultant isocitrate gets accumulated instead of being metabolized. Subsequently, the citrate stored within mitochondria when transported to the cytosol gets cleaved into acetyl-coA and oxaloacetate mediated by ATP citrate lyase. Acetyl-coA as the key intermediate is used in the biosynthesis of fatty acid and lipids in oleaginous microorganisms, and oxaloacetate is converted into malate via cytoplasmic malate dehydrogenase [4]. Different strains use acetyl-CoA in different ways during lipid accumulation [1]. The citrate (mobilized from mitochondria) is replaced by the cytosolic malate, which is catalyzed by $NADP^+$-dependent malic enzyme into pyruvate and NADPH, ultimately providing energy for the lipid accumulation process [5–7]. The activity of malic enzyme also affects lipid biosynthesis, as it regulates the generation of NADPH required by fatty acid synthase (FAS) [8]. Moreover, acetyl-CoA carboxylase (ACC) is also vital for lipid accumulation for it is truly the first enzyme of lipid biosynthesis. It was observed that the improvement of fatty acid biosynthesis in *Escherichia coli* could be achieved by overproducing the ACC activity rather than increasing the accumulation of malonyl-CoA [9, 10].

The end products of fatty acid biosynthesis are usually C16 or C18 saturated fatty acids. They are sequentially catalyzed by desaturases and elongases resulting in the production of polyunsaturated fatty acids (PUFAs), depending upon their innate genetic metabolisms [2]. After successive desaturation and elongation reactions, the saturated fatty acids could form various PUFAs [3]. So far, the pathway of PUFAs in most oleaginous yeasts uses a conventional route for the biosynthesis of fatty acids derived from acetyl-CoA and malonyl-CoA via ACC using FAS. In *Schizochytrium* sp. and related thraustochytrids, PUFA synthesis occurs through polyketide synthase (PKS) route involving two essential building blocks, viz., acetyl-CoA and malonyl-CoA, but does not involve in reducing intermediates. Meanwhile, the PKS system occurrent in bacteria like *Shewanella* sp. and *Moritella marina* produces intermediates like long-chain omega-3 fatty acids (such as EPA and DHA) [11, 12]. In prokaryotes that can synthesize DHA, the biosynthesis of long-chain fatty acids is carried out via a typical PKS route through which the growing chain of fatty acids remains unsaturated, unlike the conventional eukaryotic fatty acid synthesis system where they are reduced into saturated fatty acids [2].

Fatty acid molecules when linked to glycerol molecules via ester bonds can form lipids by the synthesis of fatty acid and/or TAG. Fatty acid chain elongation occurs through a reaction between ACC and acetyl-CoA to form malonyl-CoA, catalyzed by FAS in most microbes. The synthesized malonyl-CoA is converted into malonyl-acyl-carrier protein (malonyl-ACP) via malonyl-CoA: ACP transacetylase, one of FAS mixtures [13]. The FAS can transfer malonyl (carbon source) to acyl-carrier protein to synthesize long fatty acid chains like C16, C18, γ-linoleic acid (GLA), arachidonic acid, EPA, and DHA [14]. Differently, the vital route of TAG synthesis is glycerol-3-phosphate (G3P) or Kennedy pathway, where initial acylation of G3P with acyl-CoA results in the formation of LPA (lysophosphatidate) along with GPAT (G3P-acyltransferase). Subsequently, LPA is condensed by the catalysis through LPAT (lysophosphatidate acyltransferase) with another acyl-CoA generating PA (phosphatidic acid), which is further dephosphorylated by PAP (phosphatidic acid phosphatase) producing DAG (diacylglycerol). Finally, another acyl-CoA is introduced into DAG, and TAG is synthesized via the regulatory DGAT (diacylglycerol acyltransferase) [15–17].

It was reported that PA formation in yeasts might occur via the G3P/Kennedy pathway or DHAP (dihydroxyacetone phosphate) pathway. It is known that the G3P pathway can be found in bacteria, but the DHAP pathway is only available in yeasts and mammalian cells [15, 18, 19]. For fatty acid synthesis, it is still unclear if the FAS system operates concomitantly along with the PKS system or if fatty acids are directly generated by the TAG pathway. Genetic analysis of mutant oleaginous organisms is expected to provide a comprehensive understanding concerning PUFA biosynthesis. More details regarding the mechanism of lipid biosynthesis can be found in other Chapters of this book.

10.2 Process Development for Lipid Production Using Lignocellulosic Sugars

Biochemical routes for lipid production from lignocellulosic feedstock include biomass pretreatment to break down complex plant cell walls, enzymatic hydrolysis to release fermentable sugars, and microbial fermentation of sugars into lipids. The produced lipids can be converted through transesterification into biodiesel production [20] or through deoxygenation into hydrocarbons to produce gasoline/diesel directly [21]. Theoretical lipid yields cultured from lignocellulose derivatives like glucose and xylose can be estimated to be 0.32 g/g and 0.34 g/g, respectively, but never below 0.25 g/g even with synthetic media [20, 22]. But detected lipid yields are quite lower in lignocellulosic biomass hydrolysate ferments due to the accumulation of degradative inhibitors [23].

Many technical methods have been developed to feature the economic production of microbial lipids using lignocellulosic biomass as a potential carbohydrate feedstock. Lipid production and yields can be enhanced by process development and optimized fermentation. Different fermentation modes including batch, fed-batch, and continuous cultivation coupled with multistage strategy can be adopted for increasing cell densities of oleaginous microbes to achieve better lipid production.

10.2.1 Single-Stage Cultivation: Batch and Fed-Batch Cultures

Traditional microbial lipid production using lignocellulosic sugars can be carried out using a single-stage batch mode of fermentation. Lipid production in batch cultures has been investigated using different lignocellulosic hydrolysates for higher lipid content. Chang et al. [23] explored the batch cultivation of *Cryptococcus* using corncob hydrolysate supplementing with glucose and obtained dried cell weight of 12.6 g/L, lipid yield of 7.6 g/L, and lipid content of 60.2%, respectively. Anschau et al. [24] assessed lipid production performance of oleaginous yeast *Lipomyces starkeyi* with batch culture on sugarcane bagasse and achieved 27.7% (w/w) lipid content. Fei et al. [25] conducted a batch culture of *Rhodococcus opacus* to acquire baseline results for lipid production with C6 sugar from lignocellulosic hydrolysate, in which study lipid content of 58.7% (w/w) with lipid yield of 0.19 g lipid/g sugar was achieved. Additionally, rice hull hydrolysate was investigated for producing lipids by *Mortierella isabellina* in batch culture by Economou et al. [26], and the highest lipid content of 64.3% was reached. Though results achieved through previous reports were good enough, yet the production yields were not comparable with industrial standards.

It has been known that microbial lipid extraction is most economical when lipid productivity is above 1 g/L/h [27]. To obtain high lipid productivity, high lipid titer is desirable, which also benefits the downstream processing. High lipid titer requires high carbon source input. However, the high concentration of sugar achieved from hydrolysis will inhibit cell growth due to the high level of inhibitors from pretreatment and the sugar concentration loaded at the beginning of culture. Therefore, a fed-batch culture may reduce the inhibitory effect by adding the feeds with control [28, 29].

Fed-batch mode of cultivation is an alternative method for achieving high solid loading in fermentation processes because of the advantages it offers over batch mode. Nevertheless, the critical parameters including how and when to add nutrients to the reaction systems to ensure high conversion efficiency should be considered and explored. Various fed-batch feeding techniques have been used for obtaining high lipid productions as quoted by researchers [28, 30, 31]. Previous studies have adopted feedback control strategies that guide substrate feeding using monitored pH or dissolved oxygen (DO) provision. Although it was convenient, the tardy cellular response to pH/DO variations would arouse over or not timely feeding, thus affecting the final production [32]. The direct feedback control supervising cell growth and the residual substrate can manually be implemented through pulse feeding or automated controlled feeding modes [32, 33]. The timely monitoring of the growth condition of cells by adjusting feeding mode could avoid or relieve the inhibition of accumulated metabolites. Hassan et al. [34] reported that lipid productivity of 0.21 g/L/h along with 53% lipid content was achieved with *Cryptococcus curvatus* in the culture including the C/N ratio of 10:1 but fed glucose without nitrogen. Li et al. [30] applied discontinuous, pulse feeding approach to culture

Rhodosporidium toruloides Y4 in complex medium. The lipid content of 67.5% (w/w) and productivity of 0.54 g/L/h were achieved with a fermented batch fed with glucose solution devoid of nitrogen. In the comparative study of constant and pulse feeding mode, continuous feeding was exhibited to provide higher lipid productivity compared to pulse feeding done in the cultures of *R. toruloides* Y4 [28]. Wiebe et al. [35] conducted fed-batch cultivation of *R. toruloides* with manual feeding, which was the most efficient (75% lipid content, 0.21 g/L/h lipid productivity) using glucose as the sole carbon source. Nevertheless, these works were all carried out based on pure glucose.

To date, the lipid production studies employing carbon sources like lignocellulosic hydrolysates for the development of diverse fed-batch feeding strategies have been extensively reported [23–25]. Liu et al. [31] used corncob hydrolysates as carbon sources during fed-batch fermentation studies of *Rhodotorula glutinis* with a high C/N ratio solution feeding strategy and achieved final lipid content of 39%. Fei et al. [25] improved the lipid production of *R. toruloides* using corn stover hydrolysates as sole substrates by applying several fed-batch strategies (Fig. 10.1). The pulse mode of feeding resulted in lipid productivity of 0.35 g/L/h and lipid yield of 0.24 g/g, while the DO-stat feeding mode achieved 0.33 g/L/h lipid productivity and 0.23 g/g lipid yield. Thereafter, the highest lipid productivity of 0.4 g/L/h along with 59% lipid content and 0.29 g/g lipid yield was obtained through the application of an automated controlled sugar feeding system [22]. These findings indicate that a suitable fed-batch system with controlling the optimal sugar concentration in the culture medium could lead to an enhanced conversion efficiency of lignocellulosic-based sugar for both higher cell density and better lipid accumulation.

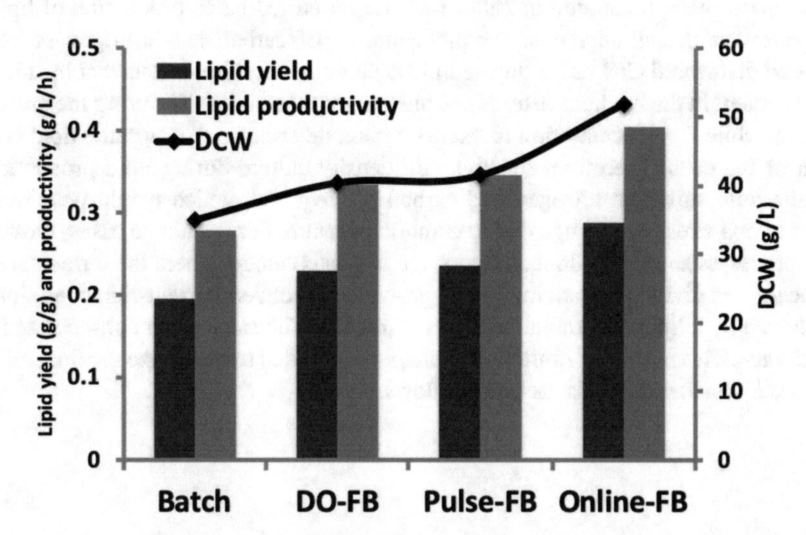

Fig. 10.1 Comparison of lipid production by using different feeding strategies in fed-batch (FB) cultures [25]. DO-FB presents DO-stat feeding fed-batch cultures, pulse-FB presents manual feeding fed-batch cultures, online-FB presents automated online sugar control fed-batch cultures

Table 10.1 Lipid production in cultures of different oleaginous microbes using various substrates

Substrates	Microbes	Culture modes	Lipid content (%)	Lipid yield (g/g)	Lipid productivity (g/L/h)	References
Corncob hydrolysates	*Cryptococcus* sp. SM5S05	Batch	60.2	0.13	0.05	[23]
Sugarcane bagasse hydrolysate	*Lipomyces starkeyi*	Batch	27.7	0.07	0.09	[24]
Corn stover hydrolysate	*Rhodococcus opacus*	Batch	58.7	0.19	0.28	[25]
Pure glucose	*C. curvatus*	Pulse fed-batch	53	0.14	0.21	[34]
Pure glucose	*R. toruloides* Y4	Pulse fed-batch	67.5	0.23	0.54	[30]
Pure glucose	*R. toruloides* Y4	Continuous fed-batch	61.8	0.23	0.57	[28]
Pure sugar mixture	*R. toruloides*	Pulse fed-batch	58	0.07	0.15	[35]
Corn stover hydrolysate	*R. toruloides*	Pulse fed-batch	62	0.24	0.35	[25]
Corn stover hydrolysate	*R. toruloides*	Online fed-batch	59	0.29	0.40	[25]
Corncob hydrolysate	*R. glutinis*	Fed-batch	39	0.15	0.15	[31]

The performances of lipid production using different substrates in various culture modes were illustrated in Table 10.1. Reported fed-batch production of lipids reveals that the simultaneous supplementation of carbon and nitrogen sources caused disfavored C/N ratios during lipid accumulation. It is obvious that the nitrogen content in the medium determines the maximum cell density during the growth phase, while the concentration of excess carbon determines the supreme lipid content of the cells. Therefore, for high cell density culture during lipid production, media with sufficient nitrogen and carbon are required, which would potentially lead to cell stress or damage due to osmotic pressure. For the above cases, most of the processes have been done in two-stage fed-batch mode, where the fermentation process was divided into an initial cell growth stage under nutrient-rich conditions, followed by a lipid accumulation under nitrogen deficiency or even absence in the 2nd stage. This nutrient's limitation strategy was applied to maximize the final product titer, which will be discussed as follows.

10.2.2 Two-Stage Cultivation: Fed-Batch and Continuous Cultures

Provision of substantial cell density, which is necessary to induce reasonable lipid yields, is a proven scope for large-scale production. Cultivations are conducted by first producing biomass and then inducing lipid accumulation, and it is possible by adopting fed-batch or continuous modes of cultivation, through which supply of nutrients can be regulated. It has been proved that applications of fed-batch or continuous cultivation could help obtain high cell density using lignocellulosic sugars meeting the industrial standards for lipid production. The two-stage cultivation that purposely separates the phases of cellular growth and lipid production in oleaginous microbes includes cell mass generated in the 1st stage with carbon-nitrogen ratios of 65:1 to stimulate cell growth, followed by amplified lipid accumulation in the 2nd stage using carbon-nitrogen hydrolysates of 500:1 for favoring lipid production and limiting cellular growth [33]. The first stage provides essential nitrogen for supporting cell growth abundantly, and the latter provides required sugars along with limited nitrogen levels to favor lipid accumulation [34]. Using the nutrient limitation strategy, a two-stage culture mode could improve lipid yield and avoid inhibitory effects in the high-solids system.

Several cultivation modes have been developed using this strategy to improve lipid productivity. The top-performing microbial lipid production yeast was successfully identified and developed through two-stage culture mode in recent studies [36]. Higher lipid titer up to 30 g/L from over 100 g/L sugars was achieved from the promising oleaginous yeast on pretreated switchgrass hydrolysates. The two-stage culture mode provides obvious benefits over single-stage fermentation for lipid production, which is because that both C/N ratio and sugar concentration can be controlled properly to enhance lipid accumulation in the second stage without concerning the dilution of hydrolysate feed and cell growth [36, 37]. Therefore, the two-staged culture mode has been considered as a basic configuration to enhance lipid production in terms of lipid titer and content.

A two-staged fed-batch process designed to achieve high cell density in *R. opacus* was reported with high lipid yields [38]. That process featured a nutrient-rich initial medium containing synthetic C6/C5-sugar mixture in the first phase and then switched into the second phase with intermittent feeding of C5-only feeds for lipid accumulation. Overall, lipid productivity of 0.29 g/L/h was recorded along with cell density of 45 g/L and lipid content of 54% w/w, and the strategy was proved significantly cost-effective. Meanwhile, the inhibitory effect caused by concentrated carbon source was also reduced in two-stage fed-batch cultivation. Liu et al. [27] also experimented with lipid production through fed-batch fermentation of *R. glutinis* adopting a two-stage feeding strategy. During the first stage, rapid biomass increase was detected with limited lipid titers of 9.42 g/L and 23.3% lipid content. However, the second stage with feeding concentrated corncob hydrolysates improved the lipid tires to 33.5 g/L and lipid content to 47.2%. Furthermore, in comparison to the batch modes of culture, the two-staged culture mode was found superior with improved biomass concentration, lipid content, and lipid titer of threefold, 29%, and fivefold, respectively [27].

Bioprocessing improvement was still required to make the fermentation more efficient. Continuous feeding mode can keep the cellular metabolism active for higher titer and productivity [39]. Increasing the supply rate of nutrients in different stages can be an efficient way to enhance cell density and lipid content, leading to high overall lipid productivities. Karamerou et al. [40] evaluated a two-stage continuous feeding strategy for the cultivation of *R. glutinis*, targeting cellular proliferation in the 1st stage followed by lipid production in the 2nd stage for favorable production of biomass and lipid yields. Continuous nutrient feeding at high dilution rates (<0.8 g/L/h) led to increases in both the cell density and lipid content (53%), reaching a high lipid yield. Thereby continuous cultivation at variable rates during each stage leads to an efficient approach for enhancement of microbial lipid production in a manner of longer timeframe and higher productivity.

10.2.3 High Cell Density Cultivation (HCDC)

Since microbial lipids are accumulated inside the cell body of oleaginous microorganisms, the productivity of lipids concerns both the cellular lipid titer and the cell density in cultures. Efficient lipid recovery from fermentation broths includes harvesting the microbial cells in the broth through biomass drying or forced cell disruption followed by lipid extraction. Cell harvesting is expensive if cellular biomass density in the fermentation broth is low; hence a high cell density is desirable in large-scale fermentation. HCDC systems refer to nearly tenfold of the cell density of a regular batch culture; for instance, compared with a 5–10 g/L of *R. glutinis* cell culture as normal cell density, a 50–100 g/L is considered high cell density [41, 42].

A major goal of any biotechnological process development is to maximize productivity by inducing the highest possible density of products within a reaction volume in the shortest period. High cell density is a requirement for higher titer and productivity, which depends not only on biocatalysts such as microbes used in the reaction but also the mode of the bioreactor operation. The HCDC system can also be modified and used in the fermentation process for microbial lipid production, which boosts the production titer of intracellular metabolic products as well as lipid productivity. Apart from the application of HCDC to produce high lipid yields [43], similar processes for multistaging continuous high cell density cultivation systems have been utilized to improve lipid production [44].

Industrial lipid production can become economically viable only when high cell densities and high lipid productivities are attained with low investments. Recently, advanced techniques for controlling and monitoring biotechnological processes are being developed to achieve high cell densities with less cost [20]. Microbes with higher productivity and cheaper feeding strategies using low-cost substrates like lignocellulosic biomass are preferable for lipid production. Co-fermentation of substrates like celluloses and hemicelluloses as raw material biomass and their conversion into cell mass and lipids could also improve lipid yield, titer, and productivities, considering cost-cutting strategies.

10.3 Lignocellulosic Inhibitor Tolerance of Lipid-Producing Microorganisms

Low lipid content and productivity are influenced by several technical challenges during the culture of oleaginous strains using lignocellulosic biomass, the most predominant bottleneck being inhibitor tolerance. In biorefineries, the pretreatment of complex lignocellulosic wastes into simpler forms is a critical prerequisite before proceeding to hydrolysis and fermentation processes. This leads to the generation of several by-products (inhibitors) like acetic acid, hydroxymethylfurfural (HMF), furfural, acidic phenolics, and other aromatic compounds that can inhibit cell growth and lipid yield [45]. *Rhodosporidium toruloides* was reported with induced tolerance against inhibitors, and it could withstand up to a 1 mM furfural content level by metabolizing it into furoic acid, but levels above that would inhibit cell growth by 45% [46]. It was noted that *Trichosporon fermentans* showed a 50% reduction of lipid production at 4.7 mM of furfural compared to 37.7 mM HMF. Besides, lipid production in oleaginous yeasts from lignocellulosic biomass can also be limited by ferulic acid due to the sensitivity. Developing microbes metabolizing lignocellulose-derived sugars that are capable of tolerating these inhibitors and applying the aforementioned fermentation modes to relieve the inhibitory effect have been efficient ways to enhance the lipid production [20, 47].

A high lipid accumulation potential and concomitant inhibitor tolerance were reported in *Trichosporon cutaneum* due to its possession of metabolic pathways to degrade eight typical inhibitors, while fermentation of corn stover hydrolysates at varying sugar concentrations. Although the strain was influenced by the presence of furfural and 4-hydroxybenzaldehyde in medium, the subsequent construction of inhibitor degradation pathways of the strain could be deciphered through genome sequencing, which indicates a potential candidate for futuristic applications in cellulosic lipid production [48]. Detoxification of furfural into furfuryl alcohol (reduction) and furoic acid (oxidation) in *Trichosporon fementans* during in vivo fermentation conditions indicated that the detoxification process can be enhanced by minimizing the transient production of furfuryl alcohol. This can be achieved by altering culture conditions, metabolic adaptation, or genetic strain modification to augment the lipid production capacity from the pretreated lignocellulosic biomass [49]. Gong and his co-workers used a co-fermentation approach to alleviate the acetate inhibitory effect on lipid accumulation and achieved improved lipid yield. In that study, the acetate and sugars were co-fermented to exploit low-cost substrates for lipid production using *Cryptococcus curvatus*. The pretreated acetate-rich corn stover hydrolysates were used as the carbon source providing a high lipid content of

61% during the cultivation, in which glucose and acetate were consumed simultaneously before the utilization of xylose [50].

The production of inhibitors was interrelated with a decrease in cell growth and lipid accumulation of the selected oleaginous yeasts, and the effect varied with the inhibitor involved. For instance, furfural could inhibit both growth and lipid accumulation in proportions greater than HMF. Hence, it is quintessential to select and modify yeasts that are strongly resistant to high concentrations of expected inhibitors that might occur in target hydrolysates chosen for lipid accumulation. Metabolic mechanisms governing microbial lipid accumulation and inhibitor degradation pathways play a vital role in improving lipid fermentation from lignocellulose feedstock [51].

10.4 Biologically Empowering Lignocellulosic-Utilizing Microorganisms for Lipid Production

Microbial lipid production using lignocellulosic biomass can be improved through genetic and/or metabolic approaches. Unlike conventional random mutagenesis for strain modification [52–54], the genetic engineering approach is more rational, rapid, and reliable resulting in microorganisms with augmented titer and yield, improved metabolic activities, tolerance to inhibitors, and adaptation to diverse conditions [55]. It is supposed that directed evolution techniques and engineered metabolic pathway approaches could enhance lipid production efficacies of microbes using lignocellulosic biomass.

Current genetic engineering strategies mainly focus on increasing the lipid content. There have been reports that engineered strains of oleaginous yeast *Yarrowia lipolytica* could accumulate up to 90% of dry cell weight as fatty acids [55] and of non-oleaginous fungal origin *Saccharomyces cerevisiae* [56] and *Ashbya gossypii* [57] as much as 50% and 70%, respectively. Engineered oleaginous bacteria like *Rhodococcus opacus* could completely and/or simultaneously utilize high concentrations of xylose or mixed glucose-xylose to produce 45.8% and 54% of lipid content [38, 58]. Furthermore, mutants of this strain were constructed with improved multiple resistance to inhibitors of phenols via adaptive evolution-based strategies, which provided more than 200% enhancement in resistance to inhibitors compared to the parent strain [59].

In wild oleaginous yeasts, the fatty acid range is somewhat limited. However, fungi and microalgae are of desirable commercial value, as the PUFAs' content comprises over 20% of the total fatty acids accumulated. Given favorable conditions, lipid content accumulated as high as 30–70% of dry cell weight could be attained in microalgae by enhancing innate lipid content levels [60]. At present, *Chlamydomonas reinhardtii* (green alga) is a potential model candidate for genetic engineering studies due to its deciphered genome, and relevant strategies have been aimed to manipulate rate-limiting steps of TAG synthesis and improve the metabolic flux [61, 62]. It has been reported that a tenfold increase in TAG production by inactivating of ADP-glucose pyrophosphorylase was achieved in the culture of *Chlamydomonas*, and the modified strain could accumulate up to 46.5% total lipid content in absence of ADP-glucose pyrophosphorylase [62].

Additionally, both overexpression and knockout of lipid synthesis genes have been applied in microbes to comprehend their metabolisms and improve lipid content levels. ACC and FAS play a crucial role as biocatalysts for biosynthesizing fatty acid as mentioned earlier [10], and improving enzyme activity through genetic engineering can significantly boost lipid metabolism [63]. In an attempt to increase lipid production without influencing growth, *Thalassiosira pseudonana* could be engineered by knock-outing mutations. This resulted in a modified strain by eliminating multifunctional lipase/phospholipase/acyltransferase, which could yield a significant increase in lipid content [64]. Biodiesel production was also reported from metabolically engineered *Escherichia coli* using glucose in fed-batch cultivation with FAEE titer of 1.28 g/L and FAEE content of 26% [10, 65]. Genetic and metabolic engineering are becoming more popular for improving the performance of lipid production by microbes, with a major long-term impact upon the commercial biofuel production scenario.

10.5 Perspectives

As the most abundant renewable and inedible feedstock, lignocellulose has been employed as the carbon source for the production of microbial lipids, which are promising precursors for biofuel production. Because of the mechanism of lipid accumulation in oleaginous microorganisms and inhibition effects of impurities from lignocellulosic hydrolysates, two-stage cultivation provides much better lipid content and yield over other culture modes due to the capability of controlling the feeding of carbon source and C/N ratio during cultures. Since lipid is an intracellular product, cell density becomes significantly important to lipid titer. Therefore, HCDC strategies coupling with hollow fiber membrane and perfusion technologies have been developed for improving both lipid titer and productivity. Overall, process development is a powerful tool for enhancing lipid production in terms of yield, titer, and productivity. Nowadays, more attention has been drawn in the field of metabolic engineering with the ability to control the expression level of genes associated with lipid accumulation. It is expected that an efficient platform for lipid production and research should integrate both micro-manipulations for the gene expression and fermentation strategies with the online control-feedback system, which may guarantee optimal conditions for a superb performance.

Acknowledgments This work is supported by the National Key R&D Program of China (2018YFA0901500) and the National Natural Science Foundation of China (21878241).

References

1. Kosa, M., & Ragauskas, A. J. (2011). Lipids from heterotrophic microbes: Advances in metabolism research. *Trends in Biotechnology, 29*(2), 53–61.
2. Ratledge, C. (2004). Fatty acid biosynthesis in microorganisms being used for single cell oil production. *Biochimie, 86*(11), 807–815.
3. Garay, L. A., Boundy-Mills, K. L., & German, J. B. (2014). Accumulation of high-value lipids in single-cell microorganisms: A mechanistic approach and future perspectives. *Journal of Agricultural and Food Chemistry, 62*(13), 2709–2727.
4. Zhang, H., Wu, C., Wu, Q., Dai, J., & Song, Y. (2016). Metabolic flux analysis of lipid biosynthesis in the yeast Yarrowia lipolytica using 13C-labeled glucose and gas chromatography-mass spectrometry. *PLoS One, 11*(7), e0159187.
5. Pereira, G., Finco, A. M., Letti, L., Karp, S., Pagnoncelli, M., Oliveira, J., Thomaz-Soccol, V., Brar, S., & Soccol, C. (2018). Microbial metabolic pathways in the production of valued-added products. In S. K. Brar, R. K. Das, & S. J. Sarma (Eds.), *Microbial sensing in fermentation* (pp. 137–167). Hoboken: Wiley.
6. Beopoulos, A., Nicaud, J.-M., & Gaillardin, C. (2011). An overview of lipid metabolism in yeasts and its impact on biotechnological processes. *Applied Microbiology and Biotechnology, 90*(4), 1193–1206.
7. Ratledge, C., & Wynn, J. P. (2002). The biochemistry and molecular biology of lipid accumulation in oleaginous microorganisms. In A. I. Laskin, J. W. Bennett, & G. M. Gadd (Eds.), *Advances in applied microbiology* (pp. 1–52). Amsterdam: Academic Press.
8. Passoth, V. (2017). Lipids of yeasts and filamentous fungi and their importance for biotechnology. In A. A. Sibirny (Ed.), *Biotechnology of yeasts and filamentous fungi* (pp. 149–204). Cham: Springer International Publishing.
9. Davis, M. S. S. J., & Cronan, J. E., Jr. (2000). Overproduction of acetyl-CoA carboxylase activity increases the rate of fatty acid biosynthesis in Escherichia coli. *Journal of Biological Chemistry, 275*(37), 28593–28598.
10. Liang, M. H. J. J. G. (2013). Advancing oleaginous microorganisms to produce lipid via metabolic engineering technology. *Progress in Lipid Research, 52*(4), 395–408.
11. Metz, J. G., Roessler, P., Facciotti, D., Levering, C., Dittrich, F., Lassner, M., Valentine, R., Lardizabal, K., Domergue, F., Yamada, A., Yazawa, K., Knauf, V., & Browse, J. (2001). Production of polyunsaturated fatty acids by polyketide synthases in both prokaryotes and eukaryotes. *Science, 293*(5528), 290–293.
12. Ferguson, J. J. A., Dias, C. B., & Garg, M. L. (2016). Omega-3 polyunsaturated fatty acids and hyperlipidaemias. In M. V. Hegde, A. A. Zanwar, & S. P. Adekar (Eds.), *Omega-3 fatty acids: Keys to nutritional health* (pp. 67–78). Cham: Springer International Publishing.
13. Ouyang, L.-L., Chen, S.-H., Li, Y., & Zhou, Z.-G. (2013). Transcriptome analysis reveals unique C4-like photosynthesis and oil body formation in an arachidonic acid-rich microalga Myrmecia incisa Reisigl H4301. *BMC Genomics, 14*(1), 396.
14. Ma Y.-L. (2006). Microbial oils and its research advance. *Chinese Journal of Bioprocess Engineering, 4*(4), 7–11.
15. Athenstaedt, K. D., & G. (1999). Phosphatidic acid, a key intermediate in lipid metabolism. *European Journal of Biochemistry, 266*(1), 1–16.
16. Coleman, R. A., & Lee, D. P. (2004). Enzymes of triacylglycerol synthesis and their regulation. *Progress in Lipid Research, 43*(2), 134–176.
17. Galán, B., Santos-Merino, M., Nogales, J., de la Cruz, F., & García, J. L. (2019). Microbial oils as nutraceuticals and animal feeds. In H. Goldfine (Ed.), *Health consequences of microbial interactions with hydrocarbons, oils, and lipids* (pp. 1–45). Cham: Springer International Publishing.
18. Minskoff, S. A., Racenis, P. V., Granger, J., Larkins, L., Hajra, A. K., & Greenberg, M. L. (1994). Regulation of phosphatidic acid biosynthetic enzymes in Saccharomyces cerevisiae. *Journal of Lipid Research (USA), 35*(12), 2254–2262.

19. Liang, M.-H., & Jiang, J.-G. (2013). Advancing oleaginous microorganisms to produce lipid via metabolic engineering technology. *Progress in Lipid Research, 52*(4), 395–408.
20. Jin, M., Slininger, P. J., Dien, B. S., Waghmode, S., Moser, B. R., Orjuela, A., Sousa, L. C., & Balan, V. (2015). Microbial lipid-based lignocellulosic biorefinery: Feasibility and challenges. *Trends in Biotechnology, 33*(1), 43–54.
21. Sousa, F. P., Silva, L. N., de Rezende, D. B., de Oliveira, L. C. A., & Pasa, V. M. D. (2018). Simultaneous deoxygenation, cracking and isomerization of palm kernel oil and palm olein over beta zeolite to produce biogasoline, green diesel and biojet-fuel. *Fuel, 223*, 149–156.
22. Huang, W.-D., & Zhang, Y. H. P. (2011). Analysis of biofuels production from sugar based on three criteria: Thermodynamics, bioenergetics, and product separation. *Energy & Environmental Science, 4*(3), 784–792.
23. Chang, Y.-H., Chang, K.-S., Lee, C.-F., Hsu, C.-L., Huang, C.-W., & Jang, H.-D. (2015). Microbial lipid production by oleaginous yeast Cryptococcus sp. in the batch cultures using corncob hydrolysate as carbon source. *Biomass and Bioenergy, 72*, 95–103.
24. Anschau, A., Xavier, M. C. A., Hernalsteens, S., & Franco, T. T. (2014). Effect of feeding strategies on lipid production by Lipomyces starkeyi. *Bioresource Technology, 157*, 214–222.
25. Fei, Q., O'Brien, M., Nelson, R., Chen, X., Lowell, A., & Dowe, N. (2016). Enhanced lipid production by Rhodosporidium toruloides using different fed-batch feeding strategies with lignocellulosic hydrolysate as the sole carbon source. *Biotechnology for Biofuels, 9*(1), 1–12.
26. Economou, C. N., Aggelis, G., Pavlou, S., & Vayenas, D. V. (2011). Single cell oil production from rice hulls hydrolysate. *Bioresource Technology, 102*(20), 9737–9742.
27. Moreton, R. S. (1988). Physiology of lipid accumulation yeast. In R. S. Moreton (Ed.), *Single cell oil*. London: Longman Higher Education Division.
28. Zhao, X., Hu, C., Wu, S., Shen, H., & Zhao, Z. K. (2011). Lipid production by Rhodosporidium toruloides Y4 using different substrate feeding strategies. *Journal of Industrial Microbiology & Biotechnology, 38*(5), 627–632.
29. Hernández-Beltrán, J. U., & Hernández-Escoto, H. (2018). Enzymatic hydrolysis of biomass at high-solids loadings through fed-batch operation. *Biomass and Bioenergy, 119*, 191–197.
30. Li, Y., Zhao, Z., & Bai, F. (2007). High-density cultivation of oleaginous yeast Rhodosporidium toruloides Y4 in fed-batch culture. *Enzyme and Microbial Technology, 41*(3), 312–317.
31. Liu, Y., Wang, Y., Liu, H., & Zhang, J. a. (2015). Enhanced lipid production with undetoxified corncob hydrolysate by Rhodotorula glutinis using a high cell density culture strategy. *Bioresource Technology, 180*, 32–39.
32. Kim, B. S., Lee, S. C., Lee, S. Y., Chang, H. N., Chang, Y. K., & Woo, S. I. (1994). Production of poly(3-hydroxybutyric acid) by fed-batch culture of Alcaligenes eutrophus with glucose concentration control. *Biotechnology and Bioengineering, 43*(9), 892.
33. Salehmin, M. N. I., Annuar, M. S. M., & Chisti, Y. (2013). High cell density fed-batch fermentations for lipase production: Feeding strategies and oxygen transfer. *Bioprocess and Biosystems Engineering, 36*(11), 1527–1543.
34. Hassan, M., Blanc, P. J., Granger, L. M., Pareilleux, A., & Goma, G. (1996). Influence of nitrogen and iron limitations on lipid production by Cryptococcus curvatus grown in batch and fed-batch culture. *Process Biochemistry, 31*(4), 355–361.
35. Wiebe, M. G., Koivuranta, K., Penttilä, M., & Ruohonen, L. (2012). Lipid production in batch and fed-batch cultures of Rhodosporidium toruloides from 5 and 6 carbon carbohydrates. *BMC Biotechnology, 12*(1), 26.
36. Slininger, P. J., Dien, B. S., Kurtzman, C. P., Moser, B. R., Bakota, E. L., Thompson, S. R., O'Bryan, P. J., Cotta, M. A., Balan, V., Jin, M., Sousa, L. C., & Dale, B. E. (2016). Comparative lipid production by oleaginous yeasts in hydrolyzates of lignocellulosic biomass and process strategy for high titers. *Biotechnology and Bioengineering, 113*(8), 1676–1690.
37. Wang, T., Tian, X., Liu, T., Wang, Z., Guan, W., Guo, M., Chu, J., & Zhuang, Y. (2017). A two-stage fed-batch heterotrophic culture of Chlorella protothecoides that combined nitrogen depletion with hyperosmotic stress strategy enhanced lipid yield and productivity. *Process Biochemistry, 60*, 74–83.

38. Fei, Q., Wewetzer, S. J., Kurosawa, K., Rha, C., & Sinskey, A. J. (2015). High-cell-density cultivation of an engineered Rhodococcus opacus strain for lipid production via co-fermentation of glucose and xylose. *Process Biochemistry, 50*(4), 500–506.
39. Fu, R., Fei, Q., Shang, L., Brigham, C. J., & Chang, H. N. (2018). Enhanced microbial lipid production by Cryptococcus albidus in the high-cell-density continuous cultivation with membrane cell recycling and two-stage nutrient limitation. *Journal of Industrial Microbiology & Biotechnology, 45*(12), 1045–1051.
40. Karamerou, E. E., Theodoropoulos, C., & Webb, C. (2017). Evaluating feeding strategies for microbial oil production from glycerol by Rhodotorula glutinis. *Engineering in Life Sciences, 17*(3), 314–324.
41. Chang, H. N., Kim, N.-J., Kang, J., Jeong, C. M., J-d-r, C., Fei, Q., Kim, B. J., Kwon, S., Lee, S. Y., & Kim, J. (2011). Multi-stage high cell continuous fermentation for high productivity and titer. *Bioprocess and Biosystems Engineering, 34*(4), 419–431.
42. Chang, H. N., Jung, K., Choi, J.-d.-r., Lee, J. C., & Woo, H.-C. (2014). Multi-stage continuous high cell density culture systems: A review. *Biotechnology Advances, 32*(2), 514–525.
43. Chang, H. N. F. Q., Choi, J. D. R., & Jung, K. S. (2011b). *Economic evaluation of heterotropic microbial lipid (C. albidus) production using low-cost volatile fatty acids in MSC-HCDC bioreactor system.* BIT's first annual congress of bioenergy, pp 0425–0429.
44. Fei, Q., Chang, H. N., Shang, L., J-d-r, C., Kim, N., & Kang, J. (2011). The effect of volatile fatty acids as a sole carbon source on lipid accumulation by Cryptococcus albidus for biodiesel production. *Bioresource Technology, 102*(3), 2695–2701.
45. Palmqvist, E., & Hahn-Hägerdal, B. (2000). Fermentation of lignocellulosic hydrolysates. II: Inhibitors and mechanisms of inhibition. *Bioresource Technology, 74*(1), 25–33.
46. Hu, C., Zhao, X., Zhao, J., Wu, S., & Zhao, Z. K. (2009). Effects of biomass hydrolysis by-products on oleaginous yeast Rhodosporidium toruloides. *Bioresource Technology, 100*(20), 4843–4847.
47. Chao, H., Hong, W., Li-ping, L., Wen-yong, L., & Min-hua, Z. (2012). Effects of alcohol compounds on the growth and lipid accumulation of oleaginous yeast Trichosporon fermentans. *PLoS One, 7*(10), 1–12.
48. Wang, J., Gao, Q., Zhang, H., & Bao, J. (2016). Inhibitor degradation and lipid accumulation potentials of oleaginous yeast Trichosporon cutaneum using lignocellulose feedstock. *Bioresource Technology, 218*, 892–901.
49. Huang, C., Wu, H., Smith, T. J., Z-j, L., Lou, W.-Y., & M-h, Z. (2012). In vivo detoxification of furfural during lipid production by the oleaginous yeast Trichosporon fermentans. *Biotechnology Letters, 34*(9), 1637–1642.
50. Gong, Z., Zhou, W., Shen, H., Yang, Z., Wang, G., Zuo, Z., Hou, Y., & Zhao, Z. K. (2016). Co-fermentation of acetate and sugars facilitating microbial lipid production on acetate-rich biomass hydrolysates. *Bioresource Technology, 207*, 102–108.
51. Sitepu, I., Selby, T., Lin, T., Zhu, S., & Boundy-Mills, K. (2014). Carbon source utilization and inhibitor tolerance of 45 oleaginous yeast species. *Journal of Industrial Microbiology & Biotechnology, 41*(7), 1061–1070.
52. Liu, J., Pei, G., Diao, J., Chen, Z., Liu, L., Chen, L., & Zhang, W. (2017). Screening and transcriptomic analysis of Crypthecodinium cohnii mutants with high growth and lipid content using the acetyl-CoA carboxylase inhibitor sethoxydim. *Applied Microbiology and Biotechnology, 101*(15), 6179–6191.
53. Cao, S., Zhou, X., Jin, W., Wang, F., Tu, R., Han, S., Chen, H., Chen, C., Xie, G.-J., & Ma, F. (2017). Improving of lipid productivity of the oleaginous microalgae Chlorella pyrenoidosa via atmospheric and room temperature plasma (ARTP). *Bioresource Technology, 244*, 1400–1406.
54. Sun, X., Li, P., Liu, X., Wang, X., Liu, Y., Turaib, A., & Cheng, Z. (2020). Strategies for enhanced lipid production of Desmodesmus sp. mutated by atmospheric and room temperature plasma with a new efficient screening method. *Journal of Cleaner Production, 250*, 119509.
55. Tai, M., & Stephanopoulos, G. (2013). Engineering the push and pull of lipid biosynthesis in oleaginous yeast Yarrowia lipolytica for biofuel production. *Metabolic Engineering, 15*, 1–9.

56. Kamisaka, Y., Kimura, K., Uemura, H., & Yamaoka, M. (2013). Overexpression of the active diacylglycerol acyltransferase variant transforms Saccharomyces cerevisiae into an oleaginous yeast. *Applied Microbiology and Biotechnology, 97*(16), 7345–7355.
57. Ledesma-Amaro, R., Santos, M. A., Jiménez, A., & Revuelta, J. L. (2014). Strain design of Ashbya gossypii for single-cell oil production. *Applied and Environmental Microbiology, 80*(4), 1237–1244.
58. Kurosawa, K., Wewetzer, S. J., & Sinskey, A. J. (2013). Engineering xylose metabolism in triacylglycerol-producing Rhodococcus opacus for lignocellulosic fuel production. *Biotechnology for Biofuels, 6*(1), 134.
59. Kurosawa, K., Laser, J., & Sinskey, A. J. (2015). Tolerance and adaptive evolution of triacylglycerol-producing Rhodococcus opacus to lignocellulose-derived inhibitors. *Biotechnology for Biofuels, 8*, 76–85.
60. Zhang, Y. W. L. W. (2012). Advances in the research of microalgae bioenergy. *Marine Sciences, 36*, 132–138.
61. Majidian, P., Tabatabaei, M., Zeinolabedini, M., Naghshbandi, M. P., & Chisti, Y. (2018). Metabolic engineering of microorganisms for biofuel production. *Renewable and Sustainable Energy Reviews, 82*, 3863–3885.
62. Li, Y., Han, D., Hu, G., Dauvillee, D., Sommerfeld, M., Ball, S., & Hu, Q. (2010). Chlamydomonas starchless mutant defective in ADP-glucose pyrophosphorylase hyper-accumulates triacylglycerol. *Metabolic Engineering, 12*(4), 387–391.
63. Sahay, S., & Braganza, V. J. (2016). Microalgae based biodiesel production-current and future scenario. *Journal of Experimental Science, 7*, 31–35.
64. Beopoulos, A., Mrozova, Z., Thevenieau, F., Le Dall, M.-T., Hapala, I., Papanikolaou, S., Chardot, T., & Nicaud, J.-M. (2008). Control of lipid accumulation in the yeast Yarrowia lipolytica. *Applied and Environmental Microbiology, 74*(24), 7779–7789.
65. Kalscheuer, R., Stölting, T., & Steinbüchel, A. (2006). Microdiesel: Escherichia coli engineered for fuel production. *Microbiology, 152*(9), 2529–2536.

Chapter 11
Fractionation, Characterization, and Valorization of Lignin Derived from Engineered Plants

Enshi Liu, Wenqi Li, Seth DeBolt, Sue E. Nokes, and Jian Shi

11.1 Introduction

Escalating concerns over global climate change and the associated environmental, societal, and economical challenges have stimulated research and development of alternative fuels, chemicals, and materials from sustainable resources [1, 2]. The use of first-generation biofuels based on starchy feedstocks is complicated by competition with food supply and for arable land. As an alternative, lignocellulosic biomass feedstocks, such as energy crops, forestry and agricultural wastes, and municipal solid wastes, have become promising feedstocks [3, 4]. However, cost-effective conversion of biomass feedstocks to value-added fuels and chemicals has not been fully realized, mainly due to the costs associated with unit operations (needed to overcome biomass recalcitrance), such as feedstock preprocessing, pretreatment, and enzymatic hydrolysis. Another challenge is the lack of infrastructure for biomass harvesting, transportation, and storage [5]. Therefore, improving the efficiency and sustainability of conversion technologies and the development of feedstocks with desirable processing traits have become the foci of the bioenergy/biochemical research community over the past few years.

A range of thermochemical and biochemical technologies have been developed for utilizing the structural sugars (i.e., cellulose and hemicellulose) from lignocellulosic biomass. In a typical conversion system, three major processes are involved in the polysaccharides conversion, that is, pretreatment/fractionation of lignocellulosic biomass, enzymatic hydrolysis, and conversion of hexose/pentose to biofuels

E. Liu · W. Li · S. E. Nokes · J. Shi (✉)
Department of Biosystems and Agricultural Engineering, University of Kentucky, Lexington, KY, USA
e-mail: j.shi@uky.edu

S. DeBolt
Department of Horticulture, University of Kentucky, Lexington, KY, USA

Z.-H. Liu, A. Ragauskas (eds.), *Emerging Technologies for Biorefineries, Biofuels, and Value-Added Commodities*,
https://doi.org/10.1007/978-3-030-65584-6_11

[6–8] (Fig. 11.1). Extensive efforts have been devoted to reducing the processing cost and improving the efficiency of conversion, such as consolidated bioprocessing of lignocellulosic biomass, reduction the costs of fungal enzymes, and developing co-product strategies from the C5 sugar stream [9, 10]. Throughout these efforts, however, lignin (as one of the major components in lignocellulosic biomass) has been considered as a waste product. However, the depolymerization and upgrading of lignin have received increased attention of late. Scientists are realizing that lignin valorization is essential to the success of the lignocellulosic biorefinery as suggested by several techno-economic analyses [11–15].

Lignin, the most abundant renewable aromatic polymer in nature, provides structural strength and protection against microbial/enzymatic degradation of plants [16]. Pretreatment is designed and widely used to overcome the lignocellulosic biomass recalcitrance and fractionate/depolymerize lignin into different streams. Moreover, characterization of the different lignin streams after pretreatment is critical to meet the requirements of different lignin valorization pathways. Increasing attention has been dedicated to elucidating the changes and variations on lignin structure and composition during its fractionation process [17–19]. Genetic modification has been used to improve crop yield and productivity and to enhance drought and salt resistance [20, 21]. Furthermore, genetically modified feedstocks can be more digestible and convertible compared to wild-type plant species [22–24]. In particular, modification of the lignin biosynthesis pathways to obtain lignocellulose with less recalcitrance has been a major research goal [25].

However, it is still not well understood on the effects of genetic modification on fractionation and characterization of lignin from genetically engineered lignocellulosic biomass feedstocks. There is a gap in linking the chemistry, biology, and

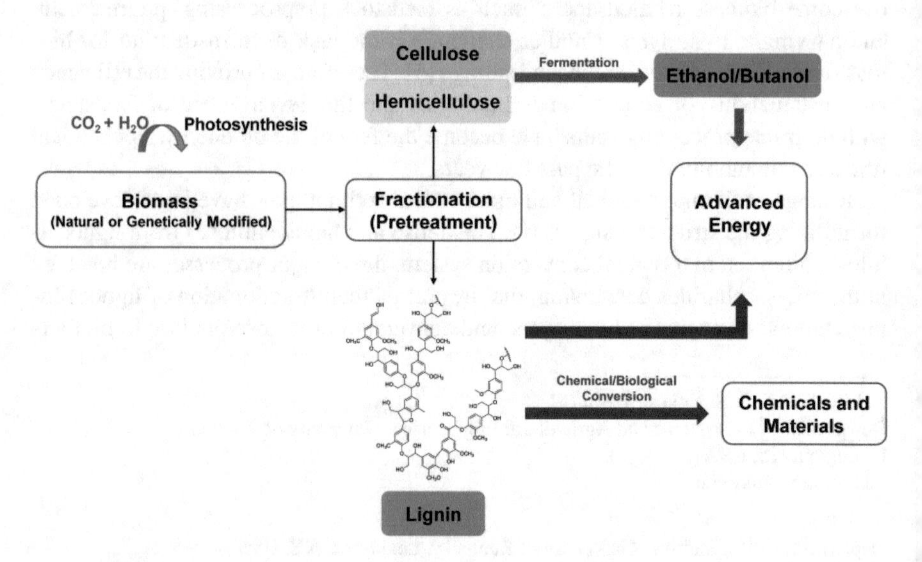

Fig. 11.1 Schematic illustration of the production of value-added fuels, chemicals, and materials from lignocellulosic biomass

conversion technology of lignin-engineered plants in the context of a biorefinery, but this knowledge is critical for rational design of lignin-engineered plants and downstream processing technologies. In the present literature review, we first summarize the recent research progress in lignin modification approaches and the current technologies for lignin fractionation and characterization. Particularly, we focus on the characteristics of lignin-modified streams from engineered biomass as a function of the biology and chemistry of the feedstocks and the conversion processes. Possible lignin valorization pathways including oxidative and reductive catalysis, electrocatalysis, and biological upgrading are also discussed. In addition, challenges and the outlook for future development are briefly reviewed. This review will provide information necessary for understanding the effects of genetic modification on lignin depolymerization and characterization using current leading lignin fractionation technologies.

11.2 Lignin Chemistry, Structure, and Biosynthesis

11.2.1 Lignin Compositions

Native lignocellulosic biomass generally contains three main structural components: cellulose (30–50%), hemicellulose (15–30%), and lignin (15–35%) [26]. Besides the three main components, lignocellulosic biomass has other extractable constituents, such as wax, protein, fat, phenolics, and minerals [27]. Major compositional components in various lignocellulosic biomass feedstocks are shown in Table 11.1, with large variations seen in different groups.

The composition and structure of lignin have been extensively investigated by wet-chemistry, microscopic, and spectroscopic methods [36–38]. Generally, lignin is considered to be a cross-linked amorphous polymer derived from the polymerization of *p*-coumaryl (H), coniferyl (G), and sinapyl (S) alcohols [39]. The chemical

Table 11.1 Structural polysaccharides and lignin in various biomass feedstocks

Biomass feedstocks	Cellulose (%)	Hemicellulose (%)	Lignin (%)	References
Poplar	42–48	16–22	21–27	[12]
Pine	30.9	19.3	29.0	[28]
Eucalyptus	24–28	39–46	19–32	[12]
Corn stover	28.7–44.4	13.4–25.0	14.3–26.0	[29]
Switchgrass	32.1–33.6	26.1–27.0	17.4–18.4	[30]
Miscanthus	43.1–52.2	27.4–34.0	9.2–12.6	[31]
Industrial hemp	43.6	15.3	21.5	[32]
Sweet sorghum	22.8	32.5	22.2	[33]
Wheat straw	33–40	15–20	20–25	[34]
Sugarcane bagasse	50	25	25	[35]

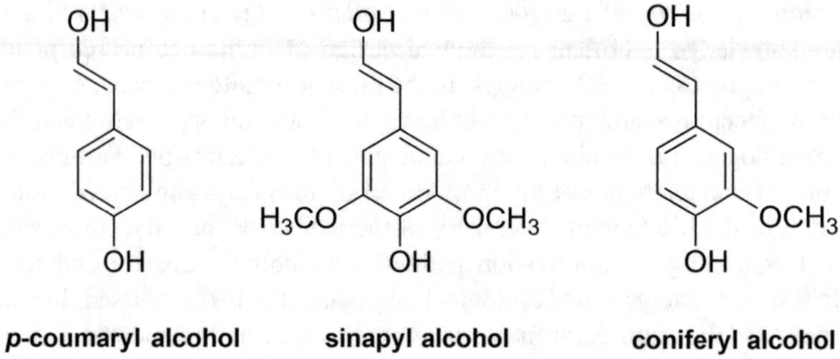

p-coumaryl alcohol sinapyl alcohol coniferyl alcohol

Fig. 11.2 Three main phenylpropanoid units in lignin structure

structures of the three main phenylpropane units are shown in Fig. 11.2. Lignin structure is generally described by the aromatic ring and the aliphatic chain [40].

Lignin contents and compositions vary among plant species, cell types, and growth stages of plants. In general, lignin in hardwood primarily consists of G and S units, while lignin in softwood is mostly composed of G units. Similar levels of G and S units plus a small amount of H units are discovered in grasses [13, 40]. Determination of lignin compositions is essential to illustrate the lignin structure and to seek appropriate routes for lignin depolymerization and valorization.

11.2.2 Lignin Inter-Unit Linkages

The common inter-unit linkages and functional groups present in lignin are ether and C–C bonds [41]. The main inter-unit linkages in lignin structure are shown in Fig. 11.3. In native lignin, two-thirds or more of the total linkages are ether bonds, and the other linkages are C–C bonds. Inter-unit linkages are named by numbering the nine C-atoms in the phenylpropanoid units, of which the aromatic carbons are marked from 1 to 6 and the aliphatic carbons are marked as α, β, and γ [42]. For instance, β-O-4 is the linkage between β carbon inside the aliphatic side chain and the oxygen atom attached to the C4 position of the aromatic moiety. The major linkages in the structural units of lignin are β-O-4 (β-aryl ether), β–β (resinol), and β-5 (phenylcoumaran) [43]. The most abundant type of inter-unit linkage is β-O-4, which accounts for more than 50% of all the linkages of lignin in grass, softwood, and hardwood. Other linkages include α-O-4 (α-aryl ether), 4-O-5 (diaryl ether), 5–5, α-O-γ (aliphatic ether), β-1 (spirodienone) [23, 44]. Lignin with relatively high numbers of C–C linkages compared to ether linkages is often referred to as condensed lignin, which is more rigid and less susceptible to degradation. The dissociation energy required to break C–C linkages, such as β-5 (125–127 kcal/mol), is higher than that of ether linkages, such as β-O-4 (54–72 kcal/mol) [41, 45].

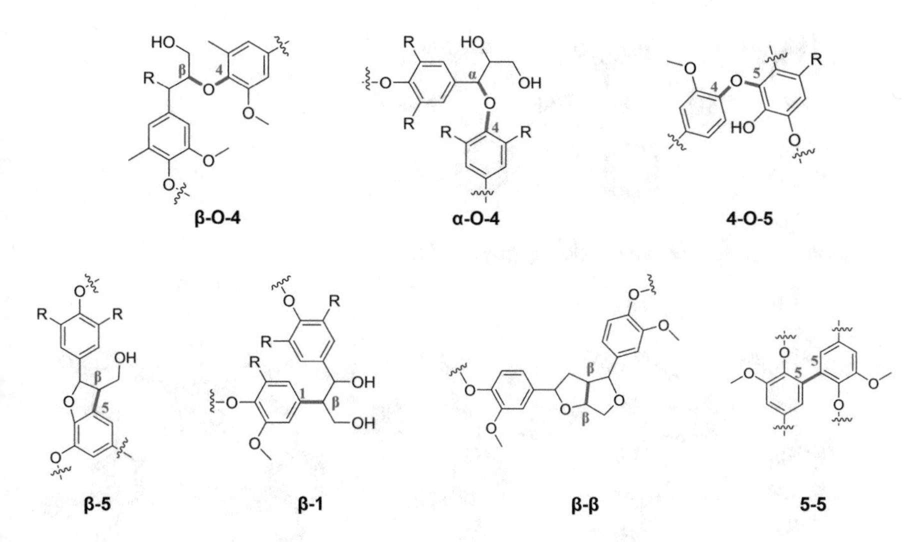

Fig. 11.3 The main inter-unit linkages in lignin structure (as indicated by the red lines and texts)

11.2.3 Lignin Biosynthesis

Lignin biosynthesis is a polymerization process where the three building blocks (i.e., *p*-coumaryl alcohol, sinapyl alcohol, and coniferyl alcohol) are linked to form *p*-hydroxyphenyl (H), guaiacyl (G), and syringyl (S) lignin, or their mixtures. As shown in Fig. 11.4, the biosynthetic pathway starts with the formation of cinnamic acid via the deamination of amino acid phenylalanine and is followed by hydroxylation reactions of aromatic rings, *O*-methylations, and different side-chain amendments [46]. Moreover, enzymes involved in lignin biosynthesis include phenylalanine ammonia-lyase (PAL), cinnamate 4-hydroxylase (C4H), 4-coumarate: CoA ligase (4CL), shikimate *p*-hydroxycinnamoyl transferase (HCT), *p*-coumarate 3-hydroxylase (C3H), caffeoyl-CoA 3-*O*-methyltransferase (CCoAOMT), cinnamoyl-CoA reductase (CCR), ferulate 5-hydroxylase (F5H), caffeic acid/5-hydroxyferulic acid *O*-methyltransferase [47], and cinnamyl alcohol dehydrogenase (CAD) [39, 46].

According to the lignin biosynthetic pathway, monolignols are derived from phenylalanine (Phe, synthesized through shikimate biosynthetic pathway), via the general phenylpropanoid and monolignol-specific pathways in plants [5]. Monolignols are generated by the conversion of Phe to cinnamic acid. Para-coumaroyl-CoA is derived from cinnamic acid by the enzymes C4H and 4CL, while *p*-coumaraldehyde and *p*-coumaroyl shikimic acid can be synthesized by CCR and HCT using *p*-coumaroyl-CoA. Subsequently, *p*-coumaroyl shikimic acid can be converted to coniferaldehyde under the function of enzymes C3H, HCT, CCoAOMT, and CCR. Furthermore, sinapaldehyde is derived from coniferaldehyde by the action of F5H and COMT. With the catalysis of CAD, *p*-coumaraldehyde, sinapaldehyde, and coniferaldehyde can be converted to *p*-coumaryl alcohol, sinapyl alcohol, and

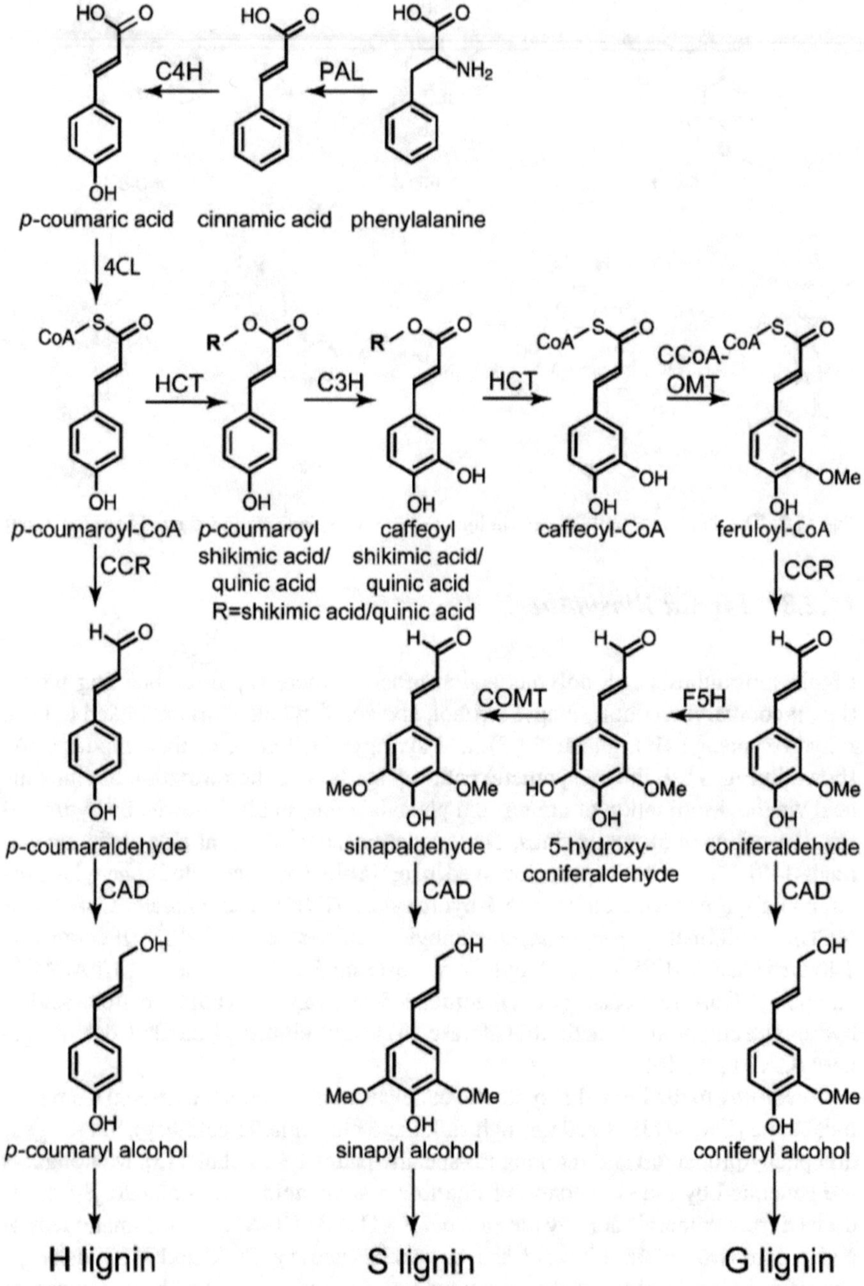

Fig. 11.4 The main biosynthetic pathway for monolignols [46]

coniferyl alcohol, respectively. After monolignol synthesis, these monomers are transported and later oxidized by laccase or peroxidases and polymerized into the cell wall structure [48]. Additionally, upstream transcription factors like *NAC*, *MYB*, and *LIM* gene families are believed to be responsible for transcriptionally regulating the genes that encode key elements of pathways for enzymatic expressions [25]. It is worth noting that investigation of lignin biosynthesis is an ongoing effort; new knowledge is being added continuously to this area. This review is not intended to be exhaustive on the topic of lignin biosynthesis. The reader is referred to the following reviews that discuss lignin biosynthesis in greater detail [12, 39, 46].

By understanding the biosynthetic pathway of lignin, it is possible to take advantage of genetic modification technologies to modify lignin content, compositions of subunits, and the inter-unit linkages by downregulation, overexpression, or suppression of certain genes regulating lignin biosynthesis. Artificial design of plants is a promising approach to develop lignocellulosic biomass with favorable characteristics for improvement of overall efficiency in a biorefinery.

11.3 Lignin-Engineered Feedstocks

Genetic modification is considered as an advanced technology to improve the crop yield and productivity and to enhance the resistance to drought, salt, and pest. Without a doubt, this technology has also been applied to various energy crops, such as switchgrass and sorghum, to enhance the convertibility of biomass feedstocks to biofuels and bioproducts [49, 50]. The manipulation on lignin biosynthesis or other relevant pathways is of particular interest. Several molecular biology methods have been developed (Table 11.2), including (a) reduction/escalation of lignin content, (b) modification of lignin composition, (c) alternation of lignin deposition in plant cell wall, and (d) alternation of inter-unit lignin linkages or linkages within LCC (lignin–carbohydrates complex) [51–53]. Some of the efforts on lignin engineering are summarized below.

11.3.1 Modification of Lignin Content

Possibly the simplest effort to date has been downregulation of the genes that are regulating the key enzymes in lignin biosynthesis. Results have demonstrated the capacity to alter the lignin structure or reduce the lignin content in biomass. Many different approaches have been developed in the past to lower the lignin content of plants. *4CL* is one of the crucial enzymes in lignin biosynthetic pathways, specially involved in the initial stage of monolignol biosynthesis. Lignin content was reduced through inhibiting one of the *4CL* gene *Pv4CL1* by RNA interference in switchgrass, and the production of fermentable sugar was significantly increased with this mutant [54]. Baucher et al. reported that 25% of lignin was reduced in tobacco

Table 11.2 Genetic modification of lignocellulosic biomass feedstocks and its effect on biomass digestibility and convertibility

Biomass feedstocks	Gene/enzymes regulation	Lignin content	Lignin composition	Digestibility	References
Alfalfa	Downregulation of HCT	Reduced	Unaffected	Increased	[23]
Switchgrass	Downregulation of 4CL	Reduced	Increased S/G	Increased	[54]
Arabidopsis	Disruption of MED5a and MED5b	Reduced	Almost entirely H	N/A	[66]
Tobacco	Downregulation of C4H	Reduced	Increased S/G	Increased	[67]
Switchgrass	Overexpression of *AtLOV1*	Increased	Increased S/G	N/A	[68]
Hybrid poplar	Downregulation of C3H	Reduced	Increased S/G	Increased	[69]
Poplar	Downregulation of 4CL	Reduced	N/A	Increased	[25]
Hybrid poplar	Downregulation of 4CL	Reduced	Increased H	N/A	[70]
Alfalfa	Downregulation of C3H or HCT	Reduced	Increased S/G	N/A	[71]
Tobacco	Downregulation of PAL	Reduced	Increased S/G	Increased	[72]

N/A Not applicable

through reduction of *4CL* by more than 90% [55]. Chen and Dixon successfully reduced lignin content in Alfalfa by downregulating *HCT* gene [23]. It was also reported that downregulation of *CCoAOMT* can cause the reduction of lignin content in alfalfa, while suppression of the ferulate 5-hydroxulase (*F5H*) can reduce the accumulation of syringyl lignin in alfalfa fibers [56]. Furthermore, downregulating *C4H* in *Populus* led to a 30% of reduction in Klason lignin content [57].

In addition, Chabannes et al. compared the downregulations of CCR and CAD, and demonstrated that both transgenic lines of tobacco exhibited a reduction of lignin content from 22% to 10.7% [58]. In another study, Kawaoka et al. suppressed the gene expression of *Ecliml* and resulted in a reduction of lignin content with 29% in Eucalyptus [59]. Researchers transferred an antisense *prxA3a* gene into a wild-type aspen to suppress the expression of *prxA3a.* Results demonstrated that aspen lignin content was reduced by ~20% because of downregulating the gene of anionic peroxidase [60]. Moreover, downregulation of C3H can cause the reduction of lignin content in alfalfa; downregulation of CCR in transgenic tobacco led to a dramatic decrease in lignin content [53, 61]. Mutations of the brown midrib *Bmr6* gene in sorghum and *Bmr1* gene in maize have been tested to decrease the lignin content, increase digestibility in forage, and enhance enzymatic saccharification efficiencies [62, 63]. By testing the convertibility of these engineered feedstock, it was generally believed that a low lignin content can help to decrease pretreatment severity, improve sugar yield, and reduce enzyme usage during enzymatic hydrolysis [52, 64].

11.3.2 Modification of Lignin Structural Subunits

Decomposition of lignin has been shown to be more effective if the subunits S or H are dominant. Biomass with a higher percentage of G lignin subunits, for example, softwood and hardwood, is more resistant to biochemical decomposition [65]. Indeed, modification of lignin structural unit ratios, such as S/G/H ratio, is one of the most common strategies of lignin engineering (Table 11.2).

Coleman et al. used RNAi to generate transgenic hybrid poplar by suppressing the *C3H* expression, and the poplar lignin content was significantly reduced. The suppression also caused an alternation of lignin monomer composition (increase of H units and decrease of G units) [69]. Sewalt et al. altered the expression of *C4H* and PAL in tobacco lines and concluded that the content of Klason lignin was reduced by downregulation of *C4H* and the syringyl to guaiacyl ratio was increased from 1.1 to 1.9 by suppressing PAL [67]. In another study, an antisense gene which can suppress the activity of CAD was introduced into the tobacco plant. The results showed that the genetic modification increased the cinnamaldehyde units in lignin, though no significant changes of overall percent lignin content were observed [73]. Reddy et al. determined the correlation between lignin content and digestibility in alfalfa and discovered that P450 enzymes, *C4H*, *C3H* and F5H, were at least partially responsible for the variations in lignin content and composition [74]. Downregulation of F5H in *Medicago sativa* (alfalfa) can reduce the accumulation of syringyl lignin in plant cells and generate more easily digestible plant materials [56]. Researchers overexpressed the *AtLOV1* gene in order to increase lignin content and monolignol composition in switchgrass cell wall, and demonstrated that lignin content increased from 20.6 to 21.8 and S/G ratio increased from 0.64 to 0.86 compared to wild-type switchgrass [68]. Bonawitz et al. disrupted the expression MED5a and MED5b and produced a mutant *Arabidopsis* that contains almost entirely H lignin subunits [66]. Results from a comparative study which examined the enzymatic digestibility of wild-type and *Arabidopsis* mutants with dominant S, G, or H subunits subjected to ionic liquid pretreatment confirmed that the *Arabidopsis* mutants with mainly S subunits (92%) were more digestible than wild-type and G-lignin (68%) dominant mutant [18].

11.3.3 Modification of Lignin Inter-Unit Linkages and Lignin Deposition

Modification of genes related to lignin biosynthesis can also cause changes in lignin deposition patterns in secondary plant cell wall, which offers another potential strategy to improve the convertibility of biomass feedstocks [75, 76]. The suppression of *C3H* in poplar reduced the lignin content and at the same time changed the distribution of lignin in secondary cell wall [69]. Rogers and Campbell surmised that the regulation of the PAL gene family provides a transcriptional regulation which may impact the timing and localization of lignification [77]. In addition, synthetic biology approaches were developed to understand the secondary cell network in *Arabidopsis thaliana*, leading to a decrease of lignin content and an increase in structural sugar depositions in fiber cells [78].

As discussed in Sect. 11.2.3, the dissociation energies needed to cleave the ether bonds are lower when compared to the C–C inter-unit lignin linkages. Engineered biomass with more ester linkages in lignin polymer is preferred. Thus, modification of inter-unit lignin linkages and lignin-hemicellulose linkages offers two potential mechanisms to enhance the convertibility of biomass. Studies on the *C3H* coding gene *ref8* showed that lack of *C3H* activity led to large differences in phenylpropanoid metabolism [79].

Modification of biomass can also decrease the ferulate and *p*-coumarate cross-linking between xylan and lignin [80]. Researchers utilized a biomimetic model to understand the reasons why reductions of ferulate lignin cross-linking can influence cell wall fermentation by rumen microbes and improve fiber fermentability [81]. In another study, higher *p*-coumarate levels in corn cell wells were proven to be correlated with higher lignin content, which indicated the *p*-coumarate-assisted lignin formation [82]. More ester linkages in poplar trees were achieved by augmenting with monolignol ferulate conjugates [83, 84]. Ralph and coworkers summarized the recent studies on monolignol-hydroxycinnamate conjugates, and demonstrated that understanding the involvement of ferulate in radical cross-coupling reactions during lignification provided potential ways to improve efficiency of biomass conversion [85].

11.3.4 CRISPR/Cas9

In addition to current genome editing tools, such as zinc-finger nucleases (ZFNs) and transcription activator-like effector nucleases (TALENs), the CRISPR/Cas9 (clustered regularly interspaced short palindromic repeats/CRISPR-associated protein 9) genome editing toolbox for aiding fundamental studies of plant gene functions and genetic improvement of crop traits has received increasing attentions [86, 87]. CRISPR/Cas9 toolbox is designed to allow targeted cleavage of genome DNA using small noncoding RNA, which stimulates the cellular DNA repair mechanisms in an efficient and precise way [88]. By disrupting the functions of specific genes (insertion or deletion), the CRISPR/Cas9 system is capable of creating variations in the plant genome and potentially generating desirable traits [89]. The development of CRISPR/Cas9 technology provided advanced methods for potential manipulation of plants, including lignocellulosic biomass. Zhou et al. reported the utilization of CRISPR/Cas9 system in woody biomass (poplar) where they altered monolignol composition by silencing *4CL1* and *4CL2* genes, and achieved reduced lignin content. Results also demonstrated that the primary role of *4CL* genes in lignin biosynthesis and revealed functional redundancy of *4CL* genes in supporting lignin and flavonoid biosynthetic pathways [90]. Park et al. established a CRISPR/Cas9 system in switchgrass to target the *4CL* gene related to lignin biosynthesis, and generated genetically engineered plants with thinner cell wall, reduced total lignin content, and increased glucose and xylose release upon saccharification [91]. Although the efficiency and range of accessible targets still require improvement, CRISPR/Cas9 holds great promise as a highly specific approach for modifying plant lignin.

11.4 Fractionation of Engineered Feedstocks

The paper and pulping industry use acidification, membrane filtration, and solvent extraction methods to isolate and remove lignin. Lignin derived from Kraft pulping is called Kraft lignin. A scheme of a typical Kraft pulping is illustrated in Fig. 11.5. In this process, softwood or hardwood is processed into paper through cooking, washing, bleaching, and drying processes. The lignin in black liquor can be precipitated by acid or CO_2 and recovered as Kraft lignin.

However, for a biorefinery focused on making biofuels, chemicals, and other bioproducts from lignocellulosic biomass, it is necessary to maximize the utilization of carbohydrates and fractionated lignin [93]. The objective of biomass pretreatment is to make the cellulose and hemicellulose more accessible to hydrolytic enzymes without destroying the sugars and without forming inhibitors to fermentation. It is the first and one of the most crucial steps for the conversion of biomass to C5 and C6 sugars [94]. Selection of pretreatment method has significant impacts on the downstream processes, such as enzymatic hydrolysis and fermentation. Here, we group pretreatment/fractionation technologies based on their chemistry and discuss in detail the pretreatment chemistry, process flow and mass flow of cellulose, hemicellulose, and lignin during each pretreatment.

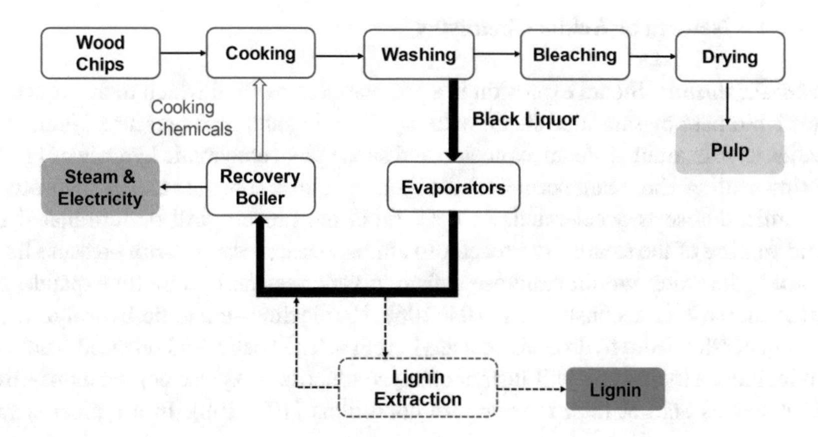

Fig. 11.5 Typical Kraft pulping process [92]

11.4.1 Physical Methods

Physical pretreatment rests on the principles to reduce biomass particle size through mechanical processes, which is typically accomplished by milling, chipping, or grinding, so that enzyme accessibility is improved by escalating the surface area to volume ratio of biomass and reducing cellulose crystallinity and/or degree of polymerization. Physical pretreatment is usually followed by other pretreatment processes because particle size reduction cannot significantly enhance sugar conversion rate by itself [95–99]. However, most subsequent pretreatments are enhanced by reducing the biomass particle size.

11.4.2 Chemical Methods

Compared to thermochemical conversion [for recent reviews see [100–102]], biochemical conversion requires less energy input and is therefore considered as a promising route for biomass utilization. Chemical pretreatments include acid pretreatment, alkaline pretreatment, steam pretreatment, and organic solvents pretreatment. Fractionation of lignin and polysaccharides is highly impacted by pretreatment chemistry. For instance, acid pretreatment can efficiently solubilize hemicellulose and leave a highly digestible solid stream containing the majority of the cellulose and lignin. As a result, lignin is mainly recovered from the solid residue after enzymatic hydrolysis/fermentation. In contrast, alkaline pretreatment solubilizes a large portion of lignin in the biomass; thus, the lignin can be recovered from liquid stream. Solvent-based pretreatment methods possess the ability to solubilize cellulose, hemicellulose, or lignin individually or simultaneously; thus, the order of recovery of the sugar and lignin streams is process dependent.

11.4.2.1 Neutral or Acidic Chemistry

Steam Explosion Steam explosion is a pretreatment method which treats lignocellulosic biomass by saturated steam with rapid heating and high pressure. Numerous studies have examined steam explosion and subsequent enzymatic hydrolysis [103]. During heating, the steam permeates the lignocellulosic substrate and the hydrolysis of hemicellulose is accelerated. At a set time, the process will be terminated by rapid venting of the steam from reactor to atmospheric pressure. With pressure flash reduced, the water within cellulose substrate vaporizes and promptly expands, by which biomass is deconstructed [104–106]. Hemicellulose can be hydrolyzed by acids generated from hydrolysis of acetyl groups. The acetic acid released from the hemicellulose hydrolysis will further catalyze glucose or xylose degradation. Also, water acts as acid at high temperature conditions [107, 108]. In a typical steam explosion process, most of the hemicellulose fraction is solubilized and removed, while majority of the lignin is preserved in the solid stream.

Hot Water Hot water pretreatment employs high temperature water, 180–250 °C, at an elevated pressure to maintain the hot water in liquid phase. Different reactors, such as co/counter-current and flow-through, have been developed to improve the mass and heat transfer between the biomass and the hot water. Also, various minerals and organic acids, acting as pretreatment catalysts, such as sulfuric, acetic, citric, maleic, and oxalic acid, have been examined [109–111]. Liquid hot water pretreatment aims to solubilize and remove hemicellulose from solids containing cellulose and lignin. Approximately 40–60% of total biomass is solubilized in the hot water solvolysis, of which 4–22% is cellulose, 35–60% is lignin and all the hemicellulose [112].

Dilute Acid Dilute acid accelerates hemicellulose hydrolysis during biomass pretreatment. The low pH facilitates hemicellulose solubilization, along with lignin relocating/precipitation to the cellulose fiber, and in turn increasing the accessible surface for enzymes to attach. As shown in Fig. 11.6a, in a typical dilute acid pretreatment process, hemicellulose is dissolved in the liquid stream, while cellulose and most of lignin fraction are fractionated into solid stream. Dilute acid pretreatment has been developed for overcoming the recalcitrance of various biomass feedstocks. Dien et al. conducted dilute acid pretreatment on switchgrass, alfalfa stems, and reed canary grass harvested at different growth stages and evaluated the sugar yields after enzymatic hydrolysis. Among the two sulfuric acid pretreatment conditions (121 °C for 1 h and 150 °C for 20 min) evaluated, 655 g/kg dry material was the highest glucose yield achieved from switchgrass [113]. Researchers evaluated the effects of dilute acid pretreatment on the supramolecular and ultrastructure of switchgrass at 160–180 °C under different sulfuric acid concentrations, and found that the crystallinity index increased whereas molecular weight decreased with pretreatment severity [114]. In another study, researchers compared transgenic and wild-type switchgrass using dilute acid pretreatment, and concluded that the proportion of acid soluble lignin was higher in transgenic switchgrass, which led to a considerable increase in sugar solubilization [115]. When compared to liquid hot water pretreatment, dilute acid pretreatment can be performed at relatively mild conditions and is effective on various biomass feedstocks including wood. However, a washing step is required to remove the excessive acids and fermentation inhibitors, such as furfural and hydroxymethyl furfural (HMF). Lignin recovered from dilute acid pretreatment may subject to recondensation, especially at severe conditions [116].

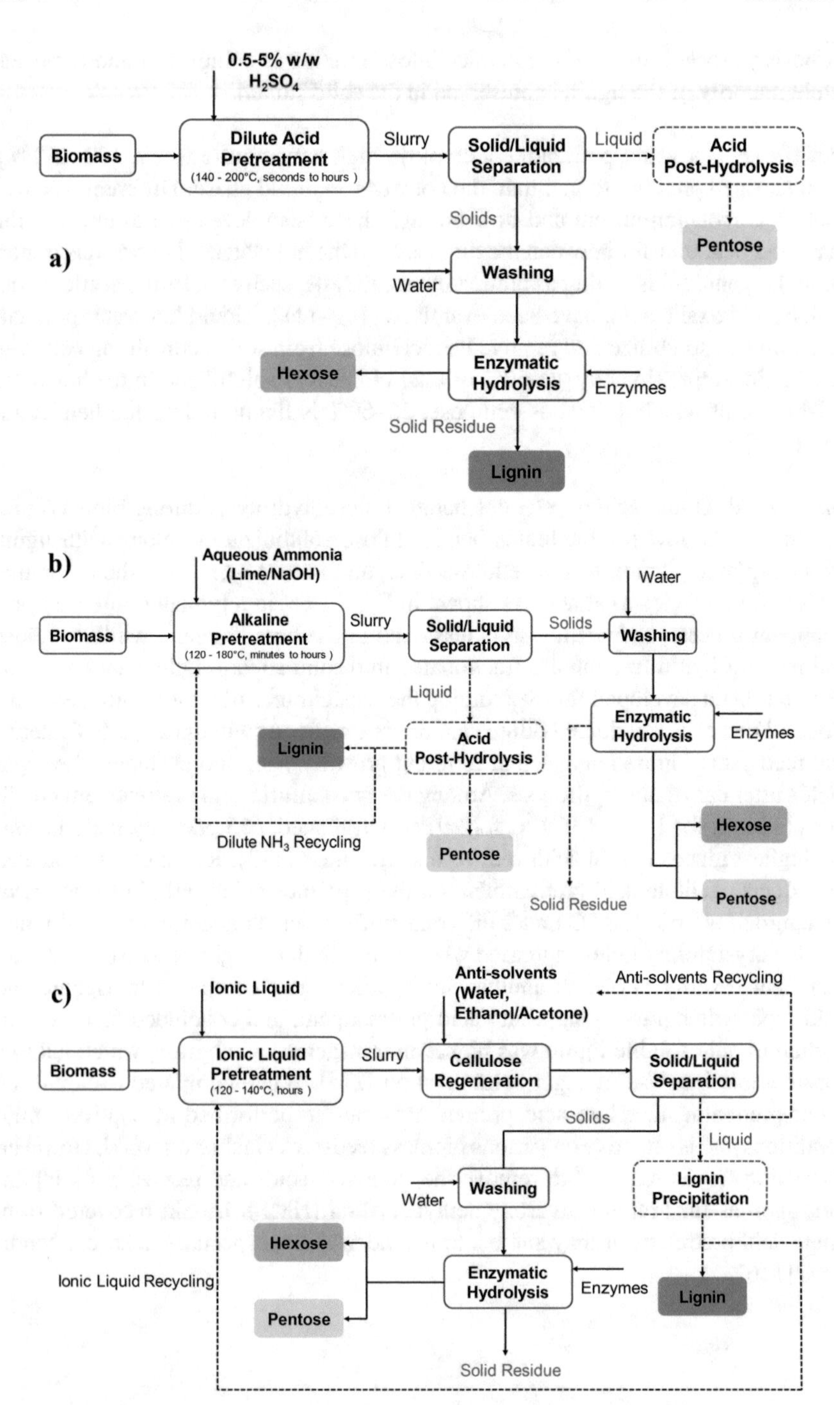

Fig. 11.6 Pretreatment process flowchart illustrating paths for structural sugars and lignin: (**a**) dilute acid pretreatment, (**b**) ammonia hydroxide pretreatment, and (**c**) ionic liquid pretreatment

11.4.2.2 Alkaline Chemistry

Alkaline (also known as high pH) pretreatment operates at high pH by adding alkalis such as lime (calcium hydroxide), sodium hydroxide, and aqueous ammonia. This type of pretreatment removes a large fraction of lignin and acetyl groups because lignin hydrophilicity/solubility is increased in alkaline solutions. In a typical alkaline or high pH (ammonia hydroxide) pretreatment process is shown in Fig. 11.6b. A high pH pretreatment breaks down the ester and glycosidic side chains and promotes the solubilization of the lignin fraction and the hemicellulose fraction. The dissolved lignin can be precipitated out at high purity by adjusting the pH to the acidic range. In a recent study, a bench-scale continuous NaOH pretreatment system for switchgrass has been developed. A glucose yield of 90.8% was achieved at a flow of 15 g/min biomass and 120 ml/min NaOH with minimum concentration [117]. Xu et al. optimized the conditions of lime pretreatment of switchgrass, with the best conditions at 50 °C for 24 h at an alkaline loading of 0.1 g lime/g raw biomass [118]. Wang et al. modified the switchgrass reducing the lignin content up to 5.8%. They pretreated the modified switchgrass with 0.5%, 1%, and 2% NaOH for 15, 30, and 60 min at 121 °C, achieving a 16–18% higher sugar yield in the transgenic switchgrass compared to wild types [116]. Salvi et al. pretreated sorghum fibers using dilute ammonia hydroxide (sorghum: ammonia: water, 1: 0.14: 8) at 160 °C for 1 h, and removed 44% of the lignin and 35% of the hemicellulose [119]. Ko et al. optimized rice straw pretreatment conditions using aqueous ammonia, and found that 69 °C, 10 h, and an ammonia concentration of 21% (w/w) were the optimal reaction conditions to achieve high sugar yields upon saccharification [120]. Gao et al. investigated acetone, butanol, and ethanol fermentation from switchgrass pretreated with 1% NaOH, acquiring a reducing-sugar yield of 365 g/kg biomass and 146 g/kg ABE yield [121].

The molecular weight distribution and structural characteristics of lignin recovered from alkaline pretreatment were significantly different. With the presence of H_2O_2, Gupta and Lee reported that NaOH pretreated switchgrass lignin had a lower content of aromatics but a greater fraction of guaiacyl-type monolignols than ammonia pretreated switchgrass lignin [122]. Qin et al. optimized the conditions of aqueous ammonia pretreatment of corn stover and recommended a condition of 180 °C, 20% ammonia hydroxide, and 30 min to optimize lignin degradation to low-molecular-weight soluble fragments [123]. Despite the high chemical cost as compared with inexpensive mineral acids and issues with waste disposal, high pH pretreatment can be a promising method for lignocellulosic biomass if an inexpensive alkali is selected or the chemical can be recycled.

Wet Oxidation In wet oxidation pretreatment, biomass is exposed to water and oxygen/air above 120 °C for a set of time period [124–126]. In a typical wet oxidation process, the lignocellulosic biomass is fractionated through dissolving lignin and hemicellulose. The presence of oxygen promotes the degradation of hemicellulose and possibly a portion of the cellulose which can boost the release of structural organic acids [126]. Attempts combining wet oxidation pretreatment and alkaline pretreatment have also been reported. Although wet oxidation can efficiently fractionate biomass and generate high sugar yield, the byproducts obtained from oxidation would have negative effects on the enzymatic hydrolysis and downstream fermentation. However, the alkaline condition can significantly reduce the formation of furfural from xylan [127–129].

11.4.2.3 Solvent-Based Fractionation

Organosolv Organosolv pretreatment involves using organic solvent or aqueous solutions of organic solvents, such as alcohol, organic acids, and acetone, to dissolve fractions of hemicellulose and lignin [130–133]. The pretreated solids, after organosolv pretreatment, preserve the cellulose component in its original form and contain fractions of hemicellulose and lignin. Organic solvents selected for the organosolv pretreatment can be recycled by distillation. The lignin in the condensed black liquor can be precipitated by diluting with water. The organosolv lignin has improved solubility in solvents and thus is amendable to producing either value-added chemicals or materials [134]. Compared to conventional acid or alkaline pretreatment, organosolv pretreatment isolates a lignin fraction with a higher quality and purity. However, the organic solvents are expensive; even though they can be recovered, energy input can be intensive [135].

Ammonia Fiber Explosion (AFEX) AFEX is performed under high pressure and moderate temperatures ranging from 60 to 100 °C and catalyzed by concentrated ammonia. In a modified version a small amount of water/moisture is introduced to allow better extraction of hemicellulose sugars and lignin. Similar to steam explosion, lignocellulosic biomass is saturated with the ammonia mixture for a set of time under a high pressure before rapidly depressurizing to atmospheric pressure in AFEX pretreatment [136, 137]. The intense expansion of ammonia vapor can break down lignin–carbohydrate linkages and improve hemicellulose hydrolysis, cellulose decrystallization, and lignin depolymerization, all of which contribute to an enhanced enzyme accessibility [138–140]. Compared to steam explosion, the lower temperature requirement significantly reduces the energy input and overall cost of biomass deconstruction. However, it is reported that AFEX pretreatment only works well on hardwood as opposed to softwood [106, 141]. In a typical AFEX process, lignin and hemicellulose fractions are retained in the pretreated solid; however, they can be recovered after enzymatic hydrolysis/fermentation removes the cellulose.

Ionic Liquid Ionic liquids (ILs) are salts which are typically in the liquid state when temperature is below 100 °C. Owing to their strong solvent power, ionic liquids can solubilize all the lignocellulosic biomass or selectively dissolve components like lignin and cellulose. The dissolved components can be re-precipitated with anti-solvent like water, methanol, and ethanol. Less-crystalline reconstituted cellulose could be obtained from this pretreatment. The typical types of ILs being used for biomass pretreatment include: EMIM (1-ethyl-3-methylimidazolium)-

based ILs (e.g., $[C_2C_1Im][OAc]$ and $[C_2C_1Im][Cl]$), BMIM (1-butyl-3-methylimidazolium)-based ILs (e.g., $[C_4C_1Im][Cl]$), and cholinium amino acids ILs (e.g., cholinium lysinate). The high hydrogen bonds basicity of IL breaks the molecular hydrogen bonds in cellulose, making the cellulose less crystalline and easier to hydrolyze. The details of a typical IL pretreatment process are illustrated in Fig. 11.6c. In a typical IL process, most of the lignin streams and part of hemicellulose fraction are fractionated into the liquid phase, and the cellulose fraction needs to be regenerated by adding water or solvents to the liquid phase.

In order to develop cost-effective IL-based pretreatment methods, recent studies have demonstrated the feasibility of using IL and water mixtures (aqueous ILs) to reduce the energy inputs and eliminate technical barriers, such as high viscosity and recycling of ILs and gel formation, during IL pretreatment [142, 143]. In addition, certain biocompatible ILs (e.g., cholinium lysinate) associated with the aqueous environment provide possible approaches to covert lignin into value-added fuels and chemicals in the liquid phase without further lignin extraction or separation [143].

11.4.3 Biological Methods

As opposed to physical and chemical pretreatments which require intense energy input, biological pretreatments employ natural microorganisms and take advantages of their enzyme systems to break down the inter-unit linkages in lignin and between lignin and structural polysaccharides. White-rot fungi, including *Pleurotus ostreatus*, *Phlebia subserialis*, *Ceriporiopsis subvermispora*, *Bjerkandera adusta*, *Cyathus bulleri*, and *Phanerochaete chrysosporium*, as well as anaerobic bacterium *Clostridium thermocellum* are believed to be the most promising microorganisms to conduct biological pretreatment [144–147]. Chinn and Nokes screened a group of thermophilic anaerobic bacteria (*Clostridium thermocellum*) for solid substrate cultivation using different biomass feedstocks (such as corn stover, sugar cane bagasse, wheat bran) [148]. Flythe et al. conducted research to study the effects of biomass particle size on switchgrass fermentation by *Clostridium thermocellum* [149]. Moreover, thermophilic anaerobe *Thermoanaerobacterium saccharolyticum* was investigated for production of *n*-butanol [150]. However, relatively longer residence times, up to days, significantly restricts biological pretreatment application in a large-scale biorefinery. In addition, the growth of microbes consumes carbohydrates, which jeopardizes the sugar yield [151].

Given the multiple pretreatment options, the pretreatment chemistry and mode of action greatly influence the pretreatment effectiveness such as lignin removal/recovery and enzymatic hydrolysis sugar yield. More importantly, different pretreatment methods lead to different characteristics of the extracted lignins. Table 11.3 recapitulates the effects of pretreatment methods on lignin fractionation and characterization. Generally, solvent-based or alkali pretreatment removes the highest fraction

Table 11.3 Effects of pretreatment methods on lignin fractionation and characterization

Pretreatments	Biomass	Lignin fractionation		Lignin characteristics			References
		Lignin removal[a]	Lignin recovered stream	Composition	Molecular weight	Inter-unit linkages	
Physical methods							
Ball milling	Corn stover/Oil Palm	Low	Solid	N/A[b]	N/A	N/A	[152, 153]
Chemical methods							
Neutral or acidic chemistry							
Steam explosion	Aspen	Low	Solid	N/A	Reduced	Breakage of ether bonds (β-O-4)	[154]
Carbon dioxide	Sugarcane bagasse	Low	Solid	N/A	N/A	N/A	[142]
Hot water	Mixed hardwoods	Low	Solid	Increased AIL/ASL ratio	N/A	Decrease in hydroxyl and methoxyl groups	[155]
Dilute acid	Switchgrass	Low	Solid	Decreased S/G ratio	Slightly decreased	Decrease of β-O-4 linkages	[156]
Alkaline chemistry							
Alkaline	Cotton stalk	Medium	Liquid	G–S units rich	Reduced	Mainly β-O-4 linkages	[157]
Wet oxidation	Grasses	Medium	Liquid	Reduced S/G ratio	N/A	Decreased β-O-4 content	[158, 159]
Solvent-based chemistry							
Organosolv	Switchgrass	High	Liquid	Increase of G–S units	Decreased	Decrease of β-O-4 linkages	[160]
Ammonia fiber explosion (AFEX)	Corn Stover	Low	Solid	N/A	N/A	N/A	[161]
Ionic liquid	Switchgrass	High	Liquid	Increased S/G ratio	Reduced	Breakage of β-O-4	[143]

Pretreatments	Biomass	Lignin fractionation		Lignin characteristics			References
		Lignin removal[a]	Lignin recovered stream	Composition	Molecular weight	Inter-unit linkages	
Biological methods							
Lignin-degrading microorganisms	Grasses/woods	Low–medium	Solid	N/A	N/A	N/A	[162, 163]

Note:
[a]Low: 0–40% lignin removal; medium: 40–60% lignin removal; high: 60–100% lignin removal
[b]*N/A* Not applicable

from the plant material, while pretreatment with neutral, acidic chemistry, or physical pretreatment, that is, ball milling, extracts lower fraction of lignin. As a result, the lignin streams can be easily recovered from either the pretreatment liquid or pretreated solids streams depending on pretreatment method. Different pretreatment chemistry alters the monolignol composition of the extract lignin by preferentially removing S or G lignins. Most of physical and chemical pretreatment methods lead to reduced lignin molecular weight and breakdown of β-*O*-4′ lignin inter-unit linkages. However, the impact of biological pretreatment on lignin characteristics is not well documented.

11.5 Lignin Characterization

11.5.1 Lignin Content and Compositions

Numerous quantification methods have been developed to measure the lignin content in different plants [164]. Near infrared spectroscopy is one method utilized to quantify lignin content in plants by comparing the intensity of several signature lignin bands representing the aromatic skeleton of lignin [165, 166]. The two most commonly used spectrometric methods involve the detection of thioglycolate and acetyl bromide, after solubilization of the lignin [167]. Spectrometric methods can then be used to determine the relative lignin abundance owing to the strong absorbance of lignin compounds in the ultraviolet region of spectrum [168]. The most commonly used method to determine lignin content is two-step acid hydrolysis, which was originally developed by Klason and improved by National Renewable Energy Laboratory (NREL) [169]. The concept that lignin is mainly generated by dehydration of three monolignols (i.e., *p*-coumaryl, coniferyl, and sinapyl alcohols) was first reviewed and proposed by Freudenberg [170]. The relative abundance of monolignols in lignin varies in different plant species.

Genetically engineered modification leads to a reduction in lignin content and an improvement of biomass digestibility, which can potentially promote the development of cost-effective pretreatment technologies in biofuels production from lignocellulosic biomass. A 40% lignin reduction and 14% cellulose escalation were achieved in genetically engineered aspen trees by silencing *4CL* gene [171]. Downregulation of cinnamyl alcohol dehydrogenase (CAD) in *Populus* spp. increased the lignin solubility in alkaline during pulping processes and facilitated the kraft delignification [172]. Genetic modification of lignin can potentially reduce the recalcitrance of biomass and further reduce the cost in biorefinery. The lignin which is broken down into low-molecular-weight fragments needs to be characterized by different technologies (e.g., GC/MS, NMR, GPC, and FTIR) before upgrading to value-added fuels and chemicals [38].

11.5.2 Lignin Molecular Weight Distribution

Lignin molecular weight distribution is a key factor that reflects the degree of lignin depolymerization and provides fundamental information for lignin upgrading. The molecular weight of lignin in native biomass varies by feedstocks. The lignin molecular weight distribution highly depends on the lignin extraction and isolation methods; thus, lignin extraction methods are required to preserve native lignin to the greatest extent. Commonly available isolated lignin, such as Milled wood lignin (MWL) [173], cellulolytic enzyme lignin (CEL) [174], and enzymatic mild acidolysis lignin (EMAL) [175, 176], is named by the mild extraction methods. Each isolation approach has merits and drawbacks in respect to the yield, integrity, and purity of the isolated lignin from biomass.

Numerous lignin molecular weight characterization techniques are available, including vapor pressure osmometry (VPO) [176], lignin ultra-filtration and nanofiltration [177], static/dynamic light scattering [178], mass spectrometry [179], and gel permeation chromatography (GPC) [180, 181], of which, the most commonly employed one is GPC or SEC (size exclusion chromatography). El Hage et al. studied and compared the molecular weight distribution between Miscanthus MWL and Miscanthus organosolv lignin, and concluded that the molecular weight of lignin on weight average (Mw) and number average (Mn) for organosolv lignin were significantly lower than that of MWL in Miscanthus [182]. Joffres et al. conducted hydroconversion of wheat straw in the presence of catalysts and characterized lignin by GPC to evaluate the conversion efficiency [183]. Salanti et al. characterized the alkaline extracted lignin from rice husk and optimized the reaction conditions based on the GPC results [184]. GPC analyses provided molecular weight distribution in the extracted lignin of waterlogged wood and determined the aging effects on the structure of archeological wood [184]. Tejado et al. explored different physical/chemical methods to characterize three different types of lignin (i.e., kraft lignin from pine, soda–anthraquinone flax lignin, and ethanol–water wild tamarind lignin) for resin production; GPC results from this research demonstrated that the molecular weight distribution of three sources of lignin were different, thus suitable for different applications [185].

Characterization of representative lignin streams from a variety of biomass feedstocks fractionated by different pretreatment methods is presented in Table 11.4. A number of studies demonstrated that genetic modification of lignin biosynthetic pathway reduced the lignin content and degree of polymerization and generated lignin with smaller molecular weight when compared to the native plants. GPC analysis of isolated ball-milled lignin from engineered alfalfa (*Medicago sativa*) by downregulation of C3H or HCT indicated significant changes in the Mw distribution between wild-type and engineered plants [71]. Lignin extracted from CAD or OMT-deficient trees denoted 11–29% lower Mw values than that of the wild-type trees [186]. The Mw of lignin created by overexpression of *PvMYB4* in alfalfa was 4400–4900 Da as compared to 5300–5500 Da in wild-type species [187]. According

Table 11.4 Lignin streams fractionated by different pretreatments and characterized by current biorefining technologies

Pretreatment	Biomass feedstocks	Lignin streams		Lignin molecular weight distribution (after pretreatment)		Lignin inter-unit linkages	References
		Liquid (%)	Solids (%)	Mw (g/mol)	Mn (g/mol)		
$[C_2C_1Im]$ [OAc]	Engineered *Arabidopsis* (*med5a med5b ref8*)	70.4	29.6	N/A	N/A	Cleavage of β-O-4 is greater than cleavage of β-β and β-5 linkages	[18]
Ethanol organosolv	Switchgrass	60.5	39.5	4200	980	Mainly breaks down β-O-4 linkages	[54]
Dilute acid	Poplar	~15	~85	8280	~2800	β-O-4 linkages are susceptible to dilute acid pretreatment	[198]
$[C_2mim]$ [OAc]	Switchgrass	40.5	59.5	N/A	N/A	Major inter-unit linkages are β-aryl ethers	[199]
Aqueous ammonia	Cotton stalk	~16	~84	~1700	~890	75.6% of β-O-4 linkages in ammonia extracted lignin	[157]
Organosolv	Miscanthus	52	48	4690	7060	~0.41 β-O-4 linkages per aromatic ring	[182]
$[C_2mim]$ [OAc]	Wheat straw	36.3	63.7	N/A	N/A	β-Aryl ethers are the major linkages	[17]
50% [Ch] [Lys]	Rice straw	58.9	41.1	N/A	N/A	N/A	[200]
[Ch][Lys]	Switchgrass	85.1	14.9	N/A	N/A	N/A	[201]
CEL followed by hydrothermal	Hybrid poplar	~44	~56	~4800	~1700	Significant reduction of β-O-4 linkages	[202]

N/A Not applicable

to lignin molecular weight distribution characterization, it is clear that genetic modification of lignin biosynthesis can produce lignocellulosic biomass feedstocks that are more susceptible to pretreatments and can easily depolymerize lignin into polymers with small molecular weights.

11.5.3 Lignin Inter-Unit Linkages and Functional Groups

Delineating the complex structure and composition of lignin is not a simple task. Traditional wet chemistry analytical methods, such as the degradation technique, will not provide sufficient information to describe the entire structure of lignin [188]. Spectroscopic techniques, such as ultraviolet–visible (UV–vis), Raman spectroscopy, and infrared (IR) spectroscopy, prove to be powerful tools as compared to wet chemistry analytical methods. Nuclear magnetic resonance (NMR) spectroscopy, in particular, can provide both compositional and structural information of lignin at high resolution [189]. Quantitative ^{13}C NMR is a useful tool for lignin characterization, and it was initiated in the early 1980s [190]. The advancement of both software and spectra quality further improved lignin quantitative ^{13}C NMR characterization methods in the past decades [191]. Recent development of multidimensional NMR has allowed spectroscopy to provide more structural information on lignin molecules [192].

Lignin structure and composition in mutant tobacco (lignin content reduced by downregulating CCR and CAD) were studied by NMR spectroscopy and results illustrated that engineered tobacco contained fewer coniferyl alcohol-derived units [175]. In another study, Marita et al. discovered an increase of syringyl unit content (determined by NMR spectroscopy) in *Arabidopsis* which was deficient in F5H by overexpression of *F5H* gene [193]. Furthermore, NMR spectroscopy provided a reliable solution to diagnose and illuminate lignin structure in CAD- and COMT-deficient poplar [194]. Pu et al. summarized the increasing attentions of ^{31}P NMR spectroscopy to quantitatively characterize lignin structure and bio-oils and biodiesels derived from biomass or lignin [195]. Yuan et al. utilized ^{1}H–^{13}C HSQC NMR spectroscopy to characterize lignin structures and lignin–carbohydrate inter-unit linkages in MWL and MAL (mild acidolysis lignin) from poplar wood and estimated the S to G ratio in these two types of lignin materials [196]. Del Rio et al. carried out the structural and compositional study of wheat straw lignin by two different methods, that is, DFRC (derivatization followed by reductive cleavage) and ^{1}H–^{13}C HSQC NMR spectroscopy, and provided information on S/G/H subunits in wheat lignin [197].

Furthermore, NMR technology (^{13}C and ^{31}P) has been developed to determine the characteristics of pyrolysis oil from softwood and to identify the cleavage of inter-unit linkages in lignin [203]. Eudes et al. expressed 3-dehydroshikimate dehydratase from *Corynebacterium glutamicum* in *Arabidopsis*, and NMR spectroscopy results revealed a significant increase of two H-lignin precursors and a reduction of G- and S-lignin, which indicated a decline of phenylcoumaran and resinol (β-5 and β-β linkages, respectively). Compared to wild-type plants, the modified species demonstrated higher sugar yields from lower amounts of enzymes used for saccharification [204]. Liu et al. fractionated genetically engineered switchgrass using aqueous cholinium lysinate and employed ^{1}H–^{13}C HSQC NMR spectroscopy to determine the lignin compositional changes and cleavage of inter-unit linkages. Results indicated that genetic modification led to compositional variations in switchgrass and generated different modes of lignin depolymerization [143]. Clearly, 2D HSQC NMR spectroscopy has become a standard tool to identify lignin monomer composition and reveal lignin structure and lignin inter-unit linkages in native and engineered plants and their changes during the lignin fractionation and valorization process. Mansfield et al. also developed a detailed protocol to characterize lignin structure and composition in plant cell wall [205].

Fourier transform infrared spectroscopy (FTIR) is an analytical and qualitative technique using infrared light to scan test materials. The resulting infrared spectrum reflects the absorption or emission of infrared from samples in solid, liquid, or gas state, by which chemical properties of organic, polymeric, and, in some cases, inorganic material can be identified [206]. Using FTIR to determine functionalities in lignin has been well documented [207–209]. Application of the infrared spectroscopy technique, especially diffuse reflectance infrared Fourier transform (DRIFT) [210], provides a snapshot of lignin samples due to the tremendous amount of information contained in the energy modes of different bonds. FTIR can reveal chemical fingerprinting of lignin and cellulosic components and monitor the chemical changes of carbohydrates/lignin during physical/chemical processing. Therefore, extensive studies have employed FTIR to qualitatively characterize lignin, cellulose, and hemicellulose components in lignocellulosic biomass before and after pretreatment [211, 212].

11.5.4 Thermal Properties

Strong interest has been shown for thermochemically depolymerizing and valorizing lignin into a potential feedstock for value-added chemicals. Thus, lignin's thermal degradation behavior needs to be better understood [213–216]. Furthermore, thermal properties provide critical information about lignin-derived materials such as carbon fiber, porous carbon, and resins. As a branch of material science, thermal analysis examines the impact of temperature change on the properties of materials. Numerous thermal analysis methods have been developed, of which DSC (differential scanning calorimetry) and TGA (thermogravimetric analysis) are the most widely used techniques [217]. TGA determines the mass gain or loss due to oxidation or decomposition properties of a sample as a function of temperature, from which information about vaporization, absorption, and desorption is revealed. In addition, TGA provides information about the chemical reactions within the material under elevated temperatures, including dehydration, decomposition, and solid–gas reactions [218]. Moreover, the heat capacity, phase transition, and glass transition temperature of amorphous polymers (e.g., cellulose and lignin) can be measured by DSC with high precision [219]. For instance, the glass transition temperature of lignins from genetically modified poplars presented differences at a temperature range from 170 to 190 °C due to the genetic modifications and the age of the plant. The results also indicated that thermal analysis was a sensitive approach to find the correlations between lignin structure and lignin properties (e.g., lignin condensation degree) [186].

11.6 Lignin Upgrading

11.6.1 Oxidative Catalysis

Catalytic oxidation of lignin aims to depolymerize lignin to value-added aromatic compounds under oxidizing condition with the aid of a catalyst. Ma et al. summarized the processes of lignin oxidation and the catalysts being applied to produce chemicals, which included phenolic compounds, dicarboxylic acids, and quinones [220]. Oxidative catalysts, including organometallic catalyst, metal-free organic catalyst, and acid or base catalysts, have been extensively investigated. The most commonly used catalysts are nitrobenzene, metal oxides, and molecular oxygen, for instance, H_2O_2/MTO, [MeRe(O_2)O_2], [Co(salen)], Mn(salen), and Cu(salen). The typical biomimetic catalysts used for lignin depolymerization are synthetic metalloporphyrins, which are similar in structure to lignin-degrading enzymes (e.g., lignin peroxidase and manganese peroxidase) and contribute to a high yield of monomeric products [221]. Photocatalytic oxidation is a method to degrade lignin using TiO_2, ZnO_2, etc. catalysts assisted by UV light to reduce reaction time and minimize the resulting organic pollutants [222].

Genetic modification of lignocellulosic biomass provides a promising method to enhance the yield and selectivity of lignin catalytic oxidation. Luo et al. reported the effects of solvent types, that is, acetic acid/formic acid, acetone, and methanol, on the organosolv extraction and catalytic upgrading of lignin from wild-type and genetically engineered poplars (high-S and low-S). The results demonstrated that genetically modified poplar with high S-lignin had higher lignin extraction yields compared to poplar with low S-lignin and wild-type poplar, and the highest yield of aromatics was obtained from methanol-extracted lignin [223]. Parsell et al. utilized bimetallic catalyst (i.e., Zn/Pd/C) to convert lignin from transgenic poplar with enhanced S-lignin to methoxyphenol products, and the remaining polysaccharides were easily hydrolyzed (over 95% glucose yield) by enzymes [224]. Oxidative catalysis is one of the most widely used catalytic methods to depolymerize lignin in native and genetically modified plants, although yield and selectivity still need to be improved.

11.6.2 Reductive Catalysis

Reductive catalysis is an efficient process to depolymerize lignin into low-molecular-weight polymers and phenols, and further to produce value-added fuels/chemicals through a thermal reduction process of lignin by hydrogen. Reductive reactions are able to break down carbon–carbon or carbon–heteroatom bonds in the presence of metal catalysts (e.g., Pt, Pd, Cu, and Ni) at an elevated temperature and hydrogen pressure. Feghali et al. evaluated an efficient method of convergent reductive catalysis (metal-free) of wood lignin, and successfully obtained high yields of phenolic compounds from lignin and woody materials [225]. Moreover, the introduction of a second metal enhanced the catalytic performance. Parsell et al. reported a bimetallic catalytic system (including Pd/C and Zn) to selectively cleave linkages in lignin model compounds [226]. Due to the complex structure and composition of native lignin, it is a big challenge to obtain pure products from lignin depolymerization. Reductive catalysis is a promising technology to deconstruct and further upgrade lignin. To our knowledge, no literature has been reported on employing reductive catalysis of lignin from genetically modified plants.

11.6.3 Electro-catalysis

Electro-catalysis is a potential method for efficiently converting lignin into fuels and chemicals, and electro-catalysis displays higher activity and stability than oxidative/reductive catalysts. For example, IrO_2-based electrodes (including Ti/RuO_2–IrO_2 and Ti/TiO_2–IrO_2) are commonly used in electro-catalysis. Milczarek et al. conducted the electro-catalysis in TRIS–HNO_3 buffers in the absence or presence of Mg^{2+} to determine the effect of pH changes on catalytic conversion [227]. Ionic liquids combined with electro-catalysis have demonstrated an efficient method to depolymerize and upgrade lignin. Ionic liquids can dissolve lignin effectively and can be an excellent mediator to selectively break down C–C and β-O-4 linkages in electronic mediator systems. Reichert et al. employed a novel approach of electro-catalysis of lignin by dissolving lignin in a special ionic liquid (triethylammonium methanesulfonate) and using a coated anode to oxidize lignin. Results demonstrated that the electro-catalysis produced various aromatic fragments from lignin [228]. Electro-catalysis is a promising lignin depolymerization method; however, the cost and the electrode fouling problems need to be addressed before widespread applications.

11.6.4 Biological Upgrading

11.6.4.1 Lignin Degrading Enzymes

Lignin is not able to be as easily and efficiently depolymerize as other polymers in nature (e.g., cellulose and starch) due to heterogeneous nature and various inter-unit linkages. However, white rot fungi have been demonstrated to produce extracellular lignin-degrading enzymes that are able to accomplish some degree of lignin depolymerization [229]. White rot fungi secrets extracellular lignin-degrading oxidative enzymes, such as LiP (lignin peroxidase, EC 1.11.1.14), MnP (manganese peroxidase, EC 1.11.1.13), AAO (aryl alcohol oxidase, EC 1.1.3.7), and laccase (EC, 1.10.3.2) [230]. Lignin peroxidase can oxidize a wide range of phenolic and non-phenolic compounds with its high redox potential and addition of hydrogen peroxide. Manganese peroxidase can oxidize Mn^{2+} to Mn^{3+}, acting as a diffusing oxidizer [231]. Lignin peroxidase can direct the interaction between lignin structure and the protein due to a highly active but short-lived catalytic site on the enzyme surface. However, the reaction is slow and ineffective, so that external H_2O_2 is required to increase the reaction rate. In the reaction the electron donor and final electron acceptor are lignin and H_2O_2, respectively [232].

Unlike lignin peroxidase, laccase uses oxygen as the electron acceptor. In a catalyzed reaction, laccase can transfer four electrons by reducing one molecule of oxygen to two molecules of H_2O. Laccase contains four copper ions with three types (i.e., T1, T2, and T3) and a trinuclear cluster formed by T2 and T3 centers. T1 is the site where oxidation of lignin takes place [233]. Laccase has a higher redox potential, which can attract electrons from lignin substrate and make it an advantage when depolymerizing lignin. Because laccase uses oxygen as the final electron acceptor, no additional H_2O_2 is needed; therefore, in theory, laccase is a more practical approach for lignin depolymerization than LiP and MnP.

Unlike LiP, which can oxidize phenolic and non-phenolic compounds, laccase alone cannot oxidize non-phenolic lignin subunits due to their high redox potential and laccase activity is inhibited by lignin depolymerizing products. However, this limitation of laccase can be overcome by using a molecule, called mediator, to act as an electron carrier between lignin subunits and laccase. The mediator can oxidize the non-phenolic compounds in lignin when the mediator is oxidized by laccase through electron abstraction [230]. The mediator can either be a natural or synthetic mediator. Natural mediators are plant/fungal metabolites (e.g., 4-hydroxybenzylic alcohol, sinapic acid, and *p*-cinnamic acid), while ABTS (2,2′-azinobis-3-ethylbenzthiazoline-6-sulfonate) and HBT (1-hydroxybenzotriazole) are commonly used as synthetic mediators [234].

Sphingobium SYK-6, an aerobic bacillus found in soil, is capable of producing β-etherases and growing on lignin-derived biaryls and monoaryls [235]. *Sphingobium* SYK-6's ability to degrade lignin model compounds was found to be structurally and biochemically related to glutathione transferases [236]. Three glutathione transferase encoding genes, that is, *ligE*, *ligF*, and *ligG*, were previously identified in *Sphingobium* SYK-6, and the gene operon was involved in lignin depolymerization [237, 238]. The β-etherases can catalyze the transfer of glutathione to an aryl ether substrate, which results in a β-thioether intermediate and further cleaves the ether linkages in lignin [239]. Pereira et al. reported detailed information of cofactor and substrate-binding sites of β-etherases in *Sphingobium* SYK-6 by conducting structural and biochemical characterizations of lignin cleavage pathway in the bacterium [240]. Mori et al. identified a transporter gene for protocatechuate in *Sphingobium* SYK-6 and found that enhanced expression of the transporter gene can improve the production and accumulation of value-added metabolites from lignin-derived aromatics [241].

McAndrew et al. discovered a lignin-degrading enzyme named NOV1, a bacterial resveratrol-cleaving dioxygenase from *Novosphingobium aromaticivorans*. The structure, iron cofactor, active sites, and degrading mechanism were reported, and potential applications of NOV1 in upgrading of solubilized fragments of lignin were discussed [242].

11.6.4.2 Lignin Fermenting Microorganisms

The environmental adaptability and ability to genetically manipulate bacteria make this a promising approach to biologically deconstructing lignin. For example, bioconversion of lignin to lipids by oleaginous *Rhodococci* has been extensively studied. Kosa and Ragauskas evaluated the possibility of using *Rhodococcus opacus* to convert lignin model compounds to lipids. Results showed that *R. opacus* can utilize different lignin model substrates as carbon source and can accumulate lipids up to 20% of the cell dry weight [243]. Wei et al. conducted research on the conversion of Kraft lignin to lipids by *R. opacus* and demonstrated that *R. opacus* were able to be utilized on oxygen-pretreated Kraft lignin and produced a significant amount of lipids [244]. Le et al. produced a high concentration of glucose and lignin oligomers with low molecular weight from corn stover using a novel two-stage alkali-peroxide pretreatment method [245]. They also converted the organic substrates to lipids by *R. opacus* and achieved a lipid concentration of 1.3 g/L in a 48 h fermentation period. Moreover, combination of *R. opacus* and engineered *R. jostii* RHA1 was conducted to reduce the chemical oxygen demand (COD) of hydrothermal liquefaction aqueous waste of algae and pine [246]. He et al. established functional modules to better understand the lipids production pathways and to enhance the production of lipids using lignin from dilute alkaline pretreated corn stover. The researchers demonstrated that co-fermentation of *R. opacus* and genetically modified *R. jostii* can produce lipids with higher yield than single strain fermentation [247]. Although clear pathways for lignin to lipids production by *R. opacus* need to be identified, the bacterial conversion of lignin is a promising approach to upgrade lignin. Moreover, other lignin fermenting microorganisms, such as *Pseudomonas putida* KT2400 [248–250], *Pandoraea* sp. ISTKB (lignin derivatives to polyhydroxyalkanoate) [251], and *Cupriavidus necator* (alkaline pretreated lignin to polyhydroxybutyrate) [252], have been recently studied.

Genetic alternation of lignin inter-unit linkages has the potential to make considerable changes in lignin properties, for example, modification of plants to have more of the dominant β-O-4 bonds will require less energy to break down the inter-unit linkages and allow easier depolymerization. Generation of homogeneous intermediates and simple monolignol solutions by structural and compositional alteration of lignin would enable efficient and selective biological upgrading of lignin decomposition products a reality, since the enzymes and microorganisms can be engineered to use a single substrate as their sole nutrient. Furthermore, simple monolignols are easier to solubilize to aid in optimizing the chemical/biological upgrading processes.

11.7 Summary and Perspectives

Despite the great potential lignin presents as a renewable resource to produce value-added fuels and chemicals, lignin is still considered as a waste stream and underutilized. It is critical to develop cost-effective biotechnologies to convert lignin wastes to value-added products to improve biofuel and chemical production in a biorefinery [11, 12, 253]. The heterogeneous nature of lignin (including the various subunits and inter-unit linkages and inconsistency among plant species), however, is the primary obstacle to lignin valorization [254]. In specific, the inter-unit linkages (e.g., β-O-4, β-β, β-5, and 5-5) and their combinations and relative abundance vary significantly among plants [222]. Development of lignin valorization approaches, such as thermochemical conversion (e.g., pyrolysis and gasification), hydrogenolysis, catalytic oxidation, and biological conversion (e.g., lignin-degrading microorganisms and/or enzymes) is currently under investigation [255].

Improving plants' characteristics through genetic modification is a promising approach to artificially design biomass feedstocks with favorable properties [5, 68, 256]. New approaches to downregulating the lignin biosynthetic key enzymes have been developed to reduce the lignin content and structurally and compositionally modify lignin in various plant species [23, 38, 54, 257]. Genetic modifications of lignin subunits ratio (i.e., S/G/H ratio) or lignin deposition patterns and artificial design of lignin linkages (including inter-unit and lignin–carbohydrate linkages) were demonstrated to be operational and effective in improving the digestibility of lignocellulosic biomass by microorganisms and/or lignin-degrading enzymes [23, 258].

Extensive research has been conducted on how pretreatment methods affect the structure and composition of biomass feedstocks and the potential upgrading approaches of lignin to value-added fuels and chemicals [156, 183]. However, research areas on impact of genetic lignin manipulation on lignin fractionation and characterization are not fully explored. It is essential to investigate the fractionation and characterization of the polysaccharides and lignin streams from wild-type and engineered plants and to build links between the biology of engineered plant, the pretreatment chemistry, and the conversion technologies if we are to be successful in building the biorefining industry. Advanced characterization methods and analytics have the potential to provide vital information about the structural and compositional changes occurring in lignin, such as lignin molecular weight distribution, inter-unit linkages, and the thermal properties of engineered biomass and fractionated lignin streams. Taken together, this advanced knowledge will help to answer fundamental questions such as (1) how do different biomass pretreatment methods affect lignin fractionation from genetically modified plants? (2) how does pretreatment chemistry affect the characteristics of lignin streams? and (3) how does lignin stream characterization inform selection of the lignin upgrading process? Answers to these questions will help inform plant scientists as to the desired plant traits needed to allow more effective and efficient lignin conversion technologies to be developed.

Acknowledgments We acknowledge the National Science Foundation under Cooperative Agreement No. 1355438 and 1632854 and the National Institute of Food and Agriculture, US Department of Agriculture, Hatch-Multistate project under accession number 1018315 and the Sustainability Challenge Area grant under accession number 1015068 for supporting this work.

References

1. Obama, B. (2017). The irreversible momentum of clean energy. *Science, 355*(6321), 126–129.
2. Ragauskas, A. J., Williams, C. K., Davison, B. H., Britovsek, G., Cairney, J., Eckert, C. A., Frederick, W. J., Hallett, J. P., Leak, D. J., & Liotta, C. L. (2006). The path forward for biofuels and biomaterials. *Science, 311*(5760), 484–489.
3. Chundawat, S. P., Beckham, G. T., Himmel, M. E., & Dale, B. E. (2011). Deconstruction of lignocellulosic biomass to fuels and chemicals. *Annual Review of Chemical and Biomolecular Engineering, 2*, 121–145.
4. Tuck, C. O., Pérez, E., Horváth, I. T., Sheldon, R. A., & Poliakoff, M. (2012). Valorization of biomass: Deriving more value from waste. *Science, 337*(6095), 695–699.
5. Hisano, H., Nandakumar, R., & Wang, Z.-Y. (2011). Genetic modification of lignin biosynthesis for improved biofuel production. In *Biofuels* (pp. 223–235). New York: Springer.
6. Carroll, A., & Somerville, C. (2009). Cellulosic biofuels. *Annual Review of Plant Biology, 60*, 165–182.
7. Wyman, C. E., Dale, B. E., Elander, R. T., Holtzapple, M., Ladisch, M. R., & Lee, Y. (2005). Coordinated development of leading biomass pretreatment technologies. *Bioresource Technology, 96*(18), 1959–1966.
8. Kim, S., & Dale, B. E. (2004). Global potential bioethanol production from wasted crops and crop residues. *Biomass & Bioenergy, 26*(4), 361–375.
9. Lynd, L. R., Laser, M. S., Bransby, D., Dale, B. E., Davison, B., Hamilton, R., Himmel, M., Keller, M., McMillan, J. D., & Sheehan, J. (2008). How biotech can transform biofuels. *Nature Biotechnology, 26*(2), 169.
10. Lynd, L. R., Van Zyl, W. H., McBride, J. E., & Laser, M. (2005). Consolidated bioprocessing of cellulosic biomass: An update. *Current Opinion in Biotechnology, 16*(5), 577–583.
11. Beckham, G. T., Johnson, C. W., Karp, E. M., Salvachúa, D., & Vardon, D. R. (2016). Opportunities and challenges in biological lignin valorization. *Current Opinion in Biotechnology, 42*, 40–53.
12. Ragauskas, A. J., Beckham, G. T., Biddy, M. J., Chandra, R., Chen, F., Davis, M. F., Davison, B. H., Dixon, R. A., Gilna, P., Keller, M., Langan, P., Naskar, A. K., Saddler, J. N., Tschaplinski, T. J., Tuskan, G. A., & Wyman, C. E. (2014). Lignin valorization: Improving lignin processing in the biorefinery. *Science, 344*(6185), 1246843.
13. Azadi, P., Inderwildi, O. R., Farnood, R., & King, D. A. (2013). Liquid fuels, hydrogen and chemicals from lignin: A critical review. *Renewable and Sustainable Energy Reviews, 21*, 506–523.
14. Linger, J. G., Vardon, D. R., Guarnieri, M. T., Karp, E. M., Hunsinger, G. B., Franden, M. A., Johnson, C. W., Chupka, G., Strathmann, T. J., & Pienkos, P. T. (2014). Lignin valorization through integrated biological funneling and chemical catalysis. *Proceedings of the National Academy of Sciences of the United States of America, 111*(33), 12013–12018.
15. Xu, C., Arancon, R. A. D., Labidi, J., & Luque, R. (2014). Lignin depolymerisation strategies: Towards valuable chemicals and fuels. *Chemical Society Reviews, 43*(22), 7485–7500.
16. Behling, R., Valange, S., & Chatel, G. (2016). Heterogeneous catalytic oxidation for lignin valorization into valuable chemicals: What results? What limitations? What trends? *Green Chemistry, 18*(7), 1839–1854.

17. Sathitsuksanoh, N., Holtman, K. M., Yelle, D. J., Morgan, T., Stavila, V., Pelton, J., Blanch, H., Simmons, B. A., & George, A. (2014). Lignin fate and characterization during ionic liquid biomass pretreatment for renewable chemicals and fuels production. *Green Chemistry, 16*(3), 1236–1247.
18. Shi, J., Pattathil, S., Parthasarathi, R., Anderson, N. A., Im Kim, J., Venketachalam, S., Hahn, M. G., Chapple, C., Simmons, B. A., & Singh, S. (2016). Impact of engineered lignin composition on biomass recalcitrance and ionic liquid pretreatment efficiency. *Green Chemistry, 18*(18), 4884–4895.
19. Tolbert, A., Akinosho, H., Khunsupat, R., Naskar, A. K., & Ragauskas, A. J. (2014). Characterization and analysis of the molecular weight of lignin for biorefining studies. *Biofuels, Bioproducts and Biorefining, 8*(6), 836–856.
20. Cabello, J. V., Lodeyro, A. F., & Zurbriggen, M. D. (2014). Novel perspectives for the engineering of abiotic stress tolerance in plants. *Current Opinion in Biotechnology, 26*, 62–70.
21. Roy, S. J., Negrão, S., & Tester, M. (2014). Salt resistant crop plants. *Current Opinion in Biotechnology, 26*, 115–124.
22. Carpita, N. C., & McCann, M. C. (2008). Maize and sorghum: Genetic resources for bioenergy grasses. *Trends in Plant Science, 13*(8), 415–420.
23. Yang, H., Yan, R., Chen, H., Lee, D. H., & Zheng, C. (2007). Characteristics of hemicellulose, cellulose and lignin pyrolysis. *Fuel, 86*(12–13), 1781–1788.
24. Liu, E., Das, L., Zhao, B., Crocker, M., & Shi, J. (2017). Impact of dilute sulfuric acid, ammonium hydroxide, and ionic liquid pretreatments on the fractionation and characterization of engineered switchgrass. *Bioenergy Research, 10*(4), 1079–1093.
25. Simmons, B. A., Loque, D., & Ralph, J. (2010). Advances in modifying lignin for enhanced biofuel production. *Current Opinion in Plant Biology, 13*(3), 312–319.
26. Menon, V., & Rao, M. (2012). Trends in bioconversion of lignocellulose: Biofuels, platform chemicals & biorefinery concept. *Progress in Energy and Combustion Science, 38*(4), 522–550.
27. Vassilev, S. V., Baxter, D., Andersen, L. K., & Vassileva, C. G. (2010). An overview of the chemical composition of biomass. *Fuel, 89*(5), 913–933.
28. Williams, C. L., Westover, T. L., Emerson, R. M., Tumuluru, J. S., & Li, C. (2016). Sources of biomass feedstock variability and the potential impact on biofuels production. *Bioenergy Research, 9*(1), 1–14.
29. Pordesimo, L., Hames, B., Sokhansanj, S., & Edens, W. (2005). Variation in corn stover composition and energy content with crop maturity. *Biomass & Bioenergy, 28*(4), 366–374.
30. Keshwani, D. R., & Cheng, J. J. (2009). Switchgrass for bioethanol and other value-added applications: A review. *Bioresource Technology, 100*(4), 1515–1523.
31. Brosse, N., Dufour, A., Meng, X., Sun, Q., & Ragauskas, A. (2012). Miscanthus: A fast-growing crop for biofuels and chemicals production. *Biofuels, Bioproducts and Biorefining, 6*(5), 580–598.
32. Kreuger, E., Sipos, B., Zacchi, G., Svensson, S.-E., & Björnsson, L. (2011). Bioconversion of industrial hemp to ethanol and methane: The benefits of steam pretreatment and co-production. *Bioresource Technology, 102*(3), 3457–3465.
33. Fernández-Fueyo, E., Ruiz-Dueñas, F. J., Ferreira, P., Floudas, D., Hibbett, D. S., Canessa, P., Larrondo, L., James, T. Y., Seelenfreund, D., Lobos, S., Polanco, R., Tello, M., Honda, Y., Watanabe, T., Watanabe, T., Ryu, J. S., Kubicek, C. P., Schmoll, M., Gaskell, J., Hammel, K. E., St. John, F. J., Vanden Wymelenberg, A., Sabat, G., Bondurant, S. S., Syed, K., Yadav, J., Doddapaneni, H., Subramanian, V., Lavín, J. L., & Oguiza, J. A. (2012). Comparative genomics of Ceriporiopisis subvermispora and Phanerochaete chrysosporium provide insight into selective ligninolysis. *Proceedings of the National Academy of Sciences of the United States of America, 109*(14), 5458–5463.
34. McKendry, P. (2002). Energy production from biomass (part 1): Overview of biomass. *Bioresource Technology, 83*(1), 37–46.

35. Pandey, A., Soccol, C. R., Nigam, P., & Soccol, V. T. (2000). Biotechnological potential of agro-industrial residues. I: Sugarcane bagasse. *Bioresource Technology, 74*(1), 69–80.
36. Adler, E. (1977). Lignin chemistry—Past, present and future. *Wood Science and Technology, 11*(3), 169–218.
37. McCarthy, J. L., & Islam, A. (2000). Lignin chemistry, technology, and utilization: A brief history. In W. G. Glasser, R. A. Northey, & T. P. Schuultz (Eds.), *Lignin: Historical, biological, and material perspectives* (ACS symposium series 742). Washington, DC: American Chemical Society.
38. Lu, F., & Ralph, J. (1997). Derivatization followed by reductive cleavage (DFRC method), a new method for lignin analysis: Protocol for analysis of DFRC monomers. *Journal of Agricultural and Food Chemistry, 45*(7), 2590–2592.
39. Boerjan, W., Ralph, J., & Baucher, M. (2003). Lignin biosynthesis. *Annual Review of Plant Biology, 54*(1), 519–546.
40. Hatakeyama, H., & Hatakeyama, T. (2009). Lignin structure, properties, and applications. In *Biopolymers* (pp. 1–63). Berlin/Heidelberg: Springer.
41. Vanholme, R., Morreel, K., Ralph, J., & Boerjan, W. (2008). Lignin engineering. *Current Opinion in Plant Biology, 11*(3), 278–285.
42. Rencoret, J., Gutierrez, A., Nieto, L., Jimenez-Barbero, J., Faulds, C. B., Kim, H., Ralph, J., Martinez, A. T., & Del Rio, J. C. (2011). Lignin composition and structure in young versus adult Eucalyptus globulus plants. *Plant Physiology, 155*(2), 667–682.
43. Pandey, M. P., & Kim, C. S. (2011). Lignin depolymerization and conversion: A review of thermochemical methods. *Chemical Engineering and Technology, 34*(1), 29–41.
44. Li, C., Zhao, X., Wang, A., Huber, G. W., & Zhang, T. (2015). Catalytic transformation of lignin for the production of chemicals and fuels. *Chemical Reviews, 115*(21), 11559–11624.
45. Chung, H., & Washburn, N. R. (2016). Extraction and types of lignin. In *Lignin in polymer composites* (pp. 13–25). New York City: William Andrew Publishing.
46. Vanholme, R., Demedts, B., Morreel, K., Ralph, J., & Boerjan, W. (2010). Lignin biosynthesis and structure. *Plant Physiology, 153*(3), 895–905.
47. Kuang, D., Walter, P., Nuesch, F., Kim, S., Ko, J., Comte, P., Zakeeruddin, S. M., Nazeeruddin, M. K., & Gratzel, M. (2007). Co-sensitization of organic dyes for efficient ionic liquid electrolyte-based dye-sensitized solar cells. *Langmuir, 23*(22), 10906–10909.
48. Weng, J. K., & Chapple, C. (2010). The origin and evolution of lignin biosynthesis. *The New Phytologist, 187*(2), 273–285.
49. Vogel, K. P., & Jung, H.-J. G. (2001). Genetic modification of herbaceous plants for feed and fuel. *Critical Reviews in Plant Sciences, 20*(1), 15–49.
50. McLaughlin, S. B., & Kszos, L. A. (2005). Development of switchgrass (Panicum virgatum) as a bioenergy feedstock in the United States. *Biomass & Bioenergy, 28*(6), 515–535.
51. Sticklen, M. B. (2007). Feedstock crop genetic engineering for alcohol fuels. *Crop Science, 47*(6), 2238.
52. Hisano, H., Nandakumar, R., & Wang, Z.-Y. (2009). Genetic modification of lignin biosynthesis for improved biofuel production. *In Vitro Cellular & Developmental Biology. Plant, 45*(3), 306–313.
53. Sticklen, M. B. (2008). Plant genetic engineering for biofuel production: Towards affordable cellulosic ethanol. *Nature Reviews. Genetics, 9*(6), 433–443.
54. Xu, B., Escamilla-Treviño, L. L., Sathitsuksanoh, N., Shen, Z., Shen, H., Percival Zhang, Y. H., Dixon, R. A., & Zhao, B. (2011). Silencing of 4-coumarate: Coenzyme A ligase in switchgrass leads to reduced lignin content and improved fermentable sugar yields for biofuel production. *The New Phytologist, 192*(3), 611–625.
55. Baucher, M., Halpin, C., Petit-Conil, M., & Boerjan, W. (2003). Lignin: Genetic engineering and impact on pulping. *Critical Reviews in Biochemistry and Molecular Biology, 38*(4), 305–350.

56. Nakashima, J., Chen, F., Jackson, L., Shadle, G., & Dixon, R. A. (2008). Multi-site genetic modification of monolignol biosynthesis in alfalfa (Medicago sativa): Effects on lignin composition in specific cell types. *The New Phytologist, 179*(3), 738–750.
57. Bjurhager, I., Olsson, A.-M., Zhang, B., Gerber, L., Kumar, M., Berglund, L. A., Burgert, I., Sundberg, B. R., & Salmén, L. (2010). Ultrastructure and mechanical properties of Populus wood with reduced lignin content caused by transgenic down-regulation of cinnamate 4-hydroxylase. *Biomacromolecules, 11*(9), 2359–2365.
58. Chabannes, M., Barakate, A., Lapierre, C., Marita, J. M., Ralph, J., Pean, M., Danoun, S., Halpin, C., Grima-Pettenati, J., & Boudet, A. M. (2001). Strong decrease in lignin content without significant alteration of plant development is induced by simultaneous down-regulation of cinnamoyl CoA reductase (CCR) and cinnamyl alcohol dehydrogenase (CAD) in tobacco plants. *The Plant Journal, 28*(3), 257–270.
59. Kawaoka, A., Nanto, K., Ishii, K., & Ebinuma, H. (2006). Reduction of lignin content by suppression of expression of the LIM domain transcription factor in Eucalyptus camaldulensis. *Silvae Genetica, 55*(6), 269–277.
60. Li, Y., Kajita, S., Kawai, S., Katayama, Y., & Morohoshi, N. (2003). Down-regulation of an anionic peroxidase in transgenic aspen and its effect on lignin characteristics. *Journal of Plant Research, 116*(3), 175–182.
61. Moura, J. C., Bonine, C. A., de Oliveira Fernandes Viana, J., Dornelas, M. C., & Mazzafera, P. (2010). Abiotic and biotic stresses and changes in the lignin content and composition in plants. *Journal of Integrative Plant Biology, 52*(4), 360–376.
62. Scully, E. D., Gries, T., Funnell-Harris, D. L., Xin, Z., Kovacs, F. A., Vermerris, W., & Sattler, S. E. (2016). Characterization of novel Brown midrib 6 mutations affecting lignin biosynthesis in sorghum. *Journal of Integrative Plant Biology, 58*(2), 136–149.
63. Sattler, S. E., Funnell-Harris, D. L., & Pedersen, J. F. (2010). Brown midrib mutations and their importance to the utilization of maize, sorghum, and pearl millet lignocellulosic tissues. *Plant Science, 178*(3), 229–238.
64. Biemelt, S., Tschiersch, H., & Sonnewald, U. (2004). Impact of altered gibberellin metabolism on biomass accumulation, lignin biosynthesis, and photosynthesis in transgenic tobacco plants. *Plant Physiology, 135*(1), 254–265.
65. Kishimoto, T., Chiba, W., Saito, K., Fukushima, K., Uraki, Y., & Ubukata, M. (2010). Influence of syringyl to guaiacyl ratio on the structure of natural and synthetic lignins. *Journal of Agricultural and Food Chemistry, 58*(2), 895–901.
66. Bonawitz, N. D., Im Kim, J., Tobimatsu, Y., Ciesielski, P. N., Anderson, N. A., Ximenes, E., Maeda, J., Ralph, J., Donohoe, B. S., & Ladisch, M. (2014). Disruption of mediator rescues the stunted growth of a lignin-deficient Arabidopsis mutant. *Nature, 509*(7500), 376–380.
67. Sewalt, V. J., Ni, W., Blount, J. W., Jung, H. G., Masoud, S. A., Howles, P. A., Lamb, C., & Dixon, R. A. (1997). Reduced lignin content and altered lignin composition in transgenic tobacco down-regulated in expression of L-phenylalanine ammonia-lyase or cinnamate 4-hydroxylase. *Plant Physiology, 115*(1), 41–50.
68. Xu, B., Sathitsuksanoh, N., Tang, Y., Udvardi, M. K., Zhang, J.-Y., Shen, Z., Balota, M., Harich, K., Zhang, P. Y.-H., & Zhao, B. (2012). Overexpression of AtLOV1 in switchgrass alters plant architecture, lignin content, and flowering time. *PLoS One, 7*(12), e47399.
69. Coleman, H. D., Park, J.-Y., Nair, R., Chapple, C., & Mansfield, S. D. (2008). RNAi-mediated suppression of p-coumaroyl-CoA 3′-hydroxylase in hybrid poplar impacts lignin deposition and soluble secondary metabolism. *Proceedings of the National Academy of Sciences of the United States of America, 105*(11), 4501–4506.
70. Voelker, S. L., Lachenbruch, B., Meinzer, F. C., Jourdes, M., Ki, C., Patten, A. M., Davin, L. B., Lewis, N. G., Tuskan, G. A., & Gunter, L. (2010). Antisense down-regulation of 4CL expression alters lignification, tree growth, and saccharification potential of field-grown poplar. *Plant Physiology, 154*(2), 874–886.
71. Ziebell, A., Gracom, K., Katahira, R., Chen, F., Pu, Y., Ragauskas, A., Dixon, R. A., & Davis, M. (2010). Increase in 4-coumaryl alcohol units during lignification in alfalfa (Medicago

sativa) alters the extractability and molecular weight of lignin. *The Journal of Biological Chemistry, 285*(50), 38961–38968.
72. Li, X., Weng, J. K., & Chapple, C. (2008). Improvement of biomass through lignin modification. *The Plant Journal, 54*(4), 569–581.
73. Hibino, T., Takabe, K., Kawazu, T., Shibata, D., & Higuchi, T. (2014). Increase of cinnamaldehyde groups in lignin of transgenic tobacco plants carrying an antisense gene for cinnamyl alcohol dehydrogenase. *Bioscience, Biotechnology, and Biochemistry, 59*(5), 929–931.
74. Reddy, M. S., Chen, F., Shadle, G., Jackson, L., Aljoe, H., & Dixon, R. A. (2005). Targeted down-regulation of cytochrome P450 enzymes for forage quality improvement in alfalfa (Medicago sativa L.). *Proceedings of the National Academy of Sciences of the United States of America, 102*(46), 16573–16578.
75. Scullin, C., Cruz, A. G., Chuang, Y. D., Simmons, B. A., Loque, D., & Singh, S. (2015). Restricting lignin and enhancing sugar deposition in secondary cell walls enhances monomeric sugar release after low temperature ionic liquid pretreatment. *Biotechnology for Biofuels, 8*, 95.
76. Bonawitz, N. D., & Chapple, C. (2013). Can genetic engineering of lignin deposition be accomplished without an unacceptable yield penalty? *Current Opinion in Biotechnology, 24*(2), 336–343.
77. Rogers, L. A., & Campbell, M. M. (2004). The genetic control of lignin deposition during plant growth and development. *The New Phytologist, 164*(1), 17–30.
78. Yang, F., Mitra, P., Zhang, L., Prak, L., Verhertbruggen, Y., Kim, J. S., Sun, L., Zheng, K., Tang, K., Auer, M., Scheller, H. V., & Loque, D. (2013). Engineering secondary cell wall deposition in plants. *Plant Biotechnology Journal, 11*(3), 325–335.
79. Franke, R., Hemm, M. R., Denault, J. W., Ruegger, M. O., Humphreys, J. M., & Chapple, C. (2002). Changes in secondary metabolism and deposition of an unusual lignin in the ref8 mutant of Arabidopsis. *The Plant Journal, 30*(1), 47–59.
80. Casler, M. D., Jung, H. G., & Coblentz, W. K. (2008). Clonal selection for lignin and etherified ferulates in three perennial grasses. *Crop Science, 48*(2), 424.
81. Grabber, J. H., Mertens, D. R., Kim, H., Funk, C., Lu, F., & Ralph, J. (2009). Cell wall fermentation kinetics are impacted more by lignin content and ferulate cross-linking than by lignin composition. *Journal of Science and Food Agriculture, 89*(1), 122–129.
82. Hatfield, R. D., & Chaptman, A. K. (2009). Comparing corn types for differences in cell wall characteristics and p-coumaroylation of lignin. *Journal of Agricultural and Food Chemistry, 57*(10), 4243–4249.
83. Wilkerson, C., Mansfield, S., Lu, F., Withers, S., Park, J.-Y., Karlen, S., Gonzales-Vigil, E., Padmakshan, D., Unda, F., & Rencoret, J. (2014). Monolignol ferulate transferase introduces chemically labile linkages into the lignin backbone. *Science, 344*(6179), 90–93.
84. Karlen, S. D., Zhang, C., Peck, M. L., Smith, R. A., Padmakshan, D., Helmich, K. E., Free, H. C., Lee, S., Smith, B. G., & Lu, F. (2016). Monolignol ferulate conjugates are naturally incorporated into plant lignins. *Science Advances, 2*(10), e1600393.
85. Ralph, J. (2010). Hydroxycinnamates in lignification. *Phytochemistry Reviews, 9*(1), 65–83.
86. Bortesi, L., & Fischer, R. (2015). The CRISPR/Cas9 system for plant genome editing and beyond. *Biotechnology Advances, 33*(1), 41–52.
87. Ma, X., Zhu, Q., Chen, Y., & Liu, Y.-G. (2016). CRISPR/Cas9 platforms for genome editing in plants: Developments and applications. *Molecular Plant, 9*(7), 961–974.
88. Rani, R., Yadav, P., Barbadikar, K. M., Baliyan, N., Malhotra, E. V., Singh, B. K., Kumar, A., & Singh, D. (2016). CRISPR/Cas9: A promising way to exploit genetic variation in plants. *Biotechnology Letters, 38*(12), 1991–2006.
89. Luo, M., Gilbert, B., & Ayliffe, M. (2016). Applications of CRISPR/Cas9 technology for targeted mutagenesis, gene replacement and stacking of genes in higher plants. *Plant Cell Reports, 35*(7), 1439–1450.

90. Zhou, X., Jacobs, T. B., Xue, L. J., Harding, S. A., & Tsai, C. J. (2015). Exploiting SNPs for biallelic CRISPR mutations in the outcrossing woody perennial Populus reveals 4-coumarate: CoA ligase specificity and redundancy. *The New Phytologist, 208*(2), 298–301.
91. Park, J.-J., Yoo, C. G., Flanagan, A., Pu, Y., Debnath, S., Ge, Y., Ragauskas, A. J., & Wang, Z.-Y. (2017). Defined tetra-allelic gene disruption of the 4-coumarate: Coenzyme A ligase 1 (Pv4CL1) gene by CRISPR/Cas9 in switchgrass results in lignin reduction and improved sugar release. *Biotechnology for Biofuels, 10*(1), 284.
92. Haddad, M., Mikhaylin, S., Bazinet, L., Savadogo, O., & Paris, J. (2017). Electrochemical acidification of Kraft black liquor by electrodialysis with bipolar membrane: Ion exchange membrane fouling identification and mechanisms. *Journal of Colloid and Interface Science, 488*, 39–47.
93. Wyman, C. E., Balan, V., Dale, B. E., Elander, R. T., Falls, M., Hames, B., Holtzapple, M. T., Ladisch, M. R., Lee, Y., & Mosier, N. (2011). Comparative data on effects of leading pretreatments and enzyme loadings and formulations on sugar yields from different switchgrass sources. *Bioresource Technology, 102*(24), 11052–11062.
94. Tao, L., Aden, A., Elander, R. T., Pallapolu, V. R., Lee, Y. Y., Garlock, R. J., Balan, V., Dale, B. E., Kim, Y., Mosier, N. S., Ladisch, M. R., Falls, M., Holtzapple, M. T., Sierra, R., Shi, J., Ebrik, M. A., Redmond, T., Yang, B., Wyman, C. E., Hames, B., Thomas, S., & Warner, R. E. (2011). Process and technoeconomic analysis of leading pretreatment technologies for lignocellulosic ethanol production using switchgrass. *Bioresource Technology, 102*(24), 11105–11114.
95. Harmsen, P., Huijgen, W., Bermudez, L., & Bakker, R. (2010). *Literature review of physical and chemical pretreatment processes for lignocellulosic biomass*. Wageningen: Wageningen UR Food & Biobased Research.
96. Sidiras, D., & Koukios, E. (1989). Acid saccharification of ball-milled straw. *Biomass, 19*(4), 289–306.
97. Tassinari, T., Macy, C., Spano, L., & Ryu, D. D. (1980). Energy requirements and process design considerations in compression-milling pretreatment of cellulosic wastes for enzymatic hydrolysis. *Biotechnology and Bioengineering, 22*(8), 1689–1705.
98. Alvo, P., & Belkacemi, K. (1997). Enzymatic saccharification of milled timothy (Phleum pratense L.) and alfalfa (Medicago sativa L.). *Bioresource Technology, 61*(3), 185–198.
99. Fan, L., Lee, Y. H., & Beardmore, D. (1981). The influence of major structural features of cellulose on rate of enzymatic hydrolysis. *Biotechnology and Bioengineering, 23*(2), 419–424.
100. Jameel, H., & Keshwani, D. R. (2017). Thermochemical conversion of biomass to power and fuels. In *Biomass to renewable energy processes* (pp. 375–422). Boca Raton: CRC Press.
101. Ong, H. C., Chen, W.-H., Farooq, A., Gan, Y. Y., Lee, K. T., & Ashokkumar, V. (2019). Catalytic thermochemical conversion of biomass for biofuel production: A comprehensive review. *Renewable and Sustainable Energy Reviews, 113*, 109266.
102. Pang, S. (2018). Advances in thermochemical conversion of woody biomass to energy, fuels and chemicals. *Biotechnology Advances, 37*(4), 589–597.
103. Ramos, L., Breuil, C., Kushner, D., & Saddler, J. (1992). Steam pretreatment conditions for effective enzymatic hydrolysis and recovery yields of Eucalyptus viminalis wood chips. *Holzforschung-International Journal of the Biology, Chemistry, Physics and Technology of Wood, 46*(2), 149–154.
104. Grous, W. R., Converse, A. O., & Grethlein, H. E. (1986). Effect of steam explosion pretreatment on pore size and enzymatic hydrolysis of poplar. *Enzyme and Microbial Technology, 8*(5), 274–280.
105. Wyman, C. (1996). *Handbook on bioethanol: Production and utilization*. Boca Raton: CRC Press.
106. Himmel, M. E., Baker, J. O., & Overend, R. P. (1994). *Enzymatic conversion of biomass for fuels production*. Washington, DC: American Chemical Society.

107. Weil, J., Sarikaya, A., Rau, S.-L., Goetz, J., Ladisch, C. M., Brewer, M., Hendrickson, R., & Ladisch, M. R. (1997). Pretreatment of yellow poplar sawdust by pressure cooking in water. *Applied Biochemistry and Biotechnology, 68*(1), 21–40.
108. Baugh, K. D., Levy, J. A., & McCarty, P. L. (1988). Thermochemical pretreatment of lignocellulose to enhance methane fermentation: II. Evaluation and application of pretreatment model. *Biotechnology and Bioengineering, 31*(1), 62–70.
109. Mosier, N. S., Ladisch, C. M., & Ladisch, M. R. (2002). Characterization of acid catalytic domains for cellulose hydrolysis and glucose degradation. *Biotechnology and Bioengineering, 79*(6), 610–618.
110. van Walsum, G. P., & Shi, H. (2004). Carbonic acid enhancement of hydrolysis in aqueous pretreatment of corn stover. *Bioresource Technology, 93*(3), 217–226.
111. Luo, C., Brink, D. L., & Blanch, H. W. (2002). Identification of potential fermentation inhibitors in conversion of hybrid poplar hydrolyzate to ethanol. *Biomass & Bioenergy, 22*(2), 125–138.
112. Mosier, N., Wyman, C., Dale, B., Elander, R., Lee, Y. Y., Holtzapple, M., & Ladisch, M. (2005). Features of promising technologies for pretreatment of lignocellulosic biomass. *Bioresource Technology, 96*(6), 673–686.
113. Dien, B., Jung, H., Vogel, K., Casler, M., Lamb, J., Iten, L., Mitchell, R., & Sarath, G. (2006). Chemical composition and response to dilute-acid pretreatment and enzymatic saccharification of alfalfa, reed canarygrass, and switchgrass. *Biomass & Bioenergy, 30*(10), 880–891.
114. Foston, M., & Ragauskas, A. J. (2010). Changes in lignocellulosic supramolecular and ultrastructure during dilute acid pretreatment of Populus and switchgrass. *Biomass & Bioenergy, 34*(12), 1885–1895.
115. Jensen, J. R., Morinelly, J. E., Gossen, K. R., Brodeur-Campbell, M. J., & Shonnard, D. R. (2010). Effects of dilute acid pretreatment conditions on enzymatic hydrolysis monomer and oligomer sugar yields for aspen, balsam, and switchgrass. *Bioresource Technology, 101*(7), 2317–2325.
116. Zhou, X., Xu, J., Wang, Z., Cheng, J. J., Li, R., & Qu, R. (2012). Dilute sulfuric acid pretreatment of transgenic switchgrass for sugar production. *Bioresource Technology, 104*, 823–827.
117. Cha, Y. L., Yang, J., Park, Y., An, G. H., Ahn, J. W., Moon, Y. H., Yoon, Y. M., Yu, G. D., & Choi, I. H. (2015). Continuous alkaline pretreatment of Miscanthus sacchariflorus using a bench-scale single screw reactor. *Bioresource Technology, 181*, 338–344.
118. Xu, J., Cheng, J. J., Sharma-Shivappa, R. R., & Burns, J. C. (2010). Lime pretreatment of switchgrass at mild temperatures for ethanol production. *Bioresource Technology, 101*(8), 2900–2903.
119. Salvi, D. A., Aita, G. M., Robert, D., & Bazan, V. (2010). Dilute ammonia pretreatment of sorghum and its effectiveness on enzyme hydrolysis and ethanol fermentation. *Applied Biochemistry and Biotechnology, 161*(1–8), 67–74.
120. Ko, J. K., Bak, J. S., Jung, M. W., Lee, H. J., Choi, I. G., Kim, T. H., & Kim, K. H. (2009). Ethanol production from rice straw using optimized aqueous-ammonia soaking pretreatment and simultaneous saccharification and fermentation processes. *Bioresource Technology, 100*(19), 4374–4380.
121. Gao, K., Boiano, S., Marzocchella, A., & Rehmann, L. (2014). Cellulosic butanol production from alkali-pretreated switchgrass (Panicum virgatum) and phragmites (Phragmites australis). *Bioresource Technology, 174*, 176–181.
122. Gupta, R., & Lee, Y. (2010). Investigation of biomass degradation mechanism in pretreatment of switchgrass by aqueous ammonia and sodium hydroxide. *Bioresource Technology, 101*(21), 8185–8191.
123. Qin, L., Liu, Z.-H., Jin, M., Li, B.-Z., & Yuan, Y.-J. (2013). High temperature aqueous ammonia pretreatment and post-washing enhance the high solids enzymatic hydrolysis of corn stover. *Bioresource Technology, 146*, 504–511.

124. Palonen, H., Thomsen, A. B., Tenkanen, M., Schmidt, A. S., & Viikari, L. (2004). Evaluation of wet oxidation pretreatment for enzymatic hydrolysis of softwood. *Applied Biochemistry and Biotechnology, 117*(1), 1–17.
125. Varga, E., Klinke, H. B., Reczey, K., & Thomsen, A. B. (2004). High solid simultaneous saccharification and fermentation of wet oxidized corn stover to ethanol. *Biotechnology and Bioengineering, 88*(5), 567–574.
126. Garrote, G., Dominguez, H., & Parajo, J. (1999). Hydrothermal processing of lignocellulosic materials. *European Journal of Wood and Wood Products, 57*(3), 191–202.
127. Bjerre, A. B., Olesen, A. B., Fernqvist, T., Plöger, A., & Schmidt, A. S. (1996). Pretreatment of wheat straw using combined wet oxidation and alkaline hydrolysis resulting in convertible cellulose and hemicellulose. *Biotechnology and Bioengineering, 49*(5), 568–577.
128. Ahring, B. K., Jensen, K., Nielsen, P., Bjerre, A., & Schmidt, A. (1996). Pretreatment of wheat straw and conversion of xylose and xylan to ethanol by thermophilic anaerobic bacteria. *Bioresource Technology, 58*(2), 107–113.
129. Martin, C., Klinke, H. B., & Thomsen, A. B. (2007). Wet oxidation as a pretreatment method for enhancing the enzymatic convertibility of sugarcane bagasse. *Enzyme and Microbial Technology, 40*(3), 426–432.
130. Curreli, N., Fadda, M. B., Rescigno, A., Rinaldi, A. C., Soddu, G., Sollai, F., Vaccargiu, S., Sanjust, E., & Rinaldi, A. (1997). Mild alkaline/oxidative pretreatment of wheat straw. *Process Biochemistry, 32*(8), 665–670.
131. Itoh, H., Wada, M., Honda, Y., Kuwahara, M., & Watanabe, T. (2003). Bioorganosolve pretreatments for simultaneous saccharification and fermentation of beech wood by ethanolysis and white rot fungi. *Journal of Biotechnology, 103*(3), 273–280.
132. Pan, X., Gilkes, N., Kadla, J., Pye, K., Saka, S., Gregg, D., Ehara, K., Xie, D., Lam, D., & Saddler, J. (2006). Bioconversion of hybrid poplar to ethanol and co-products using an organosolv fractionation process: Optimization of process yields. *Biotechnology and Bioengineering, 94*(5), 851–861.
133. Rolz, C., de Arriola, M., Valladares, J., & de Cabrera, S. (1986). Effects of some physical and chemical pretreatments on the composition and enzymatic hydrolysis and digestibility of lemon grass and citronella bagasse. *Agricultural Wastes, 18*(2), 145–161.
134. Lora, J. H., & Aziz, S. (1985). Organosolv pulping: A versatile approach to wood refining. *Tappi (United States), 68*(8), 94–97.
135. Zhao, X. (2009). Organosolv pretreatment of lignocellulosic biomass for enzymatic hydrolysis. *Applied Microbiology and Biotechnology, 82*(5), 815.
136. Dale, B. E. (1986). *Method for increasing the reactivity and digestibility of cellulose with ammonia*. United States Patent.
137. Dale, B. E., Leong, C., Pham, T., Esquivel, V., Rios, I., & Latimer, V. (1996). Hydrolysis of lignocellulosics at low enzyme levels: Application of the AFEX process. *Bioresource Technology, 56*(1), 111–116.
138. Chundawat, S. P., Venkatesh, B., & Dale, B. E. (2007). Effect of particle size based separation of milled corn stover on AFEX pretreatment and enzymatic digestibility. *Biotechnology and Bioengineering, 96*(2), 219–231.
139. Carvalheiro, F., Duarte, L. C., & Gírio, F. M. (2008). Hemicellulose biorefineries: A review on biomass pretreatments. *Journal of Scientific and Industrial Research, 67*, 849–864.
140. Lin, L., Yan, R., Liu, Y., & Jiang, W. (2010). In-depth investigation of enzymatic hydrolysis of biomass wastes based on three major components: Cellulose, hemicellulose and lignin. *Bioresource Technology, 101*(21), 8217–8223.
141. Yoon, H., Wu, Z., & Lee, Y. (1995). Ammonia-recycled percolation process for pretreatment of biomass feedstock. *Applied Biochemistry and Biotechnology, 51*(1), 5–19.
142. Shi, J., Balamurugan, K., Parthasarathi, R., Sathitsuksanoh, N., Zhang, S., Stavila, V., Subramanian, V., Simmons, B. A., & Singh, S. (2014). Understanding the role of water during ionic liquid pretreatment of lignocellulose: Co-solvent or anti-solvent? *Green Chemistry, 16*(8), 3830–3840.

143. Liu, E., Li, M., Das, L., Pu, Y., Frazier, T., Zhao, B., Crocker, M., Ragauskas, A. J., & Shi, J. (2018). Understanding lignin fractionation and characterization from engineered switchgrass treated by an aqueous ionic liquid. *ACS Sustainable Chemistry & Engineering, 6*(5), 6612–6623.
144. Kirk, T. K., & Chang, H.-M. (1981). Potential applications of bio-ligninolytic systems. *Enzyme and Microbial Technology, 3*(3), 189–196.
145. Hatakka, A. I. (1983). Pretreatment of wheat straw by white-rot fungi for enzymic saccharification of cellulose. *Applied Microbiology and Biotechnology, 18*(6), 350–357.
146. Keller, F. A., Hamilton, J. E., & Nguyen, Q. A. (2003). Microbial pretreatment of biomass. In *Biotechnology for fuels and chemicals* (pp. 27–41). Totowa: Springer.
147. Yao, W., & Nokes, S. E. (2014). Phanerochaete chrysosporium pretreatment of biomass to enhance solvent production in subsequent bacterial solid-substrate cultivation. *Biomass & Bioenergy, 62*, 100–107.
148. Chinn, M. S., & Nokes, S. E. (2006). Screening of thermophilic anaerobic bacteria for solid substrate cultivation on lignocellulosic substrates. *Biotechnology Progress, 22*(1), 53–59.
149. Flythe, M. D., Elía, N. M., Schmal, M. B., & Nokes, S. E. (2015). Switchgrass (Panicum virgatum) fermentation by Clostridium thermocellum and Clostridium beijerinckii sequential culture: Effect of feedstock particle size on gas production. *Advances in Microbiology, 5*(05), 311.
150. Bhandiwad, A., Shaw, A. J., Guss, A., Guseva, A., Bahl, H., & Lynd, L. R. (2014). Metabolic engineering of Thermoanaerobacterium saccharolyticum for n-butanol production. *Metabolic Engineering, 21*, 17–25.
151. da Costa Sousa, L., Chundawat, S. P., Balan, V., & Dale, B. E. (2009). 'Cradle-to-grave' assessment of existing lignocellulose pretreatment technologies. *Current Opinion in Biotechnology, 20*(3), 339–347.
152. Lin, Z., Huang, H., Zhang, H., Zhang, L., Yan, L., & Chen, J. (2010). Ball milling pretreatment of corn stover for enhancing the efficiency of enzymatic hydrolysis. *Applied Biochemistry and Biotechnology, 162*(7), 1872–1880.
153. Zakaria, M. R., Fujimoto, S., Hirata, S., & Hassan, M. A. (2014). Ball milling pretreatment of oil palm biomass for enhancing enzymatic hydrolysis. *Applied Biochemistry and Biotechnology, 173*(7), 1778–1789.
154. Li, J., Henriksson, G., & Gellerstedt, G. (2007). Lignin depolymerization/repolymerization and its critical role for delignification of aspen wood by steam explosion. *Bioresource Technology, 98*(16), 3061–3068.
155. Ko, J. K., Kim, Y., Ximenes, E., & Ladisch, M. R. (2015). Effect of liquid hot water pretreatment severity on properties of hardwood lignin and enzymatic hydrolysis of cellulose. *Biotechnology and Bioengineering, 112*(2), 252–262.
156. Samuel, R., Pu, Y., Raman, B., & Ragauskas, A. J. (2010). Structural characterization and comparison of switchgrass ball-milled lignin before and after dilute acid pretreatment. *Applied Biochemistry and Biotechnology, 162*(1), 62–74.
157. Kang, S., Xiao, L., Meng, L., Zhang, X., & Sun, R. (2012). Isolation and structural characterization of lignin from cotton stalk treated in an ammonia hydrothermal system. *International Journal of Molecular Sciences, 13*(11), 15209–15226.
158. Klinke, H. B., Ahring, B. K., Schmidt, A. S., & Thomsen, A. B. (2002). Characterization of degradation products from alkaline wet oxidation of wheat straw. *Bioresource Technology, 82*(1), 15–26.
159. Li, M., Foster, C., Kelkar, S., Pu, Y., Holmes, D., Ragauskas, A., Saffron, C. M., & Hodge, D. B. (2012). Structural characterization of alkaline hydrogen peroxide pretreated grasses exhibiting diverse lignin phenotypes. *Biotechnology for Biofuels, 5*(1), 38.
160. Hu, G., Cateto, C., Pu, Y., Samuel, R., & Ragauskas, A. J. (2011). Structural characterization of switchgrass lignin after ethanol organosolv pretreatment. *Energy & Fuels, 26*(1), 740–745.
161. Li, C., Cheng, G., Balan, V., Kent, M. S., Ong, M., Chundawat, S. P., daCosta, S. L., Melnichenko, Y. B., Dale, B. E., & Simmons, B. A. (2011). Influence of physico-chemical

changes on enzymatic digestibility of ionic liquid and AFEX pretreated corn stover. *Bioresource Technology, 102*(13), 6928–6936.

162. Lee, J.-W., Gwak, K.-S., Park, J.-Y., Park, M.-J., Choi, D.-H., Kwon, M., & Choi, I.-G. (2007). Biological pretreatment of softwood Pinus densiflora by three white rot fungi. *Journal of Microbiology, 45*(6), 485–491.
163. Suhara, H., Kodama, S., Kamei, I., Maekawa, N., & Meguro, S. (2012). Screening of selective lignin-degrading basidiomycetes and biological pretreatment for enzymatic hydrolysis of bamboo culms. *International Biodeterioration & Biodegradation, 75*, 176–180.
164. Martone, P. T., Estevez, J. M., Lu, F., Ruel, K., Denny, M. W., Somerville, C., & Ralph, J. (2009). Discovery of lignin in seaweed reveals convergent evolution of cell-wall architecture. *Current Biology, 19*(2), 169–175.
165. Yeh, T.-F., Chang, H.-m., & Kadla, J. F. (2004). Rapid prediction of solid wood lignin content using transmittance near-infrared spectroscopy. *Journal of Agricultural and Food Chemistry, 52*(6), 1435–1439.
166. Easty, D. B., Berben, S. A., DeThomas, F. A., & Brimmer, P. J. (1990). Near-infrared spectroscopy for the analysis of wood pulp: Quantifying hardwood-softwood mixtures and estimating lignin content. *Tappi Journal, 73*(10), 257–261.
167. Hatfield, R., & Fukushima, R. S. (2005). Can lignin be accurately measured? *Crop Science, 45*(3), 832–839.
168. Fukushima, R. S., & Hatfield, R. D. (2001). Extraction and isolation of lignin for utilization as a standard to determine lignin concentration using the acetyl bromide spectrophotometric method. *Journal of Agricultural and Food Chemistry, 49*(7), 3133–3139.
169. Sluiter, A., Hames, B., Ruiz, R., Scarlata, C., Sluiter, J., Templeton, D., & Crocker, D. (2008). Determination of structural carbohydrates and lignin in biomass. *Laboratory Analytical Procedure, 1617*(1), 1–16.
170. Freudenberg, K. (1965). Lignin: Its constitution and formation from p-hydroxycinnamyl alcohols. *Science, 148*(3670), 595–600.
171. Li, L., Zhou, Y., Cheng, X., Sun, J., Marita, J. M., Ralph, J., & Chiang, V. L. (2003). Combinatorial modification of multiple lignin traits in trees through multigene cotransformation. *Proceedings of the National Academy of Sciences of the United States of America, 100*(8), 4939–4944.
172. Pilate, G., Guiney, E., Holt, K., Petit-Conil, M., Lapierre, C., Leplé, J.-C., Pollet, B., Mila, I., Webster, E. A., & Marstorp, H. G. (2002). Field and pulping performances of transgenic trees with altered lignification. *Nature Biotechnology, 20*(6), 607.
173. Holtman, K. M., Chang, H.-M., & Kadla, J. F. (2004). Solution-state nuclear magnetic resonance study of the similarities between milled wood lignin and cellulolytic enzyme lignin. *Journal of Agricultural and Food Chemistry, 52*(4), 720–726.
174. Chang, H.-M., Cowling, E. B., & Brown, W. (1975). Comparative studies on cellulolytic enzyme lignin and milled wood lignin of sweetgum and spruce. *Holzforschung-International Journal of the Biology, Chemistry, Physics and Technology of Wood, 29*(5), 153–159.
175. Ralph, J., Hatfield, R. D., Piquemal, J., Yahiaoui, N., Pean, M., Lapierre, C., & Boudet, A. M. (1998). NMR characterization of altered lignins extracted from tobacco plants down-regulated for lignification enzymes cinnamylalcohol dehydrogenase and cinnamoyl-CoA reductase. *Proceedings of the National Academy of Sciences of the United States of America, 95*(22), 12803–12808.
176. Gosselink, R., Abächerli, A., Semke, H., Malherbe, R., Käuper, P., Nadif, A., & Van Dam, J. (2004). Analytical protocols for characterisation of sulphur-free lignin. *Industrial Crops and Products, 19*(3), 271–281.
177. Jönsson, A.-S., Nordin, A.-K., & Wallberg, O. (2008). Concentration and purification of lignin in hardwood kraft pulping liquor by ultrafiltration and nanofiltration. *Chemical Engineering Research and Design, 86*(11), 1271–1280.

178. Gidh, A. V., Decker, S. R., See, C. H., Himmel, M. E., & Williford, C. W. (2006). Characterization of lignin using multi-angle laser light scattering and atomic force microscopy. *Analytica Chimica Acta, 555*(2), 250–258.
179. Evtuguin, D., Domingues, P., Amado, F., Neto, C. P., & Correia, A. (1999). Electrospray ionization mass spectrometry as a tool for lignins molecular weight and structural characterisation. *Holzforschung, 53*(5), 525–528.
180. Gidh, A. V., Decker, S. R., Vinzant, T. B., Himmel, M. E., & Williford, C. (2006). Determination of lignin by size exclusion chromatography using multi angle laser light scattering. *Journal of Chromatography. A, 1114*(1), 102–110.
181. Baumberger, S., Abaecherli, A., Fasching, M., Gellerstedt, G., Gosselink, R., Hortling, B., Li, J., Saake, B., & de Jong, E. (2007). Molar mass determination of lignins by size-exclusion chromatography: Towards standardisation of the method. *Holzforschung, 61*(4), 459–468.
182. El Hage, R., Brosse, N., Chrusciel, L., Sanchez, C., Sannigrahi, P., & Ragauskas, A. (2009). Characterization of milled wood lignin and ethanol organosolv lignin from miscanthus. *Polymer Degradation and Stability, 94*(10), 1632–1638.
183. Joffres, B., Lorentz, C., Vidalie, M., Laurenti, D., Quoineaud, A.-A., Charon, N., Daudin, A., Quignard, A., & Geantet, C. (2014). Catalytic hydroconversion of a wheat straw soda lignin: Characterization of the products and the lignin residue. *Applied Catalysis B: Environmental, 145*, 167–176.
184. Salanti, A., Zoia, L., Orlandi, M., Zanini, F., & Elegir, G. (2010). Structural characterization and antioxidant activity evaluation of lignins from rice husk. *Journal of Agricultural and Food Chemistry, 58*(18), 10049–10055.
185. Tejado, A., Pena, C., Labidi, J., Echeverria, J., & Mondragon, I. (2007). Physico-chemical characterization of lignins from different sources for use in phenol–formaldehyde resin synthesis. *Bioresource Technology, 98*(8), 1655–1663.
186. Baumberger, S., Dole, P., & Lapierre, C. (2002). Using transgenic poplars to elucidate the relationship between the structure and the thermal properties of lignins. *Journal of Agricultural and Food Chemistry, 50*(8), 2450–2453.
187. Shen, H., Poovaiah, C. R., Ziebell, A., Tschaplinski, T. J., Pattathil, S., Gjersing, E., Engle, N. L., Katahira, R., Pu, Y., & Sykes, R. (2013). Enhanced characteristics of genetically modified switchgrass (Panicum virgatum L.) for high biofuel production. *Biotechnology for Biofuels, 6*(1), 71.
188. Chen, C. (1991). *Lignins: Occurrence in woody tissues, isolation, reactions, and structure*. New York: Wood Structure and Composition.
189. Crews, P., Rodriquez, J., Jaspars, M., & Crews, R. J. (2010). *Organic structure analysis* (Vol. 636). New York: Oxford University Press.
190. Robert, D., & Gagnaire, D. (1981). Quantitative analysis of lignins by 13C NMR. *Proceedings of the National Academy of Sciences of the United States of America, 1*, 9–12.
191. Xia, Z., Akim, L. G., & Argyropoulos, D. S. (2001). Quantitative 13C NMR analysis of lignins with internal standards. *Journal of Agricultural and Food Chemistry, 49*(8), 3573–3578.
192. Ralph, J., Marita, J. M., Ralph, S. A., Hatfield, R. D., Lu, F., Ede, R. M., Peng, J., Quideau, S., Helm, R. F., & Grabber, J. H. (1999). Solution-state NMR of lignins. In *Advances in lignocellulosics characterization* (pp. 55–108). Atlanta: TAPPI Press.
193. Marita, J. M., Ralph, J., Hatfield, R. D., & Chapple, C. (1999). NMR characterization of lignins in Arabidopsis altered in the activity of ferulate 5-hydroxylase. *Proceedings of the National Academy of Sciences of the United States of America, 96*(22), 12328–12332.
194. Ralph, J., Lapierre, C., Marita, J. M., Kim, H., Lu, F., Hatfield, R. D., Ralph, S., Chapple, C., Franke, R., & Hemm, M. R. (2001). Elucidation of new structures in lignins of CAD-and COMT-deficient plants by NMR. *Phytochemistry, 57*(6), 993–1003.
195. Pu, Y., Cao, S., & Ragauskas, A. J. (2011). Application of quantitative 31P NMR in biomass lignin and biofuel precursors characterization. *Energy & Environmental Science, 4*(9), 3154–3166.

196. Yuan, T.-Q., Sun, S.-N., Xu, F., & Sun, R.-C. (2011). Characterization of lignin structures and lignin–carbohydrate complex (LCC) linkages by quantitative 13C and 2D HSQC NMR spectroscopy. *Journal of Agricultural and Food Chemistry, 59*(19), 10604–10614.
197. Del Río, J. C., Rencoret, J., Prinsen, P., Martínez, A. T., Ralph, J., & Gutiérrez, A. (2012). Structural characterization of wheat straw lignin as revealed by analytical pyrolysis, 2D-NMR, and reductive cleavage methods. *Journal of Agricultural and Food Chemistry, 60*(23), 5922–5935.
198. Cao, S., Pu, Y., Studer, M., Wyman, C., & Ragauskas, A. J. (2012). Chemical transformations of Populus trichocarpa during dilute acid pretreatment. *RSC Advances, 2*(29), 10925–10936.
199. Shi, J., Gladden, J. M., Sathitsuksanoh, N., Kambam, P., Sandoval, L., Mitra, D., Zhang, S., George, A., Singer, S. W., & Simmons, B. A. (2013). One-pot ionic liquid pretreatment and saccharification of switchgrass. *Green Chemistry, 15*(9), 2579–2589.
200. Hou, X. D., Li, N., & Zong, M. H. (2013). Renewable bio ionic liquids-water mixtures-mediated selective removal of lignin from rice straw: Visualization of changes in composition and cell wall structure. *Biotechnology and Bioengineering, 110*(7), 1895–1902.
201. Sun, N., Parthasarathi, R., Socha, A. M., Shi, J., Zhang, S., Stavila, V., Sale, K. L., Simmons, B. A., & Singh, S. (2014). Understanding pretreatment efficacy of four cholinium and imidazolium ionic liquids by chemistry and computation. *Green Chemistry, 16*(5), 2546–2557.
202. Trajano, H. L., Engle, N. L., Foston, M., Ragauskas, A. J., Tschaplinski, T. J., & Wyman, C. E. (2013). The fate of lignin during hydrothermal pretreatment. *Biotechnology for Biofuels, 6*(1), 110.
203. Ben, H., & Ragauskas, A. J. (2011). NMR characterization of pyrolysis oils from kraft lignin. *Energy and Fuels, 25*(5), 2322–2332.
204. Eudes, A., Sathitsuksanoh, N., Baidoo, E. E., George, A., Liang, Y., Yang, F., Singh, S., Keasling, J. D., Simmons, B. A., & Loqué, D. (2015). Expression of a bacterial 3-dehydroshikimate dehydratase reduces lignin content and improves biomass saccharification efficiency. *Plant Biotechnology Journal, 13*(9), 1241–1250.
205. Mansfield, S. D., Kim, H., Lu, F., & Ralph, J. (2012). Whole plant cell wall characterization using solution-state 2D NMR. *Nature Protocols, 7*(9), 1579.
206. Griffiths, P. R., & De Haseth, J. A. (2007). *Fourier transform infrared spectrometry* (Vol. 171). Hoboken: Wiley.
207. Müller, G., Schöpper, C., Vos, H., Kharazipour, A., & Polle, A. (2008). FTIR-ATR spectroscopic analyses of changes in wood properties during particle-and fibreboard production of hard-and softwood trees. *BioResources, 4*(1), 49–71.
208. Sarkanen, K. V., & Ludwig, C. H. (1971). *Liguins. Occurrence, formation, structure, and reactions*. New York: Wiley-Interscience.
209. Faix, O. (1986). Investigation of lignin polymer models (DHP's) by FTIR spectroscopy. *Holzforschung-International Journal of the Biology, Chemistry, Physics and Technology of Wood, 40*(5), 273–280.
210. Freer, J., Ruiz, J., Peredo, M. A., Rodríguez, J., & Baeza, J. (2003). Estimating the density and pulping yield of E. globulus wood by DRIFT-MIR spectroscopy and principal components regression (PCR). *Journal of the Chilean Chemical Society, 48*(3), 19–22.
211. Kim, T. H., & Lee, Y. Y. (2005). Pretreatment and fractionation of corn stover by ammonia recycle percolation process. *Bioresource Technology, 96*(18), 2007–2013.
212. Li, C., Knierim, B., Manisseri, C., Arora, R., Scheller, H. V., Auer, M., Vogel, K. P., Simmons, B. A., & Singh, S. (2010). Comparison of dilute acid and ionic liquid pretreatment of switchgrass: Biomass recalcitrance, delignification and enzymatic saccharification. *Bioresource Technology, 101*(13), 4900–4906.
213. Caballero, J., Conesa, J., Font, R., & Marcilla, A. (1997). Pyrolysis kinetics of almond shells and olive stones considering their organic fractions. *Journal of Analytical and Applied Pyrolysis, 42*(2), 159–175.
214. Sharma, R. K., Wooten, J. B., Baliga, V. L., Lin, X., Chan, W. G., & Hajaligol, M. R. (2004). Characterization of chars from pyrolysis of lignin. *Fuel, 83*(11), 1469–1482.

215. Brebu, M., & Vasile, C. (2010). Thermal degradation of lignin—A review. *Cellulose Chemistry and Technology, 44*(9), 353.
216. Fierro, V., Torné-Fernández, V., Montané, D., & Celzard, A. (2005). Study of the decomposition of kraft lignin impregnated with orthophosphoric acid. *Thermochimica Acta, 433*(1), 142–148.
217. Erä, V., & Mattila, A. (1976). Thermal analysis of thermosetting resins. *Journal of Thermal Analysis and Calorimetry, 10*(3), 461–469.
218. Coats, A., & Redfern, J. (1963). Thermogravimetric analysis. A review. *Analyst, 88*(1053), 906–924.
219. Wunderlich, B. (2005). *Thermal analysis of polymeric materials*. New York: Springer.
220. Ma, R., Xu, Y., & Zhang, X. (2015). Catalytic oxidation of biorefinery lignin to value-added chemicals to support sustainable biofuel production. *ChemSusChem, 8*(1), 24–51.
221. Li, C., Zhao, X., Wang, A., Huber, G. W., & Zhang, T. (2015). Catalytic transformation of lignin for the production of chemicals and fuels. *Chemical Reviews, 115*(21), 11559–11624.
222. Zakzeski, J., Bruijnincx, P. C., Jongerius, A. L., & Weckhuysen, B. M. (2010). The catalytic valorization of lignin for the production of renewable chemicals. *Chemical Reviews, 110*(6), 3552–3599.
223. Luo, H., & Abu-Omar, M. M. (2018). Lignin extraction and catalytic upgrading from genetically modified poplar. *Green Chemistry, 20*(3), 745–753.
224. Parsell, T., Yohe, S., Degenstein, J., Jarrell, T., Klein, I., Gencer, E., Hewetson, B., Hurt, M., Im Kim, J., & Choudhari, H. (2015). A synergistic biorefinery based on catalytic conversion of lignin prior to cellulose starting from lignocellulosic biomass. *Green Chemistry, 17*(3), 1492–1499.
225. Feghali, E., Carrot, G., Thuéry, P., Genre, C., & Cantat, T. (2015). Convergent reductive depolymerization of wood lignin to isolated phenol derivatives by metal-free catalytic hydrosilylation. *Energy & Environmental Science, 8*(9), 2734–2743.
226. Parsell, T. H., Owen, B. C., Klein, I., Jarrell, T. M., Marcum, C. L., Haupert, L. J., Amundson, L. M., Kenttämaa, H. I., Ribeiro, F., & Miller, J. T. (2013). Cleavage and hydrodeoxygenation (HDO) of C–O bonds relevant to lignin conversion using Pd/Zn synergistic catalysis. *Chemical Science, 4*(2), 806–813.
227. Milczarek, G. (2009). Lignosulfonate-modified electrodes: Electrochemical properties and electrocatalysis of NADH oxidation. *Langmuir, 25*(17), 10345–10353.
228. Reichert, E., Wintringer, R., Volmer, D. A., & Hempelmann, R. (2012). Electro-catalytic oxidative cleavage of lignin in a protic ionic liquid. *Physical Chemistry Chemical Physics, 14*(15), 5214–5221.
229. Wen, X., Jia, Y., & Li, J. (2009). Degradation of tetracycline and oxytetracycline by crude lignin peroxidase prepared from Phanerochaete chrysosporium—A white rot fungus. *Chemosphere, 75*(8), 1003–1007.
230. Nousiainen, P., Kontro, J., Manner, H., Hatakka, A., & Sipila, J. (2014). Phenolic mediators enhance the manganese peroxidase catalyzed oxidation of recalcitrant lignin model compounds and synthetic lignin. *Fungal Genetics and Biology, 72*, 137–149.
231. Hirai, H., Sugiura, M., Kawai, S., & Nishida, T. (2005). Characteristics of novel lignin peroxidases produced by white-rot fungus Phanerochaete sordida YK-624. *FEMS Microbiology Letters, 246*(1), 19–24.
232. Thanh Mai Pham, L., Eom, M. H., & Kim, Y. H. (2014). Inactivating effect of phenolic unit structures on the biodegradation of lignin by lignin peroxidase from Phanerochaete chrysosporium. *Enzyme and Microbial Technology, 61–62*, 48–54.
233. Sitarz, A. K., Mikkelsen, J. D., Hojrup, P., & Meyer, A. S. (2013). Identification of a laccase from Ganoderma lucidum CBS 229.93 having potential for enhancing cellulase catalyzed lignocellulose degradation. *Enzyme and Microbial Technology, 53*(6–7), 378–385.
234. Shleev, S., Persson, P., Shumakovich, G., Mazhugo, Y., Yaropolov, A., Ruzgas, T., & Gorton, L. (2006). Interaction of fungal laccases and laccase-mediator systems with lignin. *Enzyme and Microbial Technology, 39*(4), 841–847.

235. Bugg, T. D., & Rahmanpour, R. (2015). Enzymatic conversion of lignin into renewable chemicals. *Current Opinion in Chemical Biology, 29*, 10–17.
236. Bugg, T. D., Ahmad, M., Hardiman, E. M., & Singh, R. (2011). The emerging role for bacteria in lignin degradation and bio-product formation. *Current Opinion in Biotechnology, 22*(3), 394–400.
237. Sato, Y., Moriuchi, H., Hishiyama, S., Otsuka, Y., Oshima, K., Kasai, D., Nakamura, M., Ohara, S., Katayama, Y., & Fukuda, M. (2009). Identification of three alcohol dehydrogenase genes involved in the stereospecific catabolism of arylglycerol-β-aryl ether by Sphingobium sp. strain SYK-6. *Applied and Environmental Microbiology, 75*(16), 5195–5201.
238. Masai, E., Kamimura, N., Kasai, D., Oguchi, A., Ankai, A., Fukui, S., Takahashi, M., Yashiro, I., Sasaki, H., Harada, T., Nakamura, S., Katano, Y., Narita-Yamada, S., Nakazawa, H., Hara, H., Katayama, Y., Fukuda, M., Yamazaki, S., & Fujita, N. (2012). Complete genome sequence of Sphingobium sp. strain SYK-6, a degrader of lignin-derived biaryls and monoaryls. *Journal of Bacteriology, 194*(2), 534–535.
239. Meux, E., Prosper, P., Masai, E., Mulliert, G., Dumarçay, S., Morel, M., Didierjean, C., Gelhaye, E., & Favier, F. (2012). Sphingobium sp. SYK-6 LigG involved in lignin degradation is structurally and biochemically related to the glutathione transferase omega class. *FEBS Letters, 586*(22), 3944–3950.
240. Pereira, J. H., Heins, R. A., Gall, D. L., McAndrew, R. P., Deng, K., Holland, K. C., Donohue, T. J., Noguera, D. R., Simmons, B. A., & Sale, K. L. (2016). Structural and biochemical characterization of the early and late enzymes in the lignin β-aryl ether cleavage pathway from Sphingobium sp. SYK-6. *The Journal of Biological Chemistry, 291*(19), 10228–10238.
241. Mori, K., Kamimura, N., & Masai, E. (2018). Identification of the protocatechuate transporter gene in Sphingobium sp. strain SYK-6 and effects of overexpression on production of a value-added metabolite. *Applied Microbiology and Biotechnology, 102*(11), 4807–4816.
242. McAndrew, R. P., Sathitsuksanoh, N., Mbughuni, M. M., Heins, R. A., Pereira, J. H., George, A., Sale, K. L., Fox, B. G., Simmons, B. A., & Adams, P. D. (2016). Structure and mechanism of NOV1, a resveratrol-cleaving dioxygenase. *Proceedings of the National Academy of Sciences of the United States of America, 113*(50), 14324–14329.
243. Kosa, M., & Ragauskas, A. J. (2012). Bioconversion of lignin model compounds with oleaginous Rhodococci. *Applied Microbiology and Biotechnology, 93*(2), 891–900.
244. Wei, Z., Zeng, G., Huang, F., Kosa, M., Huang, D., & Ragauskas, A. J. (2015). Bioconversion of oxygen-pretreated Kraft lignin to microbial lipid with oleaginous Rhodococcus opacus DSM 1069. *Green Chemistry, 17*(5), 2784–2789.
245. Le, R. K., Wells, T., Jr., Das, P., Meng, X., Stoklosa, R. J., Bhalla, A., Hodge, D. B., Yuan, J. S., & Ragauskas, A. J. (2017). Conversion of corn stover alkaline pre-treatment waste streams into biodiesel via Rhodococci. *RSC Advances, 7*(7), 4108–4115.
246. He, Y., Li, X., Xue, X., Swita, M. S., Schmidt, A. J., & Yang, B. (2017). Biological conversion of the aqueous wastes from hydrothermal liquefaction of algae and pine wood by Rhodococci. *Bioresource Technology, 224*, 457–464.
247. He, Y., Li, X., Ben, H., Xue, X., & Yang, B. (2017). Lipids production from dilute alkali corn stover lignin by Rhodococcus strains. *ACS Sustainable Chemistry & Engineering, 5*(3), 2302–2311.
248. Kohlstedt, M., Starck, S., Barton, N., Stolzenberger, J., Selzer, M., Mehlmann, K., Schneider, R., Pleissner, D., Rinkel, J., & Dickschat, J. S. (2018). From lignin to nylon: Cascaded chemical and biochemical conversion using metabolically engineered Pseudomonas putida. *Metabolic Engineering, 47*, 279–293.
249. Granja-Travez, R. S., & Bugg, T. D. (2018). Characterization of multicopper oxidase CopA from Pseudomonas putida KT2440 and Pseudomonas fluorescens Pf-5: Involvement in bacterial lignin oxidation. *Archives of Biochemistry and Biophysics, 660*, 97–107.
250. Lin, L., Wang, X., Cao, L., & Xu, M. (2019). Lignin catabolic pathways reveal unique characteristics of dye-decolorizing peroxidases in Pseudomonas putida. *Environmental Microbiology, 21*(5), 1847–1863.

251. Kumar, M., Singhal, A., Verma, P. K., & Thakur, I. S. (2017). Production and characterization of polyhydroxyalkanoate from lignin derivatives by Pandoraea sp. ISTKB. *ACS Omega, 2*(12), 9156–9163.
252. Li, M., Eskridge, K., Liu, E., & Wilkins, M. (2019). Enhancement of polyhydroxybutyrate (PHB) production by 10-fold from alkaline pretreatment liquor with an oxidative enzyme-mediator-surfactant system under Plackett-Burman and central composite designs. *Bioresource Technology, 281*, 99–106.
253. Mottiar, Y., Vanholme, R., Boerjan, W., Ralph, J., & Mansfield, S. D. (2016). Designer lignins: Harnessing the plasticity of lignification. *Current Opinion in Biotechnology, 37*, 190–200.
254. Das, L., Kolar, P., & Sharma-Shivappa, R. (2012). Heterogeneous catalytic oxidation of lignin into value-added chemicals. *Biofuels, 3*(2), 155–166.
255. Stärk, K., Taccardi, N., Bösmann, A., & Wasserscheid, P. (2010). Oxidative depolymerization of lignin in ionic liquids. *ChemSusChem, 3*(6), 719–723.
256. Himmel, M. E., Ding, S.-Y., Johnson, D. K., Adney, W. S., Nimlos, M. R., Brady, J. W., & Foust, T. D. (2007). Biomass recalcitrance: Engineering plants and enzymes for biofuels production. *Science, 315*(5813), 804–807.
257. Li, X., Ximenes, E., Kim, Y., Slininger, M., Meilan, R., Ladisch, M., & Chapple, C. (2010). Lignin monomer composition affects Arabidopsis cell-wall degradability after liquid hot water pretreatment. *Biotechnology for Biofuels, 3*(1), 27.
258. Fu, C., Mielenz, J. R., Xiao, X., Ge, Y., Hamilton, C. Y., Rodriguez, M., Chen, F., Foston, M., Ragauskas, A., & Bouton, J. (2011). Genetic manipulation of lignin reduces recalcitrance and improves ethanol production from switchgrass. *Proceedings of the National Academy of Sciences of the United States of America, 108*(9), 3803–3808.

Chapter 12
The Route of Lignin Biodegradation for Its Valorization

Weihua Qiu

12.1 Introduction

Although more complex in structure, lignin has a higher carbon content and lower oxygen content than either polysaccharides or whole-cellulose raw materials, which makes it an attractive raw material for the production of biofuels and chemicals. More importantly, the highly functional and aromatic nature of lignin offers potentiality for the direct preparation of special and fine aromatic chemicals [103]. As a result, the production of chemicals from lignin has attracted worldwide attention [73].

Researchers around the world devote increasing attention to the valorization of lignin by biological methods. In contrast to thermochemical lignin depolymerization methods which consume large amounts of energy, chemicals, and expensive catalysts, lignin biodepolymerization methods mainly use microorganisms and enzymes to break down the bonds between lignin units or bonds between lignin and carbohydrate [115]. Generally speaking, the bioprospecting of lignin-degrading microbes and enzymatic systems and the understanding of molecular and systematic degradation mechanisms and metabolic pathways of lignin and aromatics by means of systems biology analyses are two major focuses of the lignin value-added utilization [74]. At present, a large number of bacteria and fungi have been screened from environment, which contain various enzymes and have the characteristics of lignin degradation and transformation. It is very important for the cognition and accurate control of lignin biodegradation and the establishment of biological processing for lignin valorization.

W. Qiu (✉)
State Key Laboratory of Biochemical Engineering, Institute of Process Engineering, CAS, Beijing, China
e-mail: whqiu@ipe.ac.cn

Z.-H. Liu, A. Ragauskas (eds.), *Emerging Technologies for Biorefineries, Biofuels, and Value-Added Commodities*,
https://doi.org/10.1007/978-3-030-65584-6_12

12.2 Lignin-Degradable Microorganisms

12.2.1 Fungi

The degradation of lignocellulose by fungi, including white-rot fungi, brown-rot fungi, and soft-rot fungi, mainly attributes to the extracellular enzymes, including polysaccharide hydrolases and lignin-degrading enzymes [81]. Fungi can degrade lignocellulose rapidly, which makes it widely used in the fields of crop straw decay, wastewater treatment, bioconversion of lignocellulose, and so on [110]. Lignin-degrading fungi can be divided into three categories, white-rot fungi, brown-rot fungi, and soft-rot fungi. The lignin degradability is different significantly. White-rot fungi mainly degrade lignin and polysaccharides. Brown-rot fungi and soft-rot fungi mainly degrade cellulose, hemicellulose, and some polysaccharides. However, brown-rot fungi hardly degrade lignin; soft-rot fungi just degrade lignin slightly and slowly. Therefore, the degradation of lignin mainly relays on white-rot fungi.

White-rot fungi are the general name for a group of filamentous fungi that make wood white and rotten. It is the only microorganism in nature that can completely mineralize lignin. There are many white-rot fungi that can degrade lignin, mainly *Basidiomycota* and some *Ascomycetes*. The most studied white-rot fungi include *Phanerochete chrysosporium*, *Coriolus versicolor*, *Trametes versicolor*, *Phlebia radiata*, *Panus conchatus*, *Pleurotus pulmonarius*, *Pycnoporus cinnabarinus,* etc. [52]. Nowadays, *P. chrysosporium* is always used as the type strain of white-rot fungi. White-rot fungi have established unique lignocellulose degradation system in the long-term biological evolution, and they are the only known microorganism that can effectively degrade lignin into CO_2 and H_2O by pure culture. Actually, the degradation of lignin by white-rot fungi is a co-metabolic process between lignin and another extra nutrient used as carbon source. During the cultivation, these nutrients, as co-substrates, produce water-soluble extracts (such as ammonia peroxide, VA, organic acid, etc.) which could induce the aromatic oxidation and ring cleavage of lignin [61].

Therefore, the degradation of lignocellulose by white-rot fungi can be divided into two modes. The first mode is nonselective delignification, which means degrading lignin, cellulose, and hemicellulose simultaneously. The reported fungi that possessed this mode mainly included *Trametes versicolor*, *Irpex lacteus*, *Phanerochaete chrysosporium*, *Heterobasidion annosum*, *Phlebia radiate*, and some ascomycetes such as *Xylaria hypoxylon*. However, this type of degradation is limited around the hypha and degrades only a small amount of lignocellulose. The second mode is selective delignification which means lignin and hemicelluloses are attacked before cellulose. This mode mainly exists in *Ganoderma australe*, *Ceriporiopsis subvermispora*, *Phellinus pini*, *Phlebia tremellosa*, and some *Pleurotus*. In most white-rot fungi, both modes are available and are not mutually exclusive. The lignin degradability of white-rot fungus depends on its lignin-degrading enzymes, which initiate the free radical chain reaction and result in the C–C and C–O cleavage, demethylation, hydroxylation, benzyl alcohol oxidation, aromatic ring opening, etc. The products would be further degraded into CO_2 and H_2O through different pathways. In addition to white-rot fungi, other lignin degradability fungi, such as *Aspergillus oryzae* and *Aspergillus niger*, can also catalyze the degradation of lignin by ligninolytic enzymes [31, 136].

Although there had a lot of researches on the fungi degradation of lignin since 1980, the commercial application is still limited owing to the strict growth conditions and additional energy and chemical inputs [114].

12.2.2 *Bacteria*

Bacteria have many advantages over fungi on the lignin degradation, since they can tolerate a wider range of pH, temperature, and oxygen concentration, grow faster, have more biochemical functions, and have better environmental adaptability. They have attracted increasing attentions in the value-added utilization of lignin. However, the current understanding is that bacteria can only degrade low molecular weight of lignin-derived compounds. Therefore, they may play an important role in the final stage of lignin degradation [131]. In recent years, a large number of bacteria that can depolymerize lignin have been found in composted soil, rainforest, eroded bamboo slips, sludge from pulp mills, and intestines of wood-eating insects [132]. There are many types of bacteria that can degrade lignin, which are mainly *Actinobacteria*, *Proteobacteria* (mainly α-*Proteobacteria* and γ-*Proteobacteria*), and *Firmicutes*. *Actinomycota* includes *Streptomyces*, *Microbacterium*, and *Rhodococcus*. *Proteobacteria* includes *Pseudomonas*, *Burkholderia*, and *Enterobacter* [44].

12.2.2.1 *Actinomycetes*

Actinomycetes can penetrate insoluble substrates and increase the water solubility of lignocellulose. It is mainly involved in the initial degradation and humification of organic matters. They secrete a series of enzymes and convert lignin into low-molecular-weight lignin-derived compounds [43]. With the development in isolation and identification technology, more and more lignin-degrading *Actinomycetes* have been found, including *Streptomyces* sp., *Thermoactinomyces* sp., *Arthrobacter* sp., *Micromonospora* sp., *Nocardia* sp., *Thermomonospora* sp., etc. They can produce a variety of enzymes, such as lignin peroxidase, laccase, xylanase, carboxymethyl cellulase, p-nitrophenyl-β-D-glucosidase, etc., responsible for the degrading of lignocellulose [4]. With these enzymes, they can change the cell wall structure of lignocellulose effectively through selective cleavage and structure modification of lignin, thereby promoting the enzymatic hydrolysis of cellulose to release sugar [140]. It obviously modified the carbonyl and methoxy groups in the structure of lignin, especially resulting in the significant reduction of guaiacol units [141].

The advantages of *Streptomyces* in lignin degradation are becoming increasingly apparent. For example, *Streptomyces griseorubens* can degrade more than 60% of lignin by secreting extracellular laccase and lignin peroxidase, as well as exo-1,4-β-glucanase, endo-1,4-β-glucanase, and β-xylosidase [34, 130]. *Streptomyces viridosporus* T7A and *S. setonii* 75Vi2 can degrade lignin and carbohydrate in softwood, hardwood, and herbal, especially in herbs, with the generation of lignin-carbohydrate complex named acid-precipitable polymeric lignin (APPL) by cleaving p-hydroxy ether bonds in lignin [21]. The degradation rate of lignin by *S. viridosporus* T7A can reach 19.7% within 8 weeks [135]. *Streptomyces* can also degrade alkaline lignin into phenolic compounds. Meanwhile, the coculture with white-rot fungi can effectively enhance the degradation rate of alkaline lignin [137]. In particular, the recognition of some thermophilic *Actinomycetes*, such as *Thermomonospora fusca*, *T. alba*, *Micromonospora* sp., etc., makes the industrial application of *Actinomycetes* in lignin valorization possible [18, 67].

12.2.2.2 Proteobacteria

The major lignin-degrading *Proteobacteria* is α-*Proteobacteria* and γ-*Proteobacteria*. As the most important genus of α-*Proteobacteria*, *Sphingobacterium* sp. has been widely used in lignin degradation, which can use Klason lignin as the sole carbon source with the secretion of lignin peroxidase and laccase [14]. *Sphingobacterium* sp. can oxidize guaiacyl (G) and syringyl (S) units in Klason lignin to produce small molecule aromatic compounds and ketone compounds, such as guaiacol, p-hydroxybenzoic acid, vanillic acid, vanillin, 4-hydroxy-2-butanone, methyl vinyl ketone, etc. [27].

The best known lignin-degrading γ-*Proteobacteria* is *Pseudomonas fluorescens*, *Pseudomonas putida*, *Enterobacter lignolyticus*, and *E. coli*. Their strong lignin degradability could be attributed to the plentiful lignin-degrading enzymes including laccase, lignin peroxidase (LiP), and manganese-dependent peroxidase (MnP), as well as some kind of lignin-degrading auxiliary enzymes [132]. For example, *Pseudomonas fluorescens* mainly secretes LiP; dye-decolorizing peroxidase (DyP) and MnP are the major enzymes in *Pseudomonas putida*; catalase, peroxidase, and DyP are the major enzymes in *Enterobacter lignolyticus*; and laccase is the major enzymes responsible for lignin degradation of *Escherichia coli* [44].

12.2.2.3 Firmicutes

The lignin-degrading *Firmicutes* are mainly belonging to *Bacillus*, such as *Basic Bacillus*, *B. ligniniphilus*, *B. firmus*, thermophilic *B. subtilis*, and *B. licheniformis*. It was suggested that the lignin-degrading pathway in *B. ligniniphilus* included gentisate pathway, benzoic acid pathway, and β-ketoacidic acid pathway, among which β-ketoacidic acid pathway included catechuic acid branches and protocatechuic acid branches. Zhu et al. [147] reported a *B. ligniniphilus* strain with high tolerance to the extreme environments, which can grow on alkaline lignin by using it as a sole carbon or energy source. And up to 42% of monophenol aromatic compounds were identified in the fermented residues, including phenylacetic acid, 4-hydroxybenzoic acid, and vanillic acid. Thermophilic *B. subtilis* and *B. licheniformis* have also been found to degrade KL lignin significantly [29].

Other lignin-degrading *Firmicutes* include *Acetoanaerobium*, *Trabulsiella*, *Paenibacillus glucanolyticus*, etc. Duan et al. [26] reported an *Acetoanaerobium* strain that could oxidize the p-hydroxyphenyl (H), guaiacyl (G), and syringyl (S) unit of KL structure with the formation of benzene-propanoic acid, ferulic acid, syringic acid, and some low-molecular-weight acid compounds, such as adipic acid, hexanoic acid, and 2-hydroxybutyric acid. *Trabulsiella guamensis* IIPTG13 reported by Suman et al. [119] caused the degradation of over 60% of guaiacylglycerol-β-guaiacyl ether (GGE). And a variety of compounds containing polar functional groups was detected in the fermented residues, mainly including organic acids, fatty acids, and other phenolic derivatives. In particular, some compounds such as 4-hydroxy-3-methoxybenzoic acid, 3-methoxybenzoic acid, and 2-(4-hydroxy-3-methoxyphenyl) acetic acid were similar to coniferyl alcohol and sinapyl alcohol (lignin precursor).

12.3 Lignin-Degrading Enzymes

The irregularity, complexity, and specificity of the three-dimensional structure of lignin polymers determine that the lignin-degrading enzyme system is a nonspecific enzyme system. Compared with the enzymatic hydrolysis of cellulose and hemicellulose, the lignin-degrading process was oxidation rather than hydrolysis [52]. The high-molecular-weight lignin polymers generate unstable free radicals under the catalysis of enzymes, which triggers the spontaneously nonenzymatic cleavage reactions of lignin.

Generally, the lignin-degrading enzymes are divided into three categories: peroxidase, laccase, and lignin degradation auxiliary enzyme (LDA) [96]. Peroxidase mainly includes lignin peroxidase (LiP), manganese peroxidase (MnP), versatile peroxidase (VP), and dye-decolorizing peroxidase (DyP). Laccase is a type of copper-containing polyphenol oxidase which can directly use oxygen to catalyze phenolic hydroxyl groups with phenoxy radicals and water as products. LDA enzymes cannot degrade lignin directly, but it is necessary for the complete degradation of lignin. It mainly includes glyoxal oxidase, aryl alcohol dehydrogenase, heme-thiolate haloperoxidases, flavin adenine dinucleotide-dependent glucose dehydrogenase (FAD-GDH), pyranose 2-oxidase, and other enzymes like quinone reductase, superoxide dismutase, methanol oxidase, etc. Different lignin-degrading microbes usually secrete one or more lignin-degrading enzymes in different combinations, which would degrade lignin through sequential actions.

12.3.1 Lignin-Degrading Peroxidase

12.3.1.1 Manganese-Dependent Peroxidase

Manganese peroxidase (EC1. 11. 1.13, MnP) is an extracellular peroxidase, which was first found in *P. chrysosporium* more than 30 years ago, and is one of the most important lignin peroxidases [42, 95]. It was found that almost all white rot fungi contain MnP, such as *Panus tigrinus, Trametes versicolor, Lenzites betulinus, Irpex lacteus, Lenzites gibbosa, Phlebia radiata, Fomitiporia mediterranea, Agaricus bisporus, Bjerkandera sp., Nematoloma frowardii*, as well as species belong to Corticiaceae, Stereaceae, Hericiaceae, Ganodermataceae, Hymenochaetaceae, Polyporaceae, Strophariaceae, and Tricholomataceae [17]. Recently, more and more MnP was found in bacteria, yeast, and mold, such as *Bacillus pumilus*, *Paenibacillus* sp., *Azospirillum brasilense*, and *Streptomyces psammoticus* [25, 69, 90]. Generally, these microorganisms have multiple MnP isoenzymes, and there are at least two or more isomers.

The oxidation reaction of Mn^{2+} to Mn^{3+} catalyzed by MnP is H_2O_2-dependent. It takes Mn^{2+} as the preferred reducing substrate and turns it into highly active Mn^{3+}. The MnP-catalyzed lignin degradation cycle starts when H_2O_2 is bound to MnP. However, excessive H_2O_2 will affect the activity of MnP [76]. The generated Mn^{3+} would chelate with organic acids produced during the fermentation process, such as oxalic acid, to form stable low-molecular-weight substances. Then, the chelated Mn^{3+} acting as an oxidant penetrates or diffuses into the cell wall of plant and nonspecifically attacks and oxidizes the phenol structure of lignin, thus leading to the generation of unstable phenoxy radicals [17]. The phenoxy radicals are further broken down spontaneously with the reaction of demethylation, alkyl–aryl bond cleavage, Cα–Cβ cleavage, and Cα oxidation [47].

In general, secondary metabolism caused by the nutrient limitation of nitrogen and/or carbon in the growing process of microbe will cause the secretion of peroxidase and the degradation of lignin. Sufficient Mn^{2+} is necessary for the secretion of MnP by microorganisms. The addition of Mn^{2+} can increase the activity of MnP, promote the growth of fungi, shorten reaching time of the maximum MnP activity, and further enhance the degradation of lignocellulose. Meanwhile, in some microorganisms, Mn^{2+} can stimulate the transcription levels of MnP and played a post-transcriptional role in MnP production [45, 77].

12.3.1.2 Lignin Peroxidase

Lignin peroxidase (EC 1.11.1.14, LiP) is an extracellular enzyme which can oxidize the non-phenolic aromatic compounds with redox potentials higher than 1.4 V. The first LiP was found in *Phanerochaete chrysosporium*, and then many white-rot fungi, such as *Trametes versicolor*, *Bjerkandera* sp., and *Phlebia tremellosa*, were reported that can produce LiP. In recent years, it has also been detected in some bacteria, such as *Acinetobacter calcoaceticus*, *Streptomyces viridosporus*, etc. [52]. Generally, microorganisms secrete several lignin peroxidase (LiP) isoenzymes, of which the relative composition and isoelectric point (pI) are greatly affected by microbial culture conditions [116]. The crystal structure of LiPs has been recognized. It belongs to glycosylase, which possesses iron (III) protoporphyrin IX (ferriprotoporphyrin IX) as the prosthetic group [60]. It can oxidize iron cytochrome C, suggesting that the enzyme can directly act on polymerized lignin. Therefore, it plays an important role in the oxidative degradation of lignin.

Similar to MnP, the catalytic reaction of LiP requires the participation of H_2O_2. The catalytic process of LiP can be regarded as a series of free radical chain reactions initiated by H_2O_2 produced by white-rot fungi. Under the catalysis of LiP, an electron is extracted from the aromatic substrate and oxidized to form an aromatic cation group which can be used as both a reaction group and a cation. Then, the substrate would be partially or completely oxidized into various derived low-molecular-weight compounds through series branch reactions. LiP is also currently the only known oxidase that can cleave C_α–C_β and aromatic rings in an acellular state. However, different from MnP, LiP has low substrate specificity and can react with different lignin-like compounds. Meanwhile, LiP has high redox properties (about 1.2 V at pH 3), which can not only oxidize the typical phenol and aromatic amines but also can oxidize a series of other aromatic ethers and polycyclic aromatic compounds, as well as methoxy aromatic ring substrates without free phenolic group, through ionization reaction [113].

12.3.1.3 Versatile Peroxidase

Versatile peroxidase (EC 1.11.1.16, VP), considering as structural heterozygote between LiP and MnP, can directly oxidize the phenolic and non-phenolic aromatic compounds. The reported microorganisms that possess VP activities are major in genus of *Pleurotus*, *Bjerkandera*, *Phanerochaete, Trametes*, etc. [11, 19, 122, 139].

VPs have some catalytic properties of both LiP and MnP, resulting in an even wider range of substrates. It can catalyze the substrates with high and medium redox potentials, even azo dyes and other non-phenolic compounds [10, 40]. In addition, a Mn^{2+} binding site similar to MnP was in the structure of VP, which enables VP to participate in the oxidation of low and high redox potential substrates [38]. The degradation pathway of phenolic β-O-4 catalyzed by VP was identified as Cα–aryl cleavage rather than Cα–Cβ cleavage. In the VP catalyzed reaction of phenolic lignin dimers, neutral radicals were generated firstly by the oxidation of 4-OH position, and then the polymerization or depolymerization reactions were reacted depending on the functional group at the 5-position of the guaiacol group (G5). The side chain of lignin would cleave owing to the substitution of G5 with methoxy (S-O-4) [139]. Therefore, VP-producing microorganisms play an important role in the biodegradation of lignin and lignocellulose. Wen Kong et al. [64] found that although a lignin-degrading *Physisporinus vitreus* strain can secrete both laccase and VP, only VP can oxidize non-phenolic lignin compounds as well as the β-O-4 and 5–5′ dimers.

12.3.1.4 Dye-Decolorizing Peroxidase

Dye-decolorizing peroxidase (EC 1.11.1.19, DyP) was first found in *Bjerkandera adusta* (formerly named as *Geotrichum candidum*) [62]. They have low substrate specificity and can oxidize all typical peroxidase substrates, such as ABTS, DMP, adlerol, various anthraquinone dyes (e.g., Congo red, mordant black 9, reactive blue 5, etc.), and so on [20]. Presently, more and more DyP-producing fungi have been reported, such as *Pleurotus ostreatus* [32], *Termitomyces albuminosus* [54], and *Irpex lacteus* [28]. Especially, the popular reports of DyPs in bacteria, such as *Rhodococcus jostii* [104], *Bacillus subtilis* [84, 111], *Kocuria rosea* [94], and *Thermobifida fusca* [101], promote the deep researches and wide applications of DyP in lignin degradation.

DyPs also belong to the heme peroxidase family but with different structural folds from other types of heme peroxidases. They possess special reactivity to the oxidation of polycyclic dyes and phenolic compounds. The active site aspartic acid residue of DyP acts as a proton donor [118]. The catalytic characteristics on lignocellulose and lignin of DyPs are obviously similar to LiP and MnP. However, DyPs also have hydrolase or oxygenase activities [117]. Meanwhile, they can degrade different dyes at low pH (pH 3–4), especially anthraquinone dyes [20]. These characteristics make the research of VPs have significant meanings. Especially considering the high redox of VPS in fungi, the application of bacterial VPS in biotechnology will be more extensive in the future [49].

12.3.2 Lignin-Degrading Auxiliary

12.3.2.1 Glyoxal Oxidase

Glyoxal oxidase (EC 1.2.3.5, GLOX), belonging to radical copper oxidase family, is one of three known extracellular enzymes secreted by microorganisms under conditions of lignin decomposition [24]. The study of GLOX is mainly based on *Phanerochaete chrysosporium*. The catalytic mechanism of GLOX is that it can oxidize many simple dicarbonyl and hydroxycarbonyl compounds (especially glyoxal and methylglyoxal) to generate carboxylic acids coupling with the reduction of O_2 to H_2O_2 [133]. Therefore, GLOX is inactive in vitro catalysis for lignin degradation unless being coupled with peroxidase as auxiliary enzymes, since H_2O_2 is necessary for the degradation of lignin catalyzed by peroxidase such as LiPs and MnPs [52, 59].

12.3.2.2 Aryl Alcohol Oxidase

Aryl alcohol oxidase (EC 1.1.3.7, AAO), also known as glucose–methanol–choline oxidase, was originally found in *Polystictus versicolor*. At present, more and more fungi have reported possessed AAO activity, such as *Pleurotus species*, *Fusarium species*, *Geotrichum candidum*, *Amauroderma boleticeum*, *Phanerochaete chrysosporium*, and *Bjerkandera adusta*, and even some bacteria, such as *Sphingobacterium* sp. ATM [68, 120].

AAO has two domains with non-covalently bound FAD cofactors [35]. The auxiliary effect of AAO on the degradation of lignin is mainly considered to be that it can catalyze the oxidative dehydrogenation of phenol and non-phenol aromatic alcohol, the formation of corresponding aldehydes from polyunsaturated primary alcohols or aromatic secondary alcohols, and the generation of H_2O_2 as by-products [46]. In addition, AAO can promote the depolymerization of lignin by cooperating with laccase, since it can reduce the free radical intermediates (such as guaiacol, erucic acid, etc.) generated from laccase oxidation [82].

12.3.2.3 Heme-Thiolate Haloperoxidases

Heme-thiolate haloperoxidases (HTHPs) are extracellular enzymes which use cysteine residue as the proximal axial ligand of heme. For a long time, chloroperoxidase (CPO) from the ascomycete *Caldariomyces fumago* was the only known halogenating enzyme. Later, another type of fungal HTHP was found in *Agrocybe aegerita*, named as *Agrocybe aegerita* peroxidase (APO, also called aromatic peroxygenases), which showed strong brominating and weak chlorinating as well as weak iodating activities. Considering it can convert veratryl alcohol into veratryl aldehyde at pH 7, this enzyme was called "alkaline lignin peroxidase" before Ullrich et al. recognized it as HTHP [48, 123].

HTHPs represent a unique oxidoreductase sub-subclass of heme proteins with peroxygenase and peroxidase activity, but it cannot oxidize the non-phenolic β-O-4 structure of lignin. They can transfer oxygen via hydrogen peroxide to aromatic and aliphatic substrates similar to cytochrome P450, rather than directly to O_2 [98]. The auxiliary effect of fungal HTHP on lignin degradation was only found until recently. Ruiz-Dueñas et al. [107] found that the theme peroxides of *Pleurotus ostreatus* not only included cytochrome c peroxidase and lignin-degrading peroxidases (MnP and VP) but also included HTHP. This was the first description of HTHP in *Pleurotus*. Moreover, compared with the typical peroxidase, HTHP has surprisingly high reactivity against benzyl hydrogen and phenolic substrates [52, 128]. Now by accumulating the genome sequence of ascomycetes and basidiomycetes, the sequence library of HTHPs was gradually expanded [65].

12.3.2.4 Flavin Adenine Dinucleotide (FAD)-Dependent Glucose Dehydrogenase (GDH)

The flavin adenine dinucleotide (FAD)-dependent glucose dehydrogenases (EC 1.1.99.10, FADGDHs) belongs to the glucose–methanol–choline (GMC) oxidoreductase family. FAD is used as the major electron acceptor to catalyze the oxidation of first hydroxyl group of glucose and other sugar molecules [138]. The reported FADGDHs were mainly derived from fungal extracellular enzymes, membrane binding protein of Gram-negative bacteria, and cytosolic enzymes in some insects [36, 99]. The first reported FADGDHs from basidiomycetes were obtained by recombined GDH (PcGDH) from *Pycnoporus cinnabarinus* CIRM BRFM 137 in *Aspergillus niger* [105]. Compared with other glucose–methanol–choline (GMCs) oxidoreductases, the substrate binding domain of PcGDH is conserved. The enzyme does not use oxygen as the external electron acceptor and has the obvious ability of reducing the oxidized quinone or free radical intermediate, which makes the enzyme play an important role in decreasing the toxicity of quinones and protecting fungal cells from phenoxy. This provides a new way to detoxify the toxic compounds formed in the process of lignin degradation by PcGDH [105].

12.3.2.5 Pyranose 2-Oxidase

Pyranose 2-oxidase (EC 1.1.3.10, pyranose: 2-oxo oxidoreductase, P2OX) belongs to the glucose–methanol–choline oxidoreductase (GMC) family. It is a sugar oxidizing enzyme that can produce hydrogen peroxide during the glucose starvation period of *Phanerochaete chrysosporium* [30]. P2OX has been found in a variety of microorganisms, including *Phanerochaete chrysosporium*, *Trametes versicolor*, *Oudemansiella mucida*, *Phlebiopsis gigantean*, *Pleurotus ostreatus*, *Polyporus obtusus*, *Trametes multicolor*, *Peniophora* sp., *Tricholoma matsutake*, etc. [2, 23, 63, 83]. A new type bacterial P2OX (PaP2OX) was even identified from the lignin-degrading bacteria *Pantoea ananatis*. Different from other reported bacterial P2Ox, this enzyme showed homotetrameric spatial conformation that is similar to fungal P2Ox [143]. P2O was preferentially located in hyphal periplasmic space and prefer to oxidize aldopyranoses derived from cellulose and hemicellulose into ketoaldoses at C2 (also at C3) [66].

The catalytic process of P2OX can be divided into two steps, i.e., the formation of FADH intermediate by transferring hydride ion to N5 atom of isoalloxazine and the regeneration from dioxygen to FAD. It is worth noting that the benzoquinones and lignin-derived substances during lignin degradation can also be used for the regeneration of reduced FAD [2]. A large number of studies have shown that the synergistic effect of P2O and laccase can improve the degradation efficiency of lignin. It is attributed to the quinone-reducing activity of P2O which can reduce the polymerization of phenoxy radicals or quinoid intermediates in vivo. Ai et al. [2] found that during the catalysis of laccase, P2O can effectively inhibit the polymerization of lignin and even promote the lignin degradation by transferring electrons to various quinones and ABTS (2,2′-azinobis(3-ethylbenzthiazoline-6-sulfonate) cation radicals, as well as quinoids generated by laccase catalysis.

12.3.3 Laccase (EC 1.10.3.2)

Laccase, also called blue polycopper oxidase, is a kind of copper-containing polyphenol oxidase. Laccase plays an important role in the biodegradation of lignin. In plants, laccase is an intracellular enzyme involved in lignin synthesis, while in microorganisms, laccase mainly participates in the lignin degradation as an extracellular enzyme. Laccase-secreting fungi mainly include *Basidiomycota*, *Ascomycota*, and *Deuteromycota*. Compared with fungi, laccase is the most popular ligninolytic enzyme in prokaryotes. Laccase gene has been found in *Proteus*, *Actinomycetes*, and *Firmicutes* [121].

Different from lignin peroxidase taking H_2O_2 as prerequisite condition, laccase can use oxygen molecules as the final electron acceptor to catalyze the oxidation, degradation, or polymerization of phenolic and non-phenolic substrates. The utilization of O_2 instead of H_2O_2 as substrate results in a mild catalytic condition for lignin biodegradation. Meanwhile, in vitro, laccase synergistically degrades lignin with some quinone-reducing enzymes, such as pyranose oxidase, glucose oxidase, resveratrol oxidase, and cellobiose dehydrogenase [2, 33, 70, 78]. Due to its unique catalytic properties, laccase has been widely used in the high-value utilization of lignin [115].

There are four Cu^{2+} located at three binding domains in the active center of typical laccase, which have different electronic paramagnetic resonance (EPR) parameters. These copper ions synergistically mediate the catalytic oxidation of phenolic and non-phenolic substrates. However, in recent years, it has been found that there is a kind of atypical laccase in the complex laccase family. Anita et al. found a laccase that possessed two copper binding domains rather than three, which was called as small laccase or small laccase-like multicopper oxidase (LMCO) [3]. Meanwhile, it was found that in addition to copper ions, some other metal ions were combined on the binding domains of some kind of laccase. For example, the active center of *Phlebia radiate* laccase contains two Cu and one PQQ cofactor acting as T3Cu [93]. The active center of *Pleurotus ostreatus* laccase contains Cu, Fe, and Zn [85]. The active center of *Phellinus ribis* contains Cu, Mg, and Zn [5]. These differences in active centers caused the significant lignin degradability of laccase.

The oxidation lignin and its model compounds catalyzed by laccase are a one-electron reaction generating free radicals. Firstly, the substrate was catalyzed by laccase to generate unstable phenoxy radicals and continues to be converted to quinone in the consequent reactions [86]. In general, laccase can catalyze bond cleavage in low-molecular-weight phenol lignin model compounds, including ortho-diphenol and para-diphenol, aminophenol, methoxy-substituted phenol, thiol, polyphenol, polyamine, etc. In general, laccase can only oxidize the phenolic structure of lignin, accounting for 10% of the lignin structure. The rest of non-phenolic lignin structures with high redox potential could not be degraded by laccase unless by adding low-molecular-weight mediators, such as ABTS and HBT [88]. Recently, some atypical laccases have been found to directly oxidize non-phenolic lignin model compounds or even polycyclic aromatic hydrocarbons without relying on mediators. For example, a kind of atypical laccase, lacking the specific 600 nm band without the blue color (also called "yellow" or "white" laccase), was widely distributed in many microorganisms, such as *Schizophyllum commune*, *Pleurotus ostreatus*, *Leucoagaricus gongylophorus*, *Daedalea flavida*, *Aureobasidium pullulans*, and *Lentinus squarrosulus* [1, 13, 50, 87, 100, 112, 126]. These atypical laccases have important research and application significance for broadening the use of laccase and reducing the cost of laccase mediator catalytic system.

12.3.4 Other Enzymes

Besides the abovementioned ligninolytic enzymes, some enzymes reported recently would be promising options for lignin degradation. β-etherases (EC) are non-radical ligninolytic enzymes, which can selectively cleave β-O-4 aryl ether bonds in lignin. The conversion rate of lignin model compound GGE (1-(4-hydroxy-3-methoxyphenyl)-2-(2-methoxyphenoxy) propane-1,3-diol) catalyzed by β-etherase from *Sphingobium* sp. SYK-6 was close to 100% [106]. However, except for *Chaetomium* and *Sphingobium* genus, the natural existence of β-etherase was rarely reported [97]. Another lignin-degrading enzyme called superoxide dismutases has been identified recently from *Sphingomonas bacterium*, which can degrade lignin and generate low-molecular-weight aromatic derivatives, such as guaiacol and demethylated guaiacol [39].

12.4 Major Metabolic Pathway of Lignin Degradation

It is very important to study the metabolism network of lignin and construct lignin-degrading system, which is beneficial for the bioconversion of lignin into high-value intermediate metabolites. The known catabolic pathways related to lignin degradation in white-rot fungus include benzoic acid metabolism, dioxin metabolism, phenol metabolism, polycyclic aromatic hydrocarbon (PAH) metabolism, aminobenzoic acid metabolism, naphthalene metabolism, toluene metabolism, fatty acid metabolism, ether ester metabolism, xylene metabolism, carbon metabolism, etc. [75]. Due to the biochemical versatility and high environmental adaptability, increasing attention has been paid on the catabolic pathways related to lignin degradation in bacteria. Different from fungi, the metabolism of lignin and its derived aromatic compounds by bacteria is completed in vivo, taking β-ketoadipate pathway as the main pathway [131].

As bioinformatics and genomic sequencing technologies are used to screen functional genes encoding lignin-degrading enzymes, more and more genes and their encoded metabolic pathways have been reported. For example, 29 lignin-degrading enzyme genes were recognized from *Aspergillus fumigatus*, including multicopper oxidase, lignin-modifying peroxidase, glucose–methanol–choline oxidoreductase, vanillyl alcohol oxidase, galactose oxidase, 1,4-benzoquinone reductase, monooxygenase, ferulic acid esterase, etc., which degraded lignin by participating in six catabolic pathways, namely, galactose metabolism, phenylalanine metabolism, pyruvate metabolism, benzoate metabolism, toluene metabolism, and aminobenzoic acid metabolism [71].

At least 15 gene sequences related to lignin degradation, including oxidase, copper oxidase, laccase, dioxygenase, decarboxylase, and so on, were found in *Comamonas serinivorans* C35. These enzymes undergo at least four lignin-degrading pathways, i.e., benzoate pathway, phenol pathway, p-hydroxyacetophenone pathway, and β-ketoadipate pathway, to depolymerize lignin into a variety of low-molecular-weight aromatic compounds, such as 3-methylbenzaldehyde, guaiacol, vanillic acid, vanillin, syringic acid, syringaldehyde, p-hydroxybenzoic acid, ferulic acid, etc. [144]. The recognition of these metabolic pathways provides a theoretical basis for the subsequent biological transformation of lignin through the regulation of metabolic pathways. This section would introduce the major metabolic pathways related to lignin degradation in microorganisms.

12.4.1 β-Aryl Ether Catabolic Pathways

β-Aryl ether is the most abundant structure in lignin. In bacteria, the study of aryl ether catabolic pathways is mainly based on *Sphingomonas paucimobilis*, whose metabolic pathways for lignin dimmers are shown in Fig. 12.1. First, DNA-dependent lignin dehydrogenase (LigD) catalyzed the α-hydroxy groups to generate the corresponding ketones. Then, a reductive ester (ether) bond cleavage reaction took place under the action of glutathione-dependent β-esterase (LigEFG, belonging to glutathione-S-transferase superfamily) [9]. The derived ketone compounds would be consequently oxidized to vanillic acid through the oxidation of γ-hydroxyl group to carboxylic acid. In the metabolic pathway of *S. paucimobilis* SYK-6, vanillate was demethylated to protocatechuate under the catalysis of LigM (tetrahydrofolate-dependent demethylase), producing 5-methyl tetrahydrofolate as a by-product (as shown in Fig. 12.2). It can also be oxidized by C–C bond cleavage reaction in a manner similar to β-oxidation with the generation of lignin-derived products, dominating of methyl ketone (acetovanillone). The latter can be further used as the mediator of fungal laccase to promote the lignin degradation [146]. Besides *S. paucimobilis* SYK-6, β-aryl ether catabolic pathway was also found in *Pseudomonas acidovorans*, *P. putida*, and *Rhodococcus jostii* rHA1 [9].

In white-rot fungi, *P. chrysosporium* catalyzed β-aryl ether model compounds by Cα–Cβ oxidative cleavage catalyzed by lignin peroxidase catalyses, with the

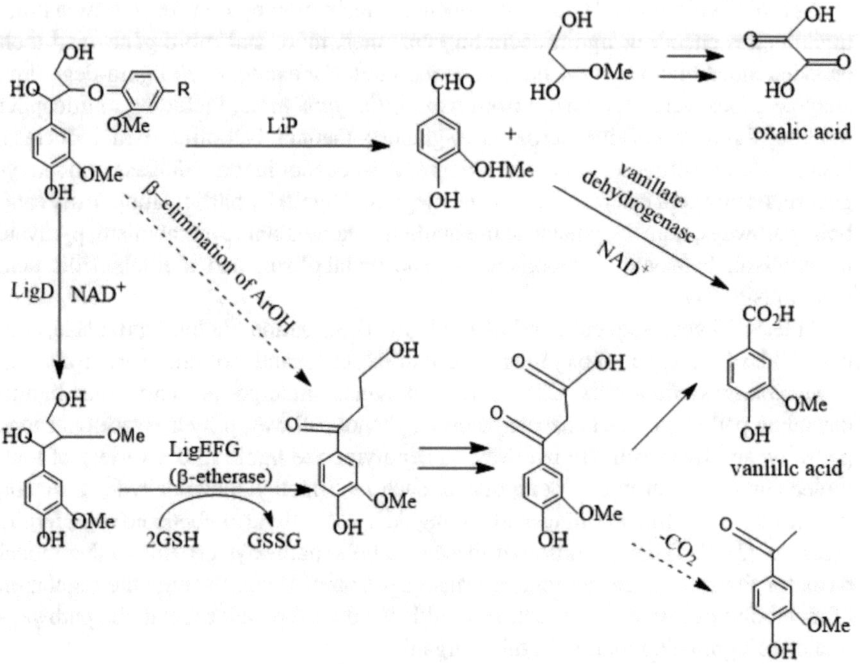

Fig. 12.1 β-aryl ether metabolic pathways in *Sphingomonas paucimobilis* for lignin and lignin-derived compounds. *LiP* lignin peroxidase, *LigD* DNA-dependent lignin dehydrogenase, *LigEFG* β-etherase (GST), *LigM* tetrahydrofolate-dependent demethylase [9]

production of vanillin as a product, as well as other oxidation products like benzyl ketone. Then vanillin was oxidized by vanillate dehydrogenase to generate vanillate and by-product 2-hydroxyacetaldehyde. The 2-hydroxyacetaldehyde might be further oxidized to oxalic acid, which would be used to complex with Mn^{2+} to promote the lignin degradation catalyzed by manganese peroxidase.

12.4.2 Biphenyl Catabolic Pathways

Biphenyl mainly exists between two guaiac units, accounting for about 10–15% in the molecular structure of lignin, only inferior to that of aryl ether. Biphenyl catabolic pathway is popular in lignin-degrading microorganisms, which mainly oxidized biphenyl and chlorinated biphenyl to 2,3-dihydroxybiphenyl by oxidative cleavage (Fig. 12.2) [145]. Firstly, non-heme iron-dependent demethylase (LigX) catalyzed the demethylation of 5,5′-dehydrodivanillate (DDVA) to 2,2′,3-trihydroxy-3′-methoxy-5,5′-dicarboxybiphenyl (OH-DDVA). Then the extradiol dioxygenase (LigZ) was catalyzed the meta-cleavage of OH-DDVA, and the ring-opened product was hydrolyzed by C–C hydrolase (LigY) to generate

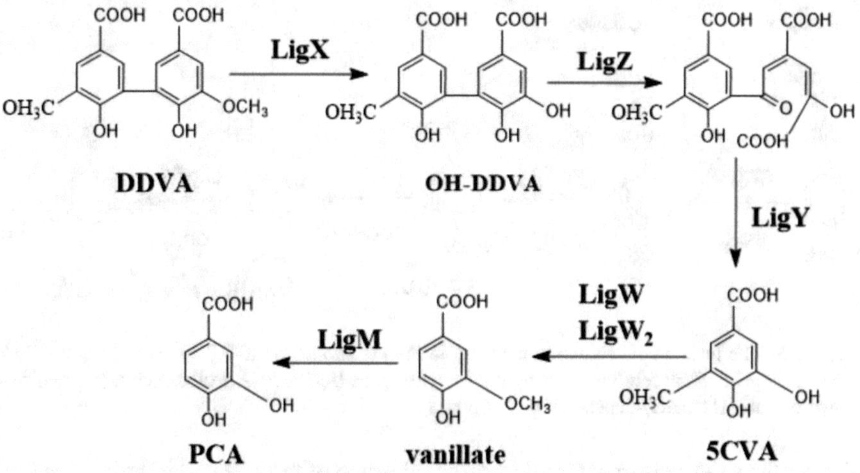

Fig. 12.2 The biphenyl metabolic pathways of lignin and lignin-derived compounds [145]. *LigZ* gene encoding OH-DDVA dioxygenase, *LigX* gene encoding non-heme iron-dependent demethylase, *LigZ* gene encoding extradiol dioxygenase, *LigY* gene encoding C–C hydrolase, *LigW/W2* gene encoding decarboxylases, *LigM* DDVA, 5,5′-dehydrodivanillate, *OH-DDVA* 2,2′,3-trihydroxy-3′-methoxy-5,5′-dicarboxybiphenyl, *5CVA* 5-carboxyvanillate, *PCA* protocatechuate

5-carboxyvanillate (5-CVA). Two decarboxylases (LigW) which can convert 5-CVA into the central intermediate vanillate were found in *Sphingomonas* sp. SYK-6 and catalyzed vanillate into protocatechuate (PCA). The metabolic pathway of biphenyls was also common in fungi. Yang and Zhang [134] found that both the laccase and MnP in *Trametes* sp. SQ01 have the ability to convert the conjugated diene of 2-hydroxy-6-oxygen −6-phenyl-2,4-hexadienoic acid (HOPDA) to monoene, so that the hydroxyl group on C disappears, resulting in the hydrolysis of HOPDA. The degradation of biphenyl by white-rot fungus mainly depends on oxidative cleavage. In vitro, using 1-hydroxybenzotriazole (HBT) as mediator, the *Trametes versicolor* laccase can significantly degrade the phenolic/non-phenolic biphenyl model compounds simulating 5–5′ type condensed lignin substructures [12].

12.4.3 Ferulic Acid Catabolic Pathway

Ferulic acid is a phenylpropane compound and is an important intermediate in the biosynthesis of lignin, which connects the benzene ring of lignin through ester bond. Ferulic acid structure accounts for about 1.5% of the total plant cell wall. It was reported that amounts of microorganisms possessed two types of ferulic acid side chain cleavage. In *Sphingomonas* SYK-6, *Pseudomonas fluorescens* AN103, *Pseudomonas* sp. HR199, *P. putida* WCS358, and *Amycolatopsis* sp. HR167, two enzymes were found participating in the cleavage of ferulic acid side chains, i.e., feruloyl-coenzyme A (CoA) synthetase and feruloyl-CoA hydratase/lyase [79].

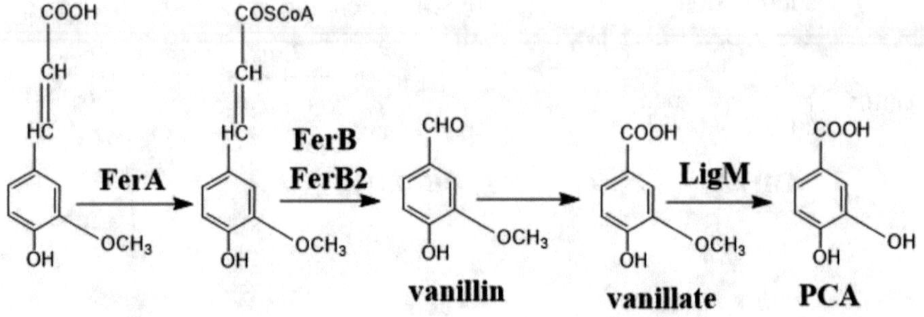

COOH CH HC OH OCH3 FerA COSCoA CH HC OH OCH3 FerB FerB2 CHO OH OCH3 vanillin COOH OH OCH3 vanillate LigM COOH OH OH PCA

Fig. 12.3 The ferulic acid catabolic pathway in *Sphingomonas* SYK-6 [79]. *FerA* feruloyl-CoA synthetase gene, *FerB* feruloyl-CoA hydratase/lyase gene (ferB), *PCA* protocatechuate, *LigM* tetrahydrofolate (H4folate)-dependent demethylase

Firstly, the COA was transferred to carboxyl group of ferulate under the catalysis of feruloyl-coenzyme A (COA) synthetase. The resulting feruloyl-CoA was then hydrated to form 4-hydroxy-3-methoxy-phenoxy-þ-hydroxypropionyl-CoA (HMPHP-CoA) under the catalysis of feruloyl-CoA hydratase/lyase (FerB). Finally, vanillin and acetyl-CoA were generated by cleaved HMPHP-CoA (Fig. 12.3). There is another ferulic acid metabolism pathway catalyzed by non-oxidative decarboxylase in some microorganisms. It removes one carbon from the side chain of ferulic acid, thereby forming 4-hydroxy-3-methoxystyrene. For example, *Cupriavidus* sp. B-8 can directly remove the carboxyl group in ferulic acid to generate corresponding aromatic compounds owing to containing of ferulic acid decarboxylase. The 4-vinyl aromatic compound generated by decarboxylation is reduced to 4-ethylphenol under the co-effect of vinylphenol reductase and NADPH. Then 4-ethylphenol was further converted to vanillic acid and consequently degraded to protocatechuic acid that enters into the ring-open reaction [142].

12.4.4 Tetrahydrofolate-Dependent O-Demethylation Catabolic Pathway

Vanillate and syringate are important intermediate metabolites in the biodegradation of lignin-derived aromatic compounds. There are two ways for microorganisms to metabolize vallinate and finally produce protocatechuate (PCA) that can be further oxidized by protocatechuic acid 3,4-dioxygenase through the β-ketoadipate pathway via intradiol oxidative cleavage (Fig. 12.4) [109].

One of the tetrahydrofolate-dependent O-demethylation catabolic pathways is vallinate metabolism depending on O-demethylase, which mainly occurs in aerobic bacteria (such as *Pseudomonas* and *Acinetobacter*). For example, the vanillate-degradation reaction catalyzed by vanillate-O-demethylase in *A. dehalogenans* firstly transferred the methyl of vanillate to the corrinoid protein by methyl

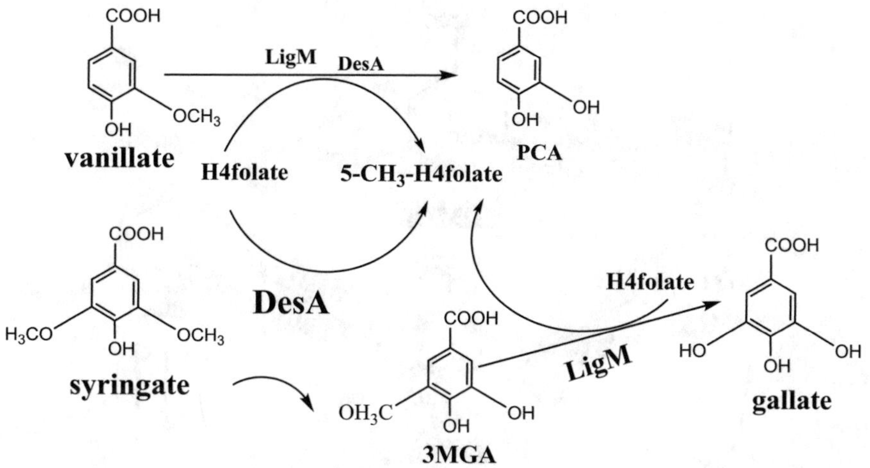

Fig. 12.4 Tetrahydrofolate-dependent O-demethylation pathway of vanillate, syringate, and 3-O-MGA linked with H4folate-mediated C1 metabolism in *S. paucimobilis* SYK-6 [109]. *DesA* gene encoding O-demethylase, *LigM* gene encoding tetrahydrofolate (H4folate)-dependent demethylase, *3-MGA* 3-O-methylgallate, *PCA* protocatechuate

transferase I; the latter then transferred methyl to tetrahydrofolate (H4folate) by methyltransferase II. The other tetrahydrofolate-dependent O-demethylation catabolic pathway is a metabolic system relying on tetrahydrofolate (H4folate)-dependent demethylase (LigM), which is mainly reported in anaerobic bacteria, including *Acetobacterium dehalogenans*, *Acetobacterium woodii*, and *Moorella thermoacetica*. In this pathway, syringate was firstly transformed into 3-O-methylgallate (3MGA) under the action of syringate O-demethylase (encoded by DesA). Then gallate was formed by catalyzing 3MGA with tetrahydrofolate (H4folate)-dependent demethylase (LigM).

12.4.5 3-Methyl Gallate Catabolic Pathway

There was three proposed 3-methyl gallate (3-MGA) catabolic pathway (Fig. 12.5): (1) converting 3-MGA to 4-oxalomesaconate (OMA) via gallate in the reaction catalyzed by vanillate/3MGA O-demethylase (LigM) and gallate dioxygenase (DesB) or PCA 4,5-dioxygenase (LigAB), which appears to be the major pathway of 3-MGA; (2) converting 3-MGA to OMA via 4-carboxy-2-hydroxy-6-methoxy-6-oxohexa-2,4-dienoate (CHMOD) by 3-MGA 3,4-dioxygenase (DesZ) and a hydrolase; and (3) converting 3-MGA to 2-pyrone-4,6-dicarboxylate (PDC) by PCA 4,5-dioxygenase (LigAB) and 3-MGA 3,4-dioxygenase (DesZ) [79, 80].

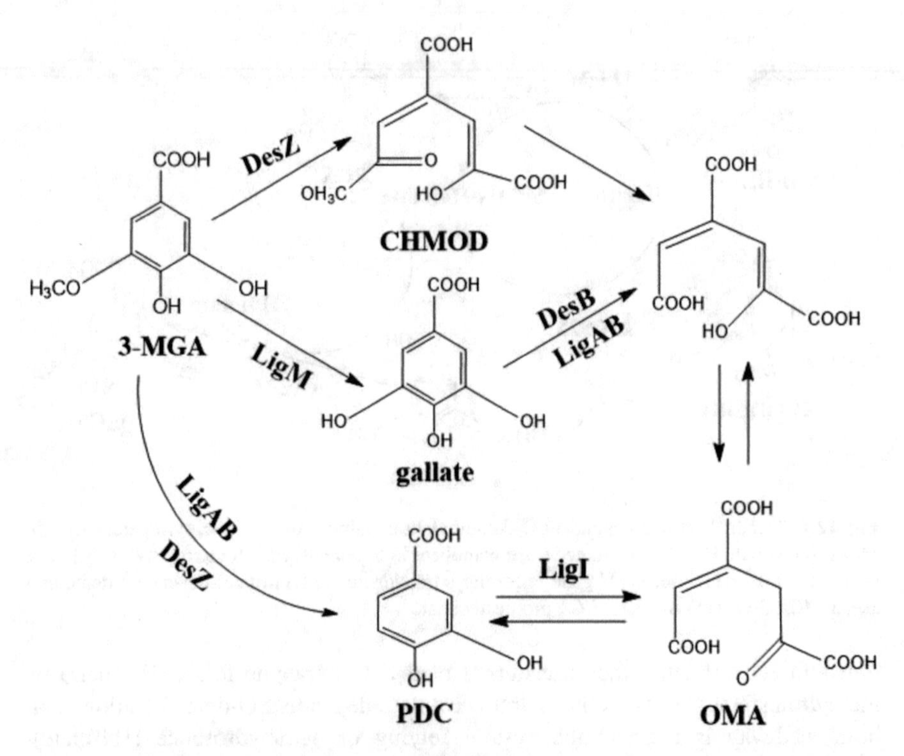

Fig. 12.5 Metabolic pathways of 3-methyl-gallic acid [79, 80]. *LigM* 3MGA O-demethylase, *DesB* gallate dioxygenase, *LigAB* PCA 4,5-dioxygenase, *DesZ* 3-MGA 3,4-dioxygenase, *3-MGA* 3-methyl gallate, *OMA* 4-oxalomesaconate, *CHMOD* 4-carboxy-2-hydroxy-6-methoxy-6-oxohexa-2,4-dienoate, *PDC* 2-pyrone-4,6-dicarboxylate

12.4.6 Protocatechuate (PCA) Catabolic Pathway

Many metabolic pathways in the lignin-degrading microbes can produce vanillin or its oxidation product vanillate. These compounds further demethylated to produce protocatechuate (PCA) which would be degraded by non-heme-o-diphenol dioxygenase through oxidative ring opening. There are three known non-heme-o-diphenol dioxygenases, i.e., PCA-3,4-dioxygenase (3,4-PCD), PCA-4,5-dioxygenase (4,5-PCD), and PCA-2,3-dioxygenase (2,3-PCD). 3,4-PCD catalyzes the meta-cleavage between two phenolic hydroxyl groups, while the other two enzymes catalyze ortho-cleavage of adjacent groups of phenolic hydroxyl groups (Fig. 12.6).

In bacteria (mainly Gram-negative bacteria), such as SYK-6 and *Bacillus Ligniniphilus* L1, protocatechuate (PCA) was converted to 4-carboxy-2-hydroxy-muconate semialdehyde (CHMS) through ortho-cleavage catalyzed by protocatechuate-4,5-dioxygenase (LigAB) [6, 129]. Then CHMS is further converted to OMA (4-oxalomesaconate) by CHMS dehydrogenase (LigC). Under the

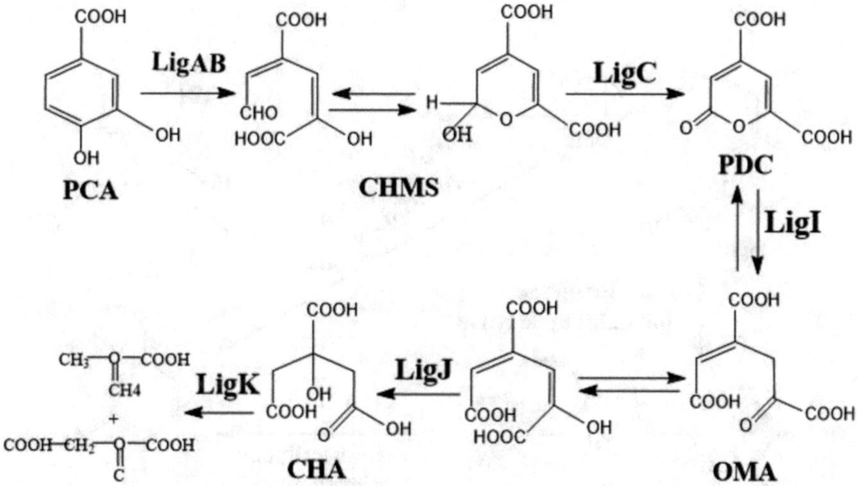

Fig. 12.6 Metabolic pathways of protocatechuic acid in SYK-6 [6, 129]. *CHMS* 4-carboxy-2-hydroxy-muconate semialdehyde, *OMA* 4-oxalomesaconate, *LigAB* protocatechuate-4,5-dioxygenase, *LigC* dehydrogenase, *LigI* PDC hydrolase, *LigJ* OMA hydratase, *LigK* CHA aldolase

catalysis of PDC hydrolase (LigI), OMA hydratase (LigJ), and CHA aldolase (LigK), CHMS was further converted to pyruvic acid and oxaloacetic acid, which would finally enter into the tricarboxylic acid cycle. In other bacteria (mainly Gram-positive bacteria), such as *Pseudomonas putida* and *calcium acetate Acinetobacter*, protocatechuate is oxidized by 3,4-PCD to generate cis-hexanedioic acid and then further degraded by β-ketoadipate metabolic pathway or ortho-cleavage pathway.

12.4.7 Diarylpropane Catabolic Pathways

The degradation of diarylpropanes by fungi has been well studied, which was attributed to the Cα–Cβ cleavage catalyzed by lignin peroxidase. In *Phanerochaete chrysosporium*, 1-(3,4-dimethoxyphenyl)-2-phenylethylene glycol (dimethoxyhydrobenzoin) (DMHB) was catalyzed by lignin peroxidase to obtain either benzaldehyde and α-hydroxy (dimethoxybenzyl) radical or resverataldehyde and α-hydroxybenzyl radicals. The diarylpropane degradation pathway in bacteria was different from that in fungi. For example, in *Pseudomonas paucimobilis* TMY1009, the interphenyl double bond in the dimeric lignin model compounds was degraded by dioxygenase with the generation of lignostilbene. Then the latter was further degraded by lignostilbene-α, β-dioxygenase to generate two vanillins which then enters the vanillin metabolic pathway (Fig. 12.7) [9, 58].

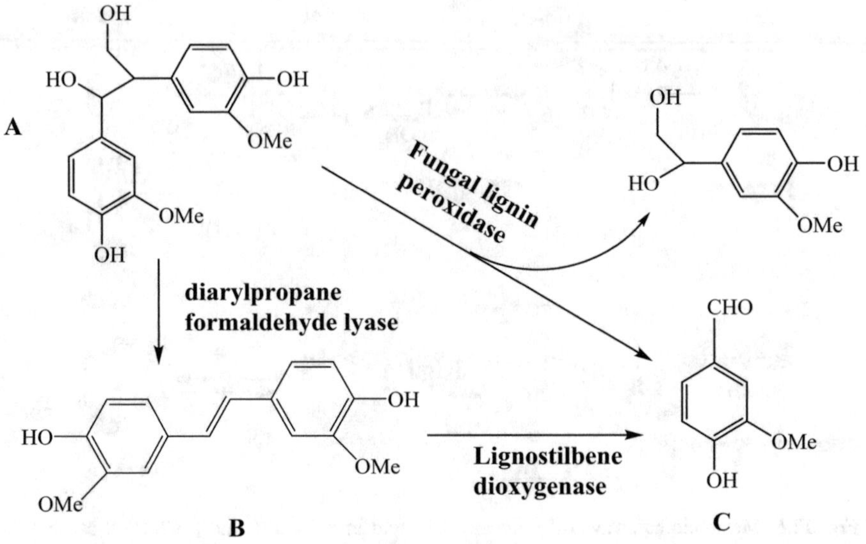

Fig. 12.7 Metabolic pathways of diarylpropane [9, 58]. (**a**) erythro-1,2-bis(4-hydroxy-3-methoxyphenyl)-1,3-propanediol; (**b**) 4,4′-dihydroxy-3,3′-dimethoxystibene; (**c**) vanillin

12.4.8 Phenylcoumarane and Pinoresinol Lignin Component Catabolic Pathway

Degradation pathways of phenylcoumarane and pinoresinol in *Phanerochaete Chrysosporium* and *Fusarium solani* have been revealed. The first step in the metabolic pathway of lignin and lignin-derived components was both referred to the α-hydroxylation [22]. The α-hydroxylated product can generate benzyl ketone by ring opening. Therefore, it was proposed that in bacteria the degradation of phenylcoumarin may be catalyzed by lignostilbene-α, β-dioxygenase. For example, *P. xanthomonas* degraded the alkylated phenylcoumarin firstly by side chain oxidation, then by heterocyclic oxidation to form a furan, and finally by oxidative cleavage of the Cα–Cβ bond [89]. The filamentous fungal *Fusarium solani* M-13-1 would break down the phenolic coumarin by directly breaking the Cα–Cβ bond to form 5-acetylvanillin [91]. The degradation of benzyl ketone can produce diarylpropane skeleton and then degrade through diarylpropane catabolic pathway. As for the degradation process of pinoresinol, in *F. solani* m-13-1, the cleavage of C–O bond was catalyzed according to the oxidation of benzyl group, thus resulting in the generation of monocyclic ketone as intermediate. Further oxidation of the aryl–alkyl group produced a carboxyl group and a corresponding lactone [57].

Sphingomonas SYK-6 can also degrade phenylcoumarin model compounds and rosinol model compounds, but the related genes were not found [9, 79].

12.5 Biocatalytic Conversion of Lignin

Systems biology provides powerful tools to reveal the basic metabolic pathways of lignin. These cognitions are applied to the construction of engineered strains using synthetic biology technology. The lignin-degrading engineered strains were used to produce various value-added products from lignin or lignin-derived aromatic compounds.

When catalyzed by lignin-degrading enzyme system, lignin is degraded by microorganisms through "upper pathways" to generate heterogeneous mixture of aromatic monomers which can be used as carbon and energy sources by many microorganisms. This process acts as a "biological funnel" to reduce the heterogeneity of carbon for catabolism [55]. It would be a great opportunity if we focus on the production of lignin-derived fuels and chemicals by selectively converting heterogeneous mixture of lignin-derived molecules instead of single intermediate. On the other hand, the lignin-derived aromatic compounds may inhibit the microbial growth, thereby hindering the value-added conversion of lignin. Metabolic engineering and synthetic biology can efficiently reduce the product inhibition. At present, the lignin-derived products through microbial degradation pathway mainly include the secondary metabolic products such as vanillin (vanillin) or cis,cis-muconic acid accumulation through the deletion of genes (blue) and the metabolic pathway of triacylglycerol lipid or polyhydroxybutyrate (green) accumulated by first metabolism (green) (as shown in Fig. 12.8). This section will introduce the typical metabolic engineering cases of lignin.

12.5.1 *Metabolic Engineering for the Production of Lignin-Derived Aromatic Compounds*

Lignin, as an aromatic polymer, has always been considered as a renewable source of aromatic chemicals in the future. In principle, the deep research and disclosure of microbial metabolic pathways and expression genes of lignin enable researchers to design biological metabolic pathways for the production of lignin-derived chemicals from renewable materials. In recent years, more and more metabolic pathways converting lignin and its derivatives into aromatic chemicals have been constructed and some engineered strains have been obtained.

12.5.1.1 Vanillin

Vanillin is a high-value chemical that has been widely used in food/condiment industry. More than 99% of vanillin produced in industry is synthesized from petrochemical products or lignin derivatives. At present, there are commercial methods to produce vanillin from lignosulfonate through chemical way. Although lignin can

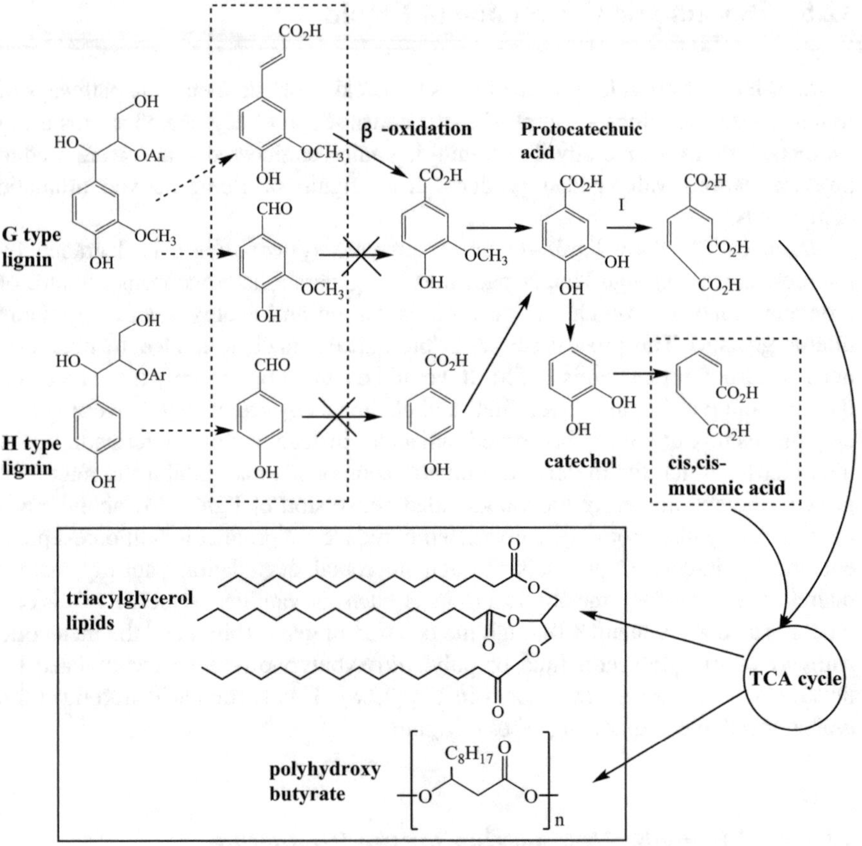

Fig. 12.8 Production of bioproducts from microbial degradation pathways [8]. The metabolic pathways responsible for accumulation of small molecule products vanillin or cis,cis-muconic acid (in dashed line) and the accumulation of triacylglycerol lipids or polyhydroxybutyrate (in solid line)

generate aromatic compounds through pyrolysis or chemical catalysis, such methods usually generate complex product mixtures with a total yield of 5–15%. Nowadays, especially in the food industry, increasing interest has been put on the natural and healthy ingredients. Therefore, the alternative source of natural vanillin from biotechnology process has attracted more and more attentions. The most attractive methods to produce vanillin include biosynthesis using natural aromatic compounds like ferulic acid and de novo biosynthesis from primary metabolites like glucose. It has reported that there were several microorganisms that can convert ferulic acid into vanillin, including *Pseudomonas*, *Rhodococcus*, *Pycnoporus cinnabarinus*, etc. The intermediate vanillin is produced by undesired catabolism caused by vanillin dehydrogenase. In order to obtain high yield and concentration of vanillin by preventing vanillin from being oxidized to vanillin acid, Fleige et al. [37] constructed the *Amycolatopsis* sp. ATCC 39116 vanillin dehydrogenase deletion mutant (ΔVDH). This strain cannot grow on vanillin anymore and cannot secrete vanillin dehydrogenase. When ferulic acid was used as substrate, vanillin concentration was enhanced by 2.3 times, while the vanillin content was greatly reduced. *Rhodococcus jostii* rHA1, a lignin decomposing bacterium, could grow to high cell density on the smallest medium containing lignocellulose and could tolerate the production of toxic aldehyde metabolites. The vdh gene (ro02986) and vanA gene (ro4165), encoding vanillin dehydrogenase and vanillin demethylase, respectively, were found in *R. jostii* RHA1. Sainsbury et al. [108] constructed *R. jostii* rHA1 mutant strain (ΔVDH gene deletion strain RHA045) which deleted vanillin dehydrogenase gene using predictive gene deletion method. The mutant strain could not grow when vanillin is used as sole organic substrate but could grow on vanillic acid. For the ΔVDH mutant, when it grows in the medium containing 2.5% wheat straw lignocellulose and 0.05% glucose, the quantity and intensity of p-hydroxybenzaldehyde, vanillin, and protocatechuic acid (4,5-dihydroxybenzoic acid) in the metabolites were all increased significantly. After being cultivated for 144 h, vanillin was accumulated up to 96 mg/L, accompanied by a small amount of ferulic acid and 4-hydroxybenzaldehyde.

12.5.1.2 Monolignols

Lignols, including coumarin alcohol, caffeinol, coniferol, and mustard alcohol, are important metabolites involved in the lignin biosynthesis of plants, and their derivatives have many physiological and pharmaceutical functions. Chemical reduction route has been developed for the production of monolignols from the corresponding phenylpropionic acid [102]. However, these methods have the disadvantages of expensive raw materials, harsh reaction conditions, low total yield, etc.

There have been some studies on the biosynthesis of coumarin by selecting effective enzymes. Then, the pathway could be further expanded to produce caffeinol and terpineol by introducing hydroxylase and methyltransferase. Jansen et al. [51], for the first time, constructed a complete artificial phenylpropane pathway of p-coumarin in *E. coli* by extending the metabolic pathway of L-tyrosine with the combination of genes from various plants and microorganisms. This pathway was co-expressed with four enzymes including L-tyrosine ammonia lyase (RsTAL) from *Rhodobacter sphaeroides*, p-coumaric acid CoA ligase (Pc4CL) from *Petroselinum crispum*, cinnamoyl-CoA reductase (ZmCCR) from *Zea mays*, and cinnamyl alcohol dehydrogenase (ZmCAD) in *E. coli*. Firstly, L-tyrosine is deaminated by RsTAL to generate p-coumaric acid which was further activated by CoA through p-coumaric acid CoA ligase (4CL) and consequently was reduced to p-coumaric acid by ZmCCR and ZmCAD. By co-expression of RsTAL, Pc4CL, ZmCCR, and ZmCAD, p-coumaryl alcohol could be directly generated from LB by the constructed strain.

However, the productivity of the artificial pathway was low which might be attributed to the low TAL activity, since the production of p-coumaryl alcohol was increased by five times when cultivating *E. coli* with 1 mM coumaric acid as substrate. Phosphorothioate-based ligase-independent gene cloning (PLICing) method has become an attractive method characterized as single-gene, enzyme-free, and sequence-independent cloning. During the assembly of DNA fragments, the Operon-PLICing method changes the interval between the Shine–Dalgarno sequence and the START codon to balance the expression of all genes in the metabolic pathway at the translation level, thereby maximizing the concentration of the target product and providing an opportunity [7]. Based on PLICing method, Jennifer et al. [53] developed Operon-PLICing method for the rapid assembly of p-coumaryl alcohol synthetic pathways. They heterologously expressed the encoding genes of the four enzymes required for the complete metabolic pathway in *E. coli*. With Operon-PLICing, 81 different clones were constructed within a few days, each of which carried a different p-coumaryl alcohol operon. The p-coumarol concentration of the selected five optimal mutants was ranging from 48 to 52 mg L^{-1}.

High catalytic activity of enzymes is the prerequisite for effective production of p-coumaryl alcohol [53]. In order to increase the production of p-coumarol, Chen et al. [15] screened TAL, 4CL1, CCR, and AD with high catalytic activity from different origins to construct a new synthetic pathway. By the combination of RgTAL from *Pseudomonas glutamicum*, the At4CL1 from *A. thaliana*, the LlCCR from *Lactobacillus leucocephala*, and the ScADH6 from *Saccharomyces cerevisiae*, the constructed strain can effectively transform the phenylpropionic acids into corresponding alcohols. The titer of p-coumaryl alcohol generated by the constructed strain was up to 656.0 ± 9.9 mg L^{-1} (Fig. 12.9).

12.5.2 Metabolic Engineering for Biomaterial Production from Lignin

12.5.2.1 Muconate/Muconic Acid

With the increase of oil price and the decrease of resources, people are increasingly interested in the production of bioplastics from renewable resources through environmentally friendly fermentation. Cis,cis-muconic acid (MA) is one of the most valuable diunsaturated dicarboxylic acids of industrial value. It can be further converted into basic material and monomer of commercial polymer materials, such as adipic acid or terephthalic acid, which are widely used in cosmetics, pharmaceutical, and food industries.

In recent years, many different functional engineering strains have been designed, such as *Pseudomonas putida* KT2440, *Amycolatopsis* species ATCC 39116, and *E. coli*, to produce MA either via biosynthesis from glucose and glycerol or via biotransformation from aromatic synthesis of biotransformed MA from glucose, glycerin, and even aromatic compounds. The production of MA from aromatic compounds is of great significance because it requires only a few biochemical reaction to achieve 100% molar yield. So far, it has been proved that different aromatic

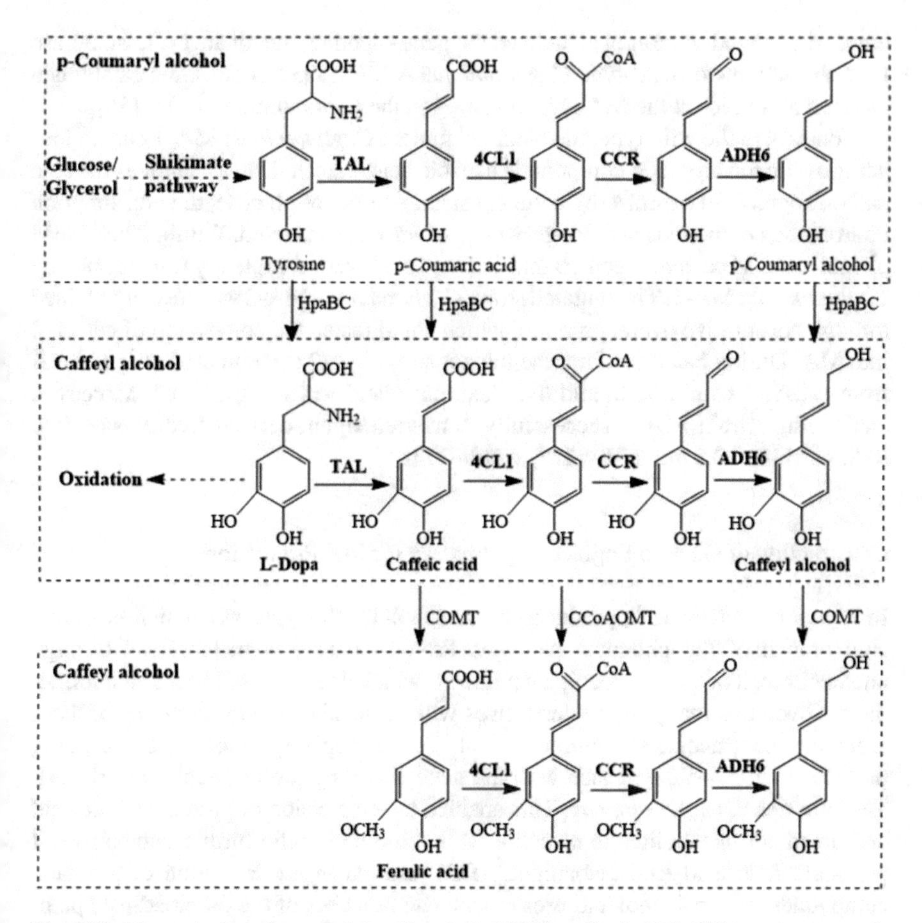

Fig. 12.9 Biosynthetic pathways of monolignols derived from lignin [15]. *TAL* tyrosine ammonia lyase, *4CL1* p-coumarate-CoA ligase, *CCR* cinnamoyl-CoA reductase, *ADH6* alcohol dehydrogenase, *HpaBC* 4-hydroxypropanoic acid 3-hydroxylase, *CCoAOMT* caffeoyl-CoA O-methyltransferase, *COMT* caffeate 3-O-methyltransferase

compounds (including benzoate, toluene, catechol, phenol, p-coumarate guaiacol) can be used in the production of MA. More importantly, the strains that can use aromatic compounds to produce MA provide the possibility for the production of MA from lignin-derived hydrolysate in the future.

Construction of Engineered *C. glutamicum* for MA Production

Most biotechnology processes that use *Corynebacterium glutamate* cell factories rely on sugar causing the competition with human nutrition. The β-ketoadipate pathway is the main pathway for assimilation of lignin-derived aromatic compounds, which indicated that engineered *C. glutamicum* strains can produce MA from lignin-based aromatic compounds if the gene encoding muconate cycloisomerase was deleted. Berker et al. deleted the gene encoding muconate cycloisomerase from the genome of wild-type *C. glutamicum* ATCC 13032 to block the catabolism of small aromatics at the MA level resulting in the accumulation of MA [56].

Contrary to the wild type, the obtained mutant *C. glutamicum* MA-1 can no longer grow on the aromatic compounds benzoic acid, catechol, and phenol as the sole carbon sources. In contrast, by adding a small amount of glucose, the consumption of aromatic compounds and the growth of strain were observed. Within 24 h, 5 mM phenol, 10 mM catechol, and 20 mM benzoic acid were completely transformed by *C. glutamicum* MA-1. The mutated strain *C. glutamicum* MA-2 was further obtained through constitutive overexpression, which could fasten the conversion of catechol into MA. During batch feeding, the mutant strain MA-2 accumulated 85 g L^{-1} mA from catechol within 60 h, and the maximum yield was 2.4 g L^{-1} h^{-1}. Moreover, engineering strain MA-2 successfully converted lignin-derived hydrolysate into MA, reaching 12.5 mm (1.8 g L^{-1}) within 27 h.

Construction of Genetic Engineering *P. putida* for MA Production

In *P. putida*, muconate is produced by catechol-1,2-dioxygenase, which is a catechol branch of the phthalate pathway. Protocatechuate is metabolized through another branch of the β-ketoadipate pathway, which does not use MA as an intermediate. Given that many lignin derivatives will be metabolized through one of these aromatic intermediates, Vardon et al. [124] replaced the gene encoding protocatechin-3,4-dioxygenase with the gene encoding protocatechin decarboxylase from *Enterobacter cloacae*. This enabled the conversion of protocatechuic acid and upstream metabolites to catechol while eliminating the further catabolism of protocatechuic acid to β-ketoadipate. This resulted in the transition of aromatic compounds from catechol and protocatechuate branches of the β-ketoadipate pathway to muconate. Acetate was used as carbon source and energy source to evaluate the muconate productivity of engineered *P. putida* KT2440-CJ103 using lignin-derived monomers, which can successfully convert catechol, phenol, and benzoate into muconate via catechol branch and convert protocatechuate, turpentine oil, ferulate, vanillin, caffeine, p-coumarate, and 4-hydroxybenzoate via protocatechuate branch.

Taking p-coumaric acid as carbon source, the cis-muconate yield of *P. putida* KT2440-CJ103 was 13.5 g/L when cultivated in a fed-batch bioreactor for 78.5 h. Alkaline pretreated liquor, mainly containing p-coumarate, ferulate, glycolate, and acetate, was used for the cultivation of *P. putida* KT2440-CJ103, which indicated that glycolate and acetate were quickly consumed and used as source of carbon and energy required for cell growth and a yield of 0.7 g L^{-1} cis-muconate was obtained after 24 h.

Construction of Genetic Engineering *E. coli* for MA Production

Microorganisms with the ability to metabolize aromatic compounds, such as *Sphingomonas* SYK-6, can take syringate and vanillin (main lignin-derived aromatic compounds) as carbon sources and metabolize them into small molecular metabolites, such as protocatechuate or catechol. The vanillin catabolic pathway is responsible for providing SYK-6 with sufficient nicotinamide adenine dinucleotide (NADH) and nicotinamide adenine dinucleotide phosphate (NADPH), which is critical for microbial growth. *Sphingomonas* SYK-6 has been proved to be extremely highly tolerant to high pH, which is a desirable characteristic for the high-value utilization of lignin. Moreover, it was found that the decomposition of lignin is a mechanism required for SYK-6 to survive [125]. The unique lignin-catabolizing characteristics of SYK-6 made it attractive in the high-value utilization of lignin. However, it was difficult for SYK-6 to be used as a host for lignin valorization since there were too many genes needed to be knocked out and overexpressed. Wu et al. [127] used vanillin as raw material which was derived from hydrogen peroxide catalyzed Kraft lignin to produce cis,cis-muconic acid by a newly developed pathway. Based on substrate consumption, *Pseudomonas putida* mt-239 containing the gene encoding catechol 1,2-dioxygenase (CatApmt2) obtained the highest yield of cis,cis-muconic acid (314 mg L^{-1}), followed by *Acinetobacter calcoaceticus* 40 containing the gene encoding catechol 1,2-dioxygenase (CatAac) (238 mg L^{-1}).

12.5.2.2 Pyruvate

Pyruvate is the key intermediate for the sugar metabolism of all biological cells and the mutual conversion of various substances in vivo. Owing to containing active ketone and carboxyl groups, pyruvate is widely used in chemical, pharmaceutical, food, agriculture, environmental protection, and other fields as a basic chemical raw material. It has been developed as the fundamental strategies for the metabolic engineering of the bio-based production of amino acids, alcohols, terpenoids, lactate, etc. In aerobic organisms, the aromatic products (such as benzoate and p-coumarate benzoate and p-coumarate) first degraded aerobically into intermediates such as protocatechuate and catechol. Then the intermediates would be cleaved by the O_2-dependent dioxygenase. However, according to the difference position of adjacent hydroxyl groups (ortho-position or meta-position), the oxygenase would catalyze the aromatic ring cleavage of catechol and protocatechuate through different degradation pathways to generate different combinations of succinate, acetyl-CoA, and pyruvate. Then the products would be metabolized through parallel pathways and finally entry into the TCA cycle. The ortho cleavage pathway of catechol and protocatechuate produces succinate and acetyl-CoA, while, the meta-cleavage pathway of catechol generates pyruvate and acetyl-CoA. The products of 2,3-meta-cleavage pathway and 4,5-meta-cleavage pathway of protocatechuate were different; the former generate pyruvate and acetyl-CoA, while the latter produce two pyruvate molecules [55].

The metabolic pathway of the target product derived from lignin degradation could be optimized. Generally, the meta-cleavage pathway of catechol and PCA is superior to the ortho-cleavage pathway. Therefore, the pyruvate yield from lignin aromatic molecules produced by a reconstructed strain by replacing the *P. putida* KT2440 endogenous catechol ortho-degradation pathway with the exogenous meta-cleavage pathway from *P. putida* MT-2 was increased by approximately 10%. Moreover, the pyruvate production would be increased by nearly five times if replacing the endogenous protocatechuate ortho-cleavage pathway with the meta-cleavage pathway from *Sphingobium* sp. SYK-6. Meanwhile, the pyruvate can further be converted into L-lactic acid aerobically, which will enhance the total pyruvate production consequently [55].

12.5.2.3 Polyhydroxyalkanoates (PHA)

Plastic products are the necessities of people's daily life and industrial and agricultural production. The widely used petroleum-based plastics cause serious environmental pollution and consume a lot of unrenewable resources. Polyhydroxyalkanoate (PHA) is a kind of environmentally friendly plastics, which can be completely biodegraded. The physical and chemical properties and processing performances of PHA are similar to traditional plastics. Therefore, PHA is an ideal substitute for petroleum-based plastics and has broad application prospects. Actually, PHA has been widely used in coatings, films, biomedical materials, biocompatible drug delivery, and organic/inorganic composite bioplastics [131].

Some microbes can produce PHA, and then PHA can be reserved as carbon source under specific nutritional constraints. Therefore, the microbial synthesis of PHAs is the major method for the industrial preparation of PHA. It has been found that PHA can be transformed from lignocellulose by microorganisms such as *Ralstonia eutropha*, *Pseudomonas putida*, and *Bacillus megaterium* [16]. Many bacteria have developed complex metabolism for lignin conversion with advanced enzyme systems for PHA production. Therefore, screening high-yield strains and choosing cheap, renewable, and wide-ranging wastes as substrates for PHA synthesis are of great significance for reducing production cost of PHA, recycling, and valorization of wastes.

The β-ketoadipate pathway also plays an important role in lignin conversion, because aromatic compounds derived from lignin can be used to synthesize high-molecular-weight compounds, including lipids and polyhydroxybenzoates (PHA). As the most thoroughly studied *Pseudomonas, P. putida* KT2440 had various metabolic functions. It had acyl-CoA synthetase that encodes fadD gene, β-oxidation pathway that provides 3-OH-aryl-CoA, and polymerization–depolymerization system that integrates two polymerases (PhaC1 and PhaC2) and a depolymerase (PhaZ). However, *P. putida* KT2440 specifically catalyzed the complete assimilation of phenylacetyl-CoA or benzoyl-CoA generated from these polyesters through two β-oxidation pathways in the cell. It was found that blocking either fadD or PhaC1 can inhibit the synthesis and accumulation of plastic polymers in cells. By disrupting PhaC2, the storage quantity of polymers could be reduced by two-thirds. The blocking of PhaZ could hinder the metabolic pathway of polymers, thus reducing its production. The productivity of plastic polymer in the mutant that absence of glyoxylate cycle or β-oxidation pathway was relatively higher than that in the wild strain [41]. The aromatic model compounds and heterogeneous lignin hydrolytes derived from alkaline pretreated lignin can be converted into medium-chain-length polyhydroxyalkanoates (mcl-PHAs). When they were used as substrate for *P. putida* KT2440, 36% of PHA was accumulated in the cell of *P. putida* after 48 h [72]. Olivera et al. [92] also proposed that the simple mutation or deletion of genes involved in the β-oxidation pathway in *P. putida* could accumulate amount of PHAs that account for more than 55% of the cell dry weight of mutants. Genetically bacteria *Pseudomonas putida* was used to produce new type of bioplastic containing aromatic or aliphatic and aromatic monomer mixture. The mutation (−) or deletion (Δ) of some genes in the β-oxidation pathway (fadA$^-$, fadB$^-$, ΔfadA, or Δfad BA mutants) can lead to the accumulation of abnormal homopolymers or copolymers in cells. The morphology of microorganisms changed significantly, since these macromolecular substances occupied 90% of the cytoplasm. Introducing blockade into the β-oxidation pathway or other related catabolic pathways allows mutant strains to synthesize polymers different from wild-type strains, to accumulate some intermediate products that are rapidly decomposed in wild type, and to accumulate the terminal decomposition products (such as phenylacetic acid, phenylbutyric acid, trans-cinnamic acid, or its derivatives) in the fermentation broth.

References

1. Ademakinwa, A. N., & Agboola, F. K. (2016). Biochemical characterization and kinetic studies on a purified yellow laccase from newly isolated *Aureobasidium pullulans* NAC8 obtained from soil containing decayed plant matter. *Journal of Genetic Engineering & Biotechnology, 14*(1), 143–151.
2. Ai, M. Q., Wang, F. F., Zhang, Y. Z., et al. (2014). Purification of pyranose oxidase from the white rot fungus *Irpex lacteus* and its cooperation with laccase in lignin degradation. *Process Biochemistry, 49*(12), 2191–2198.
3. Anita, S., Aggarwal Neeraj, K., Sharma, A., et al. (2015). Actinomycetes: A source of lignocellulolytic enzymes. *Enzyme Research, 2015*, 1–15.
4. Arias, M. E., Blanquez, A., Hernandez, M., et al. (2016). Role of a thermostable laccase produced by *Streptomyces ipomoeae* in the degradation of wheat straw lignin in solid state fermentation. *Journal of Analytical & Applied Pyrolysis, 122*, 202–208.
5. Augustin, C. M., Parvu, M., Damian, G., et al. (2012). A "yellow" laccase with "blue" spectroscopic features, from *Sclerotinia sclerotiorum*. *Process Biochemistry, 47*(6), 968–975.
6. Barry, K. P., & Taylor, E. A. (2013). Characterizing the promiscuity of LigAB, a lignin catabolite degrading extradiol dioxygenase from *Sphingomonas paucimobilis* SYK-6. *Biochemistry, 52*(38), 6724–6736.
7. Blanusa, M., Schenk, A., Sadeghi, H., et al. (2010). Phosphorothioate-based ligase-independent gene cloning (PLICing): An enzyme-free and sequence-independent cloning method. *Analytical Biochemistry, 406*(2), 141–146.

8. Bugg, T. D., & Rahmanpour, R. (2015). Enzymatic conversion of lignin into renewable chemicals. *Current Opinion in Chemical Biology, 29*, 10–17.
9. Bugg, T. D. H., Ahmad, M., Hardiman, E. M., et al. (2011). Pathways for degradation of lignin in bacteria and fungi. *Natural Product Reports, 28*(12), 1883–1890.
10. Camarero, S., Sarkar, S., Ruiz-Duenas, F. J., et al. (1999). Description of a versatile peroxidase involved in the natural degradation of lignin that has both manganese peroxidase and lignin peroxidase substrate interaction sites. *Journal of Biological Chemistry, 274*(15), 10324–10330.
11. Carabajal, M., Kellner, H., Levin, L., et al. (2013). The secretome of *Trametes versicolor* grown on tomato juice medium and purification of the secreted oxidoreductases including a versatile peroxidase. *Journal of Biotechnology, 168*(1), 15–23.
12. Castro, A. I. R. P., Evtuguin, D. V., & Xavier, A. M. B. (2003). Degradation of biphenyl lignin model compounds by laccase of *Trametes versicolor* in the presence of 1-hydroxybenzotriazole and heteropolyanion [SiW11VO40]5−. *Journal of Molecular Catalysis B: Enzymatic, 22*(1–2), 13–20.
13. Chaurasia, P. K., Yadav, R. S. S., & Yadava, S. (2014). Purification and characterization of yellow laccase from *Trametes hirsuta* MTCC-1171 and its application in synthesis of aromatic aldehydes. *Process Biochemistry, 49*(10), 1647–1655.
14. Chen, Y., Liyuan, C., Chongjian, T., et al. (2012). Kraft lignin biodegradation by *Novosphingobium* sp. B-7 and analysis of the degradation process. *Bioresour Technology, 123*, 682–685.
15. Chen, Z., Sun, X., Li, Y., et al. (2017). Metabolic engineering of *Escherichia coli* for microbial synthesis of monolignols. *Metabolic Engineering, 39*, 102–109.
16. Chen, Z., & Wan, C. (2017). Biological valorization strategies for converting lignin into fuels and chemicals. *Renewable & Sustainable Energy Reviews, 73*, 610–621.
17. Chi, Y., Wu, S., & Yu, C. (2019). Regulation of Mn2+on three manganese peroxidase genes Lg-mnp1, 2, and 3 at transcriptional level. *Journal of Jilin Agricultural University, 41*(5), 540–552.
18. Chong, G. G., Huang, X. J., Di, J. H., et al. (2018). Biodegradation of alkali lignin by a newly isolated *Rhodococcus pyridinivorans* CCZU-B16. *Bioprocess & Biosystems Engineering, 41*(4), 501–510.
19. Coconi-Linares, N., Magaña-Ortíz, D., Guzmán-Ortiz, D. A., et al. (2014). High-yield production of manganese peroxidase, lignin peroxidase, and versatile peroxidase in *Phanerochaete chrysosporium. Applied Microbiology & Biotechnology, 98*(22), 9519–9519.
20. Colpa, D. I., Fraaije, M. W., & Bloois, E. (2014). DyP-type peroxidases: A promising and versatile class of enzymes. *Journal of Industrial Microbiology & Biotechnology, 41*(1), 1–7.
21. Crowford, D. L. (1983). Lignin degradation by Streptomyces viridosporus: Isolation and characterization of a new polymeric lignin degradation intermediate. *Applied & Environmental Microbiology, 45*(3), 898–904.
22. Cui, F., & Dolphin, D. (1992). Iron porphyrin catalyzed oxidation of lignin model compounds: The oxidation of veratryl alcohol and veratryl acetate. *Canadian Journal of Chemistry, 70*(8), 2314–2318.
23. Daniel, G., Volc, J., & Kubatova, E. (1994). Pyranose oxidase, a major source of H_2O_2 during wood degradation by *Phanerochaete chrysosporium*, *Trametes versicolor*, and *Oudemansiella mucida. Applied and Environmental Microbiology, 60*(7), 2524–2532.
24. Daou, M., & Faulds, C. B. (2017). Glyoxal oxidases: Their nature and properties. *World Journal of Microbiology and Biotechnology, 33*(5), 87–97.
25. de Oliveira, P. L., Duarte, M. C. T., Ponezi, A. N., et al. (2009). Purification and partial characterization of manganese peroxidase from *Bacillus pumilus* and *Paenibacillus* sp. *Braz J Microbiol, 40*, 818–826.
26. Duan, J., Huo, X., Du, W. J., et al. (2016). Biodegradation of kraft lignin by a newly isolated anaerobic bacterial strain, *Acetoanaerobium* sp WJDL-Y2. *Letters in Applied Microbiology, 62*(1), 55–62.

27. Duan, J., Liang, J. D., Du, W. J., et al. (2014). Biodegradation of Kraft lignin by a bacterial strain *Sphingobacterium* sp. HY-H. *Advanced Materials Research, 955–959*, 548–553.
28. Duan, Z., Rui, S., Liu, B., et al. (2018). Comprehensive investigation of a dye-decolorizing peroxidase and a manganese peroxidase from *Irpex lacteus* F17, a lignin-degrading basidiomycete. *AMB Express, 8*(1), 119–134.
29. Elsalam, H. E. A., & Bahobail, A. S. (2016). Lignin biodegradation by thermophilic bacterial isolates from *Saudi Arabia. Biological Chemistry Science, 7*, 1413–1424.
30. Eriksson, K. E., Pettersson, B., Volc, J., et al. (1986). Formation and partial characterization of glucose-2-oxidase, a H2O2 producing enzyme in *Phanerochaete chrysosporium. Applied Microbiology & Biotechnology, 23*(3–4), 257–262.
31. Fan, Y., Zhang, Z., Wang, F., et al. (2019). Lignin degradation in corn stover catalyzed by lignin peroxidase from *Aspergillus oryzae* broth: Effects of conditions on the kinetics. *Renewable Energy, 130*, 32–40.
32. Faraco, V., Alessandra, P., Giovanni, S., et al. (2007). Identification of a new member of the dye-decolorizing peroxidase family from *Pleurotus ostreatus*. *World Journal of Microbiology & Biotechnology, 23*(6), 889–893.
33. Feng, H., Jing, F., Lu, X., et al. (2001). Synergistic effects of cellobiose dehydrogenase and manganese-dependent peroxidases during lignin degradation. *Chinese Science Bulletin, 46*(23), 1956–1962.
34. Feng, H., Sun, Y., Zhi, Y., et al. (2015). Lignocellulose degradation by the isolate of *Streptomyces griseorubens* JSD-1. *Functional & Integrative Genomics, 15*(2), 163–173.
35. Fernández, I. S., Ruíz-Dueñas, F. J., Santillana, et al. (2009). Novel structural features in the GMC family of oxidoreductases revealed by the crystal structure of fungal aryl-alcohol oxidase. *Acta Crystallographica, 65*(11), 1196–1205.
36. Ferri, S., Kojima, K., & Sode, K. (2011). Review of glucose oxidases and glucose dehydrogenases: A bird's eye view of glucose sensing enzymes. *Journal of Diabetes Science and Technology, 5*(5), 1068–1076.
37. Fleige, C., Hansen, G., Kroll, J., et al. (2013). Investigation of the *Amycolatopsis* sp. Strain ATCC 39116 vanillin dehydrogenase and its impact on the biotechnical production of vanillin. *Applied and Environmental Microbiology, 79*(1), 81–90.
38. Francisco, J. R.-D., & ángel, T. M. (2009). Microbial degradation of lignin: How a bulky recalcitrant polymer is efficiently recycled in nature and how we can take advantage of this. *Microbial Biotechnology, 2*(2), 164–177.
39. Gall, D. L., Ralph, J., Donohue, T. J., et al. (2017). Biochemical transformation of lignin for deriving valued commodities from lignocellulose. *Current Opinion in Biotechnology, 45*, 120–126.
40. Garcia-Ruiz, E., Mate, D., Gonzalez-Perez, D., et al. (2014). Directed evolution of ligninolytic oxidoreductases: From functional expression to stabilization and beyond. In S. Riva & W. D. Fessner (Eds.), *Cascade biocatalysis: Integrating stereoselective and environmentally friendly reactions* (pp. 1–22). Weinheim: Wiley.
41. Garcia, B., Olivera, E. R., Minambres, B., et al. (1999). Novel biodegradable aromatic plastics from a bacterial source: genetic and biochemical studies on a route of the phenylacetyl-coa catabolon. *Journal of Biological Chemistry, 274*(41), 29228–29241.
42. Glenn, J. K., & Gold, M. H. (1985). Purification and characterization of an extracellular Mn(II)-dependent peroxidase from the lignin-degrading basidiomycete, Phanerochaete chrysosporium. *Archives of Biochemistry and Biophysics, 242*, 329–341.
43. Godden, B., Ball, A. S., Helvensteian, P., et al. (1992). Towards elucidation of the lignin degradation pathway in actinomycetes. *Journal of General Microbiology, 138*, 2441–2448.
44. Grzegorz, J., Anna, P., Justyna, S., et al. (2017). Lignin degradation: microorganisms, enzymes involved, genomes analysis and evolution. *Fems Microbiology Reviews, 41*(6), 1–22.

45. Hakala, T. K., Kristiina, H., Pekka, M., et al. (2006). Differential regulation of manganese peroxidases and characterization of two variable MnP encoding genes in the white-rot fungus *Physisporinus rivulosus*. *Applied Microbiology & Biotechnology, 73*(4), 839–849.
46. Hernández-Ortega, A., Ferreira, P., & Martínez, A. T. (2012). Fungal aryl-alcohol oxidase: A peroxide-producing flavoenzyme involved in lignin degradation. *Applied Microbiology & Biotechnology, 93*(4), 1395–1410.
47. Higuchi, T. (2004). Microbial degradation of lignin: Role of lignin peroxidase, manganese peroxidase, and laccase. *Proceedings of the Japan Academy Ser B Physical and Biological Sciences, 80*(5), 204–214.
48. Hofrichter, M., & René, U. (2006). Heme-thiolate haloperoxidases: Versatile biocatalysts with biotechnological and environmental significance. *Applied Microbiology and Biotechnology, 71*(3), 276–288.
49. Hofrichter, M., Ullrich, R., Pecyna, M. J., et al. (2010). New and classic families of secreted fungal heme peroxidases. *Applied Microbiology & Biotechnology, 87*(3), 871–897.
50. Ike, P. T. L., Moreira, A. C., Almeida, d., et al. (2015). Functional characterization of a yellow laccase from *Leucoagaricus gongylophorus*. *SpringerPlus, 4*(1), 654.
51. Jansen, F., Gillessen, B., Mueller, F., et al. (2014). Metabolic engineering for p-coumaryl alcohol production in *Escherichia coli* by introducing an artificial phenylpropanoid pathway. *Applied Biochemistry and Biotechnology, 61*, 646–654.
52. Janusz, G., Pawlik, A., Sulej, J., et al. (2017). Lignin degradation: microorganisms, enzymes involved, genomes analysis and evolution. *FEMS Microbiology Reviews, 41*(6), 941–962.
53. Jennifer, A., Philana, V. S. W., Sokolowsky, S., et al. (2016). Combinatorial optimization of synthetic operons for the microbial production of p-coumaryl alcohol with *Escherichia coli*. *New Biotechnology, 33*, S26–S27.
54. Johjima, T., Ohkuma, M., & Kudo, T. (2003). Isolation and cDNA cloning of novel hydrogen peroxide-dependent phenol oxidase from the basidiomycete *Termitomyces albuminosus*. *Applied Microbiology and Biotechnology, 61*(3), 220–225.
55. Johnson, C. W., & Beckham, G. T. (2015). Aromatic catabolic pathway selection for optimal production of pyruvate and lactate from lignin. *Metabolic Engineering, 28*, 240–247.
56. Judith, B., Kuhl, M., Kohlstedt, M., et al. (2018). Metabolic engineering of *Corynebacterium glutamicum* for the production of *cis, cis*-muconic acid from lignin. *Microbial Cell Factories, 17*(1), 115.
57. Kamaya, Y., Nakatsubo, F., Higuchi, T., et al. (1981). Degradation of d,l-syringaresinol, a β-β′ linked lignin model compound, by *Fusarium solani* M-13-1. *Archives of Microbiology, 129*(4), 305–309.
58. Kamoda, S., & Saburi, Y. (1993). Cloning, expression, and sequence analysis of a lignostilbene- α,β -dioxygenase gene from *Pseudomonas paucimobilis* TMY1009. *Journal of the Agricultural Chemical Society of Japan, 57*(6), 926–930.
59. Kersten, P. J. (1990). Glyoxal oxidase of *Phanerochaete chrysosporium*: Its characterization and activation by lignin peroxidase. *Proceedings of the National Academy of Sciences of the United States of America, 87*(8), 2936–2940.
60. Khare, S., & Prakash, O. (2017). Current developments in biotechnology and bioengineering: Production, isolation and purification of industrial products. *Journal of Cleaner Production, 158*, 380–381.
61. Khatami, S., Deng, Y., Tien, M., et al. (2019). Formation of water-soluble organic matter through fungal degradation of lignin. *Organic Geochemistry, 135*, 64–70.
62. Kim, S. J., & Shoda, M. (1999). Purification and characterization of a novel peroxidase from *Geotrichum candidum* Dec1 involved in decolorization of dyes. *Applied and Environmental Microbiology, 65*(3), 1029–1035.
63. Koker, T. H. D., Mozuch, M. D., Cullen, D., et al. (2004). Isolation and purification of pyranose 2-oxidase from *Phanerochaete chrysosporium* and characterization of gene structure and regulation. *Applied & Environmental Microbiology, 70*(10), 5794–5800.

64. Kong, W., Fu, X., Wang, L., et al. (2017). A novel and efficient fungal delignification strategy based on versatile peroxidase for lignocellulose bioconversion. *Biotechnology for Biofuels, 10*(1), 218.
65. Kour, D., Rana, K. L., Yadav, N., et al. (2019). Agriculturally and industrially important fungi: Current developments and potential biotechnological applications. In A. N. Yadav, S. Singh, S. Mishra, & A. Gupta (Eds.), *Recent advancement in white biotechnology through fungi: Volume 2: Perspective for value-added products and environments* (pp. 1–64). Cham: Springer.
66. Kujawa, M., Ebner, H., Leitner, C., et al. (2006). Structural basis for substrate binding and regioselective oxidation of monosaccharides at C3 by pyranose 2-oxidase. *Journal of Biological Chemistry, 281*(46), 35104–35115.
67. Kukolya, J., Dobolyi, C., & Hornok, L. (1997). Isolation and identification of thermophilic cellulolytic actinomycetes. *Acta Phytopathologica et Entomologica Hungarica, 32*(1–2), 97–107.
68. Kumar, V. V., & Rapheal, V. S. (2011). Induction and purification by three-phase partitioning of aryl alcohol oxidase (AAO) from *Pleurotus ostreatus*. *Applied Biochemistry and Biotechnology, 163*(3), 423–432.
69. Kupryashina, M. A., Selivanov, N. Y., & Nikitina, V. E. (2012). Isolation and purification of Mn-peroxidase from *Azospirillum brasilense* SP245. *Applied Biochemistry and Microbiology, 48*(1), 17–20.
70. Leonowicz, A., Rogalski, J., Jaszek, M., et al. (1999). Cooperation of fungal laccase and glucose l-oxidase in transformation of Björkman lignin and some phenolic compounds. *Holzforschung, 53*(4), 376–380.
71. Li, X. (2019). Pathway study of lignin degradation by *Aspergillus fumigatus* from buffalo rumen, Vol. Master, Huazhong Agricultural University.
72. Linger, J. G., Vardon, D. R., Guarnieri, M. T., et al. (2014). Lignin valorization through integrated biological funneling and chemical catalysis. *Proceedings of the National Academy of Sciences of the United States of America, 111*(33), 12013–12018.
73. Liu, G., & Qu, Y. (2018). Engineering of filamentous fungi for efficient conversion of lignocellulose: Tools, recent advances and prospects. *Biotechnology Advances, 37*(4), 519–529.
74. Liu, Z. H., Le, R. K., Kosa, M., et al. (2019). Identifying and creating pathways to improve biological lignin valorization. *Renewable and Sustainable Energy Reviews, 105*, 349–362.
75. Ma, R., Zhao, F., & Zhao, H. (2017). Reveals the pathway of lignin degradation by studied the transcriptome of *Pleurotus ostreatus* fermented with corn straw. *Feed Research, 9*, 42–48.
76. Manavalan, T., Manavalan, A., & Heese, K. (2015). Characterization of lignocellulolytic enzymes from white-rot fungi. *Current Microbiology, 70*, 485–498.
77. Martínez, A. T. (2002). Molecular biology and structure-function of lignin-degrading heme peroxidases. *Enzyme & Microbial Technology, 30*(4), 425–444.
78. Marzullo, L., Cannio, R., Giardina, P., et al. (1995). Veratryl alcohol oxidase from *Pleurotus ostreatus* participates in lignin biodegradation and prevents polymerization of laccase-oxidized substrates. *Journal of Biological Chemistry, 270*(8), 3823–3827.
79. Masai, E., Katayama, Y., & Fukuda, M. (2007). Genetic and biochemical investigations on bacterial catabolic pathways for lignin-derived aromatic compounds. *Journal of the Agricultural Chemical Society of Japan, 71*(1), 1–15.
80. Masai, E., Sasaki, M., Minakawa, Y., et al. (2004). A novel tetrahydrofolate-dependent o-demethylase gene is essential for growth of *Sphingomonas paucimobilis* SYK-6 with syringate. *Journal of Bacteriology, 186*(9), 2757–2765.
81. Mathews, S. L., Pawlak, J., & Grunden, A. M. (2015). Bacterial biodegradation and bioconversion of industrial lignocellulosic streams. *Applied Microbiology and Biotechnology, 99*, 2939–2954.
82. Mathieu, Y., Piumi, F., Valli, R., et al. (2016). Activities of secreted aryl alcohol quinone oxidoreductases from *Pycnoporus cinnabarinus* provide insights into fungal degradation of plant biomass. *Applied and Environmental Microbiology, 82*, 2411–2423.

83. Michael, B., Sabine, B., Dorothee, H. P., et al. (2004). Crystal structure of pyranose 2-oxidase from the white-rot fungus *Peniophora* sp. *Biochemistry, 43*(37), 11683–11690.
84. Min, K., Gong, G., Woo, H. M., et al. (2015). A dye-decolorizing peroxidase from *Bacillus subtilis* exhibiting substrate-dependent optimum temperature for dyes and β-ether lignin dimer. *Scientific Reports, 5*, 8245–8252.
85. Min, K. L., Kim, Y. H., Kim, Y. W., et al. (2001). Characterization of a novel laccase produced by the wood-rotting fungus *Phellinus ribis*. *Archives of Biochemistry and Biophysics, 392*(2), 279–286.
86. Moreno, A. D., Ibarra, D., Eugenio, M. E., et al. (2020). Laccases as versatile enzymes: From industrial uses to novel applications. *Journal of Chemical Technology & Biotechnology, 95*(3), 481–494.
87. Mukhopadhyay, M., & Banerjee, R. (2015). Purification and biochemical characterization of a newly produced yellow laccase from *Lentinus squarrosulus* MR13. *Biotech, 5*(3), 227–236.
88. Munk, L., Sitarz, A. K., Kalyani, D. C., et al. (2015). Can laccases catalyze bond cleavage in lignin? *Biotechnology Advances, 33*(1), 13–24.
89. Nakatsubo, F., Kirk, T. K., Shimada, M., et al. (1981). Metabolism of a phenylcoumaran substructure lignin model compound in ligninolytic cultures of *Phanerochaete chrysosporium*. *Archives of Microbiology, 128*(4), 416–420.
90. Niladevi, K. N., & Mangrove, P. P. (2005). Actinomycetes as the source of ligninolytic enzymes. *Actinomycetologica, 19*, 40–47.
91. Ohta, M., Higuchi, T., & Iwahara, S. (1979). Microbial degradation of dehydrodiconiferyl alcohol, a lignin substructure model. *Archives of Microbiology, 121*(1), 23–28.
92. Olivera, E., Carnicero, D., Jodra, R., et al. (2001). Genetically engineered *Pseudomonas*: A factory of new bioplastics with broad applications. *Environmental Microbiology, 3*(10), 612–618.
93. Palmieri, G., Giardina, P., Bianco, C., et al. (2000). Copper induction of laccase isoenzymes in the ligninolytic fungus *Pleurotus ostreatus*. *Applied & Environmental Microbiology, 66*(3), 920–924.
94. Parshetti, G. K., Supriya, P., Dayanand, C. K., et al. (2012). Industrial dye decolorizing lignin peroxidase from *Kocuria rosea* MTCC 1532. *Annals of Microbiology, 62*(1), 217–223.
95. Paszczyński, A., Huynh, V. B., & Crawford, R. (1986). Comparison of ligninase-I and peroxidase-M2 from the white-rot fungus *Phanerochaete chrysosporium*. *Archives of Biochemistry and Biophysics, 244*(2), 750–765.
96. Peralta, R.M., da Silva, B.P., Gomes Côrrea, et al.., 2017. Charpt 5 Enzymes from basidiomycetes-peculiar and efficient tools for biotechnology. In G. Brahmachari (Ed.), *Biotechnology of microbial enzymes*. Springer, 1–64. Academic, 119–149.
97. Pere, P., Domínguez, D. M. P., & Anett, S. (2015). From gene to biorefinery: Microbial β-etherases as promising biocatalysts for lignin valorization. *Frontiers in Microbiology, 6*, 916.
98. Piontek, K., Ullrich, R., Liers, C., et al. (2010). Crystallization of a 45kDa peroxygenase/peroxidase from the mushroom *Agrocybe aegerita* and structure determination by SAD utilizing only the haem iron. *Acta Crystallographica, 66*(6), 693–698.
99. Piumi, F., Levasseur, A., Navarro, D., et al. (2014). A novel glucose dehydrogenase from the white-rot fungus *Pycnoporus cinnabarinus*: production in *Aspergillus niger* and physicochemical characterization of the recombinant enzyme. *Applied Microbiology and Biotechnology, 98*, 10105–10118.
100. Pozdniakova, N. N., Turkovskaia, O. V., Iudina, E. N., et al. (2006). Yellow laccase from the fungus *Pleurotus ostreatus* D1: purification and characterization. *Applied Biochemistry & Microbiology, 42*(1), 56–61.
101. Rahmanpour, R., Rea, D., Jamshidi, S., et al. (2016). Structure of *Thermobifida fusca* DyP-type peroxidase and activity towards Kraft lignin and lignin model compounds. *Archives of Biochemistry & Biophysics, 594*, 54–60.
102. Ralph, J., Helm, F. R., Quideau, S., et al. (1992). Lignin-feruloyl ester cross-links in grasses. Part 1. Incorporation of feruloyl esters into coniferyl alcohol dehydrogenation polymers. *ChemInform, 24*(21), 2961–2969.

103. Rinaldi, R., Jastrzebski, R., Clough, M. T., et al. (2016). Paving the way for lignin valorisation: Recent advances in bioengineering, biorefining and catalysis. *Angewandte Chemie International Edition, 55*(29), 8164–8215.
104. Roberts, J. N., Singh, R., Grigg, J. C., et al. (2011). Characterization of dye-decolorizing peroxidases from *Rhodococcus jostii* RHA1. *Biochemistry, 50*(23), 5108–5119.
105. Rola, B., Pawlik, A., Frąc, M., et al. (2015). The phenotypic and genomic diversity of *Aspergillus* strains producing glucose dehydrogenase. *Acta Biochimica Polonica, 62*(4), 747–755.
106. Rosini, E., Allegretti, C., Melis, R., et al. (2016). Cascade enzymatic cleavage of the beta-O-4 linkage in a lignin model compound. *Catalysis Science & Technology, 6*, 2195–2205.
107. Ruiz-Dueñas, F. J., Fernández, E., Martínez, M. J., et al. (2011). *Pleurotus ostreatus* heme peroxidases: An in silico analysis from the genome sequence to the enzyme molecular structure. *Comptes Rendus Biologies, 334*(11), 795–805.
108. Sainsbury, P. D., Hardiman, E. M., Ahmad, M., et al. (2013). Breaking down lignin to high-value chemicals: The conversion of lignocellulose to vanillin in a gene deletion mutant of *Rhodococcus jostii* RHA1. *Acs Chemical Biology, 8*(10), 2151–2156.
109. Sainsbury, P. D., Mineyeva, Y., Mycroft, Z., et al. (2015). Chemical intervention in bacterial lignin degradation pathways: Development of selective inhibitors for intradiol and extradiol catechol dioxygenases. *Bioorganic Chemistry, 60*, 102–109.
110. Sánchez, C. (2009). Lignocellulosic residues: Biodegradation and bioconversion by fungi. *Biotechnology Advances, 27*(2), 185–194.
111. Santos, A., Mendes, S., Brissos, V., et al. (2013). New dye-decolorizing peroxidases from *Bacillus subtilis* and *Pseudomonas putida* MET94: towards biotechnological applications. *Applied Microbiology & Biotechnology, 98*(5), 2053–2065.
112. Sharma, M., Chaurasia, P. K., Yadav, A., et al. (2016). Purification and characterization of a thermally stable yellow laccase from *Daedalea flavida* MTCC-145 with higher catalytic performance towards selective synthesis of substituted benzaldehydes. *Russian Journal of Bioorganic Chemistry, 42*(1), 59–68.
113. Sigoillot, J. C., Berrin, J. G., Bey, M., et al. (2012). Fungal strategies for lignin degradation. *Advances in Botanical Research, 61*, 263–308.
114. Stephanie, L. M., Joel, P., & Amy, M. G. (2015). Bacterial biodegradation and bioconversion of industrial lignocellulosic streams. *Applied Microbiology & Biotechnology, 99*(7), 2939–2954.
115. Stevens, J. C., & Shi, J. (2019). Biocatalysis in ionic liquids for lignin valorization: Opportunities and recent developments. *Biotechnology Advances, 37*(8), 107418.
116. Suderman, R. J., Dittmer, N. T., Kanost, M. R., et al. (2006). Model reactions for insect cuticle sclerotization: Cross-linking of recombinant cuticular proteins upon their laccase-catalyzed oxidative conjugation with catechols. *Insect Biochemistry and Molecular Biology, 36*(4), 353–365.
117. Sugano, Y., Matsushima, Y., Tsuchiya, K., et al. (2009). Degradation pathway of an anthraquinone dye catalyzed by a unique peroxidase DyP from *Thanatephorus cucumeris* Dec 1. *Biodegradation, 20*(3), 433–440.
118. Sugano, Y., Muramatsu, R., Ichiyanagi, A., et al. (2007). DyP, a unique dye-decolorizing peroxidase, represents a novel heme peroxidase family. *Journal of Biological Chemistry, 282*(50), 36652–36658.
119. Suman, S. K., Dhawaria, M., Tripathi, D., et al. (2016). Investigation of lignin biodegradation by *Trabulsiella* sp. isolated from termite gut. *International Biodeterioration & Biodegradation, 112*, 12–17.
120. Tamboli, D. P., Telke, A. A., Dawkar, V. V., et al. (2011). Purification and characterization of bacterial aryl alcohol oxidase from *Sphingobacterium sp*. ATM and its uses in textile dye decolorization. *Biotechnology & Bioprocess Engineering, 16*(4), 661–668.

121. Tian, J. H., Pourcher, A. M., Bouchez, T., et al. (2014). Occurrence of lignin degradation genotypes and phenotypes among prokaryotes. *Applied Microbiology & Biotechnology, 98*(23), 9527–9544.
122. Tsukihara, T., Yoichi, H., Takahito, W., et al. (2006). Molecular breeding of white rot fungus *Pleurotus ostreatus* by homologous expression of its versatile peroxidase MnP2. *Applied Microbiology and Biotechnology, 71*(1), 114–120.
123. Ullrich, R., Nüske, J., Scheibner, K., et al. (2004). Novel haloperoxidase from the agaric basidiomycete *Agrocybe aegerita* oxidizes aryl alcohols and aldehydes. *Applied & Environmental Microbiology, 70*(8), 4575.
124. Vardon, D. R., Franden, M. A., Johnson, C. W., et al. (2015). Adipic acid production from lignin. *Energy & Environmental Science, 8*(2), 617–628.
125. Varman, A. M., He, L., Follenfant, R., et al. (2016). Decoding how a soil bacterium extracts building blocks and metabolic energy from ligninolysis provides road map for lignin valorization. *Proceedings of the National Academy of Sciences, 113*(40), E5802.
126. Wang, S. N., Chen, Q. J., Zhu, M. J., et al. (2018). An extracellular yellow laccase from white rot fungus *Trametes* sp. F1635 and its mediator systems for dye decolorization. *Biochimie, 148*, 46–54.
127. Wu, W., Dutta, T., Varman, A. M., et al. (2017). Lignin valorization: Two hybrid biochemical routes for the conversion of polymeric lignin into value-added chemicals. *Scientific Reports, 7*(1), 8420.
128. Xiaoshi, W., René, U., Martin, H., et al. (2015). Heme-thiolate ferryl of aromatic peroxygenase is basic and reactive. *Proceedings of the National Academy of Sciences of the United States of America, 112*(12), 3686–3691.
129. Xie, C. (2016). The study of lignin degradation by *Bacillus ligniniphilus* L1, Master Degree, Jiangsu University.
130. Xu, J., & Yang, Q. (2010). Isolation and characterization of rice straw degrading *Streptomyces griseorubens* C-5. *Biodegradation, 21*(1), 107–116.
131. Xu, R., Zhang, K., Liu, P., et al. (2018a). Lignin depolymerization and utilization by bacteria. *Bioresource Technology, 269*, 557–566.
132. Xu, Z., Qin, L., Cai, M., Hua, W., & Jin, M. (2018b). Biodegradation of kraft lignin by newly isolated *Klebsiella pneumoniae*, *Pseudomonas putida*, and *Ochrobactrum tritici* strains. *Environmental Science and Pollution Research, 25*, 14171–14181.
133. Yamada, Y., Wang, J., Kawagishi, H., et al. (2014). Improvement of ligninolytic properties by recombinant expression of glyoxal oxidase gene in hyper lignin-degrading fungus *Phanerochaete sordida* YK-624. *Microbiology & Fermentation Technology, 78*(12), 2128–2133.
134. Yang, X., & Zhang, X. (2016). Transformation of biphenyl intermediate metabolite by manganese peroxidase from a white rot fungus SQ01. *Acta Microbiologica Sinica, 56*(6), 1044–1055.
135. Yang, Y. (2012). The screening of *Aspergillus-Streptomyces-Pleurotus* and their coimmoblized biodegradation of alkali lignin, Vol. Doctor, Dalian University of Technology.
136. Yang, Y. S., Zhou, J. T., Lu, H., et al. (2011). Isolation and characterization of a fungus *Aspergillus sp.* strain F-3 capable of degrading alkali lignin. *Biodegradation, 22*(5), 1017–1027.
137. Yang, Y. S., Zhou, J. T., Lu, H., et al. (2012). Isolation and characterization of *Streptomyces* spp. strains F-6 and F-7 capable of decomposing alkali lignin. *Environmental Technology, 33*(22–24), 2603–2609.
138. Yoshida, H., Sakai, G., Mori, K., et al. (2015). Structural analysis of fungus-derived FAD glucose dehydrogenase. *Scientific Reports, 5*(1), 13498.
139. Zeng, J., Mills, M. J., Simmons, B., et al. (2017). Understanding factors controlling depolymerization and polymerization in catalytic degradation of β-ether linked model lignin compounds by versatile peroxidase. *Green Chemistry, 19*(9), 2145–2154.

140. Zeng, J., Singh, D., Laskar, D. D., et al. (2011). Deconstruction of wheat straw lignin by *Streptomyces viridosporus* as insight into biological degradation mechanism. In AIChE Annual Meeting.
141. Zeng, J., Singh, D., Laskar, D. D., & Chen, S. (2013). Degradation of native wheat straw lignin by *Streptomyces viridosporus* T7A. *International Journal of Environmental Science & Technology, 10*(1), 165–174.
142. Zhang, H. (2012). Study on biodegradation of lignin and lignin model compounds by *Cupriavidus* sp. B-8, Master Degree, Central South University.
143. Zhang, K., Huang, M., Ma, J., et al. (2018). Identification and characterization of a novel bacterial pyranose 2-oxidase from the lignocellulolytic bacterium *Pantoea ananatis* Sd-1. *Biotechnology Letters, 40*(5), 871–880.
144. Zhang, P. (2017). Characterization and metabolic mechanism of lignin biodegradation by *Comamonas serinivorans* C35. Master Degree, Jiangsu University.
145. Zhang, X., Peng, X., & Eiji, M. (2014). Recent advances in *Sphingobium* sp. SYK-6 for lignin aromatic compounds degradation – A review. *Acta Microbiologica Sinica, 54*(8), 854–867.
146. Zhang, Y. (2013). The mechanism of lignin and related mode benzene compounds degradation by Pandoraea sp. B-6 and *Cupriavidus basilensis* B-8, Doctor Degree, Central South University.
147. Zhu, D., Zhang, P., Xie, C., et al. (2017). Biodegradation of alkaline lignin by *Bacillus ligniniphilus* L1. *Biotechnology for Biofuels, 10*(1), 44.

Chapter 13
Understanding Fundamental and Applied Aspects of Oxidative Pretreatment for Lignocellulosic Biomass and Lignin Valorization

Younghan J. Lim and Zhenglun Li

13.1 Oxidative Pretreatment: An Overview of Processes

The application of oxidation in the processing of lignocellulosic materials has a long history, with the earliest industrial use of gaseous chlorine in bleaching dating back to the eighteenth century [1]. Oxidation of lignocellulosic materials alters the physiochemical properties of plant polymers, resulting in fractionation and modification of plant cell wall components. Oxidative bleaching of plant pulp, in particular, has been the primary industrial application of lignocellulose oxidation.

Oxidation reactions have also been studied as an approach to improving the accessibility of polysaccharides during biochemical conversion of lignocellulosic biomass, and to extracting lignins from biomass for further valorization. Since these reactions precede enzymatic saccharification of plant polysaccharides, they are colloquially referred to as oxidative pretreatment. The use of oxidants in pretreatment has been demonstrated to increase the yield of sugars from enzymatic digestion of pretreated biomass, and to result in physiochemical changes in plant cell wall components. This section reviews the processing characteristics of various oxidative pretreatments, and the following section focuses on the effects of oxidative pretreatment on plant cell wall components.

Y. J. Lim · Z. Li (✉)
Department of Food Science and Technology, Oregon State University, Corvallis, OR, USA
e-mail: glen.li@oregonstate.edu

Z.-H. Liu, A. Ragauskas (eds.), *Emerging Technologies for Biorefineries, Biofuels, and Value-Added Commodities*,
https://doi.org/10.1007/978-3-030-65584-6_13

13.1.1 Pulp Bleaching Processes

Pulp bleaching is a sequence of unit operations (also referred to as stages) where wood pulp is treated with oxidants in order to whiten the pulp. A detailed overview of the process can be found in the writings of Sjöström [2], Biermann [3], and Bajpai [4]. Most pulp bleaching sequences use oxidation reactions to remove residual lignins and to convert naturally occurring wood chromophores to products that absorb little or no light in the visible range. During these reactions, color-bearing lignins and aromatic extractives are removed from wood pulp, leaving behind bleached product that is primarily cellulose. Because of the economic impacts of cellulose depolymerization during pulp bleaching, the oxidants used in pulp bleaching.

The oxidants used in pulp bleaching has evolved from early options of chlorine [5] and hypochlorite to more environmentally friendly chlorine-free oxidants including molecular oxygen, hydrogen peroxide, and ozone [6].

13.1.2 Oxidative Pretreatment for Biochemical Conversion of Lignocellulose

Biochemical conversion of lignocellulose offers potential in delivering fuel, chemicals, and materials with reduced life-cycle of greenhouse gas emissions compared to petroleum refining [7]. Pretreatment of lignocellulose disrupts the plant cell wall structure in biomass, leading to increased yields in enzymatic saccharification and sugar bioconversion. As a result, a pretreatment stage is indispensable in any economically viable biochemical biomass conversion process.

Oxidative pretreatment is a category of pretreatment methods that involves the use of oxidants as a means of disrupting and fractionating lignocellulose. Compared to other pretreatment methods, oxidative pretreatment is unique in its ability to modify and remove lignin from lignocellulosic biomass [8]. The type and extent of chemical changes to lignin, as well as other plant cell wall components, vary depending on the type of oxidant used, operation conditions, and the characteristics of biomass being pretreated.

13.1.2.1 Wet Oxidation

Wet oxidation is a pretreatment method that employs molecular oxygen or air in a pressurized reaction loaded with biomass and an alkaline aqueous solution (e.g., Na_2CO_3, NaOH). Biomass suspended in water is treated under pressurized (5–20 MPa) air or oxygen, and elevated temperature (150–350 °C). This pretreatment method has been demonstrated to be effective on various biomass feedstock including wheat straw [9, 10], reed [11], rice husk [12], rape straw [13], sugarcane bagasse [14], and wood pulp waste [15].

During wet oxidation of herbaceous biomass, lignin is converted to carbon dioxide (CO_2) as well as water-soluble carboxylic acids including acetic, formic, succinic, oxalic, and glutaconic acids. Hemicellulose is also solubilized during wet oxidation, with the extent of dissolution and degradation positively correlated with temperature [9]. Compared to steam explosion pretreatment, wet oxidation is advantageous due to lower processing temperature (<200 °C) and reduced generation of sugar dehydration products (e.g., furfural, 5-hydroxyformaldehyde) known to inhibit sugar-fermenting organisms [10].

13.1.2.2 Alkaline Hydrogen Peroxide

Hydrogen peroxide (H_2O_2) has a higher standard oxidation potential (Table 13.1) than molecular oxygen, and demonstrates effectiveness in biomass pretreatment under ambient conditions. Under alkaline conditions (pH 11–12), hydrogen peroxide dissociates and evolves into potently oxidative radicals (•OH, •OOH). These radicals oxidize lignins, resulting in change in lignin dissolution and removal from the plant cell wall matrix. The presence of alkali also promotes solubilization of hemicellulose, which further increases the accessibility of cellulose to hydrolytic enzymes.

As a pretreatment method, alkaline hydrogen peroxide (AHP) pretreatment is effective in removing both hemicellulose and lignins from the plant cell wall matrix, and thus increasing the accessibility of cellulose to hydrolytic enzymes [16, 17]. AHP has been demonstrated as an effective standalone pretreatment for a number of herbaceous feedstocks including wheat straw [18], rice hulls [19], corn stover [20, 21], bamboo [22], and agave bagasse [23], while the pretreatment is not effective on cactus pear [24] and gooseweed [25]. The efficacy of AHP pretreatment under elevated temperature (150–180 °C) on woody biomass has also been reported for Douglas fir [26, 27] and hybrid poplar [21].

AHP pretreatment has also been reported as a secondary post-treatment used in tandem with other pretreatment methods including wet oxidation [12], alkali pretreatment [28, 29], steam explosion [30], and microwave [31–33]. As part of a multi-stage pretreatment process, the delignifying AHP step reduces the cell wall recalcitrance synergistically with other pretreatments that are effective in hemicellulose removal [34]. The amount of hydrogen peroxide used in AHP post-treatment is reduced as

Table 13.1 Standard oxidation potential of oxidants used in oxidative pretreatment of biomass

Oxidant	Standard oxidation potential (V)
Hydroxyl radical (•OH)	2.86
Ozone (O_3)	2.07
Hydrogen peroxide (H_2O_2)	1.77
Oxygen (O_2)	1.23

compared to using AHP as a standalone pretreatment process, and may lead to an improvement in the economic feasibility of the entire pretreatment process [29].

13.1.2.3 Catalytic Hydrogen Peroxide Pretreatment

The application of metal catalysts in AHP pretreatment has been reported as an approach to improving both pretreatment efficacy and reducing oxidant use. Examples of effective catalysts include manganese acetate [35], copper-diimine complexes [36], magnetite nanoparticles [37], and titanium dioxide [38]. The effects of catalysts include accelerated oxidation reactions, reduced time needed for effective pretreatment, and improved efficacy of recalcitrance reduction. In presence of metal catalysts, oxidation reactions between hydrogen peroxide and plant cell wall components can be accelerated, or go through pathways different from those under catalyst-free conditions. Although the specific reaction mechanisms of catalytic biomass oxidation by hydrogen peroxide is not fully understood, a number of possible reaction schemes have been proposed. Transitional metal catalysts including manganese, copper, and iron can catalyze Fenton reactions that generate hydroxyl (•OH) and superoxide (O2•–) radicals, which in turn oxidizes lignins and other aromatic structures in biomass [39].

13.1.2.4 Ozone

Ozone is an environmentally friendly oxidant that generates hydroxyl radicals while dissolved in water and thus has potential applications in lignin oxidation and biomass pretreatment. Recent developments in ozone pretreatment of biomass have been detailed in a recent review by Travaini and coauthors [40]. The efficacy of ozone pretreatment is associated with the moisture content of biomass during pretreatment process, as dissolved ozone is more effective in lignin oxidation compared to gaseous ozone [41]. Currently, the high energy and financial cost of ozone generation has limited the use of ozone in biomass pretreatment as part of cellulosic biofuel production. To overcome these limitations, ozone pretreatment with low severity may be used in tandem with other synergistic pretreatment methods [42].

13.1.2.5 Enzymatic Pretreatment

Oxidoreductases have been associated with both the plant biosynthesis and fungal biodegradation of lignins. Laccase and peroxidases have been shown to affect lignin synthesis and accumulation in *Arabidopsis* [43–45], and their use in biomass pretreatment has also been reported. Deng et al. independently reported a *Trametes versicolor* laccase pretreatment mediated with 1-hydroxybenzotriazole that reduces non-productive enzymatic binding on lignin during wheat straw hydrolysis, and the pretreatment also results in a 26% increase in the yield of reducing sugars during enzymatic saccharification [46]. Heap et al. used *T. versicolor* laccase pretreatment in tandem with alkaline peroxide pretreatment to further improve saccharification yield from wheat straw [47]. Enhanced saccharification yield was also observed in Eucalyptus [48] and bamboo [49] pretreated with a laccase-mediator system.

13.2 Effects of Oxidative Pretreatment Processes on Plant Cell Wall Components

Physical and chemical changes to plant cell wall components, as well as an increase in enzymatic digestibility, have been associated with oxidation reactions during oxidative pretreatment. The extent of these changes is associated with the type of the pretreatment, as well as pretreatment process parameters including pH temperature, oxygen partial pressure, residence time, and oxidant–biomass stoichiometry. Some examples of pretreatment effects on enzymatic conversion are summarized in Table 13.2.

Via cleavage of covalent intramolecular linkages [39] and oxidation of lignin side chains [50], oxidative pretreatment is effective in hemicellulose solubilization and lignin removal. Redistribution and removal of recalcitrant cell wall poly-

Table 13.2 Effect of oxidative pretreatment on enzymatic digestibility of biomass

Feedstock	Pretreatment conditions	Enzyme loading	Cellulose conversion (of theoretical maximum)	References
Common reed (*Phragmites australis*)	12 bar O_2, 2 g/L Na_2CO_3, 200 °C, 12 min	25 FPU per g dry matter	82.4% after 48 h	[56]
Norway spruce (*Picea abies*)	12 bar O_2, initial pH = 7, 200 °C, 10 min	30 FPU per g dry matter	50% after 24 h, 79% after 72 h	[57]
Sugarcane bagasse	12 bar O_2, 2 g/L Na_2CO_3, 190 °C, 15 min	25 FPU per g dry matter	57.4% after 24 h	[58]
Wood pulp waste	12 bar O_2, pH = 10, 195 °C, 15 min	35 FPU per g dry matter	42.9% after 48 h	[15]
Corn stover	500 mg H_2O_2/g dry biomass, pH = 11.5, 30 °C, 24 h	20 FPU per g dry matter	88.2% after 48 h	[16]
Cashew apple bagasse	4.3% v/v H_2O_2, biomass loading at 2% w/v, pH 11.5, 35 °C, 24 h	60 FPU per g cellulose	50% after 48 h	[59]
Douglas fir (*Pseudotsuga menziesii*)	0.1 g H_2O_2/g biomass, pH 11.9, 180 °C, 30 min	20 FPU per g cellulose	95% after 96 h	[27]
Hybrid poplar (*Populus nigra* var. *charkoviensis* × *caudina* cv. NE-19)	Alkaline extraction (120 °C, 1 h) followed by AHP catalyzed by Cu-2,2′-bipyridyl (30 °C, 23 h)	30 mg protein per g dry matter	94% after 72 h	[60]

mer is associated with improved accessibility of hydrolytic enzymes. Wu et al. reported the reduced binding affinity of oxidized lignins to cellulases, which leads to improved enzymatic hydrolysis rate of cellulose [51]. The surface morphology of plant cell wall matrix is also affected by oxidative pretreatment, with an increase in surface roughness [52], pore volume [16], and pore surface area [22, 53]. Oxidative pretreatment also increases the water retention value [54], which is an indicator of enzymatic digestibility of pretreated biomass [55].

13.3 Knowledge Gaps

Oxidative pretreatment has demonstrated potential in reducing plant cell recalcitrance to biochemical conversion, although additional research is still needed in this area to better understand the physiological changes in biomass during oxidative pretreatment. Specifically, the oxidation chemistry during depolymerization and repolymerization of lignin polymers is not yet fully understood, and the effect of catalysts during oxidative pretreatment has not been well studied. The role of naturally occurring, cell wall–associated transition metals in biomass has been reported [61], which reveals the potential of breeding biomass crops with metal profiles suitable to oxidative pretreatment. In addition, more research is warranted in identifying strategies in reducing costs related to pretreatment chemical consumption and product separations during oxidative pretreatment [29].

References

1. Torén, K., & Blanc, P. D. (1997). The history of pulp and paper bleaching: Respiratory-health effects. *The Lancet, 349*, 1316–1318.
2. Sjöström, E. (1993). Chapter 8 – Pulp bleaching. In E. Sjöström (Ed.), *Wood Chem* (2nd ed., pp. 165–203). San Diego: Academic Press.
3. Biermann, C. J. (1996). Chapter 5 – Pulp bleaching. In: C. J. Biermann (Ed.), *Biermann's handbook of pulp and paper: Raw material and pulp making* (2nd ed., pp. 123–136). San Diego: Academic.
4. Bajpai, P. (2018). Chapter 19 – Pulp bleaching. In: P. Bajpai (Ed.), *Biermann's handbook of pulp and paper: Raw material and pulp making* (3rd ed., pp. 465–491). San Diego: Elsevier.
5. Koda, K., Shintani, H., Matsumoto, Y., & Meshitsuka, G. (1999). Evaluation of the extent of the oxidation reaction during chlorine bleaching of pulp. *Journal of Wood Science, 45*, 149–153.
6. Kaur, D., Bhardwaj, N. K., & Lohchab, R. K. (2019). Effect of incorporation of ozone prior to ECF bleaching on pulp, paper and effluent quality. *Journal of Environmental Management, 236*, 134–145.
7. Cai, H., Han, J., Wang, M., Davis, R., Biddy, M., & Tan, E. (2018). Life-cycle analysis of integrated biorefineries with co-production of biofuels and bio-based chemicals: Co-product handling methods and implications. *Biofuels, Bioproducts and Biorefining, 12*, 815–833.
8. Chaturvedi, V., & Verma, P. (2013). An overview of key pretreatment processes employed for bioconversion of lignocellulosic biomass into biofuels and value added products. *3 Biotech, 3*, 415–431.

9. Schmidt, A. S., & Thomsen, A. B. (1998). Optimization of wet oxidation pretreatment of wheat straw. *Bioresource Technology, 64*, 139–151.
10. Bjerre, A. B., Olesen, A. B., Fernqvist, T., Plöger, A., & Schmidt, A. S. (1996). Pretreatment of wheat straw using combined wet oxidation and alkaline hydrolysis resulting in convertible cellulose and hemicellulose. *Biotechnology and Bioengineering, 49*, 568–577.
11. Szijarto, N., Kadar, Z., Varga, E., Thomsen, A. B., Costa-Ferreira, M., & Reczey, K. (2009). Pretreatment of reed by wet oxidation and subsequent utilization of the pretreated fibers for ethanol production (Report). *Applied Biochemistry and Biotechnology, 157*, 83.
12. Banerjee, S., Sen, R., Pandey, R. A., Chakrabarti, T., Satpute, D., Giri, B. S., & Mudliar, S. (2009). Evaluation of wet air oxidation as a pretreatment strategy for bioethanol production from rice husk and process optimization. *Biomass and Bioenergy, 33*, 1680–1686.
13. Arvaniti, E., Bjerre, A. B., & Schmidt, J. E. (2012). Wet oxidation pretreatment of rape straw for ethanol production. *Biomass and Bioenergy, 39*, 94–105.
14. Martín, C., Marcet, M., & Thomsen, A. B. (2008). Comparison between wet oxidation and steam explosion as pretreatment methods for enzymatic hydrolysis of sugarcane bagasse. *BioResources, 3*, 670–683.
15. Ji, X., Liu, S., Wang, Q., Yang, G., Chen, J., & Fang, G. (2015). Wet oxidation pretreatment of wood pulp waste for enhancing enzymatic saccharification. *BioResources, 10*, 2184.
16. Li, J., Lu, M., Guo, X., Zhang, H., Li, Y., & Han, L. (2018). Insights into the improvement of alkaline hydrogen peroxide (AHP) pretreatment on the enzymatic hydrolysis of corn stover: Chemical and microstructural analyses. *Bioresource Technology, 265*, 1–7.
17. Ho, M. C., Ong, V. Z., & Wu, T. Y. (2019). Potential use of alkaline hydrogen peroxide in lignocellulosic biomass pretreatment and valorization – A review. *Renewable and Sustainable Energy Reviews, 112*, 75–86.
18. Saha, B. C., & Cotta, M. A. (2006). Ethanol production from alkaline peroxide pretreated enzymatically saccharified wheat straw. *Biotechnology Progress, 22*, 449–453.
19. Saha, B. C., & Cotta, M. A. (2007). Enzymatic saccharification and fermentation of alkaline peroxide pretreated rice hulls to ethanol. *Enzyme and Microbial Technology, 41*, 528–532.
20. Mittal, A., Katahira, R., Donohoe, B. S., Black, B. A., Pattathil, S., Stringer, J. M., & Beckham, G. T. (2017). Alkaline peroxide delignification of corn stover. *ACS Sustainable Chemistry & Engineering, 5*, 6310–6321.
21. Gupta, R., & Lee, Y. Y. (2010). Pretreatment of corn stover and hybrid poplar by sodium hydroxide and hydrogen peroxide. *Biotechnology Progress, 26*, 1180–1186.
22. Song, X., Jiang, Y., Rong, X., Wei, W., Wang, S., & Nie, S. (2016). Surface characterization and chemical analysis of bamboo substrates pretreated by alkali hydrogen peroxide. *Bioresource Technology, 216*, 1098–1101.
23. Perez-Pimienta, J. A., Poggi-Varaldo, H. M., Ponce-Noyola, T., Ramos-Valdivia, A. C., Chavez-Carvayar, J. A., Stavila, V., & Simmons, B. A. (2016). Fractional pretreatment of raw and calcium oxalate-extracted agave bagasse using ionic liquid and alkaline hydrogen peroxide. *Biomass and Bioenergy, 91*, 48–55.
24. de Souza Filho, P. F., Ribeiro, V. T., dos Santos, E. S., & de Macedo, G. R. (2016). Simultaneous saccharification and fermentation of cactus pear biomass – Evaluation of using different pretreatments. *Industrial Crops and Products, 89*, 425–433.
25. Vu, P. T., Unpaprom, Y., & Ramaraj, R. (2018). Impact and significance of alkaline-oxidant pretreatment on the enzymatic digestibility of Sphenoclea zeylanica for bioethanol production. *Bioresource Technology, 247*, 125–130.
26. Alvarez-Vasco, C., & Zhang, X. (2013). Alkaline hydrogen peroxide pretreatment of softwood: Hemicellulose degradation pathways. *Bioresource Technology, 150*, 321–327.
27. Alvarez-Vasco, C., & Zhang, X. (2017). Alkaline hydrogen peroxide (AHP) pretreatment of softwood: Enhanced enzymatic hydrolysability at low peroxide loadings. *Biomass and Bioenergy, 96*, 96–102.
28. Liu, T., Williams, D. L., Pattathil, S., Li, M., Hahn, M. G., & Hodge, D. B. (2014). Coupling alkaline pre-extraction with alkaline-oxidative post-treatment of corn stover to enhance enzymatic hydrolysis and fermentability. *Biotechnology for Biofuels, 7*, 48.

29. Stoklosa, R. J., del Pilar Orjuela, A., da Costa Sousa, L., Uppugundla, N., Williams, D. L., Dale, B. E., Hodge, D. B., & Balan, V. (2017). Techno-economic comparison of centralized versus decentralized biorefineries for two alkaline pretreatment processes. *Bioresource Technology, 226*, 9–17.
30. Yang, B., Boussaid, A., Mansfield, S. D., Gregg, D. J., & Saddler, J. N. (2002). Fast and efficient alkaline peroxide treatment to enhance the enzymatic digestibility of steam-exploded softwood substrates. *Biotechnology and Bioengineering, 77*, 678–684.
31. Wu, C. J., Zhao, C. S., Li, J., & Chen, K. F. (2011). The effect of microwave treatment on the hydrogen peroxide bleaching of soda-AQ wheat straw pulp. *Advances in Materials Research, 236–238*, 1307–1312.
32. Ayyasamy, S., Venkatachalam, S., Sangeetha, V., & Priyenka, D. (2016). Improving enzymatic saccharification of cassava stem using peroxide and microwave assisted pre-treatment techniques. *Chemical Industry and Chemical Engineering Quarterly, 23*, 50–50.
33. Huang, X., De Hoop, C. F., Li, F., Xie, J., Hse, C.-Y., Qi, J., Jiang, Y., & Chen, Y. (2017). Dilute alkali and hydrogen peroxide treatment of microwave liquefied rape straw residue for the extraction of cellulose nanocrystals. *Journal of Nanomaterials, 2017*, 4049061.
34. Zhu, M.-Q., Wen, J.-L., Wang, Z.-W., Su, Y.-Q., Wei, Q., & Sun, R.-C. (2015). Structural changes in lignin during integrated process of steam explosion followed by alkaline hydrogen peroxide of Eucommia ulmoides Oliver and its effect on enzymatic hydrolysis. *Applied Energy, 158*, 233–242.
35. Lucas, M., Hanson, S. K., Wagner, G. L., Kimball, D. B., & Rector, K. D. (2012). Evidence for room temperature delignification of wood using hydrogen peroxide and manganese acetate as a catalyst. *Bioresource Technology, 119*, 174–180.
36. Bhalla, A., Bansal, N., Pattathil, S., et al. (2018). Engineered lignin in poplar biomass facilitates cu-catalyzed alkaline-oxidative pretreatment. *ACS Sustainable Chemistry & Engineering, 6*, 2932–2941.
37. Koo, H., Salunke, B., Iskandarani, B., Oh, W.-G., & Kim, B. (2017). Improved degradation of lignocellulosic biomass pretreated by Fenton-like reaction using Fe 3 O 4 magnetic nanoparticles. *Biotechnology and Bioprocess Engineering, 22*, 597–603.
38. Kuznetsov, B., Kuznetsova, S., Danilov, V., & Yatsenkova, O. (2009). Influence of UV pretreatment on the abies wood catalytic delignification in the medium "acetic acid–hydrogen peroxide–TiO 2". *Reaction Kinetics and Catalysis Letters, 97*, 295–300.
39. Gierer, J. (1982). The chemistry of delignification – A general concept – Part II. *Holzforschung, 36*, 55–64.
40. Travaini, R., Martín-Juárez, J., Lorenzo-Hernando, A., & Bolado-Rodríguez, S. (2016). Ozonolysis: An advantageous pretreatment for lignocellulosic biomass revisited. *Pretreat Biomass, 199*, 2–12.
41. Den, W., Sharma, V. K., Lee, M., Nadadur, G., & Varma, R. S. (2018). Lignocellulosic biomass transformations via greener oxidative pretreatment processes: Access to energy and value-added chemicals. *Frontiers in Chemistry, 6*, 141.
42. Mulakhudair, A. R., Hanotu, J., & Zimmerman, W. (2017). Exploiting ozonolysis-microbe synergy for biomass processing: Application in lignocellulosic biomass pretreatment. *Biomass and Bioenergy, 105*, 147–154.
43. Berthet, S., Demont-Caulet, N., Pollet, B., et al. (2011). Disruption of LACCASE4 and 17 results in tissue-specific alterations to lignification of Arabidopsis thaliana stems. *Plant Cell, 23*, 1124.
44. Shigeto, J., Itoh, Y., Hirao, S., Ohira, K., Fujita, K., & Tsutsumi, Y. (2015). Simultaneously disrupting AtPrx2, AtPrx25 and AtPrx71 alters lignin content and structure in Arabidopsis stem. *Journal of Integrative Plant Biology, 57*, 349–356.
45. Cosio, C., Ranocha, P., Francoz, E., Burlat, V., Zheng, Y., Perry, S. E., Ripoll, J.-J., Yanofsky, M., & Dunand, C. (2017). The class III peroxidase PRX17 is a direct target of the MADS-box transcription factor AGAMOUS-LIKE15 (AGL15) and participates in lignified tissue formation. *The New Phytologist, 213*, 250–263.

46. Deng, Z., Xia, A., Liao, Q., Zhu, X., Huang, Y., & Fu, Q. (2019). Laccase pretreatment of wheat straw: Effects of the physicochemical characteristics and the kinetics of enzymatic hydrolysis. *Biotechnology for Biofuels, 12*, 159.
47. Heap, L., Green, A., Brown, D., van Dongen, B., & Turner, N. (2014). Role of laccase as an enzymatic pretreatment method to improve lignocellulosic saccharification. *Catalysis Science & Technology, 4*, 2251–2259.
48. Rico, A., Rencoret, J., del Río, J. C., Martínez, A. T., & Gutiérrez, A. (2014). Pretreatment with laccase and a phenolic mediator degrades lignin and enhances saccharification of Eucalyptus feedstock. *Biotechnology for Biofuels, 7*, 6.
49. Gaikwad, A., & Meshram, A. (2019). Effect of particle size and mixing on the laccase-mediated pretreatment of lignocellulosic biomass for enhanced saccharification of cellulose. *Chemical Engineering Communications, 207*, 1–11.
50. Gould, J. M. (1985). Studies on the mechanism of alkaline peroxide delignification of agricultural residues. *Biotechnology and Bioengineering, 27*, 225–231.
51. Wu, K., Ying, W., Shi, Z., Yang, H., Zheng, Z., Zhang, J., & Yang, J. (2018). Fenton reaction-oxidized bamboo lignin surface and structural modification to reduce nonproductive cellulase binding and improve enzyme digestion of cellulose. *ACS Sustainable Chemistry & Engineering, 6*, 3853–3861.
52. Selig, M. J., Vinzant, T. B., Himmel, M. E., & Decker, S. R. (2009). The effect of lignin removal by alkaline peroxide pretreatment on the susceptibility of corn stover to purified cellulolytic and xylanolytic enzymes. *Applied Biochemistry and Biotechnology, 155*, 94–103.
53. Shimizu, F. L., de Azevedo, G. O., Coelho, L. F., Pagnocca, F. C., & Brienzo, M. (2020). Minimum lignin and xylan removal to improve cellulose accessibility. *Bioenergy Research, 13*, 775–785.
54. Bhalla, A., Bansal, N., Stoklosa, R. J., Fountain, M., Ralph, J., Hodge, D. B., & Hegg, E. L. (2016). Effective alkaline metal-catalyzed oxidative delignification of hybrid poplar. *Biotechnology for Biofuels, 9*, 34.
55. Weiss, N. D., Felby, C., & Thygesen, L. G. (2018). Water retention value predicts biomass recalcitrance for pretreated lignocellulosic materials across feedstocks and pretreatment methods. *Cellulose, 25*, 3423–3434.
56. Szijártó, N., Kádár, Z., Varga, E., Thomsen, A. B., Costa-Ferreira, M., & Réczey, K. (2009). Pretreatment of reed by wet oxidation and subsequent utilization of the pretreated fibers for ethanol production. *Applied Biochemistry and Biotechnology, 155*, 83–93.
57. Palonen, H., Thomsen, A. B., Tenkanen, M., Schmidt, A. S., & Viikari, L. (2004). Evaluation of wet oxidation pretreatment for enzymatic hydrolysis of softwood. *Applied Biochemistry and Biotechnology, 117*, 1–17.
58. Martin, C., Marcet, M., & Thomsen, A. B. (2008). Comparison between wet oxidation and steam explosion as pretreatment methods for enzymatic hydrolysis of sugarcane bagasse. *BioResources, 3*, 670–683.
59. Correia, J. A. d. C., Júnior, J. E. M., Gonçalves, L. R. B., & Rocha, M. V. P. (2013). Alkaline hydrogen peroxide pretreatment of cashew apple bagasse for ethanol production: Study of parameters. *Bioresource Technology, 139*, 249–256.
60. Bhalla, A., Fasahati, P., Particka, C. A., Assad, A. E., Stoklosa, R. J., Bansal, N., Semaan, R., Saffron, C. M., Hodge, D. B., & Hegg, E. L. (2018). Integrated experimental and technoeconomic evaluation of two-stage Cu-catalyzed alkaline–oxidative pretreatment of hybrid poplar. *Biotechnology for Biofuels, 11*, 143.
61. Bansal, N., Bhalla, A., Pattathil, S., Adelman, S. L., Hahn, M. G., Hodge, D. B., & Hegg, E. L. (2016). Cell wall-associated transition metals improve alkaline-oxidative pretreatment in diverse hardwoods. *Green Chemistry, 18*, 1405–1415.

Chapter 14
Lignin Valorization in Biorefineries Through Integrated Fractionation, Advanced Characterization, and Fermentation Intensification Strategies

Zhi-Min Zhao, Yan Chen, Xianzhi Meng, Siying Zhang, Jingya Wang, Zhi-Hua Liu, and Arthur J. Ragauskas

14.1 Introduction

The valorization of lignin has attracted increasing attention as it is now considered to be an essential part to achieve cost-effective and sustainable biorefineries [1, 2]. Lignin is more energy-dense in comparison with cellulose and hemicellulose due to its higher carbon-to-oxygen ratio [3]. However, it is also much more difficult to depolymerize due to its complex three dimensional molecular structure and the presence of several C–C interunit linkages. Furthermore, the presence of lignin–carbohydrate complex (LCC) results in difficulties in obtaining carbohydrate and lignin without structural alterations [4]. Recently, various pretreatment methods have been developed to fractionate lignin, cellulose, and hemicellulose in biomass.

Z.-M. Zhao
School of Ecology and Environment, Inner Mongolia Key Laboratory of Environmental Pollution Controlling and Wastes Recycling, Inner Mongolia University, Hohhot, China

Department of Chemical and Biomolecular Engineering, The University of Tennessee, Knoxville, TN, USA

Y. Chen · S. Zhang · J. Wang
School of Ecology and Environment, Inner Mongolia Key Laboratory of Environmental Pollution Controlling and Wastes Recycling, Inner Mongolia University, Hohhot, China

X. Meng
Department of Chemical and Biomolecular Engineering, The University of Tennessee, Knoxville, TN, USA

Z.-H. Liu
Synthetic and Systems Biology Innovation Hub (SSBiH), Texas A&M University, College Station, TX, USA

Department of Plant Pathology and Microbiology, Agriculture & Life Sciences, Texas A&M University, College Station, TX, USA

Z.-H. Liu, A. Ragauskas (eds.), *Emerging Technologies for Biorefineries, Biofuels, and Value-Added Commodities*,
https://doi.org/10.1007/978-3-030-65584-6_14

However, the properties of lignin obtained from different methods vary significantly. Table 14.1 lists the different sources, characteristics, and applications of different types of lignin.

Even in the same pretreatment process, the obtained lignin exhibits inhomogeneous physicochemical properties such as different molecular weights and functional groups due to the complexity of bond cleavage and repolymerization reactions. Lignin chemistry determines its functions, while its structural heterogeneity significantly restricts its valorization. To address these challenges, lignin fractionation provides an effective approach to obtain lignin fractions with relatively uniform properties. The resulting different lignin fractions can be utilized rationally based

Table 14.1 Sources, characteristics, and applications of different types of lignin

Lignin types	Sources	Characteristics and applications
Milled wood lignin (MWL)	Extraction with neutral solvents (dioxane/water) from milled wood [5]	Slight change on lignin structure; Used for lignin structure comparison
Klason lignin	Lignocellulosics treated by concentrated acid (72% (w/w) H_2SO_4)	Applied for lignin content measurement [6]
Kraft lignin (KL)	Kraft pulping	Macromolecular destroyed, low molecular weight lignin, main kind of industrial lignin, account for 85% of whole industrial lignin; Mainly burned for energy, low utilization efficiency [7]
Lignosulfonate	Sulfite pulping	Soluble in water, higher molecular weight than kraft lignin; Relatively wide applications in dispersing agents, surfactants, cement additives, etc. [8]
Organosolv lignin	Extraction with organic solvents from lignocellulosics	Relatively high processability and purity; Application in adhesive, filler, etc. [9, 10]
Alkaline lignin	Extraction with alkaline or alkaline–anthraquinone	Degradation of β-O-4 linkages and aromatic ring openings, high content of phenolic hydroxyl; Application in surfactants, phenol substitute, nanoparticles, etc. [11]
Enzymatic hydrolysis lignin (EHL)	Isolated from enzymatic hydrolysis residues of lignocellulosics	Relatively high processability and reactivity; Application in adsorbents, modification of polymer materials, etc. [9]
Steam exploded lignin	Isolation by steam explosion process	Relatively high processability and purity, high content of phenolic hydroxyl; Application in phenol substitute, filler, etc. [9, 12]

A. J. Ragauskas (✉)
Department of Chemical and Biomolecular Engineering, The University of Tennessee, Knoxville, TN, USA

Center for Bioenergy Innovation, Joint Institute of Biological Science, Biosciences Division, Oak Ridge National Laboratory (ORNL), Oak Ridge, TN, USA

Center for Renewable Carbon, Department of Forestry, Wildlife, and Fisheries, The University of Tennessee Institute of Agriculture, Knoxville, TN, USA
e-mail: aragausk@utk.edu

on their special properties, which could contribute to the efficient valorization of the parent lignin. Regardless of the source of lignin and different pretreatment or fractionation technologies, lignin characterization is vital to understand the structural changes and further provides useful guidance for the lignin processing. Therefore, characterization also plays an essential role in lignin valorization.

At present, there are two major ways for lignin valorization, including thermochemical and biological conversion [8, 13]. The thermochemical processes rely heavily breaks down large molecules of lignin with high temperature, high pressure, extreme pH, etc. While obtaining high reaction rate and efficiency, thermochemical processes often have disadvantages such as intensive energy input, severe operating conditions, and complex products [13, 14]. Recently, biological conversion of lignin has begun to open an alternative pathway for lignin valorization with significant potentials [15]. Typically, lignin streams are selected as a feedstock, which can be eventually converted to value-added compounds such as lipids, polyhydroxyalkanoates (PHAs), muconic acids, and vanillin through the microbial metabolism [8]. Although the reaction time is relatively longer, these biological conversions of lignin show potential green and economic process benefits, which have emerged as promising approaches for lignin valorization.

This chapter illustrates recent lignin fractionation methods to overcome the heterogeneity of lignin. Characterization techniques as well as their important roles for understanding lignin structure and improving lignin valorization are also systematically summarized. Besides these, as an emerging technology platform, bioconversion and intensification strategies of lignin valorization through fermentation are reviewed systematically from the aspects of microbial strains, substrates, and processes. Finally, the perspectives for future research on more directional lignin valorization routes are discussed.

14.2 Lignin Fractionation

Fractionation has been employed to deconstruct biomass and decrease the structural heterogeneity of lignin and thus produces the lignin with specific structures, which is beneficial to its rational utilization for material applications and microbial conversion. At present, there are principally three methods for lignin fractionation, including membrane technology, sequential precipitation, and organosolv sequential dissolution [16, 17].

14.2.1 Fractionation by Membrane Technology

Based on the difference of lignin molecular size (i.e., molecular weight), membrane technology is commonly applied for lignin fractionation. Wang and Chen [18] used ultrafiltration membrane with different cut-offs (20, 10, and 6 kDa) to fractionate alkali-extracted lignin from steam-exploded corn straw. Considering the pH tolerance of the membrane materials, dilute acid was added in the alkali-extracted lignin solution to ensure that the pH value was acceptable for membrane operation. Results showed that four lignin fractions were obtained after the ultrafiltration membrane treatment, of which the weight-average molecular weight (M_w) was 15,867, 7332, 4575, and 2882 g/mol, respectively. Lignin concentration and purity increased while inorganic salts content decreased in the lignin fractions. Moreover, lignin fractions with different molecular weight showed various acid-precipitation behaviors, which justified the necessity of the fractionation. Compared with traditional polymer membranes, a ceramic membrane can be used under a wide range of pH and temperature values, which enhances the applicability of membrane fractionation. Toledano et al. [17] applied ceramic membrane with different cut-offs (15, 10, and 5 kDa) to fractionate black liquor, which was generated from alkaline pulping of the *Miscanthus sinensis* [7.5% (w/w) NaOH, 90 min and 90 °C]. Four fractions with different lignin molecular weights were efficiently obtained. Generally, membrane technology is suitable for fractionation of dissolved lignin, which can remove inorganic salts, oligosaccharides, monosaccharides, and improve the purity of the fractionated lignin [16].

14.2.2 Fractionation by Sequential Precipitation

Sequential precipitation method typically fractionates lignin based on its molecular weight as well as functional groups. Gradient acid precipitation is commonly used for alkaline lignin. Lignin was dissolved and formed a stable colloid structure in alkaline solutions. Due to the phenolic hydroxyl and carboxyl groups in lignin, the surface of colloid lignin is negatively charged in an alkaline solution, which prevents colloid aggregation and precipitation formation. When acid (i.e., HCl, H_2SO_4) is added, the free protons neutralize the negative groups on the lignin colloid surface, which facilitate precipitation formation. Lourençon et al. [19] applied sequential acid precipitation to fractionate lignin from hardwood and softwood kraft black liquors, respectively. Lignin precipitation was made sequentially by acidification at pH 9, 7, 5, 3, and 1, and five lignin fractions were obtained accordingly. Regardless of the lignins from hardwood or softwood, lignin with higher molecular weight precipitated at higher pH, which may due to the relatively fewer contents of acidic groups compared with low molecular weight lignin fractions. Meanwhile, lignin precipitation using antisolvents is a promising strategy for the separation of lignin from organic solvents. The addition of antisolvents with marginal solubility for lignin to organic phase forms solid lignin, which can be filtered from the remaining liquid. Holtz et al. [20] studied the separation of lignin from 2-methyltetrahydrofuran using antisolvent precipitation. Antisolvent screening showed that high lignin precipitation yield and efficient antisolvent recovery were achieved when using *n*-pentane as an antisolvent. Wang et al. [21] applied sequential precipitation with hexane as an antisolvent to fractionate both Kraft lignin (KL) and cosolvent enhanced lignocellulosic fractionation (CELF) lignin. Characterization results showed that lignin fractions with high molecular weight possessed a relatively higher frequency of aliphatic hydroxyl groups on the macromolecular chains, while fractions with low molecular weight contained more phenolic hydroxyl groups due to the cleavage of alkyl aryl ether bonds. The lignin fractions obtained from sequential precipitation showed narrowly distributed molecular weights and tunable chemical structure, which was beneficial to the subsequent specific application. The disadvantage of the sequential precipitation fractionation method is that coprecipitation might occur. Santos et al. [22] detected the presence of Na_2SO_4, NaCl in lignins precipitated with sulfuric acid, or hydrochloric acid, respectively, which decreased the purity of the obtained lignins.

14.2.3 Fractionation by Organosolv Sequential Dissolution

Organosolv sequential dissolution method also fractionates lignin based on its molecular weight and functional groups. Lignin after organosolv dissolution fractionation is not only more homogeneous in molecular weight, but also contains similar functional groups, which is advantageous to the subsequent specific application [23]. Mörck's research group [24–26] fractionated hardwood KL from birch by successive extraction with organic solvents including dichloromethane and methanol. Characterization results of high performance size exclusion chromatography and ^{13}C nuclear magnetic resonance spectrometry (NMR) showed that the low molecular weight fractions had lower polydispersity but higher contents of phenolic hydroxyl groups and higher syringyl/guaiacyl (S/G) ratio. Song et al. [27] applied solvents of ethyl acetate, ethanol, acetone, and dioxane/water (95:5, v/v) to fractionate lignin that was extracted by *p*-toluenesulfonic acid from hybrid poplar. As shown in Table 14.2, four fractions were obtained accordingly. Lignin fractions that were dissolved in ethanol and acetone possessed high contents of phenolic hydroxyls, which might be attributed to the formation of numerous hydrogen bonds. Meanwhile, 2D heteronuclear single quantum coherence (2D HSQC) NMR results showed that the lignin fraction dissolved in ethyl acetate contained more C–C bonds, mainly β–β units, compared with other fractions. A single solvent system consisting of ethanol and water for lignin fractionation was established as an attractive green system. Wang et al. [28] applied a simple one-step ethanol fractionation method to fractionate bamboo KL. Soluble and insoluble fractions were subdivided. The soluble fraction (F_s) showed enhanced antibacterial activity compared to bamboo KL, while the insoluble fraction (F_i) barely showed any inhibition on bacteria growth, and even promoted bacterial growth. As shown in Table 14.2, molecular weight was much lower while phenolic hydroxyls content was higher for F_s compared with those of F_i. It was deduced that the better solubility and higher phenolic hydroxyls content in the F_s fraction contributed to its enhanced antimicrobial performance [29].

Furthermore, organosolv sequential dissolution could also fractionate lignins directly from raw lignocellulosics biomass. In a biorefinery concept, Liu et al. [23]

Table 14.2 Molecular weight and functional group content of different lignin fractions

	Successive organic solvents fractionation of *p*-toluenesulfonic acid-extracted lignin from hybrid poplar [27]				One-step ethanol fractionation of bamboo kraft lignin [28]	
	F_1 (ethyl acetate soluble)	F_2 (ethanol soluble)	F_3 (acetone soluble)	F_4 (dioxane/water soluble)	F_s (95% ethanol soluble)	F_i (95% ethanol insoluble)
M_w (g/mol)	1903	2275	4571	45,420	2518	5216
PDI[a]	1.81	1.44	1.78	4.88	1.79	1.65
Syringyl OH[b]	1.15	1.67	1.53	0.90	0.89	0.43
Guaiacyl OH[b]	0.55	0.85	0.75	0.52	0.51	0.27
p-Hydroxyphenyl OH[b]	0.30	0.52	0.45	0.28	0.54	0.29
COOH[b]	0.68	0.29	0.16	0.14	0.77	0.41

[a]PDI represents polydispersity index
[b]The unit of the functional group content is mmol/g

applied ethanol plus different-stage catalysts to selectively dissolve lignin from corn stover for producing multiple uniform lignin streams, and to tailor their chemistry and reactivity for fabricating lignin nanoparticles (LNPs). As a result, polydispersity index (PDI) of LNPs from dissolved lignin by ethanol plus sulfuric acid was less than 0.08, indicating good uniformity and stability of the LNPs. Lignin characterization results showed that S/G ratio, β-O-4, and β–β linkage abundance of the obtained lignin decreased while more phenolic hydroxyl groups were present, which helped to enhance the stability of the LNPs. By tailoring the lignin chemistry using sequential dissolution, high-quality LNPs of a spherical shape, small effective diameters, and good stability have been fabricated. Moreover, the glucose and xylose yields were also increased due to the improved hydrolysis performance. Therefore, fractionation by organosolv dissolution could provide an effective route for specific utilization of the lignin fractions, which shows great potential for upgrading the low-value lignin and thus contributes to the profitability of biorefineries. The disadvantages of organosolv sequential dissolution fractionation are generally attributed to complicated processes and difficulties in the recycling of organic solvents [30].

Overall, as shown in Table 14.3, each fractionation method has its own special characteristics. The options of fractionation method should depend not only on the lignin rich substrates but also on the further utilization of the fractionated lignin. Clearly knowing the further valorization route of the obtained lignin fractions can facilitate the options of lignin fractionation method.

Table 14.3 Comparison of different lignin fractionation methods

Fractionation methods	Principles	Characteristics	Applications
Membrane technology	Based on lignin molecular weight strictly	Remove inorganic salts, oligosaccharides, monosaccharides, improve lignin concentration and purity, single function [9]	Suitable for dissolved lignin, relatively mature in industry [31]
Sequential precipitation	Based on lignin solubility, involving lignin molecular weight and functional groups	Obtain solid lignin fractions with narrowly distributed molecular weights and tunable chemical structure, lignin purity is not always guaranteed [21, 22]	Always applied for lignin in alkaline or organic solvents system, could be applied in industry
Organosolv sequential dissolution	Based on lignin solubility in organic solvents, involving lignin molecular weight and polar groups [27, 32]	Remove polysaccharide, improve lignin purity and activity obviously, complicated processes, difficulties in organosolv recycling [33]	Suitable for solid lignin and raw lignocellulosics biomass, relatively high cost, generally applied in laboratory analysis at present

14.3 Lignin Characterization

Lignin conversion to fuels and value-added chemicals requires a sound understanding of the lignin structure [34]. Lignin characterization not only provides a detailed understanding of lignin structure, but also provides guidance for enhancing lignin valorization, which is of great importance for improving the efficiency of biorefineries.

14.3.1 Lignin Characterization Methods

14.3.1.1 Molecular Weight Analysis

Molecular weight is one of the critical properties of lignin. Various analytical methods including electrospray ionization mass spectrometry, vapor pressure osmometry (VPO), light scattering, and gel permeation chromatography (GPC) have been applied to measure lignin molecular weights [35]. In particular, GPC is commonly used to characterize weight-average molecular weight (M_w) and number-average molecular weight (M_n) of lignin because of its relatively short processing time and broad range of detected molecular weights. The PDI, which indicates the distribution of molecular weights, can be calculated as M_w/M_n. Tetrahydrofuran (THF) is commonly used as the mobile phase in GPC, and lignin acetylation using acetic anhydride and pyridine mixture is typically required to improve the solubility of lignin in the mobile phase. Table 14.4 shows the molecular weights and PDI of

Table 14.4 Molecular weight and PDI of different kinds of lignin

Lignins	M_w (g/mol)	M_n (g/mol)	PDI
Hardwood kraft lignin [9, 36]	3300	1000	3.30
Softwood kraft lignin [36]	6500	1600	4.06
Spruce Na-lignosulfonate [37]	64,000	7200	8.89
Organosolv lignin from wheat straw [38]	8680	1702	5.10
Alkaline lignin from wheat straw [38]	8000	1667	4.80
Enzymatic hydrolysis lignin from corn straw [29]	12,293	7210	1.70

different lignins, which demonstrate that the lignin molecular weights varied depending on their species as well as the processing methods and chemicals employed [39].

14.3.1.2 Chemical Structure Analysis of Lignin Polymer

Lignin structure analysis is the basis of lignin research, which is related to lignin biosynthesis and regulation, lignin separation and purification, and the subsequent lignin utilization [2, 9]. Chemical degradation processes can be used for lignin structure analysis. However, these traditional wet-chemical techniques are usually time-consuming and complex. Among all the chromatographic and spectroscopic approaches, NMR characterization remains the most comprehensive technique and has been widely used to offer structural insight into lignin by far. NMR characterization techniques mainly include ^{1}H NMR, ^{13}C NMR, ^{31}P NMR, and 2D HSQC NMR. ^{1}H NMR has been regarded as a valuable technique for the characterization of lignin chemical structure, by which acetate derivatives or underivatized lignin can be quantitatively analyzed. It provides information on some key functionalities such as alkyl groups, carboxyl group, methoxyl group, phenolic hydroxyl groups, aldehydes, and ketones. Both qualitative and quantitative ^{13}C NMR can be applied in lignin characterization, of which the quantitative ^{13}C NMR is usually time consuming. The addition of a relaxation agent such as chromium acetylacetonate to the solution reduces the acquisition time by relaxing all the nuclei in the lignin. ^{13}C NMR can provide information of lignin functional groups and structures, such as aryl ether bonds, condensation and noncondensation structural units, and methoxyl. Meanwhile, the content of aryl ether and C–C linkages can be also calculated quantitatively based on the ^{13}C NMR characterization data [40]. The hydroxyl groups affect the physical and chemical properties of lignin significantly, which play a significant role in defining the lignin reactivity. Lignin phosphitylation followed by ^{31}P NMR can quantitatively determine the major hydroxyl groups, including aliphatic, carboxylic, and different types of phenolic hydroxyls in short time with only small amount of samples. However, the samples are required to be phosphitylated prior to the analysis, and an appropriate internal standard needs to be selected for the analysis [41]. 2D HSQC NMR is a powerful method for lignin structural analysis. It is widely used as a semiquantitative method to analyze the relative abundance of interunit linkages. Besides, the information of monolignol ratios in lignin from native and genetically engineered plants as well as pretreated biomass can be detected [42, 43]. Detailed principles, characteristics, and applications of different lignin characterization methods were illustrated in Table 14.5.

14.3.2 Application of Lignin Characterization

14.3.2.1 Lignin Characterization Enriches Understanding of Lignin Properties

Characterization technologies to analyze lignin provide detailed knowledge of lignin structure and its properties. Meng et al. [55] isolated hololignin, which was the fraction of poplar lignin that remained after acid chlorite delignification. Hololignin properties were analyzed for the first time with various techniques including GPC, quantitative ^{13}C NMR, ^{31}P NMR, and 2D HSQC NMR. Compared with milled wood lignin (MWL), the representative native lignin isolated from poplar, hololignin has a significantly lower molecular weight and higher oxygenated aromatic carbon content. Meanwhile, hololignin is relatively enriched in guaiacyl units and has a lower S/G ratio, fewer aliphatic and phenolic hydroxyl groups, more carboxylic acid groups, and lower β-O-4 ether linkages than the poplar MWL. Moreover, it was noted that hololignin is enriched in condensed structures, which probably contributes to the resistance of residual lignin to acid chlorite. Meng et al. [56] examined a relatively new poplar lignin stream called CELF lignin, which was acquired from an acidic aqueous THF pretreatment (CELF: cosolvent enhanced lignocellulosic fractionation process). GPC results showed that CELF lignin presented a decreased molecular weight. 2D HSQC NMR analysis illustrated that β-O-4 interunit linkages were extensively cleaved after CELF pretreatment at elevated temperatures (>160 °C). Meanwhile, ^{31}P NMR analysis showed a decrease in aliphatic hydroxyl groups that was attributed, in part, due to dehydration reactions. The content of total phenolic hydroxyl groups significantly increased due to the cleavage of interunit linkages.

Pseudo-lignin with lignin-like structures, which is generated under acidic pretreatment conditions from hemicelluloses, has been shown to retard biological biomass conversion through either unproductive binding to enzymes and microbes or a physical hindrance to enzymes and microbes by blocking the active cellulose surface binding sites. Different analytical techniques were used for pseudo-lignin characterization to reveal the formation mechanism of pseudo-lignin [53]. Pseudo-lignin was observed as spherical balls or droplets on the surface of the materials after severe acid and hydrothermal pretreatment by using scanning electron microscopy (SEM). The generation of pseudo-lignin increased with an increase of pretreatment severity [57]. Ma et al. [54] applied XPS to trace pseudo-lignin formation and characterized the structure of isolated pseudo-lignin during hydrothermal pretreatment of holocellulose. It was also found that more pseudo-lignin would appear on the

Table 14.5 Principles, characteristics, and applications of different lignin characterization methods

Classifications	Methods	Principles	Characteristics	Applications	Usage frequency[a]
Chromatography	GPC	Separated according to molecular size on a column filled with porous filler [44]	Broad range of molecular weights, small amount of samples required, relatively short processing time, determination is simple and accurate	Wide applications in lignosulfonate, kraft lignin, alkaline lignin, lignin after pretreatment and enzymatic hydrolysis, etc.	*****
Physical spectroscopy methods	^{1}H NMR	Qualitative analysis according to different chemical shift of hydrogen proton, quantitative analysis of functional groups according to proportional relation of peak area and hydrogen atom number	Strong signal, relatively short processing time, signal overlap often occurs [45]	Qualitative and quantitative analysis of basic structures including alkyl groups, carboxyl group, methoxyl group, phenolic hydroxyl groups, aldehydes, and ketones	****
	^{13}C NMR	^{13}C atom has spin phenomena and nuclear magnetic resonance, the principle is similar as ^{1}H NMR	Broad spectral window, high resolution, less signal overlap, low natural isotopic abundance of ^{13}C and low sensitivity [46]	Lignin compositions, linkage types, condensation and noncondensation structures, methoxyl contents can be determined	*****
	^{31}P NMR	Phosphoric derivatives are formed by phosphorylation reagent and unstable hydroxyl protons in functional groups of lignin	High resolution, high abundance of ^{31}P nucleus, small amount of samples required, shot experimental time, derivatization of samples required	Contents of aliphatic, carboxylic, and phenolic hydroxyls can be determined [41]	*****

Classifications	Methods	Principles	Characteristics	Applications	Usage frequency[a]
	2D HSQC NMR	Combining ^{1}H NMR and ^{13}C NMR into a spectrum, cross signals to specifically determine lignin structure	High sensitivity, wide range spectrum, high resolution, avoid signal overlap effectively, relative contents determined [43]	Relative abundance of interunit linkages and monolignol ratios can be obtained [47]	*****
	Fourier-transform infrared spectroscopy (FTIR)	Different absorption at different wavelengths of various groups	High spectral resolution, short analysis time, relatively easy sample preparation	Qualitative and quantitative analysis of lignin functional groups [48]	****
	Ultraviolet absorption spectroscopy	Aromatic ring structure and various chromophoric groups in lignin have strong absorption to ultraviolet, so the functional groups of lignin can be analyzed	Simple and rapid operation, determination would be affected by degradation products from glycan [45]	Qualitative and quantitative analysis of kraft lignin, alkali lignin, organic solvent lignin, MWL, lignosulfonate, etc.	****
Chemical degradation processes	Nitrobenzene oxidation	β-aryl ether bonds are broken to form aromatic ring monomers, then lignin compositions can be analyzed [49]	Large amount of samples required, time consuming, unable to cleave C–C bonds	Determination of noncondensed lignin	***
	Potassium permanganate oxidation	Degradation products are analyzed to achieve information about lignin benzene ring structure, linkages between phenylpropane structural units	Side chain of lignin is selectively oxidized, qualitative analysis [45]	Mainly applied in qualitative analysis of lignin in wood and pulping [10]	*
	Ozone oxidization	Oxidation cleavage of aromatic ring, stereochemistry of side chains was detected by comparing with model compounds	Strong oxidation ability, less affected by pH value, high cost	Stereoscopic structure of lignin can be analyzed [50]	**

(continued)

Table 14.5 (continued)

Classifications	Methods	Principles	Characteristics	Applications	Usage frequency[a]
	Mercaptan acidolysis	Selective β-O-4 ether bonds cleavage to release lignin monomers after hydrolysis	Selective cleavage reaction, high yield, strict operation conditions, ethyl mercaptan smells foul [50]	Cleave β-O-4 bonds, further analysis of lignin monomers	*
	Derivatization reduction degradation	Effectively destroy β-aryl ether bonds and release lignin monomers [51]	Mild reaction conditions, high yield, reaction mechanism involving some structural units remains unclear	Structure of most lignin can be analyzed	**
Mass spectrometry	Gas chromatography-Mass spectrometry (GC-MS)	Samples are ionized and arranged according to charge–mass ratio, composition, structure and molecular weight can be detected	High specificity, sensitivity and efficiency	Analyze proportion of G, S, and H, and structure of dehydrogenation products of macromolecules [52]	****
Photoelectric methods	Scanning electron microscopy (SEM)	Observe surface morphology using secondary electronic signal imaging	Relatively easy to operate in morphology observation and size detection	Widely used in ultrastructural imaging of biomass to elucidate morphological changes after treatment [53]	****
	X-ray photoelectron spectroscopy (XPS)	Photoelectron spectroscopy is produced by radiating samples with X-rays to excite inner layer electrons or valence electrons	Observe chemical shift, high sensitivity, quantitative analysis [53]	Characterize surface chemistry and disorder degree of aromatic compounds, quantitatively determine element compositions and atoms location in samples [54]	***

Classifications	Methods	Principles	Characteristics	Applications	Usage frequency[a]
Thermal analysis	Thermogravimetric analysis (TGA)	Relationship between substance mass and temperature or time is measured under the program-controlled temperature, lignin decomposition and components can be analyzed	Quantitative analysis, accurately measure mass change and the change rate	Characterize lignin compositions and properties [21]	***

[a]Symbol "*" was applied to represent usage frequency of the characterization method. The more "*", the higher usage frequency

biomass surface with extended pretreatment time. Hu et al. [58] characterized pseudo-lignin using Fourier-transform infrared spectroscopy (FTIR) during severe acid and hydrothermal pretreatment. Spectra results showed that pseudo-lignin consisted of hydroxyl, carbonyl, and aromatic structure, and further revealed that pseudo-lignin was generated through dehydration, rearrangement, aromatization, and condensation reactions from carbohydrates. These findings have also been confirmed by ^{13}C NMR analysis. Therefore, through a series of characterization techniques, a better understanding of pseudo-lignin structure and formation mechanism was accomplished, which is of great significance for the future development of pretreatment technology that avoids the formation of pseudo-lignin.

14.3.2.2 Lignin Characterization Reveals Mechanisms and Provides Guidance for Improving Biomass Valorization

Lignin characterization is also beneficial for revealing pretreatment/extraction mechanisms and providing guidance for optimization of lignin manipulation in plants, biomass pretreatment, fractionation, and conversion processes. Liu et al. [59] investigated lignin structural and compositional changes of wild-type and engineered switchgrass before and after aqueous ionic liquid (IL) pretreatment. Their results indicated that the engineered switchgrass mutant *4CL* was more susceptible to aqueous IL pretreatment and more digestible during enzymatic saccharification due to its lower lignin content, higher S/G ratio, higher amounts of β-O-4 linkages, *p*-coumarate, and ferulate. Advanced characterization techniques provided insight into the impact of lignin manipulation on biomass pretreatment and lignin depolymerization, which helped to develop a more efficient biomass utilization.

Softwood KL was applied to produce microbial lipids with oleaginous *Rhodococcus opacus*. Poor bacterial growth was observed when KL was used directly as a substrate [60, 61]. Wei et al. [61] carried out O_2 pretreatment of KL under alkali conditions to improve the bioconversion efficiency. Results showed that the yield of lipids utilizing alkaline O_2-pretreated KL as a sole carbon source was significantly improved. GPC characterization illustrated that M_w and PDI of KL decreased significantly after O_2 pretreatment. 2D HSQC NMR demonstrated that part of signals corresponding to the aliphatic-ether and aliphatic alcohol (lignin-interunit) moieties greatly dimished after the O_2 pretreatment, whereas aromatic structures were fairly resistant to degradation. Quantitative ^{31}P NMR showed that the amount of noncondensed structures including aliphatic hydroxyl and guaiacyl hydroxyl decreased. On the other hand, the C-5 type condensed structures increased slightly after O_2 pretreatment, which indicated a selective enrichment of condensed structures during the pretreatment process. It was also noted that KL after O_2 pretreatment possessed more carboxylic acid functionalities. Moreover, SEM observed that the surface morphology of KL changed from the initially smooth surface into multilayer "eroded" regions with broken particles. These changes might increase the surface area and reduce the recalcitrance of KL, making it easier to be accessed by the bacteria. The characterization results above clearly indicated the degradation and depolymerization of KL after O_2 pretreatment. Based on these characterization results, the mechanisms of enhanced lignin bioconversion efficiency were analyzed. Das et al. [62] characterized deep eutectic solvent (DES, 1:2 choline chloride:lactic acid) extracted sorghum lignin using 2D HSQC NMR, ^{13}C NMR, and ^{31}P NMR. Results showed that DES cleaved nearly all ether linkages in native lignin, resulting in significant molecular weight reduction and benefiting the preparation of low molecular weight phenolic compounds. The advanced characterization techniques provided a mechanistic understanding of lignin depolymerization in DES and a guidance for rational valorization of the obtained lignin.

Wang et al. applied ethanol–water dissolution to fractionate EHL and bamboo KL [28, 29]. It was interesting to note that different lignin fractions had totally different antimicrobial activity against both Gram-positive and Gram-negative bacteria. Systematic characterization including GPC, FTIR, 2D HSQC NMR, and ^{31}P NMR were performed. GPC results showed that the lignin polydispersity significantly decreased after fractionation. Moreover, it was observed that the antimicrobial activity was centralized in the lignin fraction with lower molecular weight, better solubility, and higher phenolic hydroxyl contents, whereas the fractions with higher molecular weight had lower phenolic hydroxyl contents and poor water solubility, which barely showed inhibition on bacteria growth, and even promoted bacterial growth. The mechanism was deduced that the poor water-solubility resulted in the formation of insoluble particles, which acted as a carrier for bacteria and promoted the bacteria growth. Therefore, advanced characterization confirmed the efficiency of fractionation and further provided an explanation for the different antibacterial activity in different lignin fractions.

Li et al. [63] revealed different inhibitory mechanisms for different types of lignin upon cellulose enzymatic hydrolysis using NMR techniques. It was found that lignin structural features determined the behaviors of lignin in cellulose enzymatic hydrolysis. Organosolv lignin from loblolly pine could adsorb enzyme nonproductively, reducing the available enzyme for cellulose, which decreased the hydrolysis rate and ultimate sugar yield. Kraft pine lignin mainly precipitated on the cellulose surface and negatively affected hydrolysis by steric repulsion, which limited the productive contact between cellulose and enzyme. On the other hand, Zhou et al. [64] found that sodium lignosulfonate from poplar wood could enhance enzymatic saccharification of lignocelluloses, which was attributed to lignosulfonate groups acting as a surfactant in the enzymatic hydrolysis process. The advanced characterization provided insights into the effects of different lignin structure on lignin–enzyme interaction and cellulose enzymatic hydrolysis, which could be helpful to biomass pretreatment design and process optimization. Overall, the advanced characterization could provide guidance for optimization of lignin manipulation in plants, biomass pretreatment, fractionation, and conversion processes, which should further improve the biomass valorization.

Overall, lignin characterization techniques are developing in progress with analytical advances. Due to the disadvantages of complicated operation and noncomprehensive information obtained, traditional chemical degradation processes including nitrobenzene oxidation, potassium permanganate oxidation, and mercaptan acidolysis are gradually being replaced or supplemented by more advanced chromatography and spectroscopy analytical techniques, such as GPC, NMR, FTIR, and GC-MS. These advanced techniques provide more information on lignin properties including molecular weight, functional groups, and other structures. However, accurate quantitative characterization and structure–activity relationships of lignin have still not been completely achieved. Therefore, combining various advanced technologies to characterize lignin and developing more powerful and suitable analysis techniques still need further exploration.

14.4 Fermentation Process Intensification Strategies

Bioconversion of lignin through fermentation is an emerging approach with significant potentials. Advanced fermentation utilizing lignin is a systematic project, requiring optimal microbial stains, substrates, and processes [15]. Figure 14.1 illustrated the outline of the fermentation process intensification strategies involving the aspects of microbial stains, substrates, and processes design.

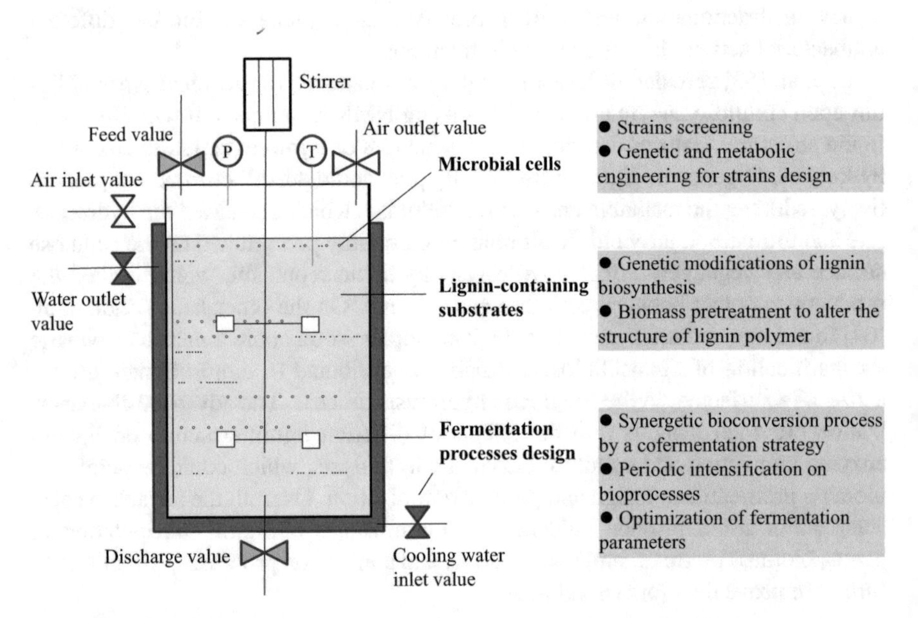

Fig. 14.1 Fermentation process intensification strategies involving microbial stains, substrates, and processes design

14.4.1 Exploration on Microbial Strains

14.4.1.1 Strains Screening

The screening and selection of promising microbial strains with high growth rates and productivities is a crucial step for improving lignin bioconversion. Traditional methods for screening and identifying strains utilizing lignin are typically achieved through monitoring cell growth with lignin as carbon source and detecting ligninolytic enzyme activities and products' yield [13, 65]. Salvachúa et al. [66] utilized alkaline pretreatment liquor from corn stover as a substrate to screen strains. Results illustrated that *Pseudomonas putida* KT2440 and *Pseudomonas putida* mt-2 could produce medium-chain length PHAs while *Cupriavidus necator* accumulated mainly short-chain PHAs. *Rhodococcus* strains are widely used for lipids generation utilizing lignin-derived aromatic molecules due to the pathways for oxidizing ring opening of central aromatic intermediates via β-ketoadipate pathway and enabling shuttling of aromatic-derived carbon into central carbon metabolism via the tricarboxylic acid (TCA) cycle [67]. *Rhodococcus opacus* PD630 was isolated from a soil sample at a gas-works plant in Germany and was shown to be able to produce lipids utilizing lignin substrates effectively [68, 69]. In addition to the traditional methods, various emerging techniques, such as genomics/metagenomics analysis, biosensor, and the spectrometric assay of lignin degradation products, contribute to the fast screening for promising strains. Lin et al. [70] conducted comparative genomics analysis between a novel strain *P. putida* A514 with 12 other existing *P. putida* strains. Results showed that genes encoding a broad range of oxidoreductases, enzymes required for aromatic compound funneling and degradation as well as a PHA synthetic gene cluster had been discovered in *P.* putida A514, which demonstrated its potential on lignin bioconversion. Sana et al. [71] designed a biosensor that could detect vanillin at a concentration as low as 200 μM. The vanillin biosensor could be used for screening lignin-degrading strains that can convert lignin to vanillin, which provided a fast method for screening strains.

14.4.1.2 Genetic and Metabolic Engineering for Strains Design

In addition to screening strains from environment samples, genetic and metabolic engineering is another rational way to improve strains' ability on lignin valorization through regulating gene expression and metabolic pathway. From a metabolic flux perspective, multiple steps of lignin degradation pathways can be engineered and optimized to produce valuable products [14]. Lin et al. [70] performed a system biology-guided design for KL bioconversion by *P. putida* A514. It was found that *P. putida* A514 used a DyP-based enzymatic system for lignin depolymerization and utilized a variety of peripheral and central catabolism pathways to metabolize aromatic compounds. Based on these discoveries, an effective multifunctional DyP was introduced to *P. putida* A514 via genetic engineering. By overexpressing the key enzymes of aromatic compound funneling, the monomers from lignin depolymerization were consumed more efficiently by the engineered *P. putida* A514, which resulted in higher cell growth rate and cell mass concentration. To maximize carbon flux into PHAs synthesis, the biosynthesis of fatty acids was down-regulated while the β-oxidation of fatty acids was up-regulated, which significantly increased PHAs content up to 73% of cell dry weight. Xie et al. [72] conducted proteomics study on *R. opacus* PD630 with lignin as a sole carbon source. The proteomics results revealed unique mechanisms of lipids accumulation in *R. opacus*. The high correlation between FASI expressions with lipids accumulation along with the phenotype in engineered strains indicated that FASI was one of the key enzymes driving the carbon partition to the storage of lipids through channeling acetyl-CoA to fatty acids biosynthesis. This discovery thus provided effective guidance for design of *R. opacus* PD630 for high lipids productivity. With the development of synthesis and system biology, more defined and controllable genetic and metabolic processes should provide more suitable and promising strains for lignin bioconversion [73].

14.4.2 Modifications of Lignin Substrates

14.4.2.1 Genetic Modifications of Lignin Biosynthesis

The regulation of lignin biosynthesis though genetic engineering represents a promising strategy to improve lignin properties for fermentation, but this is a nontrivial task [15]. Lignin pathway-related metabolites vary among species and individuals within a population [74]. Genome-wide association study (GWAS) has been applied to help to link genes to phenotypes [75, 76]. Wei et al. [77] identified a glycosyl hydrolase, *CYT1*, *SHINE1*, and *DAR6* as novel regulators for lignin biosynthesis in *Brassica napus* by using GWAS approach combined with transcriptomics. Morris et al. [78] found genes related to lignin content and agroclimatic adaptation in *Sorghum bicolor* also by using GWAS. S/G ratio is an important parameter affecting lignin properties [79]. Takeda et al. [80] regulated S/G lignin unit ratio by manipulation of a gene *CAld5H* in rice. It was noted that RNAi-mediated downregulation of *CAld5H* generated lignins enriched in G units, whereas upregulation of *CAld5H* resulted in lignins enriched in S units, which was revealed by a series of wet-chemical and NMR structural analyses. In addition, it was found that lignin biosynthetic genes coexpress with genes related to cellulose and hemicellulose biosynthesis, suggesting that biosynthesis of the overall plant cell wall may be in part controlled by common regulatory networks [74]. Genetic modifications of lignin biosynthesis provide an approach to regulate lignin structure from the original point. As the developments in sequencing technologies, high-throughput cellular phenotyping platforms, and lignin characterization techniques, effective strategies to reduce the recalcitrance of native lignin and improve lignin properties for fermentation would become more feasible [74, 81, 82].

14.4.2.2 Biomass Pretreatment to Alter the Structure of Lignin Polymer

Biomass pretreatment could alter lignin structure and therefore influence lignin properties. Optimization of pretreatment process is an important way to make lignin more suitable for fermentation. Traditional biorefineries emphasized carbohydrates utilization, while leaving lignin as a waste stream from the processes targeted at improving carbohydrates output. Nowadays, lignin valorization has attracted increasing attentions to obtain high carbon conversion of the entire process of biorefineries [83]. Therefore, the pretreatment should focus not only on increasing carbohydrates output but also on improving lignin processability [14, 15]. Wei et al. [61] applied O_2 pretreatment to modify kraft lignin from black liquor. Physical and chemical properties of KL were improved for fermentation by *R. opacus* DSM 1069. As a result, lipids production was significantly increased. Liu et al. [84] developed a combinatorial pretreatment strategy aiming not only to increase carbohydrates and lignin output but also to improve the lignin processability. By applying the optimized combinatorial pretreatment with 1% H_2SO_4 followed by 1% NaOH, glucose and xylose yields increased by 10.0% and 8.1%, respectively, over a single NaOH pretreatment while lignin yield increased by 33.4%. Moreover, lignin processability was obviously improved, which was indicated by the NMR, GPC, and FTIR characterization results. A record level of PHA concentration was then achieved using the obtained lignin as a carbon source by *P. putida* KT2440. Therefore, the pretreatment maximized the carbohydrates output and lignin processability, and thus improved the fermentation performance utilizing lignin as a substrate, which could make a significant contribution to improve the efficiency of whole biorefineries.

14.4.3 Advanced Processes Design and Intensification

14.4.3.1 Synergetic Bioconversion Process by a Cofermentation Strategy

Cofermentation with different strains has been employed for synergetic bioconversion of lignin rich substrates to enhance fermentation performance. He et al. [67] reported that the alkali-extracted lignin could be successfully utilized by a cofermentation strategy with the wild-type *R. opacus* PD630 and engineered *R. jostii* RHA1 $VanA^-$. Results showed that the cofermentation strategy could depolymerize lignin into aromatics effectively and produce higher yield of lipids than single strain fermentation. In order to further enhance the biological conversion efficiency and clarify the mechanism, cofermentation strategy was further developed with strains including *R. opacus* PD630, *R. jostii* RHA1, and *R. jostii* RHA1 $VanA^-$ as well as substrates containing both lignin and carbohydrates. Meanwhile, fermentation kinetics and proteomics analysis were conducted to identify potential catabolic pathways during cofermentation for funneling biosynthesis of triacylglycerol (TAG) from both lignin and carbohydrates. Results demonstrated that coculture of the three strains achieved the highest activity of lignin conversion compared with single culture of each strain or coculture of *R. opacus* PD630 and *R. jostii* RHA1 VanA$^-$, which may attribute to the synergy among *R. opacus* PD630, *R. jostii* RHA1, and *R. jostii* RHA1 VanA$^-$. Proteomics analysis suggested that *R. opacus* PD630 had lower extracellular activity during lignin fermentation compared with *R. jostii* RHA1 and *R. jostii* RHA1 VanA$^-$. The cofermentation with *R. opacus* PD630 and other two strains may help *R. opacus* PD630 get access to lignin-derived products as carbon sources for cell growth and lipids production [83]. Based on the proteomics study and microorganism engineering, Xie et al. [72] designed a cofermentation process utilizing lignin-containing biorefineries stream from the ammonia fiber expansion pretreated corn stover for lipids production. The strains engineered with laccase secretion module (*R. opacus* PD630_La) and lipids synthesis module (*R. opacus* PD630_Fa) were cocultivated. Results showed that the yield of lipids in cofermentation increased by 48.4% than that in a single strain fermentation. The substrates composition analysis before and after fermentation clearly showed that lignin was significantly consumed by the consortium strains. Therefore, the cofermentation showed an effective strategy to improve fermentation performance utilizing lignin. However, more detailed study is needed to confirm the synergic effect and mechanism of coculturing strains [83].

14.4.3.2 Periodic Intensification on Bioprocesses

Periodic intensification has been shown to be an effective strategy to enhance bioprocess performance [85, 86]. Chen and Li [87] invented a bioreactor with pressure pulsation through feeding and exhaling sterile air in the reactor periodically. Results showed that the periodic pressure pulsation improved fermentation performance significantly. A systematic study was further performed to reveal the fermentation enhancement mechanisms. First, periodic pressure pulsation strengthened heat and mass transfer in the fermentation substrates [88]. Second, it was found that the pressure pulsation enhanced microbial metabolic key enzyme ATPase activity, which hydrolyzes ATP to produce energy and provides transmembrane proton and ionic gradients for the growth of cells. Thus, the microbial metabolism was stimulated [89, 90]. Third, it was noted that microbes grew and metabolized synchronously due to the periodic stimulation by pressure pulsation, which was beneficial to shorten the fermentation time [88]. Moreover, the pressure pulsation frequency was regarded as a key parameter determining fermentation performance and energy consumption. Zhao et al. [90] optimized the pressure pulsation frequency systematically based on a heat balance model in fermentation. Results indicated that the variable frequency was the preferred operation mode rather than the constant one, which further promoted the fermentation performance. Although lignin was not applied as fermentation substrate in these studies above, periodic intensification strategy provides a promising approach for lignin bioconversion.

14.4.3.3 Optimization of Fermentation Parameters

Microbial growth and product yield are sensitive to a number of environmental factors [91]. Lignin bioconversion is no exception and is determined by the strains, substrates properties as well as the culture parameters. Liu et al. [1] optimized fermentation conditions systematically using lignin as the carbon source by *R. opacus* PD630. Parameters including inoculation density, nitrogen sources, and soluble substrate concentration were optimized. Meanwhile, different fermentation modes including batch and fed-batch fermentation were investigated. Results showed that both the cell dry weight and lipids concentration in fed-batch fermentation mode were higher than those in batch fermentation mode. A record of the yield of lipids was obtained in fed-batch fermentation under the optimized culture parameters. Therefore, the lipids fermentation performance using lignin as a carbon source can be favorably improved through the optimization of fermentation conditions.

14.5 Conclusions and Future Perspectives

Lignin valorization requires a systematic design involving lignin substrates, processing, and lignin-based products. Advanced characterization techniques are expected to provide more exact lignin structure–activity relationships. From the point of reverse thinking, value-added products (i.e., lignin nanoparticles) should be further developed and expanded. Focus on the lignin characteristics required by the target products, suitable raw materials should be selected or even bio-designed, and moreover the pretreatment and fractionation methods should be designed directionally. Meanwhile, conversion processes should be stimulated by exploring various intensification ways to improve the production. Overall, with the increase of fundamental understanding of how different chemical structures will affect lignin processability illustrated by the advanced characterization, lignin valorization path should be more directional and controllable to improve the profitability and carbon efficiency of the entire biorefineries.

Acknowledgments This manuscript has been authored (AJR) in part, by UT-Battelle, LLC under Contract No. DE-AC05-00OR22725 with the U.S. Department of Energy. The United States Government retains and the publisher, by accepting the article for publication, acknowledges that the United States Government retains a non-exclusive, paid-up, irrevocable, world-wide license to publish or reproduce the published form of this manuscript, or allow others to do so, for United States Government purposes. The Department of Energy will provide public access to these results of federally sponsored research in accordance with the DOE Public Access Plan (http://energy.gov/downloads/doe-public-access-plan). The views and opinions of the authors expressed herein do not necessarily state or reflect those of the United States Government or any agency thereof. Neither the United States Government nor any agency thereof, nor any of their employees, makes any warranty, expressed or implied, or assumes any legal liability or responsibility for the accuracy, completeness, or usefulness of any information, apparatus, product, or process disclosed, or represents that its use would not infringe privately owned rights.

References

1. Liu, Z. H., Xie, S., Lin, F., Jin, M., & Yuan, J. S. (2018b). Combinatorial pretreatment and fermentation optimization enabled a record yield on lignin bioconversion. *Biotechnology for Biofuels, 11*(1), 21.
2. Ragauskas, A. J., Beckham, G. T., Biddy, M. J., Chandra, R., Chen, F., Davis, M. F., Davison, B. H., Dixon, R. A., Gilna, P., & Keller, M. (2014). Lignin valorization: Improving lignin processing in the biorefinery. *Science, 344*(6185), 1246843.
3. Wang, M., & Liu, C. (2016). Theoretic studies on decomposition mechanism of o-methoxy phenethyl phenyl ether: Primary and secondary reactions. *Journal of Analytical and Applied Pyrolysis, 117*, 325–333.
4. Giummarella, N., Pu, Y., Ragauskas, A. J., & Lawoko, M. (2019). A critical review on the analysis of lignin carbohydrate bonds. *Green Chemistry, 21*(7), 1573–1595.
5. Björkman, A. (1956). Studies on finely divided wood. Part 1. Extraction of lignin with neutral solvents. *Svensk Papperstidning, 59*(13), 477–485.
6. Sluiter, A., Hames, B., Ruiz, R., Scarlata, C., Sluiter, J., Templeton, D., & Crocker, D. (2008). Determination of structural carbohydrates and lignin in biomass. *Laboratory Analytical Procedure, 1617*, 1–16.
7. Tejado, A., Pena, C., Labidi, J., Echeverria, J., & Mondragon, I. (2007). Physico-chemical characterization of lignins from different sources for use in phenol–formaldehyde resin synthesis. *Bioresource Technology, 98*(8), 1655–1663.
8. Ragauskas, A. J. (2016). Challenging/interesting lignin times. *Biofuels, Bioproducts and Biorefining, 10*(5), 489–491.
9. Wang, G. (2015). *Fractionation of lignin from steam-exploded corn stalk and lignin-based materials preparation*. Beijing: University of Chinese Academy of Sciences.
10. Xue, B. (2015). *Preparation of lignin-based polyurethane and its performance characterization*. Beijing: Beijing Forestry University.
11. Mahan, K. M., Le, R. K., Yuan, J., & Ragauskas, A. J. (2017). A review on the bioconversion of lignin to microbial lipid with oleaginous *Rhodococcus opacus*. *Journal of Biotechnology & Biomaterials, 7*, 02.
12. Sasaki, C., Wanaka, M., Takagi, H., Tamura, S., Asada, C., & Nakamura, Y. (2013). Evaluation of epoxy resins synthesized from steam-exploded bamboo lignin. *Industrial Crops and Products, 43*, 757–761.
13. Li, X., & Zheng, Y. (2019). Biotransformation of lignin: Mechanisms, applications and future work. *Biotechnology Progress, 36*, e2922.
14. Xie, S., Ragauskas, A. J., & Yuan, J. S. (2016). Lignin conversion: Opportunities and challenges for the integrated biorefinery. *Industrial Biotechnology, 12*(3), 161–167.
15. Beckham, G. T., Johnson, C. W., Karp, E. M., Salvachúa, D., & Vardon, D. R. (2016). Opportunities and challenges in biological lignin valorization. *Current Opinion in Biotechnology, 42*, 40–53.
16. Fernández-Rodríguez, J., Erdocia, X., Hernández-Ramos, F., Alriols, M. G., & Labidi, J. (2019). Lignin separation and fractionation by ultrafiltration. In *Separation of functional molecules in food by membrane technology* (pp. 229–265). London: Elsevier.
17. Toledano, A., García, A., Mondragon, I., & Labidi, J. (2010). Lignin separation and fractionation by ultrafiltration. *Separation and Purification Technology, 71*(1), 38–43.
18. Wang, G., & Chen, H. Z. (2013). Fractionation of alkali-extracted lignin from steam-exploded stalk by gradient acid precipitation. *Separation and Purification Technology, 105*, 98–105.
19. Lourençon, T. V., Hansel, F. A., da Silva, T. A., Ramos, L. P., de Muniz, G. I., & Magalhães, W. L. (2015). Hardwood and softwood kraft lignins fractionation by simple sequential acid precipitation. *Separation and Purification Technology, 154*, 82–88.
20. Holtz, A., Weidener, D., Leitner, W., Klose, H., Grande, P. M., & Jupke, A. (2020). Process development for separation of lignin from OrganoCat lignocellulose fractionation using anti-solvent precipitation. *Separation and Purification Technology, 236*, 116295.

21. Wang, Y. Y., Li, M., Wyman, C. E., Cai, C. M., & Ragauskas, A. J. (2018b). Fast fractionation of technical lignins by organic cosolvents. *ACS Sustainable Chemistry & Engineering, 6*(5), 6064–6072.
22. Santos, P. S., Erdocia, X., Gatto, D. A., & Labidi, J. (2014). Characterisation of Kraft lignin separated by gradient acid precipitation. *Industrial Crops and Products, 55*, 149–154.
23. Liu, Z. H., Hao, N., Shinde, S., Pu, Y., Kang, X., Ragauskas, A. J., & Yuan, J. S. (2019a). Defining lignin nanoparticle properties through tailored lignin reactivity by sequential organosolv fragmentation approach (SOFA). *Green Chemistry, 21*(2), 245–260.
24. Mörck, R., Reimann, A., & Kringstad, K. P. (1988). Fractionation of kraft lignin by successive extraction with organic solvents. III. Fractionation of kraft lignin from birch. *Holzforschung, 42*(2), 111–116.
25. Mörck, R., Yoshida, H., Kringstad, K. P., & Hatakeyama, H. (1986). Fractionation of kraft lignin by successive extraction with organic solvents. I. Functional groups (13) C-NMR-spectra and molecular weight distributions. *Holzforschung, 40*, 51–60.
26. Yoshida, H., Mörck, R., Kringstad, K. P., & Hatakeyama, H. (1987). Fractionation of Kraft lignin by successive extraction with organic solvents. II. Thermal properties of kraft lignin fractions. *Holzforschung, 41*(3), 171–176.
27. Song, Y., Shi, X., Yang, X., Zhang, X., & Tan, T. (2019). Successive organic solvent fractionation and characterization of heterogeneous lignin extracted by *p*-Toluenesulfonic acid from hybrid poplar. *Energy & Fuels, 34*, 557–567.
28. Wang, G., Pang, T., Xia, Y., Liu, X., Li, S., Parvez, A. M., Kong, F., & Si, C. (2019a). Subdivision of bamboo kraft lignin by one-step ethanol fractionation to enhance its water-solubility and antibacterial performance. *International Journal of Biological Macromolecules, 133*, 156–164.
29. Wang, G., Xia, Y., Liang, B., Sui, W., & Si, C. (2018a). Successive ethanol–water fractionation of enzymatic hydrolysis lignin to concentrate its antimicrobial activity. *Journal of Chemical Technology and Biotechnology, 93*(10), 2977–2987.
30. Meng, X., Parikh, A., Seemala, B., Kumar, R., Pu, Y., Wyman, C. E., Cai, C. M., & Ragauskas, A. J. (2019b). Characterization of fractional cuts of co-solvent enhanced lignocellulosic fractionation lignin isolated by sequential precipitation. *Bioresource Technology, 272*, 202–208.
31. Sultan, Z., Graça, I., Li, Y., Lima, S., Peeva, L. G., Kim, D., Ebrahim, M. A., Rinaldi, R., & Livingston, A. G. (2019). Membrane fractionation of liquors from lignin-first biorefining. *ChemSusChem, 12*(6), 1203–1212.
32. Bär, J., Phongpreecha, T., Singh, S. K., Yilmaz, M. K., Foster, C. E., Crowe, J. D., & Hodge, D. B. (2018). Deconstruction of hybrid poplar to monomeric sugars and aromatics using ethanol organosolv fractionation. *Biomass Conversion and Biorefinery, 8*(4), 813–824.
33. Ramakoti, B., Dhanagopal, H., Deepa, K., Rajesh, M., Ramaswamy, S., & Tamilarasan, K. (2019). Solvent fractionation of organosolv lignin to improve lignin homogeneity: Structural characterization. *Bioresource Technology Reports, 7*, 100293.
34. Das, P., Stoffel, R. B., Area, M. C., & Ragauskas, A. J. (2019). Effects of one-step alkaline and two-step alkaline/dilute acid and alkaline/steam explosion pretreatments on the structure of isolated pine lignin. *Biomass & Bioenergy, 120*, 350–358.
35. Tolbert, A., Akinosho, H., Khunsupat, R., Naskar, A. K., & Ragauskas, A. J. (2014). Characterization and analysis of the molecular weight of lignin for biorefining studies. *Biofuels, Bioproducts and Biorefining, 8*(6), 836–856.
36. Asikkala, J., Tamminen, T., & Argyropoulos, D. S. (2012). Accurate and reproducible determination of lignin molar mass by acetobromination. *Journal of Agricultural and Food Chemistry, 60*(36), 8968–8973.
37. Fredheim, G. E., Braaten, S. M., & Christensen, B. E. (2002). Molecular weight determination of lignosulfonates by size-exclusion chromatography and multi-angle laser light scattering. *Journal of Chromatography. A, 942*(1–2), 191–199.

38. Wörmeyer, K., Ingram, T., Saake, B., Brunner, G., & Smirnova, I. (2011). Comparison of different pretreatment methods for lignocellulosic materials. Part II: Influence of pretreatment on the properties of rye straw lignin. *Bioresource Technology, 102*(5), 4157–4164.
39. Yoo, C. G., Ragauskas, A. J., & Pu, Y. (2019). Measurement of physicochemical properties of lignin. In *Understanding lignocellulose: Synergistic computational and analytic methods* (pp. 33–47). Washington, DC: ACS Publications.
40. Froass, P. M., Ragauskas, A. J., & Jiang, J. E. (1998). NMR studies part 3: Analysis of lignins from modern kraft pulping technologies. *Holzforschung, 52*(4), 385–390.
41. Meng, X., Crestini, C., Ben, H., Hao, N., Pu, Y., Ragauskas, A. J., & Argyropoulos, D. S. (2019a). Determination of hydroxyl groups in biorefinery resources via quantitative ^{31}P NMR spectroscopy. *Nature Protocols, 14*(9), 2627–2647.
42. Chen, W., McClelland, D. J., Azarpira, A., Ralph, J., Luo, Z., & Huber, G. W. (2016). Low temperature hydrogenation of pyrolytic lignin over Ru/TiO2: 2D HSQC and 13C NMR study of reactants and products. *Green Chemistry, 18*(1), 271–281.
43. Wang, H., Pu, Y., Ragauskas, A. J., & Yang, B. (2019b). From lignin to valuable products–strategies, challenges, and prospects. *Bioresource Technology, 271*, 449–461.
44. Lange, H., Rulli, F., & Crestini, C. (2016). Gel permeation chromatography in determining molecular weights of lignins: Critical aspects revisited for improved utility in the development of novel materials. *ACS Sustainable Chemistry & Engineering, 4*(10), 5167–5180.
45. Cao, S., Hu, W., & Fan, L. (2012). Progress in the structure of lignin and its analyzing methods. *Polymer Bulletin, 3*, 1.
46. Balakshin, M. Y., Capanema, E. A., Santos, R. B., Chang, H.-M., & Jameel, H. (2016). Structural analysis of hardwood native lignins by quantitative 13C NMR spectroscopy. *Holzforschung, 70*(2), 95–108.
47. Jensen, A., Cabrera, Y., Hsieh, C.-W., Nielsen, J., Ralph, J., & Felby, C. (2017). 2D NMR characterization of wheat straw residual lignin after dilute acid pretreatment with different severities. *Holzforschung, 71*(6), 461–469.
48. Moghaddam, L., Rencoret, J., Maliger, V. R., Rackemann, D. W., Harrison, M. D., Gutiérrez, A., del Río, J. C., & Doherty, W. O. (2017). Structural characteristics of bagasse furfural residue and its lignin component. An NMR, Py-GC/MS, and FTIR study. *ACS Sustainable Chemistry & Engineering, 5*(6), 4846–4855.
49. Iiyama, K., & Lam, T. B. T. (1990). Lignin in wheat internodes. Part 1: The reactivities of lignin units during alkaline nitrobenzene oxidation. *Journal of Science and Food Agriculture, 51*(4), 481–491.
50. Wen, J. L., Chen, T. Y., & Sun, R. C. (2017). Research progress on separation and structural analysis of lignin in lignocellulosic biomass. *International Journal of Forest Engineering, 2*(5), 76–84.
51. Lu, F., & Ralph, J. (1997). Derivatization followed by reductive cleavage (DFRC method), a new method for lignin analysis: Protocol for analysis of DFRC monomers. *Journal of Agricultural and Food Chemistry, 45*(7), 2590–2592.
52. Zhang, Y., Zhang, R., Zhang, Y., Ai, M., & Huang, F. (2011). Research progress of analysis methods of lignin structure. *Journal of Anhui Agricultural Sciences, 2011*(36), 120.
53. Shinde, S. D., Meng, X., Kumar, R., & Ragauskas, A. J. (2018). Recent advances in understanding the pseudo-lignin formation in a lignocellulosic biorefinery. *Green Chemistry, 20*(10), 2192–2205.
54. Ma, X., Yang, X., Zheng, X., Chen, L., Huang, L., Cao, S., & Akinosho, H. (2015). Toward a further understanding of hydrothermally pretreated holocellulose and isolated pseudo lignin. *Cellulose, 22*(3), 1687–1696.
55. Meng, X., Pu, Y., Sannigrahi, P., Li, M., Cao, S., & Ragauskas, A. J. (2017). The nature of hololignin. *ACS Sustainable Chemistry & Engineering, 6*(1), 957–964.
56. Meng, X., Parikh, A., Seemala, B., Kumar, R., Pu, Y., Christopher, P., Wyman, C. E., Cai, C. M., & Ragauskas, A. J. (2018). Chemical transformations of poplar lignin during cosolvent

enhanced lignocellulosic fractionation process. *ACS Sustainable Chemistry & Engineering, 6*(7), 8711–8718.
57. Kumar, R., Hu, F., Sannigrahi, P., Jung, S., Ragauskas, A. J., & Wyman, C. E. (2013). Carbohydrate derived-pseudo-lignin can retard cellulose biological conversion. *Biotechnology and Bioengineering, 110*(3), 737–753.
58. Hu, F., Jung, S., & Ragauskas, A. (2012). Pseudo-lignin formation and its impact on enzymatic hydrolysis. *Bioresource Technology, 117*, 7–12.
59. Liu, E., Li, M., Das, L., Pu, Y., Frazier, T., Zhao, B., Crocker, M., Ragauskas, A. J., & Shi, J. (2018a). Understanding lignin fractionation and characterization from engineered switchgrass treated by an aqueous ionic liquid. *ACS Sustainable Chemistry & Engineering, 6*(5), 6612–6623.
60. Kosa, M. (2012). *Direct and multistep conversion of lignin to biofuels*. Atlanta: Georgia Institute of Technology.
61. Wei, Z., Zeng, G., Huang, F., Kosa, M., Huang, D., & Ragauskas, A. J. (2015). Bioconversion of oxygen-pretreated Kraft lignin to microbial lipid with oleaginous *Rhodococcus opacus* DSM 1069. *Green Chemistry, 17*(5), 2784–2789.
62. Das, L., Li, M., Stevens, J., Li, W., Pu, Y., Ragauskas, A. J., & Shi, J. (2018). Characterization and catalytic transfer hydrogenolysis of deep eutectic solvent extracted sorghum lignin to phenolic compounds. *ACS Sustainable Chemistry & Engineering, 6*(8), 10408–10420.
63. Li, X., Li, M., Pu, Y., Ragauskas, A. J., Klett, A. S., Thies, M., & Zheng, Y. (2018). Inhibitory effects of lignin on enzymatic hydrolysis: The role of lignin chemistry and molecular weight. *Renewable Energy, 123*, 664–674.
64. Zhou, H., Lou, H., Yang, D., Zhu, J., & Qiu, X. (2013). Lignosulfonate to enhance enzymatic saccharification of lignocelluloses: Role of molecular weight and substrate lignin. *Industrial and Engineering Chemistry Research, 52*(25), 8464–8470.
65. Xu, Z., Qin, L., Cai, M., Hua, W., & Jin, M. (2018). Biodegradation of kraft lignin by newly isolated *Klebsiella pneumoniae*, *Pseudomonas putida*, and *Ochrobactrum tritici* strains. *Environemental Science and Pollution Research, 25*(14), 14171–14181.
66. Salvachúa, D., Karp, E. M., Nimlos, C. T., Vardon, D. R., & Beckham, G. T. (2015). Towards lignin consolidated bioprocessing: Simultaneous lignin depolymerization and product generation by bacteria. *Green Chemistry, 17*(11), 4951–4967.
67. He, Y., Li, X., Ben, H., Xue, X., & Yang, B. (2017). Lipid production from dilute alkali corn stover lignin by *Rhodococcus* strains. *ACS Sustainable Chemistry & Engineering, 5*(3), 2302–2311.
68. Alvarez, H. M., Mayer, F., Fabritius, D., & Steinbüchel, A. (1996). Formation of intracytoplasmic lipid inclusions by *Rhodococcus opacus* strain PD630. *Archives of Microbiology, 165*(6), 377–386.
69. Zhang, D. (2016). *Carbon metabolic flux analysis of lipid accumulation mechanism and key genetic modification in Rhodococcus Opacus*. Wuxi: Jiangnan University.
70. Lin, L., Cheng, Y., Pu, Y., Sun, S., Li, X., Jin, M., Pierson, E. A., Gross, D. C., Dale, B. E., & Dai, S. Y. (2016). Systems biology-guided biodesign of consolidated lignin conversion. *Green Chemistry, 18*(20), 5536–5547.
71. Sana, B., Chia, K. H. B., Raghavan, S. S., Ramalingam, B., Nagarajan, N., Seayad, J., & Ghadessy, F. J. (2017). Development of a genetically programed vanillin-sensing bacterium for high-throughput screening of lignin-degrading enzyme libraries. *Biotechnology for Biofuels, 10*(1), 32.
72. Xie, S., Sun, S., Lin, F., Li, M., Pu, Y., Cheng, Y., Xu, B., Liu, Z. H., da Costa Sousa, L., Dale, B. E., Ragauskas, A. J., Dai, S. Y., & Yuan, J. S. (2019). Mechanism-guided design of highly efficient protein secretion and lipid conversion for biomanufacturing and biorefining. *Advancement of Science, 6*, 1801980.
73. Liu, Z. H., Le, R. K., Kosa, M., Yang, B., Yuan, J. S., & Ragauskas, A. J. (2019b). Identifying and creating pathways to improve biological lignin valorization. *Renewable and Sustainable Energy Reviews, 105*, 349–362.

74. Tuskan, G. A., Muchero, W., Tschaplinski, T. J., & Ragauskas, A. J. (2019). Population-level approaches reveal novel aspects of lignin biosynthesis, content, composition and structure. *Current Opinion in Biotechnology, 56*, 250–257.
75. Adamski, J. (2012). Genome-wide association studies with metabolomics. *Genome Medicine, 4*(4), 34.
76. Visscher, P. M., Brown, M. A., McCarthy, M. I., & Yang, J. (2012). Five years of GWAS discovery. *American Journal of Human Genetics, 90*(1), 7–24.
77. Wei, L., Jian, H., Lu, K., Yin, N., Wang, J., Duan, X., Li, W., Liu, L., Xu, X., & Wang, R. (2017). Genetic and transcriptomic analyses of lignin-and lodging-related traits in *Brassica napus*. *Theoretical and Applied Genetics, 130*(9), 1961–1973.
78. Morris, G. P., Ramu, P., Deshpande, S. P., Hash, C. T., Shah, T., Upadhyaya, H. D., Riera-Lizarazu, O., Brown, P. J., Acharya, C. B., & Mitchell, S. E. (2013). Population genomic and genome-wide association studies of agroclimatic traits in sorghum. *Proceedings of the National Academy of Sciences of the United States of America, 110*(2), 453–458.
79. Umezawa, T. (2018). Lignin modification *in planta* for valorization. *Phytochemistry Reviews, 17*(6), 1305–1327.
80. Takeda, Y., Koshiba, T., Tobimatsu, Y., Suzuki, S., Murakami, S., Yamamura, M., Rahman, M. M., Takano, T., Hattori, T., & Sakamoto, M. (2017). Regulation of *CONIFERALDEHYDE 5-HYDROXYLASE* expression to modulate cell wall lignin structure in rice. *Planta, 246*(2), 337–349.
81. Rinaldi, R., Jastrzebski, R., Clough, M. T., Ralph, J., Kennema, M., Bruijnincx, P. C., & Weckhuysen, B. M. (2016). Paving the way for lignin valorisation: Recent advances in bioengineering, biorefining and catalysis. *Angewandte Chemie. International Edition, 55*(29), 8164–8215.
82. Simmons, B. A., Loqué, D., & Ralph, J. (2010). Advances in modifying lignin for enhanced biofuel production. *Current Opinion in Plant Biology, 13*(3), 312–319.
83. Li, X., He, Y., Zhang, L., Xu, Z., Ben, H., Gaffrey, M. J., Yang, Y., Yang, S., Yuan, J. S., Qian, W.-J., & Yang, B. (2019). Discovery of potential pathways for biological conversion of poplar wood into lipids by co-fermentation of *Rhodococci* strains. *Biotechnology for Biofuels, 12*(1), 60.
84. Liu, Z. H., Olson, M. L., Shinde, S., Wang, X., Hao, N., Yoo, C. G., Bhagia, S., Dunlap, J. R., Pu, Y., Kao, K. C., Ragauskas, A. J., Jin, M., & Yuan, J. S. (2017). Synergistic maximization of the carbohydrate output and lignin processability by combinatorial pretreatment. *Green Chemistry, 19*(20), 4939–4955.
85. Chen, H. Z. (2013). *Modern solid state fermentation: Theory and practice*. Dordrecht: Springer.
86. Liu, Z. H., & Chen, H. Z. (2016). Periodic peristalsis enhancing the high solids enzymatic hydrolysis performance of steam exploded corn stover biomass. *Biomass & Bioenergy, 93*, 13–24.
87. Chen, H. Z., & Li, Z. H. (2007). *Gas dual-dynamic solid state fermentation technique and apparatus*. US7183074.
88. Chen, H. Z., Zhao, Z. M., & Li, H. Q. (2014b). The effect of gas double-dynamic on mass distribution in solid-state fermentation. *Enzyme and Microbial Technology, 58*, 14–21.
89. Chen, H. Z., Shao, M. X., & Li, H. Q. (2014a). Effects of gas periodic stimulation on key enzyme activity in gas double-dynamic solid state fermentation (GDD-SSF). *Enzyme and Microbial Technology, 56*, 35–39.
90. Zhao, Z. M., Wang, L., & Chen, H. Z. (2015). Variable pressure pulsation frequency optimization in gas double-dynamic solid-state fermentation (GDSSF) based on heat balance model. *Process Biochemistry, 50*(2), 157–164.
91. Chatterjee, A., DeLorenzo, D. M., Carr, R., & Moon, T. S. (2020). Bioconversion of renewable feedstocks by *Rhodococcus opacus*. *Current Opinion in Biotechnology, 64*, 10–16.

Chapter 15
Novel and Efficient Lignin Fractionation Processes for Tailing Lignin-Based Materials

Chuanling Si, Jiayun Xu, Lin Dai, and Chunlin Xu

15.1 Introduction

Due to resource shortage and the energy crisis, biorefinery has attracted much attention. Lignin is the most abundant aromatic renewable resource in nature, accounting for 15–30% by weight and 40% by the energy of lignocellulosic biomass [1]. However, in the pulp and paper industry, lignin has long been treated as a waste byproduct and commonly burned to supply heat and energy due to its chemical recalcitrance and structural complexity. In recent years, with the better understanding of lignin physics and chemistry, more and more high performance materials are being developed from lignin resources, which would open not only a new design for functional materials but also a promising way for biorefinery [2].

The efficient fractionation strategies for lignin are the key and basis for biorefinery [3]. The intricate relationship among lignin, cellulose, and hemicellulose and the similarity between some components of lignin and high glycan increases the difficulty of separation [4]. So, the harsh treatment was put to issue this question, which causes an irreversible degradation and condensation reactions of lignin

---

Authors Chuanling Si and Jiayun Xu have equally contributed to this chapter.

C. Si (✉) · L. Dai (✉)
Tianjin Key Laboratory of Pulp and Paper, Tianjin University of Science and Technology, Tianjin, P. R. China
e-mail: sichli@tust.edu.cn; dailin@tust.edu.cn

J. Xu
Tianjin Key Laboratory of Pulp and Paper, Tianjin University of Science and Technology, Tianjin, P. R. China

Laboratory of Natural Materials Technology, Åbo Akademi University, Turku/Åbo, Finland

C. Xu
Laboratory of Natural Materials Technology, Åbo Akademi University, Turku/Åbo, Finland

Z.-H. Liu, A. Ragauskas (eds.), *Emerging Technologies for Biorefineries, Biofuels, and Value-Added Commodities*,
https://doi.org/10.1007/978-3-030-65584-6_15

during traditional carbohydrate's first conception of lignocellulose biorefinery. Generally speaking, lignin presents the heterogeneous structure and broad molecular weight distribution, which would severely affect its potential applications in materials [5]. In addition, the phenolic hydroxyl (PhOH) group content of lignin is one of the most important factors for chemical modification and physical interaction. Many studies showed that the β-O-4 content closely related to the reactivity of lignin [6]. However, during the cleavage of ether bonds of lignin, the stable carbon-carbon (C-C) bonds instead of the hydroxyl are formed, thus the lignin exhibits lower reactivity than native lignin. That has brought up the difficulty of catalytic depolymerization of lignin in order to modify or functionalize lignin for downstream application.

Therefore, the single target product, the serious degradation of lignin, and some other disadvantages of the traditional fractionation method need to be overcome urgently. Moreover, the presence of sulfur in lignin makes them more problematic for chemical modification and further application. Although sulfite lignin is the most widely used industrial lignin, sulfur-free lignin has greater potential applications because it is more environmentally friendly and odor-free [7]. According to this philosophy, it is very important to obtain high purity lignin with commendable chemical reactivity. In addition, the exploitation of a lignin-based coproduct also depends on the inherent structures and chemical reactivity of the lignin polymers.

Hence, numerous researches have been made to develop novel fractionation approaches to isolate lignin from biomass and to obtain chemicals economically valuable for several applications. Based on the concept to utilize the entire lignocellulose substrate, alkali-based fractionation, organosolv fractionation, ionic liquids (ILs) fractionation, and deep eutectic solvents (DESs) fractionation approaches are recently gradually developed and have received the attention among academia and industries. Enzymes can also be used to assist the separation of lignin by, for example, hydrolyzing carbohydrates. What's more, several methods for further fractionation of lignin have been studied to obtain specific molecular lignin fractions with defined properties.

The main objectives of the lignin fractionation process are to obtain relatively high-purity lignin besides carbohydrates (hemicelluloses and celluloses) and to reduce the condensed and degraded nature of conventional biorefinery lignin. Importantly, fractionation should meet the following requirements:

- Basing on all-components separation
- Avoiding further condensation of lignin
- Increasing the stability and uniformity of lignin

15.2 Lignin Properties

15.2.1 The Basic Structural Features of Lignin

Lignin, as one of the three major components of plant raw materials, has a complex structure. Numerous studies have confirmed that the chemical structure of lignin consists of phenylpropane units, originating from three aromatic alcohol precursors (monolignols), p-coumaryl, coniferyl, and sinapyl alcohols. The phenolic substructures that originate from these monolignols are called p-hydroxyphenyl (H, from coumaryl alcohol), guaiacyl (G, from coniferyl alcohol), and syringyl (S, from sinapylalcohol) moieties, as shown in Fig. 15.1 [8]. Softwood lignins mainly consist of G units whereas hardwood lignins are mainly composed of G and S units. The lignins from non-wood such as wheat straw are mainly composed of G, S, and H units.

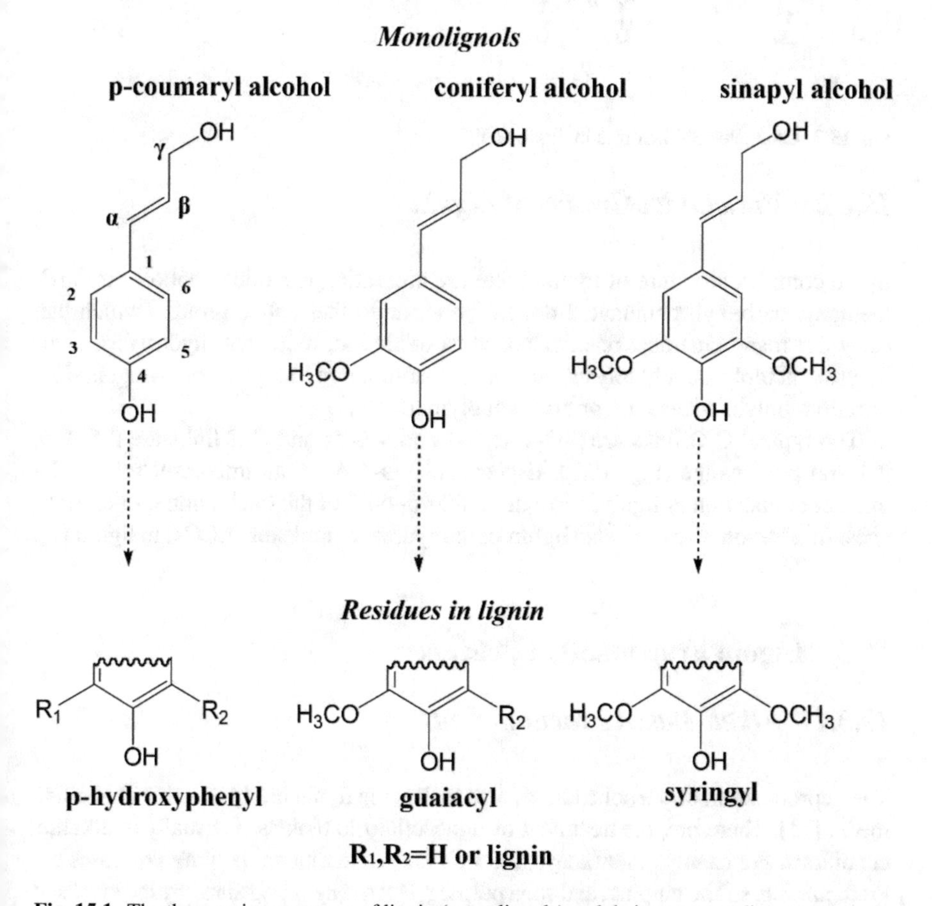

Fig. 15.1 The three main precursors of lignin (monolignols) and their corresponding structures in lignin polymers

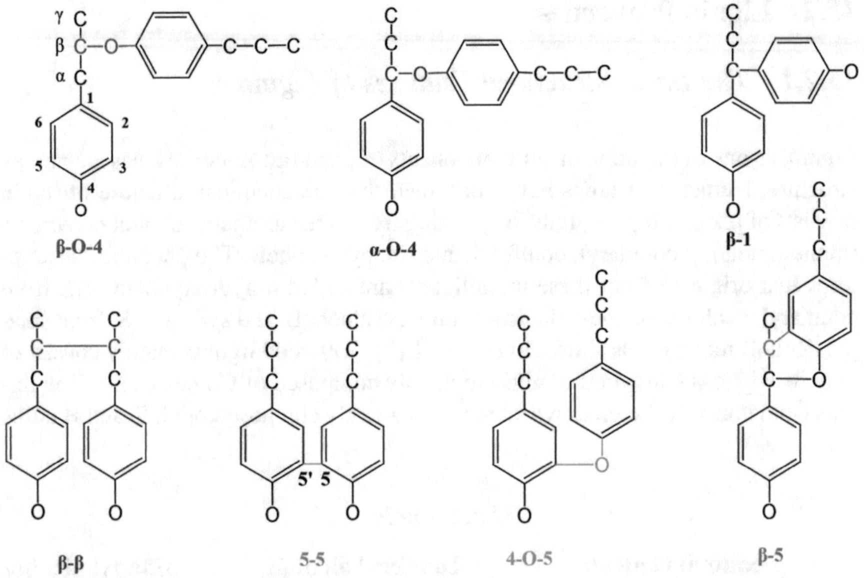

Fig. 15.2 Characteristic linkages in lignin [10]

15.2.2 Functional Groups of Lignin

In the complex structure of lignin, there are aromatic, phenolic, alcohol, carbonyl, methoxy, carboxyl, conjugated double bond, and other active groups, which can carry out many chemical reactions such as oxidation, reduction, hydrolysis, alcoholysis, acidolysis, photolysis, acylation, sulfonation, alkylation, halogenation, nitration, polycondensation or graft copolymerization.

The typical C-O links are β-O-4, α-O-4 and 4-O-5; and C-C links are β-5, 5-5, β-1 and β-β linkage (Fig. 15.2). Especially, β-O-4 plays an important role in the internal connection of lignin, accounting for 45–60% of the total number of all linkages. In addition, there are also lignin-carbohydrate complexes (LCCs) in lignin [9].

15.3 Lignin Fractionation Process

15.3.1 Alkali-Based Fractionation

The deprotonation of phenolic OH-groups makes the lignin easily dissolve in alkaline media [11]. Therefore, the treatment of lignocellulosic biomass is usually in alkaline conditions. For example, nowadays, the well-known traditional pulping processes are kraft pulping, sulfite pulping, and soda pulping. But the lignin obtained by the kraft and sulfite pulping process containing the sulfur and the structure of the lignin was severely damaged after soda pulping, which can complicate lignin downstream valorization.

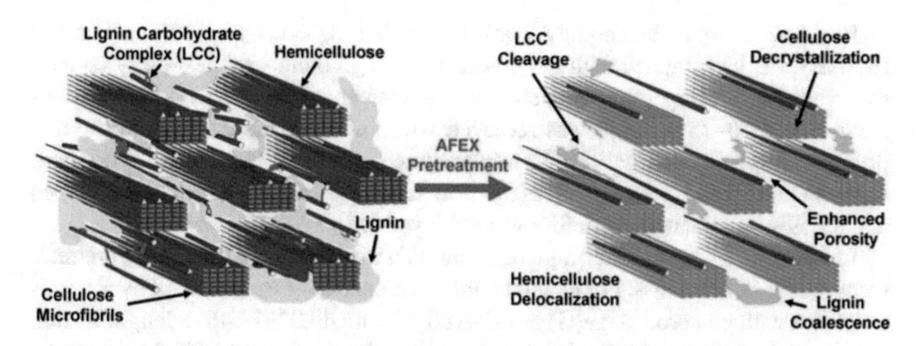

Fig. 15.3 Schematic models for lignocellulosic cell walls depicting overall nanoscale ultrastructural modifications as a result of ammonia-based pretreatments. Different components of the cell wall depicted are cellulose (green lines), hemicelluloses (blue strings), lignin (yellow matrix) [18]

Alkali-based pretreatment was investigated, which utilizes various kinds of alkaline (e.g., NaOH, $Ca(OH)_2$, ammonia). Typically, it can efficiently treat the graminaceous feedstocks due to its lower severity of the treatment compared with traditional soda pulping [12]. Obtaining the lignin from wood after alkali-based pretreatment may need harsher conditions that risk altering the lignin structure [13]. The common reagent used in alkali-based fractionation was NaOH, followed by ammonia, which with several well-known benefits, such as the easy recovery of solvent, the preserve of lignin structure and alters cellulose crystalline structure [14]. Up to now, some ammonia based on extraction and isolation procedures are available, mainly including ammonia fiber explosion (AFEX) [15], anhydrous ammonia pretreatment (AAP), [16] and recycled aqueous ammonia expansion (RAAE) [17].

Ammonia fiber expansion (AFEX) is a promising thermochemical pretreatment, which only changes the cell walls to enhance cellulase accessibility without the lignin and hemicelluloses removal. The mechanism of AFEX was discussed by Chundawat et al.. They found that the ammonia penetrates the cell walls from the outer walls facing the lumen and middle lamella, which would facilitate various ester linkages at first, then decomposition products could be allowed onto outer wall surfaces with the rapid pressure release and hence results in increased wall porosity (white spaces) (Fig. 15.3). During the process, the biomass matrix cannot fractionate to the lignin fraction, but the following lignin extractions were benefiting a lot from AFEX, which can up to 50% removal of the AFEX-pretreated feedstocks by organic or alkaline solution [18]. Subsequently, the mechanism and the causes for improved digestibility of AFEX-pretreated materials were further studied; the results show that it is the disruption of lignin-carbohydrate linkages of mainly polymeric lignin that contribute to the efficiency of AFEX pretreatment, and the process shows minimum change to feedstocks, which can preserve the lignin structure [19].

Another method is anhydrous ammonia pretreatment (AAP). Mittal et al. used this method followed by a mild NaOH extraction (0.1 M NaOH, 25 °C) to treat corn stover with the result of more than 65% lignin removal [16]. The extracts of lignin show considerable antimicrobial and antioxidant activities, when corn stover treated by low-moisture anhydrous ammonia (LMAA), subsequently extracted its lignin by 4% (w/v) sodium hydroxide [20].

Besides, a novel technology named recycled aqueous ammonia expansion (RAAE) was investigated with the object to get high biomass digestibility under a relatively lower temperature and shorter pretreatment time. After RAAE pretreatment, about 50–75% lignin was removed while most of the carbohydrates were preserved. Specifically, only by 11 min treatment at 85 °C, 80% water to dry corn stalks loading and 1.5 L/min ammonia flow rate, up to 68.3% lignin was removed and 85.69% of glucan digestibility was achieved [17].

In addition, a few modified aqueous ammonia methods were used in the pretreatment process of biomass. For example, the sugarcane bagasse is treated by aqueous ammonia with glycerol (AAWG) at 120–180 °C with 49.34–77.46% delignification rates, which indicates that the level of delignification is directly related to temperature [21]. A modified aqueous ammonia soaking (AAS) method was used to treat the eucalyptus, and the results show that the addition of H_2O_2 to the AAS could further promote the delignification of the eucalyptus [22].

The ammonia-based fractionation process is sustainable for large-scale applications without containing carbon or release greenhouse gas (GHG) contain [23]. Although the ammonia-based fractionation is more preferable to limit ether bond cleavage due to milder treatment conditions, ammonia lignin residue showed higher cellulase adsorption affinity compared to organosolv lignin, which reduced cellulose hydrolysis [24].

15.3.2 *Organosolv Fractionation*

Organosolv fractionation is a promising process using an organic solvent to dissolve lignin from the plant cell walls for effective separation of lignin. The lignin was separated in the recovery process to achieve the purpose of the comprehensive utilization of components, which fits well into the concept of sustainable biorefinery (Fig. 15.4). Unlike impure lignin obtained by the traditional pulping process, for example, kraft and sulfite process, the lignin obtained from organosolv fractionation is free of sulfur and the molecular weight is low compared with other lignin production methods. Moreover, nontoxic reagent organic solvent can be recovered during

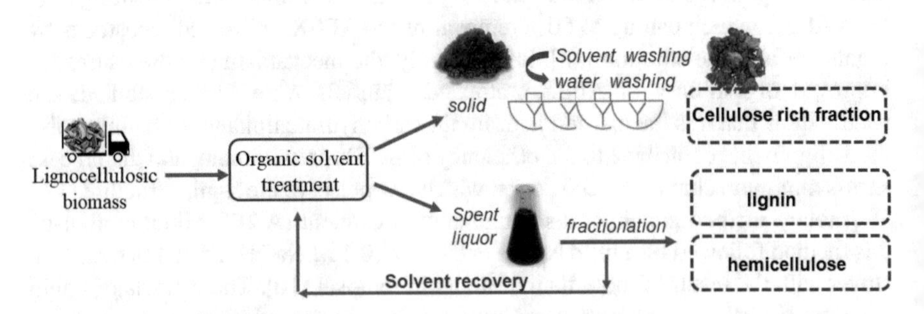

Fig. 15.4 Flowchart of organic solvent treatment of lignocellulosic biomass

organosolv fractionation. A wide range of organic solvents, mainly including organic alcohols (i.e., methanol, ethanol), organic acids (i.e., acetic acid, formic acid), esters, and combined solvents with and without acid or alkaline catalysts are used in biomass fractionation. The catalysts used are acid, a base, or a salt.

These solvents provide selective solubilization of lignin from various lignocellulosic biomass including poplar wood, wheat straw, and eucalyptus, etc. The excellent solubility to dissolve lignin results in slight chemical modification and the mechanisms are considered as the cleavage of aryl ether bonds. During the organosolv delignification, the β-O-aryl ether bonds are more hardly broken than the α-O-aryl ether bonds, which normally need more severe conditions, especially at elevated acid concentrations [25].

Despite organosolv fractionation or pulping being utilized for more than 100 years, the application is still not widely used in industrial production due to high equipment requirements and the dangers, toxic, and costs-increase posed by the use of solvents [26]. In addition, organosolv fractionation for non-wood feedstocks (e.g., corn stover, wheat straw, rice straw, bamboo) exhibit better effect relative to woody feedstocks. These feedstocks represent a small fraction of the global chemical pulping market but may represent a significant feedstock for biorefining processes. Up to now, several organosolv fractionations based on extraction and isolation procedures are available. The main fractionations include alcohol-based fractionation and organic acid-based fractionation.

15.3.2.1 Alcohol-Based Fractionation

For alcohol-based fractionation, ethanol pulping is the earliest and has the most reports. Methanol is cheaper and yields higher pulp than ethanol, but it is toxic. So, the ethanol has been considered one potential solvent to treat lignocellulosic biomass. Generally, the ethanol fractionation process is carried out under elevated temperatures without or with the addition of the acidic or alkaline catalysts.

Temperatures used in the process can be as high as 200 °C, but lower temperatures can also be used depending on the type of biomass and the use of a catalyst. The process produces three main fractions including high purity lignin, a relatively pure cellulose fraction, and hemicellulose-derived products (mainly xylose).

Wheat straw is fractionated by using autocatalytic and acid-catalyzed ethanol organosolv and the lignin yield of 84% was reached without the use of a catalyst (organosolv at 210 °C, 50% w/w aqueous EtOH). Similar results were obtained at 190 °C using 30 mM H_2SO_4 as the catalyst and 60% w/w aqueous EtOH [27].

Lignin condensation is unavoidable reaction during acidic or alkaline ethanol fractionation process. It is formed via formation of the carbocation since it easily forms bond with an electron-rich carbon atom in the aromatic ring of another lignin units [28]. The obtained solid residues are separated by filtration and washed with warm organosolv solvent avoiding dissolved lignin redeposition. Then, the lignin was obtained by precipitation with water. The treatment of willow uses 2-PrOH/H_2O (7:3) at 190 °C for 10 h, which shows that the obtained lignin yield can reach 61% with high purity (93%) [29].

Obviously, ethanol fractionation yields relatively clean and impurity-free lignin that can be directly used for the production of value-added materials. But there are some factors limiting the application, such as limited delignification (high kappa number), safety problems caused by high temperature and pressure.

Based on flow-through fractionation development, the flow-through setup provided lignin with a β-O-4 content in the range of 45–60 β-O-4 linking motifs, whereas in the batch system the β-O-4 content decreased to 8–30 β-O-4 linking motifs for the extractions performed with over 80% EtOH [30].

15.3.2.2 Organic Acid-Based Fractionation

Because of effective hydrolysis and extensive delignification organic acids, primarily acetic acid, and formic acid, have been extensively employed in the biomass fractionation process. Moreover, the ash content is lower than traditional pulping, for example, approximately 64% of ash or 83% of silica in wheat straw remained in the pulp after atmospheric acetic acid (AcOH) pulping [31], lower impurities (ash 0–0.1%) in formic acid lignins [32]. Due to a lower boiling point than acetic acid, formic acid had been recognized as a promising agent for organosolv pulping and fractionation. As a volatile weak organic acid, formic acid can be converted directly from the fraction of lignocelluloses or directly recycled for delignification [33].

Chempolis and CIMV processes involve using formic acid for the treatment of non-wood fiber sources in a single-stage process. DapíA et al. demonstrated that beech wood could be fractionated using 80% formic acid at a temperature from 110 to 130 °C [34]. Zhang et al. obtained the FA-lignin (FAL) with high guaiacyl content by using one-step mild formic acid pretreatment, and FAL fraction exhibited a loose structure, which is prominent for its further catalytic conversion into chemicals and energy [35]. Similarly, after one-step formic acid hydrolysis (85 °C, 5 h), about 80% of lignin was removed from corn husk, and the obtained lignin with high-purity (>99%) would have potential for phenol-formaldehyde resin and concrete water-reducer applications [36].

In order to realize delignification, severe experimental conditions were needed such as increasing temperature and pressure. So, the β-O-4 bonds are broken during the formic acid fractionation, which results in the increase of phenol hydroxyl content, the formation of C-C condensation structure, and the side chain formylation of lignin. Although the yield of lignin separated by this method is high, the high value potential use of lignin is greatly limited due to a large number of β-O-4 broken.

For example, bamboo was fractionated using formic acid under high pressure (at 145 °C for 45 min, 0.3 MPa), presents a quick and efficient delignification method by enhancing the cleavage of interunitary bonds in lignin (β-O-4′, β-β, and β-5′), and dissolved lignin also enabled condensation reaction [37].

Formic acid fractionation can be conducted under low temperatures and at atmospheric pressure [33], but it still needs more treatment or long reaction time to avoid lignin condensation, for example, 60 °C for 8 h [38], 105 °C for 3 h [39]. Li et al. reported bamboo using the Milox method at 101 °C for 2 h achieved the highest delignification (88.9%) [32]. After acetic and formic acid treatment, M. x giganteus showed a low S/G ratio (0.7) and the β-O-4 linkages lost during organosolv fractionation up to 21% and 32%, respectively [40]. It can conclude that extensive degradation of lignin occurred as a result of long reaction time as indicated by a significant reduction of aryl-ether linkages.

These fractionations using formic acid-based organosolv processes are in a batch operation. In order to develop organosolv processes for commercialization, developing a continuous treatment system suitable for flexible biomass feedstocks should be considered [41]. Recently, the flow-through strategy was successively applied in lignin-first biorefinery [42] and hemicelluloses extraction [43]. The flow-through strategy had been verified to be an alternative for effective biomass fractionation with insignificant degradation of lignin and carbohydrate. Wang et al. used 72 wt% aqueous formic acid by flow-through fractionation of biomass, which shows similar lignin yields compared with batch extractions but retains high β-O-4 ether bond [44, 45].

15.3.3 Ionic Liquid (IL)-Assisted Fractionation

Ionic liquids (ILs) have received considerable attention and have been extensively exploited for biomass fractionation due to their excellent ability to destroy the crystalline structure of cellulose or remove lignin/hemicelluloses, the low equipment, and energy costs requirement. ILs are mainly composed of cations (generally organic) and anions (organic or inorganic) [46]. The use of ILs as biomass processing solvents started with the discovery of cellulose dissolving ILs [47].

Ionic liquids (ILs) can selectively break the bond between cellulose, hemicellulose, and lignin. According to selectivity of ILs, the methods of lignin extraction from lignocellulosic biomass can be classfied into partial and total dissolution systems (Fig. 15.5). In the former IL-based system, only lignin and hemicellulose can be extracted by ILs while the cellulose fraction remains in the form of a solid pulp, then the lignin can be precipitated by antisolvent [48]. In the latter system, the entire lignocellulose substrate can be dissolved, then cellulose and lignin were precipitated, respectively, from the product mixture by the addition of an antisolvent (organic or aqueous–organic solution) [49]. Recent researches on lignin fractionation in ionic liquids have been listed in Table 15.1.

Among numerous ILs, 1-ethyl-3-methylimidazolium acetate ([C_2mim]-[OAc]) was demonstrated to effectively extract lignin from lignocellulose raw material, such as wood [56] and wheat straw. Bogel-Łukasik et al. investigated the fractionation process of the wheat straw in 1-ethyl-3-methylimidazolium acetate ([emim] [CH_3COO]) and reported that 87 wt% purity of the carbohydrate-free lignin-rich fraction can be recovered [57]. And the chemical transformations involved in the [C_2mim]-[OAc] pretreatment were more deeply investigated [58].

Various process parameters would affect the dissolution of lignocellulosic material, such as the effect of IL anion or cation, solvation properties of ILs, and viscosity. These relevant parameters were summarized by Badgujar and Bhanage [59],

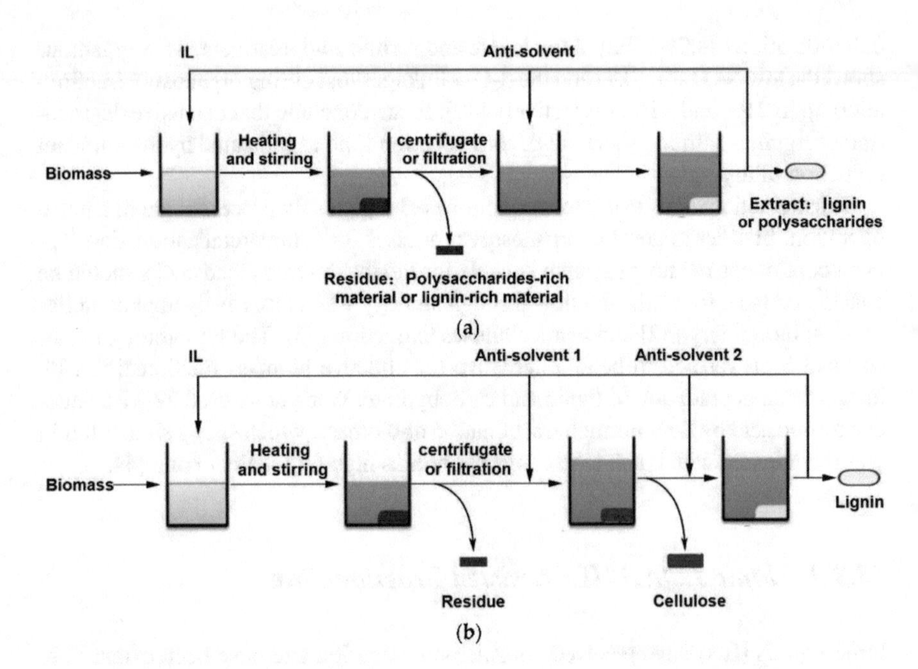

Fig. 15.5 Fractionation procedure of lignocellulose with ILs: (**a**) extraction of lignin or polysaccharide using selective ILs (partial dissolution systems) and (**b**) separation of cellulose and lignin with non-selective ILs (full dissolution systems) [50]

Table 15.1 Ionic liquids used for lignin fractionation

Substrate	Ionic liquids[a]	Conditions	Results	References
Corn stover	[HMIM]Cl	70 °C, 3 h	Lignin yield 60.48%	[48]
Wheat straw	[emim][HSO_4]	131.0 °C, 58.7 wt.% H_2O and 88.0 min, followed by alkaline extraction and acidified	Lignin yield 42.6 wt.%	[51]
Miscanthus x giganteus	[TEA][HSO_4]	120 °C, 8 h, 1:10	88% delignification	[52]
Poplar	[Hpy]Cl and [Hmim]Cl	100 °C for 30 min	Lignin yield 61.0% and 60.4%	[53]
Bamboo	[Hpy]Cl and [Hmim]Cl	100 °C for 30 min	Lignin yield 51.7% and 50.3%	[53]
Eucalyptus	[bmim]OAc	120 °C, 4 h	29.7–43.3% delignification	[54]
Alkaline lignin	[C_2mim]-[OAc]	110–170 °C, 1–16 h	Lignin yield (60–93%)	[55]

[a]Ionic liquids: [HMIM]Cl: 1-*H*-3-methy-limidazolium chloride, [emim][HSO_4]:1-ethyl-3-methylimidazolium hydrogen sulfate, [TEA][HSO_4]:triethylammonium hydrogen sulfate, [Hpy]Cl: Pyridinium chloride, [bmim]OAc:1-butyl-3-methylimidazolium acetate, [C_2mim]-[OAc]: 1-ethyl-3-methylimidazolium acetate

who suggested that IL should in the context of green sustainable chemistry and IL properties should be optimized for biomass processing. A key example is the high viscosity of most ionic liquids, so several studies try to solve this issue. For example, increasing the temperature may be a method to reduce IL viscosity [55]. In addition, the currently popular ionic liquids in lignocellulosic biomass treatment such as the imidazolium-based ionic liquids are likely to remain expensive and their preparation is often very difficult. Therefore, the recyclability of ionic liquids is necessary. Up to now, numerous investigations of the use of IL have focused on the cost and recyclable problems. Up to 85% of the lignin can be solubilized into the IL solution 80% lignin recovery can be achieved by using the low-cost ionic liquid triethylammonium hydrogen sulfate [TEA][HSO_4] at mild temperature (120 °C) [52]. The lignin-rich solid fractions were obtained with distinct purities and yields by aqueous ionic liquid solution ([emim][HSO_4]/H_2O), and the IL was successfully separated, recovered, and reused [51].

15.3.4 Deep Eutectic Solvents (DESs) Fractionation

Issues such as toxicity, poor biodegradability, and high cost have nevertheless restricted the implementation of ILs. In order to overcome these problems, a new solvent-deep eutectic solvent (DES, also known as the third generation of ionic liquids) has entered the field of vision of researchers. The term deep eutectic solvents (DESs) were coined by Abbott and co-workers in 2003 [60]. It has important significance as a substitute for traditional ionic liquids for the isolation and extraction of lignin. Deep eutectic solvents (DESs) are solvents that are synthesized in a liquid state at room temperature by at least one hydrogen bond donor (HBD) usually quaternary ammonium salts and one hydrogen bond acceptor (HBA) such as alcohols or carboxylic acids.

Francisco et al. found that DESs show high lignin solubility and very poor or negligible cellulose solubility [61], and the properties of DESs can be easily tuned by changing the HBDs and HBAs [62]. Mechanism investigation revealed that the functional groups in the DESs significantly affected their ability to dissolve lignin [63], and it can evaluate the potential of DES for lignin processing more closely. In addition, parameters such as temperature, pH, and DES viscosity play a role in the fractionation of lignocellulosic components by DES. Chen et al. found that the molar ratio of ChCl–PCA (1:1) would allow the formation of more hydrogen bonds, which is helpful for the stability of the solvents and lignin removal [64].

Several notable studies have highlighted the benefits of DESs for the preparation of lignin. Yu et al. have demonstrated that both the hydrogen bonds and ether bonds in lignin–carbohydrate complexes could be cleaved during DESs pretreatment, thereby facilitating the selective extraction of lignin [65]. The lignin yield by ChCl and HBD (acetic acid, lactic acid, levulinic acid, and glycerin) could reach 78% and 58% from poplar and Douglas (D) fir, respectively, and the extracted lignin has high purity (95%) with unique structural properties [66]. Although the main delignifica-

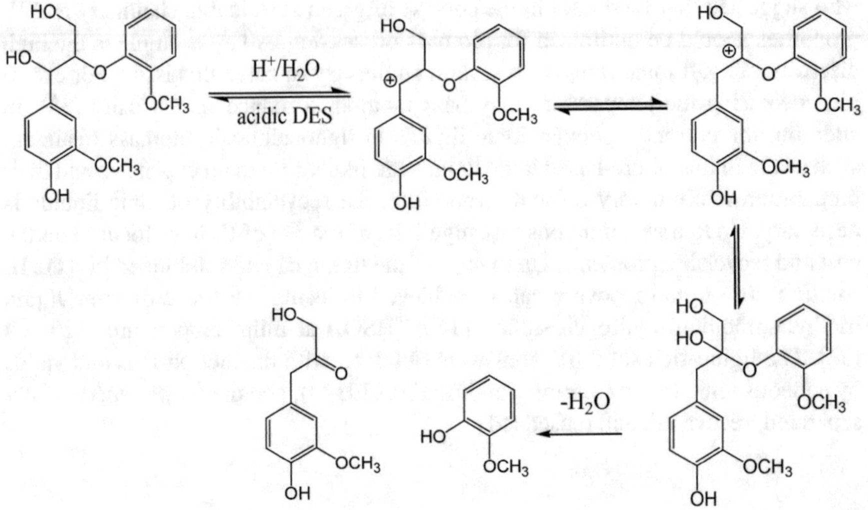

Fig. 15.6 Main mechanisms of β-O-4 linkage cleavage reaction during DES treatment of lignin based on study of model compound guaiacylglycerol-β-guaiacyl ether [66]

tion mechanism employed by DES treatment is the cleavage of ether linkages, little re-condensation of lignin fragments occurred (Fig. 15.6). Singh et al. used lignin-derived compounds as potential raw materials for DES preparation to biomass processes and the lignin removal (60.8%) provided by ChCl-PCA. Higher lignin yield (up to 80%) could be obtained from prairie cordgrass (PCG) and switchgrass (SWG) biomass [67].

Generally speaking, there is still much room for improvement in the fractionation of lignocellulosic biomass and the extraction of lignin using DES. The integration of DES pretreatment with other technologies show good results. For example, the lignin removal can reach 65–80% using acidic ChCl: lactic acid pretreatment with ultrafast 45 s microwave heating at 800 W and the purity reach 85–87% [68]. Recently, the green processing of lignocellulosic biomass and its derivatives in the DESs system has been comprehensively and critically reviewed. The authors pointed out that DESs are attractive solvents for the fractionation of lignocelluloses and the valorization of lignin [69]. Another review stressed three key parts: performance of varying types of DESs and pretreatment schemes for biopolymer fractionation, properties, and conversion of fractionated saccharides, as well as DES, extracted lignin [70].

In spite of the rapid development in the past 10 years, some drawbacks still restricted the practical applications of these solvent systems, such as the thermal instability, susceptibility to contaminants [69]. In another aspect, the ideal DES type for biomass conversion is still uncertain, and the studies about the application using the DES lignin still not enough. There is still much research to be done for the DES fractionation technology.

15.3.5 Enzyme Assisted Fractionation

Usually, Milled wood lignin (MWL) was prepared for the lignin structural characterization research, which was considered to be the most similar to native lignin. The isolation of lignin has a detailed introduction [71]. The obtained lignin has low yield, which maybe half that of the crude preparation and its residual carbohydrate content is about 4%. Subsequently, enzyme assisted extraction to obtain lignin was developed, which mainly uses cellulase to hydrolyze cellulose aiming to remove carbohydrates. The assistant function of enzyme improves the yield of lignin and the obtained lignin was known as cellulolytic enzyme lignin (CEL).

The addition of enzymes into lignin fractionation would reduce the use of toxic chemicals with economic and environmental protection advantages. But it may extend treat time, so enzymatic mild acidolysis lignin (EMAL), one modified method was investigated to improve lignin recovery and purity [72]. This strategy is usually used in laboratory research. Li et al. compared the pyrolysis of wood and herbaceous EMAL and found that high phenolics yield can be obtained at mild pyrolysis temperature of 450–650 °C [73].

There is no reason not to acknowledge that the enzymatic treatment has successfully assisted in lignin extraction, such as mild reaction conditions, the use of renewable and inexpensive biocatalysts, and reduction in the use of toxic chemicals. Enzymatic hydrolysis was applied after mild alkaline treatments in situ to the ball-milled swollen cell wall and the yield of swollen residual enzyme lignin (SREL) can reach 95% [74]. The novel method combined the alkaline treatment and enzymatic hydrolysis, which not only has little change on the lignin structures but also obtained the syringyl-rich lignin macromolecules as compared with CEL.

The enzyme was also used in further fractionation of the lignin. For example, two common lignin (alkaline lignin: AL, and hydrolysis lignin: HL) were treated with laccase, and the results show that the structure, especially the Ph-OH contents and molecular weight of lignin could be effectively controlled, which benefits for the antioxidant application [75].

15.3.6 Strategies to Quench Reaction During Lignin Fractionation

During lignin fractionation, ether bonds are cleaved and a new stable carbon-carbon (C-C) bonds are formed, which is an undesirable reaction, limiting the lignin activity for added-value using. It is well acknowledged that the use of acid and/or high temperatures during lignin fractionation leads to severe and irreversible condensation, so the mild treatment conditions can avoid condensation to some extent. Wang et al. obtained the lignin (up to 90% of original lignin removal) using one-pot method for poplar fractionation by acidic water/phenol pretreatment at mild temperature (120 °C), and the phenol phase, which contains depolymerized and phenolated lignin, is directly used to prepare lignin-based phenolic foam, with satisfactory properties [76]. In order to develop a new solvent with mild treatment conditions, the separation of lignin with high yield and high activity can also be achieved by appropriate physical and chemical quenching methods.

In order to avoid condensation, strategies to prevent structural degradation is the direct hydrogenolysis of native lignin in biomass. Currently, A"lignin-first" biorefinery concept was proposed, which mainly included tandem depolymerization–stabilization of native lignin and active preservation of β-O-4 bonds [77–79]. However, there are still several important obstacles, such as catalyst recovery and mass transfer limitations, which have nevertheless restricted the large-scale implementation of this method. So, Shuai et al. [80] reported a lignin stabilization strategy by forming 1,3-dioxane structure using formaldehyde, which can block reactive positions without extra catalysts.

However, these yields uncondensed or less condensed strategies require extra chemicals or catalysts. Different from these strategies, a simple but effective strategy to preserve lignin structures by employing rapid flow-through fractionation was applied over the last few decades. The diffusive flux remains high in a flow-through setup and the lignin concentration is relatively low, as a fresh solvent is constantly added to the system. In a batch system, the lignin concentration is increased during the extraction, resulting in a decrease in diffusive flux as a function of time. In another word, when operating in flow-through mode, the dissolved lignin fragments are removed from the heating zone, which limits the extent of structural alteration and redeposition. Wang et al. used 72 wt% aqueous formic acid by flow-through fractionation of biomass, which shows similar lignin yields compared with batch extractions but retains high β-O-4 ether bond [81] (Fig. 15.7a). Using acidic alcohol–water mixtures (120 °C) by flow-through fractionation show the similar results that extraction efficiencies of over 55% were achieved, yielding lignin with good structural quality in terms of β-O-4 linking motifs (typically over 60 per 100 aromatic units from Fig. 15.7b) [30]. In addition, sequential extraction (SE) of hemicellulose and lignin to avoid their excessive degradation have been described [82, 83].

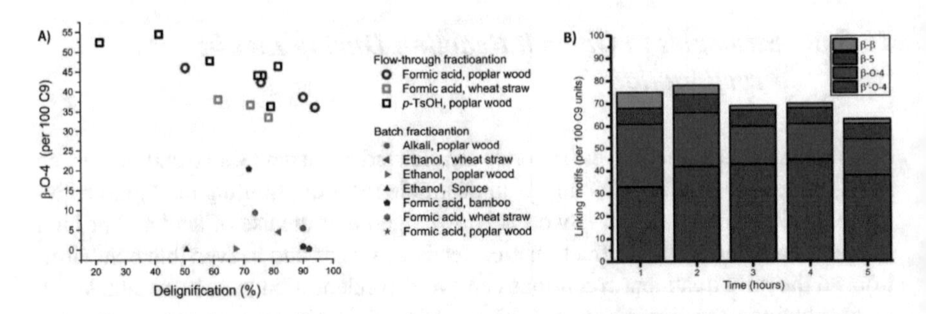

Fig. 15.7 (**a**) Correlation between delignification and β-O-4 linkages during fractionation [44]. (**b**) Linking motifs distribution during the mild organosolv extraction (80,20 EtOH/H_2O, 0.18 M H_2SO_4, 120 °C, 5 h) [30]

The second step of SF was characterized by mild conditions of 100 °C, 40 wt% p-TsOH, and 1, and achieved 83% lignin dissolution with a well-preserved structure.

The given discussion showed that even if the cleavage of β-O-4 ether bond could not be prevented during the lignin fractionation, further C-C bonds condensation reaction could be prevented by appropriate quenching methods for improving the reactivity of lignin after fractionation.

15.4 Further Fractionation of Lignin

Although numerous studies have showed to obtain the lignin for further application, the lignin presents a broad molecular weight distribution that limits the comprehensive and efficient utilization. Therefore, the further fractionation of lignin seems to be a preferential way as it can achieve specific molecular lignin fractions with defined properties.

Until now, many studies have reported about the methods to fractionate lignin. Table 15.1 highlights the major fractionation conditions, molecular weights distribution, and potential applications. After further fractionation of lignin, the differences in their chemical structure were investigated. The lignin obtained by ultrafiltration present narrow molecular weight distribution in the results of the size exclusion chromatography (Table 15.1). After the ultrafiltration, the obtained lignin fractions can be used as bio-based components in blends with polyethylene due to low molecular weight (Mw) and a high quantity of phenolic hydroxyl groups [84]. As mature technology is widely used in industry, but there are some disadvantages, for example, the existence of partial polysaccharides, fouling of the membranes and expensive instrumentation. Compared with ultrafiltration, differential precipitation was an easier and simpler technique and less energy consuming. Labidi et al. divided kraft lignin into three fractions by precipitation at different pH conditions with various acids (sulfuric acid and hydrochloric acid), and the effect of the process and the precipitation technique on lignin composition were analyzed. The fraction of lignin with higher Mw were precipitated at higher pH due to this fraction is easier to destabilize (Table 15.2) [85]. Subsequently, sequential acid precipitation method showed the similar results that lignins with a higher molecular mass were obtained at higher pH values [86].

Solvent fractionation is based on partial solubility of polymers into solvents. Jiang et al. using the solvent assisted fractionation method of kraft lignin yielded four fractions with varied molecular weights and polydispersities [87]. These solvents are inexpensive, but toxic. GVL is arousing the interest of researchers, due to its better stability, harmlessness, biodegradability, and recyclability. Wang et al. fractionated enzymatic hydrolysis lignin (EHL) into three parts only by GVL/water solvent (GWS), the GVL subsequent reuse for lignin fractionation, and they found that the obtained fractions presented significantly low molecular weight polydispersity and structural heterogeneity [88]. Another renewable solvent, ethanol–water solution, was successfully used to fractionate EHL. The high-molecular-weight lignin obtained by 80% ethanol–water (v/v) fractionation of EHL, and its antioxidant

Table 15.2 A list of major methods for further fractionation of lignin

Lignin type	Fractionation conditions	Molecular weight distribution				References
			$\bar{M}w$	$\bar{M}n$	D($\bar{M}w$/$\bar{M}n$)	
Enzymatic hydrolysis lignin (EHL)	Gradient precipitation 40%, 30%, and 5% aqueous GVL solutions	F1	11,135	7131	1.56	[88]
		F2	8310	4704	1.77	
		F3	4235	2277	1.86	
		EHL	8930	4420	2.02	
Bamboo kraft lignin (BKL)	95% ethanol, solid–liquid ratio of 1:20, ambient temperature	Fs	2518		1.79	[93]
		Fi	5216		1.65	
		BKL	4032		2.04	
Black liquor (alkaline pulping)	Ultrafiltration	Rough	5654	1879	3.01	[94]
		>15 kDa	6300	2032	3.10	
		15 kDa	3544	1891	1.87	
		10 kDa	20,221	946	2.14	
		5 kDa	1806	940	1.92	
Black liquor (kraft pulp)	Acid precipitation	LKS 2	3522	695	5.06	[85]
		LKS 4	5741	852	6.73	
		LKS 6	4575	750	6.10	
		LKH 2	4993	866	5.76	
		LKH 4	5798	895	6.47	
		LKH 6	6760	1054	6.41	

activity improved followed by depolymerization [89]. Also, the antimicrobial activity of EHL was improved after ethanol–water fractionation [90]. Based on the concept of green environmental protection, glycerol was applied with ethanol for lignin fractionation and the results showed that this approach could realize the efficient fractionation of EHL with relatively narrow polydispersity [91]. In addition, Dai et al. developed a novel hydrogel-assisted fractionation approach using lignin-containing cellulose (Cell-AL) hydrogel to fractionate lignin to produce uniform lignins [92]. Depending on the future use of the lignin, the right technique to obtain the fractions has to be chosen.

15.5 Lignin-Based Material and Applications

Based on the richness of its functional groups and high carbon contents, lignin has been known as a very useful macromolecule in material industries [95]. Generally, lignin can be utilized as a replacement in different polymeric materials by modification or not [96], such as polyurethane products that include rigid and flexible foams, adhesives, coatings, and elastomers [96, 97]. A variety of polymers can be derived from lignin by simple chemical modification in order to overcome the brittle nature and its incompatibility with other polymer systems, for example, the esterification of lignin [98] Dai et al. developed a novel functional lignin-based filler and found that lignin-graft-poly(D-lactic acid) (LG-g-PDLA) could accelerate crystallization in PLLA systems [99]. The hydroxyl group acts as a reaction site for the given chemical reaction. But there are several factors limited to its application, including some technical lignin that may contain sulfur leading to products yellowing [100]; polydisperse lignin show low solubility, which may lead to the difficult application; low reactivity leads to the poor incorporation into a polymer matrix. So, obtaining relative purity and high reactivity lignin is essential for lignin valorization.

Up to now, the application of lignin-based functional materials has changed from the direct physical blending and they are being incorporated into other polymer materials to be used as a new functional material for high value-added applications. There are two main ways to prepare lignin-based functional materials including preparing lignin nanostructured functional materials and lignin carbon functional materials. The former materials are based on its own structure and performance characteristics. The latter materials are based on high carbon contents (up to 60%) and aromatic monomers [101].

Currently, a variety of nanomaterials would be prepared by different methods in different fields. Therefore, it is worthwhile to prepare nanomaterials combined with the characteristics of lignin. Lignin-based nanomaterials have been used in adsorption [102, 103], catalysis [104], packaging [105], drug delivery, and sustained release system [106] to name a few. In the biomedical field, Dai et al. created a lignin-containing self-nanoemulsifying drug delivery system (SNEDDS) that can effectively improve the stability of trans-RSV [107]. The delivery and storage of trans-RSV could also be achieved via Pickering emulsion stabilized by lignin-based nanoparticles [108]. In wastewater treatment field, Wang and his co-workers evaluated the behavior of dye adsorption from wastewater using lignin-based Fe_3O_4@ lignosulfonate/phenolic core-shell microspheres and their results demonstrated that it could be very efficient in removing methylene blue from aqueous solutions [109]. However, in the aquatic environment, the aggregation and deactivation of nanomaterials have hindered its development and practical application. Currently, using organosolv lignin for the modification of nanoscale zero-valent iron (nZVI) was investigated to issue this problem [102]. Ma et al. acquired super long-term stable lignin nanoparticles via an acid precipitation method directly from black liquor, which also offer the possibility to overcome the poor stability problem [110].

In addition, many studies have reported to use lignin as carbon materials [111, 112] including activated carbons [113], carbon fibers [101], and graphitic carbon or carbon dots [114]. It has brought forth a wide array of promising applications, ranging from advance materials, energy storage to as adsorbents for organic pollutants or heavy metals.

Approaches to preparing carbon sorbents derived from lignin are summarized by Chistyakov [111]. Zhu et al. summarized the applications of lignin-derived materials in rechargeable batteries and supercapacitors including their use as binders and electrodes for rechargeable batteries and electrodes and electrolytes for supercapacitors with a focus on the mechanisms behind their operation [70]. Lawoko et al. reviewed both the topics of the direct use of lignin and of the chemical modifications of lignin toward micro- and nanostructured materials. But there are various factors that limit its application such as the heterogeneity of lignin molecule, ash content, and thermoplastic foaming behavior [12].

In the last decade, lignin-based materials development has picked significant momentum due to the bio-refinery concept as aging pulp and paper mills need to diversify their product portfolio to maintain their vitality. However, most of these studies have not yet reached an industrial scale. So far, there are many studies needed to do about lignin-based materials including the major lignin-derived materials, functional products, and their potential applications.

15.6 Summary and Outlook

Lignin is the only renewable aromatic native biopolymer in nature, and lignin-based materials avoiding the waste of resources become one of the hot topics recently. The application can lead to higher profits for pulp and paper industries and second-generation biorefineries and better environmental performance. Because the fractionation conditions, such as temperature and robust solvent can destroy the structure of lignin and affect the purity of lignin, accordingly, the main challenge in the development of lignin for materials is that the stable carbon-carbon (C-C) bonds are formed once β-O-4 linkages break, which can influence the activities of lignin. That gives the opportunities to investigate novel effective fractionation technologies for the next application of lignin.

It has now been widely demonstrated that a higher quantity of phenolic hydroxyl, higher β-O-4 linkages, narrow polydispersity, and purity of lignin often determines the lignin activities for downstream application. With this in mind, we sought to further improve the quality of fractionated lignin by using novel solvents, mild conditions or novel technologies, such as IL-assisted fractionation, EDS-assisted fractionation, and flow-through fractionation. In view of the huge varieties of fractionation technologies available, most focus on lignin yield and purity. However, the ideal fractionation for lignin obtained is still uncertain. Aside from the ideal fractionation, the fractioned lignin can be integrated into a variety of applications as it can also be treated as raw materials for further use. But there is a lack of studies in the downstream application of fractionated lignin.

In addition, consideration has to be taken on the prospect of applying the fractionation technologies on a large scale and eventually achieving the commercial utilization process. Techno-economic analysis is also needed to constantly evaluate the progress of the developed technology and scale-up production and application.

Acknowledgments Financial support from National Key Research and Development Program of China (2017YFB0307903), Key Technology Research and Development Program of Tianjin (19YFZCSN00950), Tianjin Enterprise Technology Commissioner Project (19JCTPJC52800), National Natural Science Foundation of China (21706193), and Young Elite Scientists Sponsorship Program by Tianjin (TJSQNTJ-2017-19) are greatly appreciated. Jiayun Xu acknowledges the financial support from the China Scholarship Council.

References

1. Zakzeski, J., Bruijnincx, P. C., Jongerius, A. L., & Weckhuysen, B. M. (2010). The catalytic valorization of lignin for the production of renewable chemicals. *Chemical Reviews, 110*(6), 3552–3599.
2. Ragauskas, A. J., Beckham, G. T., Biddy, M. J., Chandra, R., Chen, F., Davis, M. F., Davison, B. H., Dixon, R. A., Gilna, P., Keller, M., Langan, P., Naskar, A. K., Saddler, J. N., Tschaplinski, T. J., Tuskan, G. A., & Wyman, C. E. (2014). Lignin valorization: Improving lignin processing in the biorefinery. *Science, 344*(6185), 1246843.
3. Yoo, C. G., Meng, X., Pu, Y., & Ragauskas, A. J. (2020). The critical role of lignin in lignocellulosic biomass conversion and recent pretreatment strategies: A comprehensive review. *Bioresource Technology, 301*, 122784.
4. Tarasov, D., Leitch, M., & Fatehi, P. (2018). Lignin–carbohydrate complexes: Properties, applications, analyses, and methods of extraction: A review. *Biotechnology for Biofuels, 11*(1), 269.
5. Li, Q., Serem, W. K., Dai, W., Yue, Y., Naik, M. T., Xie, S., Karki, P., Liu, L., Sue, H.-J., Liang, H., Zhou, F., & Yuan, J. S. (2017). Molecular weight and uniformity define the mechanical performance of lignin-based carbon fiber. *Journal of Materials Chemistry A, 5*(25), 12740–12746.
6. Lancefield, C. S., Rashid, G. M. M., Bouxin, F., Wasak, A., Tu, W.-C., Hallett, J., Zein, S., Rodríguez, J., Jackson, S. D., Westwood, N. J., & Bugg, T. D. H. (2016). Investigation of the chemocatalytic and biocatalytic valorization of a range of different lignin preparations: The importance of β-O-4 content. *ACS Sustainable Chemistry & Engineering, 4*(12), 6921–6930.
7. Mandlekar, N., Cayla, A., Rault, F., Giraud, S., Salaün, F., Malucelli, G., & Guan, J.-P. (2018). An overview on the use of lignin and its derivatives in fire retardant polymer systems. *Lignin-Trends and Applications, 9*, 207–231.
8. Laurichesse, S., & Avérous, L. (2014). Chemical modification of lignins: Towards biobased polymers. *Progress in Polymer Science, 39*(7), 1266–1290.
9. Koshijima, T., Watanabe, T., & Yaku, F. (1989). Structure and properties of the lignin—Carbohydrate complex polymer as an amphipathic substance. In *Lignin* (ACS symposium series) (pp. 11–28). Washington, DC: ACS Publication.
10. Windeisen, E., & Wegener, G. (2012). Lignin as building unit for polymers. In *Polymer science: A comprehensive reference* (pp. 255–265). Amsterdam: Elsevier.
11. Schutyser, W., Renders, T., Van den Bosch, S., Koelewijn, S. F., Beckham, G. T., & Sels, B. F. (2018). Chemicals from lignin: An interplay of lignocellulose fractionation, depolymerisation, and upgrading. *Chemical Society Reviews, 47*(3), 852–908.
12. Wang, H., Pu, Y., Ragauskas, A., & Yang, B. (2019). From lignin to valuable products-strategies, challenges, and prospects. *Bioresource Technology, 271*, 449–461.
13. Bouxin, F. P., David Jackson, S., & Jarvis, M. C. (2014). Isolation of high quality lignin as a by-product from ammonia percolation pretreatment of poplar wood. *Bioresource Technology, 162*, 236–242.
14. Kim, J. S., Lee, Y. Y., & Kim, T. H. (2016). A review on alkaline pretreatment technology for bioconversion of lignocellulosic biomass. *Bioresource Technology, 199*, 42–48.
15. Meyer, J. R., Waghmode, S. B., He, J., Gao, Y., Hoole, D., da Costa Sousa, L., Balan, V., & Foston, M. B. (2018). Isolation of lignin from Ammonia Fiber Expansion (AFEX) pretreated biorefinery waste. *Biomass and Bioenergy, 119*, 446–455.
16. Mittal, A., Katahira, R., Donohoe, B. S., Pattathil, S., Kandemkavil, S., Reed, M. L., Biddy, M. J., & Beckham, G. T. (2017). Ammonia pretreatment of corn stover enables facile lignin extraction. *ACS Sustainable Chemistry & Engineering, 5*(3), 2544–2561.
17. Zhang, C., Pang, F., Li, B., Xue, S., & Kang, Y. (2013). Recycled aqueous ammonia expansion (RAAE) pretreatment to improve enzymatic digestibility of corn stalks. *Bioresource Technology, 138*, 314–320.

18. Chundawat, S. P. S., Donohoe, B. S., da Costa Sousa, L., Elder, T., Agarwal, U. P., Lu, F., Ralph, J., Himmel, M. E., Balan, V., & Dale, B. E. (2011). Multi-scale visualization and characterization of lignocellulosic plant cell wall deconstruction during thermochemical pretreatment. *Energy & Environmental Science, 4*(3), 973–984.
19. Singh, S., Cheng, G., Sathitsuksanoh, N., Wu, D., Varanasi, P., George, A., Balan, V., Gao, X., Kumar, R., Dale, B. E., Wyman, C. E., & Simmons, B. A. (2015). Comparison of different biomass pretreatment techniques and their impact on chemistry and structure. *Frontiers in Energy Research, 2*, 62.
20. Guo, M., Jin, T., Nghiem, N. P., Fan, X., Qi, P. X., Jang, C. H., Shao, L., & Wu, C. (2018). Assessment of antioxidant and antimicrobial properties of lignin from corn stover residue pretreated with low-moisture anhydrous ammonia and enzymatic hydrolysis process. *Applied Biochemistry and Biotechnology, 184*(1), 350–365.
21. Shi, T., Lin, J., Li, J., Zhang, Y., Jiang, C., Lv, X., Fan, Z., Xiao, W., Xu, Y., & Liu, Z. (2019). Pre-treatment of sugarcane bagasse with aqueous ammonia-glycerol mixtures to enhance enzymatic saccharification and recovery of ammonia. *Bioresource Technology, 289*, 121628.
22. Huo, D., Yang, Q., Fang, G., Liu, Q., Si, C., Hou, Q., & Li, B. (2018). Improving the efficiency of enzymatic hydrolysis of Eucalyptus residues with a modified aqueous ammonia soaking method. *Nordic Pulp & Paper Research Journal, 33*(2), 165–174.
23. Zhao, C., Shao, Q., & Chundawat, S. P. S. (2020). Recent advances on ammonia-based pretreatments of lignocellulosic biomass. *Bioresource Technology, 298*, 122446.
24. Yoo, C. G., Li, M., Meng, X., Pu, Y., & Ragauskas, A. J. (2017). Effects of organosolv and ammonia pretreatments on lignin properties and its inhibition for enzymatic hydrolysis. *Green Chemistry, 19*(8), 2006–2016.
25. Brosse, N., Hussin, M. H., & Rahim, A. A. (2019). Organosolv Processes. *Advances in Biochemical Engineering/Biotechnology, 166*, 153–176.
26. Zhao, X., Li, S., Wu, R., & Liu, D. (2017). Organosolv fractionating pre-treatment of lignocellulosic biomass for efficient enzymatic saccharification: Chemistry, kinetics, and substrate structures. *Biofuels, Bioproducts and Biorefining, 11*(3), 567–590.
27. Wildschut, J., Smit, A. T., Reith, J. H., & Huijgen, W. J. (2013). Ethanol-based organosolv fractionation of wheat straw for the production of lignin and enzymatically digestible cellulose. *Bioresource Technology, 135*, 58–66.
28. McDonough, T. J. (1992). *The chemistry of organosolv delignification*.
29. Wu, Y. L., Saeed, H. A. M., Li, T. F., Lyu, G. J., Wang, Z. W., Liu, Y., Yang, G. H., & Lucia, L. A. (2018). Organic solvent isolation and structural characterization of willow lignin. *BioResources, 13*(4), 7957–7968.
30. Zijlstra, D. S., Analbers, C. A., Korte, J., Wilbers, E., & Deuss, P. J. (2019). Efficient mild organosolv lignin extraction in a flow-through setup yielding lignin with high beta-O-4 content. *Polymers (Basel), 11*(12), 1913.
31. Pan, X., & Sano, Y. (2005). Fractionation of wheat straw by atmospheric acetic acid process. *Bioresource Technology, 96*(11), 1256–1263.
32. Li, M.-F., Sun, S.-N., Xu, F., & Sun, R.-C. (2012). Formic acid based organosolv pulping of bamboo (Phyllostachys acuta): Comparative characterization of the dissolved lignins with milled wood lignin. *Chemical Engineering Journal, 179*, 80–89.
33. Zhao, X., & Liu, D. (2012). Fractionating pretreatment of sugarcane bagasse by aqueous formic acid with direct recycle of spent liquor to increase cellulose digestibility – The Formiline process. *Bioresource Technology, 117*, 25–32.
34. Dapia, S., Santos, V., & Parajo, J. C. (2002). Study of formic acid as an agent for biomass fractionation. *Biomass & Bioenergy, 22*(3), 213–221.
35. Jin, C., Yang, M., Shuang, E., Liu, J., Zhang, S., Zhang, X., Sheng, K., & Zhang, X. (2020). Corn stover valorization by one-step formic acid fractionation and formylation for 5-hydroxymethylfurfural and high guaiacyl lignin production. *Bioresource Technology, 299*, 122586.
36. Hu, L., Du, H., Liu, C., Zhang, Y., Yu, G., Zhang, X., Si, C., Li, B., & Peng, H. (2018). Comparative evaluation of the efficient conversion of corn husk filament and corn husk pow-

der to valuable materials via a sustainable and clean biorefinery process. *ACS Sustainable Chemistry & Engineering, 7*(1), 1327–1336.
37. Zhang, Y., Hou, Q., Xu, W., Qin, M., Fu, Y., Wang, Z., Willför, S., & Xu, C. (2017). Revealing the structure of bamboo lignin obtained by formic acid delignification at different pressure levels. *Industrial Crops and Products, 108*, 864–871.
38. Zhang, M., Qi, W., Liu, R., Su, R., Wu, S., & He, Z. (2010). Fractionating lignocellulose by formic acid: Characterization of major components. *Biomass and Bioenergy, 34*(4), 525–532.
39. Snelders, J., Dornez, E., Benjelloun-Mlayah, B., Huijgen, W. J., de Wild, P. J., Gosselink, R. J., Gerritsma, J., & Courtin, C. M. (2014). Biorefining of wheat straw using an acetic and formic acid based organosolv fractionation process. *Bioresource Technology, 156*, 275–282.
40. Villaverde, J. J., Li, J., Ek, M., Ligero, P., & de Vega, A. (2009). Native lignin structure of Miscanthus x giganteus and its changes during acetic and formic acid fractionation. *Journal of Agricultural and Food Chemistry, 57*(14), 6262–6270.
41. Zhang, K., Pei, Z., & Wang, D. (2016). Organic solvent pretreatment of lignocellulosic biomass for biofuels and biochemicals: A review. *Bioresource Technology, 199*, 21–33.
42. Anderson, E. M., Stone, M. L., Hülsey, M. J., Beckham, G. T., & Román-Leshkov, Y. (2018). Kinetic studies of lignin solvolysis and reduction by reductive catalytic fractionation decoupled in flow-through reactors. *ACS Sustainable Chemistry & Engineering, 6*(6), 7951–7959.
43. Tarasov, D., Leitch, M., & Fatehi, P. (2018). Flow through autohydrolysis of spruce wood chips and lignin carbohydrate complex formation. *Cellulose, 25*(2), 1377–1393.
44. Zhou, H., Xu, J. Y., Fu, Y. J., Zhang, H. G., Yuan, Z. W., Qin, M. H., & Wang, Z. J. (2019). Rapid flow-through fractionation of biomass to preserve labile aryl ether bonds in native lignin. *Green Chemistry, 21*(17), 4625–4632.
45. Zhou, H., Tan, L., Fu, Y., Zhang, H., Liu, N., Qin, M., & Wang, Z. (2019). Rapid nondestructive fractionation of biomass (</=15 min) by using flow-through recyclable formic acid toward whole valorization of carbohydrate and lignin. *ChemSusChem, 12*(6), 1213–1221.
46. Dai, J., Patti, A. F., & Saito, K. (2016). Recent developments in chemical degradation of lignin: Catalytic oxidation and ionic liquids. *Tetrahedron Letters, 57*(45), 4945–4951.
47. Swatloski, R. P., Spear, S. K., Holbrey, J. D., & Rogers, R. D. (2002). Dissolution of cellulose [correction of cellulose] with ionic liquids. *Journal of the American Chemical Society, 124*(18), 4974–4975.
48. Zhang, P., Dong, S.-J., Ma, H.-H., Zhang, B.-X., Wang, Y.-F., & Hu, X.-M. (2015). Fractionation of corn stover into cellulose, hemicellulose and lignin using a series of ionic liquids. *Industrial Crops and Products, 76*, 688–696.
49. Leskinen, T., King, A. W. T., & Argyropoulos, D. S. (2014). Fractionation of lignocellulosic materials with ionic liquids. In *Production of biofuels and chemicals with ionic liquids. Biofuels and biorefineries* (pp. 145–168). New York: Springer.
50. Hou, Q., Ju, M., Li, W., Liu, L., Chen, Y., & Yang, Q. (2017). Pretreatment of lignocellulosic biomass with ionic liquids and ionic liquid-based solvent systems. *Molecules, 22*(3), 490.
51. da Costa Lopes, A. M., Lins, R. M. G., Rebelo, R. A., & Łukasik, R. M. (2018). Biorefinery approach for lignocellulosic biomass valorisation with an acidic ionic liquid. *Green Chemistry, 20*(17), 4043–4057.
52. Brandt-Talbot, A., Gschwend, F. J. V., Fennell, P. S., Lammens, T. M., Tan, B., Weale, J., & Hallett, J. P. (2017). An economically viable ionic liquid for the fractionation of lignocellulosic biomass. *Green Chemistry, 19*(13), 3078–3102.
53. Wang, F.-L., Li, S., Sun, Y.-X., Han, H.-Y., Zhang, B.-X., Hu, B.-Z., Gao, Y.-F., & Hu, X.-M. (2017). Ionic liquids as efficient pretreatment solvents for lignocellulosic biomass. *RSC Advances, 7*(76), 47990–47998.
54. Xu, J., Liu, B., Hou, H., & Hu, J. (2017). Pretreatment of eucalyptus with recycled ionic liquids for low-cost biorefinery. *Bioresource Technology, 234*, 406–414.
55. Tan, H. T., & Lee, K. T. (2012). Understanding the impact of ionic liquid pretreatment on biomass and enzymatic hydrolysis. *Chemical Engineering Journal, 183*, 448–458.

56. Sun, N., Rahman, M., Qin, Y., Maxim, M. L., Rodríguez, H., & Rogers, R. D. (2009). Complete dissolution and partial delignification of wood in the ionic liquid 1-ethyl-3-methylimidazolium acetate. *Green Chemistry, 11*(5), 646–655.
57. da Costa Lopes, A. M., Joao, K. G., Rubik, D. F., Bogel-Lukasik, E., Duarte, L. C., Andreaus, J., & Bogel-Lukasik, R. (2013). Pre-treatment of lignocellulosic biomass using ionic liquids: Wheat straw fractionation. *Bioresource Technology, 142*, 198–208.
58. Wen, J.-L., Yuan, T.-Q., Sun, S.-L., Xu, F., & Sun, R.-C. (2014). Understanding the chemical transformations of lignin during ionic liquid pretreatment. *Green Chemistry, 16*(1), 181–190.
59. Badgujar, K. C., & Bhanage, B. M. (2015). Factors governing dissolution process of lignocellulosic biomass in ionic liquid: Current status, overview and challenges. *Bioresource Technology, 178*, 2–18.
60. Abbott, A. P., Capper, G., Davies, D. L., Rasheed, R. K., & Tambyrajah, V. (2003). Novel solvent properties of choline chloride/urea mixtures. *Chemical Communications (Cambridge, England), 1*, 70–71.
61. Francisco, M., van den Bruinhorst, A., & Kroon, M. C. (2012). New natural and renewable low transition temperature mixtures (LTTMs): Screening as solvents for lignocellulosic biomass processing. *Green Chemistry, 14*(8), 2153–2157.
62. Škulcová, A., Kamenská, L., Kalman, F., Ház, A., Jablonský, M., Čížová, K., & Šurina, I. (2016). Deep eutectic solvents as medium for pretreatment of biomass. *Key Engineering Materials, 688*, 17–24.
63. Liu, Q., Zhao, X., Yu, D., Yu, H., Zhang, Y., Xue, Z., & Mu, T. (2019). Novel deep eutectic solvents with different functional groups towards highly efficient dissolution of lignin. *Green Chemistry, 21*(19), 5291–5297.
64. Chen, L., Yu, Q., Wang, Q., Wang, W., Qi, W., Zhuang, X., Wang, Z., & Yuan, Z. (2019). A novel deep eutectic solvent from lignin-derived acids for improving the enzymatic digestibility of herbal residues from cellulose. *Cellulose, 26*(3), 1947–1959.
65. Liu, Y., Chen, W., Xia, Q., Guo, B., Wang, Q., Liu, S., Liu, Y., Li, J., & Yu, H. (2017). Efficient cleavage of lignin-carbohydrate complexes and ultrafast extraction of lignin oligomers from wood biomass by microwave-assisted treatment with deep eutectic solvent. *ChemSusChem, 10*(8), 1692–1700.
66. Alvarez-Vasco, C., Ma, R., Quintero, M., Guo, M., Geleynse, S., Ramasamy, K. K., Wolcott, M., & Zhang, X. (2016). Unique low-molecular-weight lignin with high purity extracted from wood by deep eutectic solvents (DES): A source of lignin for valorization. *Green Chemistry, 18*(19), 5133–5141.
67. Degam, G. (2017). *Deep eutectic solvents synthesis, characterization and applications in pretreatment of lignocellulosic biomass*.
68. Chen, Z., & Wan, C. (2018). Ultrafast fractionation of lignocellulosic biomass by microwave-assisted deep eutectic solvent pretreatment. *Bioresource Technology, 250*, 532–537.
69. Tang, X., Zuo, M., Li, Z., Liu, H., Xiong, C., Zeng, X., Sun, Y., Hu, L., Liu, S., Lei, T., & Lin, L. (2017). Green processing of lignocellulosic biomass and its derivatives in deep eutectic solvents. *ChemSusChem, 10*(13), 2696–2706.
70. Zhu, J., Yan, C., Zhang, X., Yang, C., Jiang, M., & Zhang, X. (2020). A sustainable platform of lignin: From bioresources to materials and their applications in rechargeable batteries and supercapacitors. *Progress in Energy and Combustion Science, 76*, 100788.
71. John, R., & Obst, T. K. K. (1988). Isolation of lignin, methods in enzymology. In *Isolation of lignin* (Vol. 161). San Diego: Academic.
72. Wang, Y., Liu, W., Zhang, L., & Hou, Q. (2019). Characterization and comparison of lignin derived from corncob residues to better understand its potential applications. *International Journal of Biological Macromolecules, 134*, 20–27.
73. Li, T., Lyu, G., Saeed, H. A., Liu, Y., Wu, Y., Yang, G., & Lucia, L. A. (2018). Analytical pyrolysis characteristics of enzymatic/mild acidolysis lignin (EMAL). *BioResources, 13*(2), 4484–4496.

74. Wen, J.-L., Sun, S.-L., Yuan, T.-Q., & Sun, R.-C. (2015). Structural elucidation of whole lignin from Eucalyptus based on preswelling and enzymatic hydrolysis. *Green Chemistry, 17*(3), 1589–1596.
75. Li, Z., Zhang, J., Qin, L., & Ge, Y. (2018). Enhancing antioxidant performance of lignin by enzymatic treatment with laccase. *ACS Sustainable Chemistry & Engineering, 6*(2), 2591–2595.
76. Wang, G., Qi, S., Xia, Y., Parvez, A. M., Si, C., & Ni, Y. (2020). Mild one-pot lignocellulose fractionation based on acid-catalyzed biphasic water/phenol system to enhance components processability. *ACS Sustainable Chemistry & Engineering, 8*(7), 2772–2782.
77. Huang, Y., Duan, Y., Qiu, S., Wang, M., Ju, C., Cao, H., Fang, Y., & Tan, T. (2018). Lignin-first biorefinery: A reusable catalyst for lignin depolymerization and application of lignin oil to jet fuel aromatics and polyurethane feedstock. *Sustainable Energy & Fuels, 2*(3), 637–647.
78. Rinaldi, R. (2017). A tandem for lignin-first biorefinery. *Joule, 1*(3), 427–428.
79. Renders, T., Van den Bosch, S., Koelewijn, S. F., Schutyser, W., & Sels, B. F. (2017). Lignin-first biomass fractionation: The advent of active stabilisation strategies. *Energy & Environmental Science, 10*(7), 1551–1557.
80. Shuai, L., Amiri, M. T., Questell-Santiago, Y. M., Heroguel, F., Li, Y., Kim, H., Meilan, R., Chapple, C., Ralph, J., & Luterbacher, J. S. (2016). Formaldehyde stabilization facilitates lignin monomer production during biomass depolymerization. *Science, 354*(6310), 329–333.
81. Tian, G., Xu, J., Fu, Y., Guo, Y., Wang, Z., & Li, Q. (2019). High β-O-4 polymeric lignin and oligomeric phenols from flow-through fractionation of wheat straw using recyclable aqueous formic acid. *Industrial Crops and Products, 131*, 142–150.
82. Wu, X., Zhang, T., Liu, N., Zhao, Y., Tian, G., & Wang, Z. (2020). Sequential extraction of hemicelluloses and lignin for wood fractionation using acid hydrotrope at mild conditions. *Industrial Crops and Products, 145*, 112086.
83. Olsson, J., Novy, V., Nielsen, F., Wallberg, O., & Galbe, M. (2019). Sequential fractionation of the lignocellulosic components in hardwood based on steam explosion and hydrotropic extraction. *Biotechnology for Biofuels, 12*(1), 1.
84. Huang, C., He, J., Narron, R., Wang, Y., & Yong, Q. (2017). Characterization of kraft lignin fractions obtained by sequential ultrafiltration and their potential application as a biobased component in blends with polyethylene. *ACS Sustainable Chemistry & Engineering, 5*(12), 11770–11779.
85. Santos, P. S. B. D., Erdocia, X., Gatto, D. A., & Labidi, J. (2014). Characterisation of Kraft lignin separated by gradient acid precipitation. *Industrial Crops and Products, 55*, 149–154.
86. Lourençon, T. V., Hansel, F. A., da Silva, T. A., Ramos, L. P., de Muniz, G. I. B., & Magalhães, W. L. E. (2015). Hardwood and softwood kraft lignins fractionation by simple sequential acid precipitation. *Separation and Purification Technology, 154*, 82–88.
87. Jiang, X., Savithri, D., Du, X., Pawar, S., Jameel, H., Chang, H.-m., & Zhou, X. (2016). Fractionation and characterization of kraft lignin by sequential precipitation with various organic solvents. *ACS Sustainable Chemistry & Engineering, 5*(1), 835–842.
88. Wang, G., Liu, X., Yang, B., Si, C., Parvez, A. M., Jang, J., & Ni, Y. (2019). Using green γ-valerolactone/water solvent to decrease lignin heterogeneity by gradient precipitation. *ACS Sustainable Chemistry & Engineering, 7*(11), 10112–10120.
89. An, L., Si, C., Wang, G., Sui, W., & Tao, Z. (2019). Enhancing the solubility and antioxidant activity of high-molecular-weight lignin by moderate depolymerization via in situ ethanol/acid catalysis. *Industrial Crops and Products, 128*, 177–185.
90. Wang, G., Xia, Y., Liang, B., Sui, W., & Si, C. (2018). Successive ethanol–water fractionation of enzymatic hydrolysis lignin to concentrate its antimicrobial activity. *Journal of Chemical Technology & Biotechnology, 93*(10), 2977–2987.
91. Liu, C., Si, C., Wang, G., Jia, H., & Ma, L. (2018). A novel and efficient process for lignin fractionation in biomass-derived glycerol-ethanol solvent system. *Industrial Crops and Products, 111*, 201–211.
92. Dai, L., Zhu, W., Lu, J., Kong, F., Si, C., & Ni, Y. (2019). A lignin-containing cellulose hydrogel for lignin fractionation. *Green Chemistry, 21*(19), 5222–5230.

93. Wang, G., Pang, T., Xia, Y., Liu, X., Li, S., Parvez, A. M., Kong, F., & Si, C. (2019). Subdivision of bamboo kraft lignin by one-step ethanol fractionation to enhance its water-solubility and antibacterial performance. *International Journal of Biological Macromolecules, 133*, 156–164.
94. Toledano, A., García, A., Mondragon, I., & Labidi, J. (2010). Lignin separation and fractionation by ultrafiltration. *Separation and Purification Technology, 71*(1), 38–43.
95. Rico-García, D., Ruiz-Rubio, L., Pérez-Alvarez, L., Hernández-Olmos, S. L., Guerrero-Ramírez, G. L., & Vilas-Vilela, J. L. (2020). Lignin-based hydrogels: Synthesis and applications. *Polymers, 12*(1), 81.
96. Alinejad, M., Henry, C., Nikafshar, S., Gondaliya, A., Bagheri, S., Chen, N., Singh, S. K., Hodge, D. B., & Nejad, M. (2019). Lignin-based polyurethanes: Opportunities for bio-based foams, elastomers, coatings and adhesives. *Polymers (Basel), 11*(7), 1202.
97. Li, B., Wang, Y., Mahmood, N., Yuan, Z., Schmidt, J., & Xu, C. (2017). Preparation of bio-based phenol formaldehyde foams using depolymerized hydrolysis lignin. *Industrial Crops and Products, 97*, 409–416.
98. An, L., Si, C., Wang, G., Choi, C. S., Yu, Y. H., Bae, J. H., Lee, S. M., & Kim, Y. S. (2020). Efficient and green approach for the esterification of lignin with oleic acid using surfactant-combined microreactors in water. *BioResources, 15*(1), 89–104.
99. Dai, L., Liu, R., & Si, C. (2018). A novel functional lignin-based filler for pyrolysis and feedstock recycling of poly(l-lactide). *Green Chemistry, 20*(8), 1777–1783.
100. Vishtal, A., & Kraslawski, A. (2011). Challenges in industrial applications of technical lignins. *BioResources, 6*(3), 3547–3568.
101. Li, Q., Xie, S., Serem, W. K., Naik, M. T., Liu, L., & Yuan, J. S. (2017). Quality carbon fibers from fractionated lignin. *Green Chemistry, 19*(7), 1628–1634.
102. Chi, Z., Hao, L., Dong, H., Yu, H., Liu, H., Wang, Z., & Yu, H. (2020). The innovative application of organosolv lignin for nanomaterial modification to boost its heavy metal detoxification performance in the aquatic environment. *Chemical Engineering Journal, 382*, 122789.
103. Zhang, Y., Ni, S., Wang, X., Zhang, W., Lagerquist, L., Qin, M., Willför, S., Xu, C., & Fatehi, P. (2019). Ultrafast adsorption of heavy metal ions onto functionalized lignin-based hybrid magnetic nanoparticles. *Chemical Engineering Journal, 372*, 82–91.
104. Chen, S., Wang, G., Sui, W., Parvez, A. M., Dai, L., & Si, C. (2020). Novel lignin-based phenolic nanosphere supported palladium nanoparticles with highly efficient catalytic performance and good reusability. *Industrial Crops and Products, 145*, 112164.
105. Zhang, Y., Xu, W., Wang, X., Ni, S., Rosqvist, E., Smått, J.-H., Peltonen, J., Hou, Q., Qin, M., Willför, S., & Xu, C. (2019). From biomass to nanomaterials: A green procedure for preparation of holistic bamboo multifunctional nanocomposites based on formic acid rapid fractionation. *ACS Sustainable Chemistry & Engineering, 7*(7), 6592–6600.
106. Dai, L., Liu, R., Hu, L.-Q., Zou, Z.-F., & Si, C.-L. (2017). Lignin nanoparticle as a novel green carrier for the efficient delivery of resveratrol. *ACS Sustainable Chemistry & Engineering, 5*(9), 8241–8249.
107. Dai, L., Zhu, W. Y., Liu, R., & Si, C. L. (2018). Lignin-containing self-nanoemulsifying drug delivery system for enhance stability and oral absorption of trans-resveratrol. *Particle & Particle Systems Characterization, 35*(4), 1700447.
108. Dai, L., Li, Y., Kong, F., Liu, K., Si, C., & Ni, Y. (2019). Lignin-based nanoparticles stabilized pickering emulsion for stability improvement and thermal-controlled release of trans-resveratrol. *ACS Sustainable Chemistry & Engineering, 7*(15), 13497–13504.
109. Wang, G., Liu, Q., Chang, M., Jang, J., Sui, W., Si, C., & Ni, Y. (2019). Novel Fe3O4@lignosulfonate/phenolic core-shell microspheres for highly efficient removal of cationic dyes from aqueous solution. *Industrial Crops and Products, 127*, 110–118.
110. Ma, M., Dai, L., Si, C., Hui, L., Liu, Z., & Ni, Y. (2019). A facile approach to prepare super long-term stable lignin nanoparticles from black liquor. *ChemSusChem, 12*, 5239.
111. Chistyakov, A. V., & Tsodikov, M. V. (2018). Methods for preparing carbon sorbents from lignin (review). *Russian Journal of Applied Chemistry, 91*(7), 1090–1105.

112. Zhang, X., Yan, Q., Li, J., Chu, I. W., Toghiani, H., Cai, Z., & Zhang, J. (2018). Carbon-based nanomaterials from biopolymer lignin via catalytic thermal treatment at 700 to 1000 degrees C. *Polymers (Basel), 10*(2), 183.
113. Martin-Martinez, M., Barreiro, M. F. F., Silva, A. M. T., Figueiredo, J. L., Faria, J. L., & Gomes, H. T. (2017). Lignin-based activated carbons as metal-free catalysts for the oxidative degradation of 4-nitrophenol in aqueous solution. *Applied Catalysis B: Environmental, 219*, 372–378.
114. Niu, N., Ma, Z., He, F., Li, S., Li, J., Liu, S., & Yang, P. (2017). Preparation of carbon dots for cellular imaging by the molecular aggregation of cellulolytic enzyme lignin. *Langmuir, 33*(23), 5786–5795.

Index

Z.-H. Liu, A. Ragauskas (eds.), *Emerging Technologies for Biorefineries, Biofuels, and Value-Added Commodities*,
https://doi.org/10.1007/978-3-030-65584-6

D

E

F

T

Printed in the United States
by Baker & Taylor Publisher Services